ANNUAL EDITIONS

Sustainability 12/13

First Edition

EDITOR

Nicholas J. Smith-Sebasto
Kean University

Nicholas J. Smith-Sebasto is Executive Director of the Center for Sustainability Studies at Kean University where he also supervises the B.S. in Sustainability Science program. He received a B.S. in Biology from Trenton State College (now The College of New Jersey), a M.S. in Fishery and Wildlife Biology from Colorado State University, and a Ph.D. in Environmental Communication, Education, and Interpretation from The Ohio State University. For the past two decades, he has been recognized as one of the global leaders in the field of environmental education. He is the only person ever to serve simultaneously on the editorial boards of the *Journal of Environmental Education, Environmental Education Research,* and *International Research in Geographical and Environmental Education.* He served on the Board of Directors of the North American Association for Environmental Education (NAAEE), the largest professional society dedicated to environmental education in the world. He is the recipient of the NAAEE's award for *Outstanding Contributions to Research in Environmental Education.* He has authored more original research articles in the *Journal of Environmental Education,* five of which have included his students as co-authors, than any other author in the 40-year history of the journal. He is the recipient of seven teaching excellence recognitions, appearing in *Who's Who Among America's Teachers* twice. He is currently a Technical Advisor for the STARS program of the Association for the Advancement of Sustainability in Higher Education.

His new research and teaching interests concern sustainability science. His specific interest is on food and organics recycling. He has quickly become an authority on aerobic in-vessel digestion of food and organic wastes. He has operated one of the only university-based food and organics composting demonstration projects in the nation for the past five years, including an expanded effort at Kean University where a system he custom-designed is being used to compost more than 50 tons of food scraps generated on the campus annually. His other interests include sustainability education and sustainability technologies, especially Earth-sheltering architecture.

The McGraw·Hill Companies

ANNUAL EDITIONS: SUSTAINABILITY, FIRST EDITION

Published by McGraw-Hill, a business unit of The McGraw-Hill Companies, Inc., 1221 Avenue of the Americas, New York, NY 10020.
Annual Editions is published by the **Contemporary Learning Series** group within The McGraw-Hill Higher Education division.

1 2 3 4 5 6 7 8 9 0 QDB/QDB 1 0 9 8 7 6 5 4 3 2 1

ISBN: 978–0–07–352869–4
MHID: 0–07–352869–2
ISSN: 2162–5689 (print)
ISSN: 2162–5697 (online)

Managing Editor: *Larry Loeppke*
Developmental Editor II: *Debra A. Henricks*
Senior Permissions Coordinator: *Shirley Lanners*
Marketing Specialist: *Alice Link*
Project Manager: *Connie Oertel*
Design Coordinator: *Margarite Reynolds*
Buyer: *Susan K. Culbertson*
Cover Designer: *Kristine Jubeck*
Media Project Manager: *Sridevi Palani*

Compositor: Laserwords Private Limited
Cover Image Credits: Ingram Publishing (inset); Photo by Ron Nichols, USDA Natural Resources Conservation Service (background)

www.mhhe.com

Editors/Academic Advisory Board

Members of the Academic Advisory Board are instrumental in the final selection of articles for each edition of ANNUAL EDITIONS. Their review of articles for content, level, and appropriateness provides critical direction to the editors and staff. We think that you will find their careful consideration well reflected in this volume.

ANNUAL EDITIONS: Sustainability 12/13
1st Edition

EDITOR

Nicholas J. Smith-Sebasto
Kean University

ACADEMIC ADVISORY BOARD MEMBERS

Preface

In publishing ANNUAL EDITIONS we recognize the enormous role played by the magazines, newspapers, and journals of the public press in providing current, first-rate educational information in a broad spectrum of interest areas. Many of these articles are appropriate for students, researchers, and professionals seeking accurate, current material to help bridge the gap between principles and theories and the real world. These articles, however, become more useful for study when those of lasting value are carefully collected, organized, indexed, and reproduced in a low-cost format, which provides easy and permanent access when the material is needed. That is the role played by ANNUAL EDITIONS.

"...there is nothing wrong with the planet.... The planet is fine.... Compared to the people, the planet is doing great. [It has] been here 4½ billion years. Do you ever think about the arithmetic? The planet has been here 4½ billion years; we've been here what, 100,000, maybe 200,000? And we've only been engaged in heavy industry for a little over 200 years. Two hundred years versus 4½ billion and we have the conceit to think that somehow we're a threat? That somehow, we're going to put in jeopardy this beautiful little blue-green ball that's just a-floatin' around the Sun? The planet has been through a lot worse than us. Been through all kinds of things worse than us; been through earthquakes, volcanoes, plate tectonics, continental drifts, solar flares, sunspots, magnetic storms, the magnetic reversal of the poles, hundreds of thousands of years of bombardment by comets and asteroids and meteors, worldwide floods, tidal waves, worldwide fires, erosion, cosmic rays, recurring ice ages, and we think some plastic bags and aluminum cans are going to make a difference? The planet isn't going anywhere, we are! We're going away!"

These words are from the late comedian George Carlin's HBO special in 1992. I think they represent a very astute assessment of sustainability. Sustainability is about humanity behaving in a manner that does not compromise the life support systems of Earth. To be sure, it is about making sure that we don't go away, but it is also about making sure that our actions do not so degrade the life support systems of the planet that other species are adversely affected.

In order to achieve sustainability, humanity is going to have to change many of its behaviors. Before such changes are made, essential learnings are going to have to occur. These learnings are characterized by four essential questions:

What are the unique characteristics of Earth that have allowed life to evolve and develop to the extent that it has?

As near as we know, and we might very well be wrong on this, Earth is the only planetary body in the solar system, and perhaps the universe, where life exists. Clearly, Earth must have some unique characteristics. The problem is that most people know painfully little about these characteristics. If people are not knowledgeable about the life support systems of the planet, how can they be expected to value them and/or to behave in ways that sustain them?

What is humanity doing to compromise or degrade these unique characteristics?

There is something very close to a scientific consensus that human actions are compromising the life support systems of the planet. What most people don't know or understand is the specific ways in which we are doing so. It is important that we move beyond simplistic issues such as littering and begin to consider larger issues such as climate change, landscape change, accelerated extinction, and other issues. More people need to know precisely how humans are degrading the life support systems of Earth before any substantive movement in the direction of sustainability will occur.

Why do humans behave in unsustainable ways?

It is so utterly illogical that a species would knowingly and intentionally destroy its own life support system that there must be a rational explanation for such actions. Before sustainability will be achieved, it is essential that we understand what exactly we are doing that is unsustainable.

What changes are necessary to achieve sustainability?

Once the answers to the first three questions are identified, then, and only then, will it be possible to address the question of what changes to our actions and behaviors are necessary.

Clearly, education has an enormous role to play in the transition to sustainability. Increased student interest in the subject has been documented. For example, On Tuesday, March 24, 2009, *NBC Nightly News* with Brian Williams closed its broadcast with a 2 minute 23 second segment that Mr. Williams began with,

> *If you know somebody heading off to college, there's an increasingly good chance they might want to major in something green. As you know, all things environmental are very hot right now, and colleges are taking notice.*

Early in the segment, a student is observed stating, "I feel that we need to protect our environment." Chief Environmental Correspondent Anne Thompson followed with,

To do that, schools are moving beyond science, creating a new major called sustainability; bringing together everything from architecture to engineering to urban planning in an effort to find solutions to environmental issues.

The August 3, 2009 edition of *USA Today* included the headline, "College students are flocking to sustainability degrees, careers." Similarly, the August 19, 2009 edition of the *New York Times* included the headline, "Sustainability field booms on campus." In the August 31, 2009 edition of *The Chronicle of Higher Education,* "Sustainability" was identified as one of "5 college majors on the rise." More recently, in the June 24, 2011 edition of the *New York Times,* an article with the title "Green jobs attract graduates," suggested, "Suddenly, "sustainability" seems to resonate with the sex appeal of "docom" or "start-up,". . ." Such news seems to suggest that institutions of higher education have recognized the notion that they "bear a profound moral responsibility to increase the awareness, knowledge, skills and values needed to create a just and sustainable future."

This volume in the *Annual Editions* series represents another step in the direction of sustainability. The readings included provide an introduction to the questions identified above with the expectation that they will instigate a desire to explore them in greater detail and that they will inspire readers to move to the front of the line in the march toward sustainability.

Also included in this volume are a number of features that are designed to make it useful for students, researchers, and professionals in the field of sustainability. The *Topic Guide* can be used to establish specific reading assignments tailored to the needs of a particular course of study. Other helpful features include the *Table of Contents* abstracts, which summarize each article and present key concepts in bold italics. In addition, each unit is preceded by an overview, which provides a background for informed reading of the articles, emphasizes critical issues, and presents *Learning Outcomes* that are tied to *Critical Thinking* questions at the end of each article. The *Internet References* section can be used to further explore the topics online. Instructors will appreciate a password-protected online *Instructor's Resource Guide,* and students will find online quizzing to further test their understanding of the material. These tools are available at www.mhhe.com/cls via the *Online Learning Center* for this book.

Annual Editions: Sustainability will be updated annually. Those involved in producing this volume wish to make the next one as useful and effective as possible. Your criticism and advice are always welcome.

Nicholas J. Smith-Sebasto
Editor

The Annual Editions Series

VOLUMES AVAILABLE

Adolescent Psychology
Aging
American Foreign Policy
American Government
Anthropology
Archaeology
Assessment and Evaluation
Business Ethics
Child Growth and Development
Comparative Politics
Criminal Justice
Developing World
Drugs, Society, and Behavior
Dying, Death, and Bereavement
Early Childhood Education
Economics
Educating Children with Exceptionalities
Education
Educational Psychology
Entrepreneurship
Environment
The Family
Gender
Geography
Global Issues
Health
Homeland Security
Human Development
Human Resources
Human Sexualities
International Business
Management
Marketing
Mass Media
Microbiology
Multicultural Education
Nursing
Nutrition
Physical Anthropology
Psychology
Race and Ethnic Relations
Social Problems
Sociology
State and Local Government
Sustainability
Technologies, Social Media, and Society
United States History, Volume 1
United States History, Volume 2
Urban Society
Violence and Terrorism
Western Civilization, Volume 1
Western Civilization, Volume 2
World History, Volume 1
World History, Volume 2
World Politics

Contents

UNIT 1
Sustainability—An Indictment of Human Actions?

The concepts in bold italics are developed in the article. For further expansion, please refer to the Topic Guide.

UNIT 2
Sustainability—A New Paradigm?

The concepts in bold italics are developed in the article. For further expansion, please refer to the Topic Guide.

UNIT 3
Earth's Life Support Systems and Ecosystem Services

UNIT 4
Why Do Humans Behave in Unsustainable Ways?

The concepts in bold italics are developed in the article. For further expansion, please refer to the Topic Guide.

UNIT 5
What Are the Impacts of Our Actions?

The concepts in bold italics are developed in the article. For further expansion, please refer to the Topic Guide.

UNIT 6
How Do We Correct Our Actions and Embrace Sustainability?

The concepts in bold italics are developed in the article. For further expansion, please refer to the Topic Guide.

The concepts in bold italics are developed in the article. For further expansion, please refer to the Topic Guide.

Correlation Guide

The *Annual Editions* series provides students with convenient, inexpensive access to current, carefully selected articles from the public press. **Annual Editions: Sustainability 12/13** is an easy-to-use reader that presents articles on important topics such as *biodiversity, climate change, pro-sustainability actions/behaviors,* and many more. For more information on *Annual Editions and other McGraw-Hill Contemporary Learning Series* titles, visit www.mhhe.com/cls.

This convenient guide matches the units in **Annual Editions: Sustainability 12/13** with the corresponding chapters in two of our best-selling McGraw-Hill Environmental Science textbooks by Cunningham/Cunningham and Enger/Smith.

Annual Editions: Sustainability 12/13	Environmental Science: A Global Concern, 12/e, by Cunningham/ Cunningham	Environmental Science, 13/e, by Enger/Smith
Unit 1: Sustainability—An Indictment of Human Actions?	**Chapter 1:** Understanding Our Environment **Chapter 5:** Biomes: Global Patterns of Life **Chapter 6:** Population Biology	
Unit 2: Sustainability—A New Paradigm?	**Chapter 1:** Understanding Our Environment	
Unit 3: Earth's Life Support Systems and Ecosystem Services	**Chapter 1:** Understanding Our Environment **Chapter 4:** Evolution, Biological Communities, and Species Interactions **Chapter 5:** Biomes: Global Patterns of Life **Chapter 11:** Biodiversity: Preserving Species	**Chapter 5:** Interactions: Environments and Organisms **Chapter 11:** Biodiversity Issues **Chapter 12:** Land-Use Planning
Unit 4: Why Do Humans Behave In Unsustainable Ways?	**Chapter 1:** Understanding Our Environment	**Chapter 2:** Environmental Ethics
Unit 5: What Are The Impacts of Our Actions?	**Chapter 3:** Matter, Energy, and Life **Chapter 10:** Farming: Conventional and Sustainable Practices **Chapter 11:** Biodiversity: Preserving Species **Chapter 25:** What Then Shall We Do?	**Chapter 2:** Environmental Ethics **Chapter 5:** Interactions: Environments and Organisms **Chapter 11:** Biodiversity Issues **Chapter 13:** Soil and Its Uses
Unit 6: How Do We Correct Our Actions and Embrace Sustainability?	**Chapter 10:** Farming: Conventional and Sustainable Practices **Chapter 16:** Air Pollution **Chapter 20:** Sustainable Energy **Chapter 23:** Ecological Economics **Chapter 25:** What Then Shall We Do?	**Chapter 3:** Environmental Risk: Economics, Assessment, and Management **Chapter 9:** Energy Sources **Chapter 14:** Agricultural Methods and Pest Management **Chapter 16:** Air Quality Issues

Topic Guide

This topic guide suggests how the selections in this book relate to the subjects covered in your course. You may want to use the topics listed on these pages to search the Web more easily.

On the following pages a number of websites have been gathered specifically for this book. They are arranged to reflect the units of this Annual Editions reader. You can link to these sites by going to www.mhhe.com/cls.

All the articles that relate to each topic are listed below the bold-faced term.

Internet References

The following Internet sites have been selected to support the articles found in this reader. These sites were available at the time of publication. However, because websites often change their structure and content, the information listed may no longer be available. We invite you to visit www.mhhe.com/cls for easy access to these sites.

Annual Editions: Sustainability 12/13

Unit 1: Sustainability—An Indictment of Human Actions?

Convention on Biological Diversity (CBD)
www.cbd.int

Has produced three editions of Global Biodiversity Outlook, an effort that resulted from the increasing recognition of the importance of biodiversity and facilitated by the United Nations Environment Program. The CBD was "inspired by the world community's growing commitment to sustainable development."

The H. John Heinz, III Center for Science, Economics and the Environment
www.heinzctr.org

Named in honor of the late Senator H. John Heinz, III, the Center is "dedicated to improving the scientific and economic foundation for environmental policy through multisectoral collaboration." It has produced 21 major reports related to sustainability.

Millennium Ecosystem Assessment
www.maweb.org/en/index.aspx

An international effort of nearly 1,400 scientists to assess the status of the ecosystems of Earth and the services provided by them that contribute to human well-being. Recommendations for needed behavioral changes to assure the sustainability of both are provided.

Union of Concerned Scientists (UCS)
www.ucsusa.org/about/1992-world-scientists.html

Since 1969, the UCS has been providing science-based information about the environment and global issues. For the past four decades, it has consistently demonstrated how "thoughtful action based on the best available science can help safeguard our future and the future of our planet."

World Wildlife Fund (WWF)
www.worldwildlife.org

With more than 5 million members, including over 1 million in the United States, the World Wildlife Fund is the largest conservation organization in the world. Its mission is clear: the conservation of nature and reversing environmental degradation.

Unit 2: Sustainability—A New Paradigm?

International Union for the Conservation of Nature (IUCN)
www.iucn.org

What the WWF is to individual membership in an organization dedicated to sustainability, the IUCN is to governments and large organizations, with over 1,000 members in the organization.

United States Environmental Protection Agency (USEPA)
www.epa.gov/sustainability

USEPA's website concerning sustainability.

United States National Park Service (NPS)
www.nps.gov/sustainability

The NPS's website concerning sustainability.

World Resources Institute (WRI)
www.wri.org

The mission of the WRI "is to move human society to live in ways that protect Earth's environment and its capacity to provide for the needs and aspirations of current and future generations."

Unit 3: Earth's Life Support Systems and Ecosystem Services

EcosystemMarketplace (EM)
www.ecosystemmarketplace.com

EM "is a leading source of news, data, and analytics on markets and payments for ecosystem services (such as water quality, carbon sequestration, and biodiversity)."

International Union for the Conservation of Nature (IUCN), Commission on Environmental Management (CEM)
www.iucn.org/about/union/commissions/cem/cem_work/cem_services

The objective of the CEM is "to improve the knowledge base on ecosystem services and values and stimulate the integration of this knowledge in planning and decision making for sustainable Ecosystem Management through development of case studies, guidelines, and dissemination."

United States Environmental Protection Agency (USEPA)
www.epa.gov/ecology

The USEPA's website regarding ecosystem services.

United States Forest Service (USFS)
www.fs.fed.us/ecosystemservices

The USFS's website regarding ecosystem services.

World Resources Institute (WRI)
www.wri.org/project/mainstreaming-ecosystem-services

"WRI is helping governments, businesses, and multilateral development banks include . . . ecosystem services in their decision-making, with the ultimate goal of reducing ecosystem degradation around the world."

Unit 4: Why Do Humans Behave In Unsustainable Ways?

Post Carbon Institute–Energy Bulletin
www.energybulletin.net/node/46276

This site contains a very informative article titled, *Our American Way of Life is Unsustainable–Evidence.*

Sustainability Institute (SI)
www.sustainer.org

Founded by the late noted biophysicist and author Donella Meadows, the SI is an organization dedicated to identifying and addressing the root causes of unsustainable behaviors. The philosophy of the SI is: ". . .unsustainability does not arise out of ignorance, irrationality or greed. It is largely the collective consequence of rational, well-intended decisions. . . ."

Internet References

Unit 5: What Are the Impacts of Our Actions?

Center for Science in the Public Interest (CSPI)

www.cspinet.org/EatingGreen/calculator.html

"Since 1971, CSPI has been a strong advocate for nutrition and health, food safety, alcohol policy, and sound science." Its Eating Green Calculator allows you to examine the impact of your diet on your health and the environment.

Conservation International (CI)

www.conservation.org/act/live_green/carboncalc/Pages/default.aspx

"Building upon a strong foundation of science, partnership and field demonstration, CI empowers societies to responsibly and sustainably care for nature, our global biodiversity, for the well-being of humanity." It is one of several organizations that offer a carbon footprint calculator.

Global Footprint Network (GFN)

www.footprintnetwork.org

GFN is a nonprofit organization that was founded in 2003 "to enable a sustainable future where all people have the opportunity to live satisfying lives within the means of one planet." A major focus of the GFN is to "accelerate the use of the Ecological Footprint—a resource accounting tool that measures how much nature we have, how much we use, and who uses what."

The Nature Conservancy (TNC)

www.nature.org/greenliving/carboncalculator

Formed in 1951, the mission of TNS is "to preserve the plants, animals and natural communities that represent the diversity of life on Earth by protecting the lands and waters they need to survive." Its carbon footprint calculator allows you to "to measure your impact on our climate."

United States Department of Transportation (USDOT), Federal Transit Administration (FTA)

www.fta.dot.gov/planning/planning_environment_8523.html

The carbon calculator on this site allows you to calculate how much carbon you contribute to the atmosphere as a result of your driving habits.

Unit 6: How Do We Correct Our Actions and Embrace Sustainability?

GoingGreen

www.goinggreen.com

Provides suggestions for simple behavior changes that support sustainability.

The Green Guide

http://environment.nationalgeographic.com/environment/green-guide

This National Geographic site provides assorted information on sustainable living.

Original Green

www.originalgreen.org

This site focuses predominantly on architecture, but it also provides links to other useful sites.

Sustainable Communities Online

www.sustainable.org

This site provides a "pool information on sustainability to make it more readily accessible to the public."

YouSustain

www.yousustain.com/footprint/actions

This calculator allows users to calculate how changes in 24 behaviors will affect the environment.

UNIT 1

Sustainability—An Indictment of Human Actions?

Unit Selections

1. **World Scientists' Warning to Humanity,** Henry Kendall
2. **Population and the Environment: The Global Challenge,** Don Hinrichsen and Bryant Robey
3. **Ecosystems and Human Well-Being: Summary for Decision-Makers,** *Millennium Ecosystem Assessment*
4. **The Anthropocene: Are Humans Now Overwhelming the Great Forces of Nature?,** Will Steffen, Paul J. Crutzen, and John R. McNeill
5. **The State of the Nation's Ecosystems 2008: What the Indicators Tell Us,** The H. John Heinz
6. **Global Biodiversity Outook 3: Executive Summary,** *Secretariat of the Convention on Biological Diversity*

Learning Outcomes

After reading this unit, you should be able to:

- Identify, by listing, the components of the World Scientists' Warning to Humanity that are consistent with the components of an indictment.
- Indentify, by listing, examples of how human actions are detrimental to the life support systems of Earth.
- Classify, by using a definition, the concept of the Anthropocene.
- Classify, by using a definition, the concept of biodiversity.
- Generate, by synthesizing the provided evidence, a paragraph describing the cumulative impact of humanity on the life support systems of Earth.

Student Website

www.mhhe.com/cls

Internet References

Convention on Biological Diversity (CBD)
www.cbd.int

The H. John Heinz, III Center for Science, Economics and the Environment
www.heinzctr.org/ecosystems/2008report/pdf_files/Highlights_Final_low_res.pdf

Millennium Ecosystem Assessment
www.maweb.org/documents/document.356.aspx.pdf

Union of Concerned Scientists
www.ucsusa.org/about/1992-world-scientists.html

World Wildlife Fund (WWF)
www.worldwildlife.org

An indictment is defined as "a formal written statement framed by a prosecuting authority and found by a jury (as a grand jury) charging a person with an offense."[1] An arraignment is "a formal reading of a criminal complaint in the presence of the defendant to inform the defendant of the charges against her/him."[2] Where sustainability is concerned, a document that could be considered to at least partially meet the definition of an indictment of humanity was provided nearly 15 years ago. Several documents that would likely be part of an arraignment have also been provided.

The World Without Us, a *New York Times* bestselling book; the *Life After People* series on The History Channel; and *Aftermath-Population Zero* on the National Geographic Channel are a few examples of how the issue of what would Earth be like if humans went extinct has been explored. The general thinking is that humans have so degraded the life support systems of Earth that it is only a matter of time before we experience the same fate as 99 percent of the species that have ever existed on this planet. Is this assessment accurate? Is it possible to indict humanity for crimes against the planet and the life that lives on it? Is this thinking new, or has it existed for some time? Is there evidence to support such a notion or are these efforts simply more propaganda from the lunatic fringe of environmentalism? A modest pamphlet titled the "World Scientists' Warning to Humanity" was released by the Union of Concerned Scientists in 1997. It reads like an indictment of humanity insofar as it is a formal written statement of a prosecuting authority of nearly 1,700 of the most pre-eminent scientists in the world, including half of the living Nobel laureates in the sciences at the time of its release, all of whom agreed with the charge that human actions are reducing the life support systems of the planet to an extent that might prevent life as we know it from continuing to exist.

Article 2 corroborates the indictment of humanity by providing an update and elaboration of the charges against it. Importantly, this document was produced by another exceptionally credible source: The Johns Hopkins University School of Public Health. After careful examination, it offered the observation that, "In the past decade in every environmental sector, conditions have either failed to improve, or they are worsening."

Article 3 establishes even more conclusively the evidence in support of the warning. Called for by the former Secretary General of the United Nations and the authorized by governments by four conventions, it presents the efforts of nearly 1,400 experts worldwide. The major finding was that humanity has altered ecosystems more in the past 50 years than at any other period in the history of our existence on the planet.

© Ingram Publishing

Article 4 explores a question posed by the late Dolores LaChappelle, namely, how did we reach such a state of insanity? The Anthropocene is a term used to describe a period in Earth's history when human actions have come to rival more so-called natural forces, such as geological ones, insofar as their ability to alter the actual appearance of the planet, the distribution of life on it, the climate of it, etc. It turns out that humans have been having such an effect for over 200 years.

Article 5 continues the support of the indictment offered by the warning with evidence specific to the United States. It provides data to support the conclusion that "The United States landscape has changed substantially in the last half-century." This finding is consistent with those provided by the MEA. Still, what these changes mean for the future is unclear. Continued monitoring and presentation of data such as is provided in this report is essential.

Article 6 examines a specific sustainability concern, namely the loss of biodiversity. Alfred Russell Wallace once suggested that the loss of biodiversity (even though that term was not used during his lifetime) is like the loss of a few letters in a sentence, which may make the sentence unreadable. Many consider species a metaphor for letters; the loss of a few of them may make life of Earth, at least in the form that we know it, impossible. To understand this matter, we must understand the diversity of life on Earth, which we don't. We don't even know with any real precision the number of species that inhabit the planet and estimates vary by an order of magnitude.

[1]www.merriam-webster.com/dictionary/indictment
[2]www.en.wikipedia.org/wiki/Arraignment

World Scientists' Warning to Humanity

Some 1,700 of the world's leading scientists, including the majority of Nobel laureates in the sciences, issued this appeal in November 1992. The World Scientists' Warning to Humanity was written and spearheaded by the late Henry Kendall, former chair of UCS's board of directors.

HENRY KENDALL

Introduction

Human beings and the natural world are on a collision course. Human activities inflict harsh and often irreversible damage on the environment and on critical resources. If not checked, many of our current practices put at serious risk the future that we wish for human society and the plant and animal kingdoms, and may so alter the living world that it will be unable to sustain life in the manner that we know. Fundamental changes are urgent if we are to avoid the collision our present course will bring about.

The Environment

The environment is suffering critical stress:

The Atmosphere

Stratospheric ozone depletion threatens us with enhanced ultraviolet radiation at the earth's surface, which can be damaging or lethal to many life forms. Air pollution near ground level, and acid precipitation, are already causing widespread injury to humans, forests, and crops.

Water Resources

Heedless exploitation of depletable ground water supplies endangers food production and other essential human systems. Heavy demands on the world's surface waters have resulted in serious shortages in some 80 countries, containing 40 percent of the world's population. Pollution of rivers, lakes, and ground water further limits the supply.

Oceans

Destructive pressure on the oceans is severe, particularly in the coastal regions, which produce most of the world's food fish. The total marine catch is now at or above the estimated maximum sustainable yield. Some fisheries have already shown signs of collapse. Rivers carrying heavy burdens of eroded soil into the seas also carry industrial, municipal, agricultural, and livestock waste—some of it toxic.

Soil

Loss of soil productivity, which is causing extensive land abandonment, is a widespread by-product of current practices in agriculture and animal husbandry. Since 1945, 11 percent of the earth's vegetated surface has been degraded—an area larger than India and China combined—and per capita food production in many parts of the world is decreasing.

Forests

Tropical rain forests, as well as tropical and temperate dry forests, are being destroyed rapidly. At present rates, some critical forest types will be gone in a few years, and most of the tropical rain forest will be gone before the end of the next century. With them will go large numbers of plant and animal species.

Living Species

The irreversible loss of species, which by 2100 may reach one-third of all species now living, is especially serious. We are losing the potential they hold for providing medicinal and other benefits, and the contribution that genetic diversity of life forms gives to the robustness of the world's biological systems and to the astonishing beauty of the earth itself. Much of this damage is irreversible on a scale of centuries, or permanent. Other processes appear to pose additional threats. Increasing levels of gases in the atmosphere from human activities, including carbon dioxide released from fossil fuel burning and from deforestation, may alter climate on a global scale. Predictions of global warming are still uncertain—with projected effects ranging from tolerable to very severe—but the potential risks are very great.

Our massive tampering with the world's interdependent web of life—coupled with the environmental damage inflicted by deforestation, species loss, and climate change—could trigger widespread adverse effects, including unpredictable collapses of critical biological systems whose interactions and dynamics we only imperfectly understand.

Uncertainty over the extent of these effects cannot excuse complacency or delay in facing the threats.

Population

The earth is finite. Its ability to absorb wastes and destructive effluent is finite. Its ability to provide food and energy is finite. Its ability to provide for growing numbers of people is finite. And we are fast approaching many of the earth's limits. Current economic practices which damage the environment, in both developed and underdeveloped nations, cannot be continued without the risk that vital global systems will be damaged beyond repair.

Pressures resulting from unrestrained population growth put demands on the natural world that can overwhelm any efforts to achieve a sustainable future. If we are to halt the destruction of our environment, we must accept limits to that growth. A World Bank estimate indicates that world population will not stabilize at less than 12.4 billion, while the United Nations concludes that the eventual total could reach 14 billion, a near tripling of today's 5.4 billion. But, even at this moment, one person in five lives in absolute poverty without enough to eat, and one in ten suffers serious malnutrition.

No more than one or a few decades remain before the chance to avert the threats we now confront will be lost and the prospects for humanity immeasurably diminished.

Warning

We the undersigned, senior members of the world's scientific community, hereby warn all humanity of what lies ahead. A great change in our stewardship of the earth and the life on it is required, if vast human misery is to be avoided and our global home on this planet is not to be irretrievably mutilated.

What We Must Do

Five inextricably linked areas must be addressed simultaneously:

- **We must bring environmentally damaging activities under control to restore and protect the integrity of the earth's systems we depend on.**
- We must, for example, move away from fossil fuels to more benign, inexhaustible energy sources to cut greenhouse gas emissions and the pollution of our air and water. Priority must be given to the development of energy sources matched to Third World needs—small-scale and relatively easy to implement.
- We must halt deforestation, injury to and loss of agricultural land, and the loss of terrestrial and marine plant and animal species.
- **We must manage resources crucial to human welfare more effectively.**
- We must give high priority to efficient use of energy, water, and other materials, including expansion of conservation and recycling.
- **We must stabilize population.**
- This will be possible only if all nations recognize that it requires improved social and economic conditions, and the adoption of effective, voluntary family planning.
- **We must reduce and eventually eliminate poverty.**
- **We must ensure sexual equality, and guarantee women control over their own reproductive decisions.**

Developed Nations Must Act Now

The developed nations are the largest polluters in the world today. They must greatly reduce their overconsumption, if we are to reduce pressures on resources and the global environment. The developed nations have the obligation to provide aid and support to developing nations, because only the developed nations have the financial resources and the technical skills for these tasks.

Acting on this recognition is not altruism, but enlightened self-interest: whether industrialized or not, we all have but one lifeboat. No nation can escape from injury when global biological systems are damaged. No nation can escape from conflicts over increasingly scarce resources. In addition, environmental and economic instabilities will cause mass migrations with incalculable consequences for developed and undeveloped nations alike.

Developing nations must realize that environmental damage is one of the gravest threats they face, and that attempts to blunt it will be overwhelmed if their populations go unchecked. The greatest peril is to become trapped in spirals of environmental decline, poverty, and unrest, leading to social, economic, and environmental collapse.

Success in this global endeavor will require a great reduction in violence and war. Resources now devoted to the preparation and conduct of war—amounting to over $1 trillion annually—will be badly needed in the new tasks and should be diverted to the new challenges.

A new ethic is required—a new attitude towards discharging our responsibility for caring for ourselves and for the earth. We must recognize the earth's limited capacity to provide for us. We must recognize its fragility.

We must no longer allow it to be ravaged. This ethic must motivate a great movement, convincing reluctant leaders and reluctant governments and reluctant peoples themselves to effect the needed changes.

The scientists issuing this warning hope that our message will reach and affect people everywhere. We need the help of many.

- We require the help of the world community of scientists—natural, social, economic, and political.
- We require the help of the world's business and industrial leaders.
- We require the help of the world's religious leaders.
- We require the help of the world's peoples.
- **We call on all to join us in this task.**

Critical Thinking

1. If this warning were written today, how would it differ from what was originally written?
2. What vital action is missing from the list of "What We Must Do?"
3. While 1,700 scientists endorsed the Warning, not all scientists did. Is this fact important?

Population and the Environment

The Global Challenge

As the century begins, natural resources are under increasing pressure, threatening public health and development. Water shortages, soil exhaustion, loss of forests, air and water pollution, and degradation of coastlines afflict many areas. As the world's population crows, improving living standards without destroying the environment is a global challenge.

DON HINRICHSEN AND BRYANT ROBEY

Most developed economies currently consume resources much faster than they can regenerate. Most developing countries with rapid population growth face the urgent need to improve living standards. As we humans exploit nature to meet present needs, are we destroying resources needed for the future?

Environment Getting Worse

In the past decade in every environmental sector, conditions have either failed to improve, or they are worsening:

Public Health Unclean water, along with poor sanitation, kills over 12 million people each year, most in developing countries. Air pollution kills nearly 3 million more. Heavy metals and other contaminants also cause widespread health problems.

Food Supply Will there be enough food to go around? In 64 of 105 developing countries studied by the UN Food and Agriculture Organization, the population has been growing faster than food supplies. Population pressures have degraded some 2 billion hectares of arable land—an area the size of Canada and the US.

Freshwater The supply of freshwater is finite, but demand is soaring as population grows and use per capita rises. By 2025, when world population is projected to be 8 billion, 48 countries containing 3 billion people will face shortages.

Coastlines and Oceans Half of all coastal ecosystems are pressured by high population densities and urban development. A tide of pollution is rising in the world's seas. Ocean fisheries are being overexploited, and fish catches are down.

Forests Nearly half of the world's original forest cover has been lost, and each year another 16 million hectares are cut, bulldozed, or burned. Forests provide over US$400 billion to the world economy annually and are vital to maintaining healthy ecosystems. Yet, current demand for forest products may exceed the limit of sustainable consumption by 25 percent.

Biodiversity The earth's biological diversity is crucial to the continued vitality of agriculture and medicine—and perhaps even to life on earth itself. Yet human activities are pushing many thousands of plant and animal species into extinction. Two of every three species is estimated to be in decline.

Global Climate Change The earth's surface is warming due to greenhouse gas emissions, largely from burning fossil fuels. If the global temperature rises as projected, sea levels would rise by several meters, causing widespread flooding. Global warming also could cause droughts and disrupt agriculture.

Toward a Livable Future

How people preserve or abuse the environment could largely determine whether living standards improve or deteriorate. Growing human numbers, urban expansion, and resource exploitation do not bode well for the future. Without practicing sustainable development, humanity faces a deteriorating environment and may even invite ecological disaster.

Taking action Many steps toward sustainability can be taken today. These include using energy more efficiently; managing cities better; phasing out subsidies that encourage waste; managing water resources and protecting freshwater sources; harvesting forest products rather than destroying forests; preserving arable land and increasing food production through a second Green Revolution; managing coastal zones and ocean fisheries; protecting biodiversity hotspots; and adopting an international convention on climate change.

Stabilizing Population While population growth has slowed, the absolute number of people continues to increase—by about 1 billion every 13 years. Slowing population growth would help improve living standards and would buy time to protect natural resources. In the long run, to sustain higher living standards, world population size must stabilize.

The Earth and Its People

As the 21st century begins, growing numbers of people and rising levels of consumption per capita are depleting natural resources and degrading the environment. In many places chronic water shortages, loss of arable land, destruction of natural habitats, and widespread pollution undermine public health and threaten economic and social progress. Many experts think that current trends cannot continue much longer without dire consequences.

In most developed countries population is growing slowly or no longer growing at all, but levels of per capita consumption are so high that the environment is under pressure. Most developing countries face even greater pressures, however. Population is growing rapidly, while consumption is increasing as living standards improve. Every person has an equal right to achieve a high standard of living. But, if every person in the world consumed as much as the average American or Western European, the demand for natural resources would exceed nature's supply.

"There is no question that improving standards of living for the current poor of the world, plus providing for the billions still to come, will increase global demand for food, water, energy, wood, housing, sanitation, and disposal of wastes," writes Richard E. Benedick, former US assistant secretary of state responsible for population and environmental policies. One of the world's main challenges is practicing sustainable development—that is, improving living standards today without foreclosing the opportunities of future generations to meet their needs.

An Environmental Scorecard

In 1992, concerned about worsening environmental conditions, delegates to the UN Conference on Environment and Development (UNCED) in Rio de Janeiro, Brazil, stressed the need for action. The Rio "Earth Summit" set specific goals for environmental improvements. Then in 1997 a Special Session of the UN General Assembly—popularly known as the "Rio Plus Five Conference"—met to assess progress toward these goals.

The conclusions were discouraging. In such sectors as land, freshwater, forests, biodiversity, and climate change, the 1997 UN assessment found that conditions either were no better than in 1992 or had worsened.

Despite lower poverty rates, the number of poor people had increased—in large part because of rapid population growth in developing countries, as well as uneven development, and increasing concentration of wealth.

Arable land At the beginning of the 1990s, about 560 million hectares of cropland worldwide were degraded, of a total 1.5 billion hectares. At the end of the decade about 610 million hectares were degraded (265). Soils can become degraded rapidly when they are overworked and thus become more exposed to erosion.

Freshwater Worldwide, the percentage of the population with access to clean freshwater increased during the 1990s. Nevertheless, due to rapid population growth, currently an estimated 1.2 billion people lack potable water—20 percent more than in 1990. Also, about 3 billion people lack adequate sanitation facilities compared with 2 billion in 1990.

Forests Half the world's original forest cover—over 3 billion hectares—has been lost, largely during the past five decades. Deforestation has accelerated since 1990. For instance, tropical forests declined from 1.7 billion hectares in 1990 to 1.4 billion in 1999.

Globally, about 16 million hectares of forest, an area roughly the size of Nepal, are cut, bulldozed, or burned each year. In the Brazilian Amazon the annual deforestation rate has increased by about one-third since 1992.

Biodiversity Human activities already have pushed many plant and animal species into extinction. While no one knows the exact number, there is wide agreement that the rate of extinction will accelerate as population growth and development put more pressure on prime habitats of other species.

Pollution Air pollution, already a serious problem in many cities, is becoming worse as urban populations grow and the number of motor vehicles rises. Water pollution is a serious problem almost everywhere. Biologist Peter Vitousek and colleagues have warned that human numbers and actions risk fundamentally disrupting nature's basic cycling of water, nitrogen, phosphorus, and carbon among the ecosystems. Largely by releasing carbon dioxide into the atmosphere and by destroying or altering biological resources, humanity is causing "rapid, novel, and substantial" changes to the environment

Climate Change At the Rio Earth Summit in 1992, whether the global climate was changing was still a matter of debate. Since then, the evidence has mounted. In 1990 atmospheric concentrations of carbon dioxide—the main climate-changing gas—were measured at about 355 parts per million. In 1997 concentrations were measured at about 364 parts per million. Since 1950 carbon dioxide emissions have increased fourfold.

Poverty During the 1990s the number of people in poverty increased by about 1 billion. In 1990 about 2 billion people were subsisting on the equivalent of US$2 a day or less. By 2000 that number had risen to about 3 billion—half of the world's population.

Population and Sustainable Development

Environmentalists and economists increasingly agree that efforts to protect the environment and to achieve better living standards can be closely linked and are mutually reinforcing. Slowing the

Measuring Population's Impact

There is no easy way to measure the overall impact of human activities on the environment. Nevertheless, several approaches have been developed, as follows:

Environmental Resource Accounting

Environmental resource accounting attempts to place an economic value on "environmental goods and services" used—natural resources that conventionally have been regarded as free and used in common. These include unpolluted freshwater, clean air, ocean life, forests, and wetlands. A recent study by Robert Costanza of the University of Maryland estimated the total value of ecosystem services and products at US$33 trillion per year—an amount that exceeds the total value of the global economy as conventionally measured (US$29 trillion in 1998).

Some economists argue that the value of environmental goods and services should be incorporated into estimates of Gross Domestic Product (GDP), as are manufactured assets. Unlike manufactured capital, which depreciates in value over time, environmental capital (such as forests, fisheries, and unpolluted air and water) currently is not considered to depreciate, and no charge is made against current income as it is used. "A country could exhaust its mineral resources, cut down its forests, erode its soils, pollute its aquifers, and hunt its wildlife and fisheries to extinction, but measured income would not be affected as these natural assets disappeared," notes Robert Repetto of the World Resources Institute.

If natural resources were valued in the same way that manufactured assets are valued, it might help economies learn to use them more efficiently and to conserve them in order to assure continued use in the future. Such valuations also might help indicate the economic benefits of protecting the environment, as well as the ecological benefits. In other terms, instead of continuing to draw down their "environmental capital" until it is gone, economies could begin to live on its interest, maintaining the capital for use indefinitely in the future.

I = P × A × T

The equation $I = P \times A \times T$ represents another effort to describe the overall impact of humanity on the environment. In the equation:

- I is environmental impact,
- P is population (including size, growth, and distribution),
- A is the level of affluence (consumption per capita), and
- T is the level of technology.

Despite its limitations—for instance, inability to assign actual values to each component or to depict changes in the factors over time—the equation is valuable. In particular, it emphasizes that developing countries with large and rapidly growing populations affect the environment, even though their levels of affluence may be low, while at the same time countries in the developed world with little or no population growth have a substantial environmental impact because consumption per capita is so high.

The equation makes clear that slowing population growth is a key part of any strategy to reduce humanity's impact on the environment. For example, even if per capita resource consumption (A) declined or technologies (T) improved enough to reduce the environmental impact (I) of humanity by 10 percent, this gain would be wiped out in less than a decade because world population (P) is growing at over 1 percent per year. Since per capita consumption of resources is expected to increase as living standards rise, protecting the environment requires more efficient production technologies, less waste, and ultimately a stable world population size.

Ecological Footprints of Nations

In 1997, as part of the five-year review of environmental conditions following the Rio Earth Summit, the Earth Council of Costa Rica sponsored a major "Ecological Footprints of Nations" study. The chief researcher was Mathis Wackernagel of the University of Anahuac de Xalapa in Mexico.

Wackernagel's study calculated, nation by nation, the biologically productive areas needed to provide the resources consumed by the population and to absorb their wastes, given prevailing levels of technology. As Wackernagel explained, "Everybody has an impact on the Earth, because they consume the products and services of nature. Their ecological impact corresponds to the amount of nature they occupy to keep them going. In other words, we calculate the 'ecological footprints' of these countries."

Wackernagel and his group calculated the ecological footprints of 52 nations containing 80 percent of the global population and accounting for 95 percent of the World Domestic Product. The researchers concluded that the world's people are using about one-third more of the earth's biological productivity than can be regenerated.

Carrying Capacity

The term "carrying capacity" refers to the number of people the earth can support. Logically, population growth must stop at some point, or the earth would become overcrowded and its resources eventually would be depleted. But what is this maximum human population?

This question has been debated since 1798, when English economist Thomas Malthus predicted that population growth inevitably would outstrip the food and water supply at some point. Since then, estimates of carrying capacity have varied a great deal depending on what assumptions are made about technology, consumption levels, and other factors that are not easily forecast. Some have even argued that the earth's carrying capacity may already have been exceeded in the sense that that the world could support only 2 billion people if the entire world consumed at the rate that Americans and Western Europeans consume.

While nobody can know how many people the earth could support, few would want to find out the hard way—by reaching this theoretical limit. Calculating the maximum number of people who could exist on earth seems less important than determining how resources can be used wisely and managed sustainably to improve living standards without eventually destroying the natural environment that supports life itself.

increase in population, especially in the face of rising per capita demand for natural resources, can take pressure off the environment and buy time to improve living standards on a sustainable basis.

Although it is not clear whether in the long run rapid population growth causes poverty, "it is clear that high fertility leading to rapidly growing population will increase the number of people in poverty in the short run and, in some cases, make escape from poverty more difficult," observes researcher Dennis A. Ahlburg. It is difficult to make investments for the future when resources are already fully used trying to keep up with the current needs of rapidly growing populations.

As population growth slows, countries can invest more in education, health care, job creation, and other improvements that help boost living standards. In turn, as individual income, savings, and investment rise, more resources become available that can boost productivity. This dynamic process has been identified as one of the key reasons that the economies of many Asian countries grew rapidly between 1960 and 1990.

In recent years fertility has been falling in many developing countries and, as a result, annual world population growth has fallen to about 1.4 percent in 2000 compared with about 2 percent in 1960. The UN estimated recently that population is growing by about 78 million per year, down from about 90 million estimated early in the 1990s (243). Still, at the current pace world population increases by about 1 billion every 13 years. World population surpassed 6 billion in 1999 and is projected to rise to over 8 billion by 2025.

Globally, fertility has fallen by half since the 1960s, to about three children per woman (243). In 65 countries, including 9 in the developing world, fertility rates have fallen below replacement level of about two children per woman. Nonetheless, fertility is above replacement level in 123 countries, and in some countries it is substantially above replacement level. In these countries the population continues to increase rapidly.

About 1.7 billion people live in 47 countries where the fertility rate averages between three and five children per woman. Another 730 million people live in 44 countries where the average woman has five children or more.

Almost all population growth is in the developing world. As a result of differences in population growth, Europe's population will decline from 13 percent to 7 percent of world population over the next quarter century, while that of sub-Saharan Africa will rise from 10 percent to 17 percent. The shares of other regions are projected to remain about the same as today.

As population and demand for natural resources continue to grow, environmental limits will become increasingly apparent. Water shortages are expected to affect nearly 3 billion people in 2025, with sub-Saharan Africa worst affected. Many countries could avoid environmental crises if they took steps now to conserve and manage supplies and demand better, while slowing population growth by providing families and individuals with information and services needed to make informed choices about reproductive health.

Family planning programs play a key role. When family planning information and services are widely available and accessible, couples are better able to achieve their fertility desires. "Even in adverse circumstance—low incomes, limited education, and few opportunities for women—family planning programs have meant slower population growth and improved family welfare," the World Bank has noted.

If every country made a commitment to population stabilization and resource conservation, the world would be better able to meet the challenges of sustainable development. Practicing sustainable development requires a combination of wise public investment, effective natural resource management, cleaner agricultural and industrial technologies, less pollution, and slower population growth.

Better resource management protects the environment and preserves nature's productive capacity. Stronger economies can afford to invest more in protecting the environment. Slower population growth can speed economic growth and conserve natural resources.

Critical Thinking

1. It has been more than a decade since this report was published. How have the situations changed?
2. What are the implications of the developing world living a developed world lifestyle?
3. Of the four ways to measure the overall impact of human activities on the environment identified in this report, which one do you think is the most reasonable? Why?

Ecosystems and Human Well-Being

Summary for Decision-Makers

MILLENNIUM ECOSYSTEM ASSESSMENT

Everyone in the world depends completely on Earth's ecosystems and the services they provide, such as food, water, disease management, climate regulation, spiritual fulfillment, and aesthetic enjoyment. Over the past 50 years, humans have changed these ecosystems more rapidly and extensively than in any comparable period of time in human history, largely to meet rapidly growing demands for food, fresh water, timber, fiber, and fuel. This transformation of the planet has contributed to substantial net gains in human well-being and economic development. But not all regions and groups of people have benefited from this process—in fact, many have been harmed. Moreover, the full costs associated with these gains are only now becoming apparent.

Three major problems associated with our management of the world's ecosystems are already causing significant harm to some people, particularly the poor, and unless addressed will substantially diminish the long-term benefits we obtain from ecosystems:

- First, approximately 60 percent (15 out of 24) of the ecosystem services examined during the Millennium Ecosystem Assessment are being degraded or used unsustainably, including fresh water, capture fisheries, air and water purification, and the regulation of regional and local climate, natural hazards, and pests. The full costs of the loss and degradation of these ecosystem services are difficult to measure, but the available evidence demonstrates that they are substantial and growing. Many ecosystem services have been degraded as a consequence of actions taken to increase the supply of other services, such as food. These trade-offs often shift the costs of degradation from one group of people to another or defer costs to future generations.
- Second, there is *established but incomplete* evidence that changes being made in ecosystems are increasing the likelihood of nonlinear changes in ecosystems (including accelerating, abrupt, and potentially irreversible changes) that have important consequences for human well-being. Examples of such changes include disease emergence, abrupt alterations in water quality, the creation of "dead zones" in coastal waters, the collapse of fisheries, and shifts in regional climate.
- Third, the harmful effects of the degradation of ecosystem services (the persistent decrease in the capacity of an ecosystem to deliver services) are being borne disproportionately by the poor, are contributing to growing inequities and disparities across groups of people, and are sometimes the principal factor causing poverty and social conflict. This is not to say that ecosystem changes such as increased food production have not also helped to lift many people out of poverty or hunger, but these changes have harmed other individuals and communities, and their plight has been largely overlooked. In all regions, and particularly in sub-Saharan Africa, the condition and management of ecosystem services is a dominant factor influencing prospects for reducing poverty.

The degradation of ecosystem services is already a significant barrier to achieving the Millennium Development Goals agreed to by the international community in September 2000 and the harmful consequences of this degradation could grow significantly worse in the next 50 years. The consumption of ecosystem services, which is unsustainable in many cases, will continue to grow as a consequence of a likely three- to sixfold increase in global GDP by 2050 even while global population growth is expected to slow and level off in mid-century. Most of the important direct drivers of ecosystem change are unlikely to diminish in the first half of the century and two drivers—climate change and excessive nutrient loading—will become more severe.

Already, many of the regions facing the greatest challenges in achieving the MDGs coincide with those facing significant problems of ecosystem degradation. Rural poor people, a primary target of the MDGs, tend to be most directly reliant on ecosystem services and most vulnerable to changes in those services. More generally, any progress achieved in addressing the MDGs of poverty and hunger eradication, improved health, and environmental sustainability is unlikely to be sustained if most of the ecosystem services on which humanity relies continue to be degraded. In contrast, the sound management of ecosystem services provides cost-effective opportunities for addressing multiple development goals in a synergistic manner.

There is no simple fix to these problems since they arise from the interaction of many recognized challenges, including climate change, biodiversity loss, and land degradation, each of which is complex to address in its own right. Past actions to slow or reverse the degradation of ecosystems have yielded significant benefits, but these improvements have generally not kept pace with growing pressures and demands. Nevertheless, there is tremendous scope

Four Main Findings

- Over the past 50 years, humans have changed ecosystems more rapidly and extensively than in any comparable period of time in human history, largely to meet rapidly growing demands for food, fresh water, timber, fiber, and fuel. This has resulted in a substantial and largely irreversible loss in the diversity of life on Earth.
- The changes that have been made to ecosystems have contributed to substantial net gains in human well-being and economic development, but these gains have been achieved at growing costs in the form of the degradation of many ecosystem services, increased risks of nonlinear changes, and the exacerbation of poverty for some groups of people. These problems, unless addressed, will substantially diminish the benefits that future generations obtain from ecosystems.
- The degradation of ecosystem services could grow significantly worse during the first half of this century and is a barrier to achieving the Millennium Development Goals.
- The challenge of reversing the degradation of ecosystems while meeting increasing demands for their services can be partially met under some scenarios that the MA has considered, but these involve significant changes in policies, institutions, and practices that are not currently under way. Many options exist to conserve or enhance specific ecosystem services in ways that reduce negative trade-offs or that provide positive synergies with other ecosystem services.

for action to reduce the severity of these problems in the coming decades. Indeed, three of four detailed scenarios examined by the MA suggest that significant changes in policies, institutions, and practices can mitigate some but not all of the negative consequences of growing pressures on ecosystems. But the changes required are substantial and are not currently under way.

An effective set of responses to ensure the sustainable management of ecosystems requires substantial changes in institutions and governance, economic policies and incentives, social and behavior factors, technology, and knowledge. Actions such as the integration of ecosystem management goals in various sectors (such as agriculture, forestry, finance, trade, and health), increased transparency and accountability of government and private-sector performance in ecosystem management, elimination of perverse subsidies, greater use of economic instruments and market-based approaches, empowerment of groups dependent on ecosystem services or affected by their degradation, promotion of technologies enabling increased crop yields without harmful environmental impacts, ecosystem restoration, and the incorporation of nonmarket values of ecosystems and their services in management decisions all could substantially lessen the severity of these problems in the next several decades.

The remainder of this Summary for Decision-makers presents the four major findings of the Millennium Ecosystem Assessment on the problems to be addressed and the actions needed to enhance the conservation and sustainable use of ecosystems.

Finding #1: Over the past 50 years, humans have changed ecosystems more rapidly and extensively than in any comparable period of time in human history, largely to meet rapidly growing demands for food, fresh water, timber, fiber, and fuel. This has resulted in a substantial and largely irreversible loss in the diversity of life on Earth.

The structure and functioning of the world's ecosystems changed more rapidly in the second half of the twentieth century than at any time in human history.

- More land was converted to cropland in the 30 years after 1950 than in the 150 years between 1700 and 1850. Cultivated systems (areas where at least 30 percent of the landscape is in croplands, shifting cultivation, confined livestock production, or freshwater aquaculture) now cover one quarter of Earth's terrestrial surface.
- Approximately 20 percent of the world's coral reefs were lost and an additional 20 percent degraded in the last several decades of the twentieth century, and approximately 35 percent of mangrove area was lost during this time (in countries for which sufficient data exist, which encompass about half of the area of mangroves).
- The amount of water impounded behind dams quadrupled since I960, and three to six times as much water is held in reservoirs as in natural rivers. Water withdrawals from rivers and lakes doubled since I960; most water use (70 percent worldwide) is for agriculture.
- Since 1960, flows of reactive (biologically available) nitrogen in terrestrial ecosystems have doubled, and flows of phosphorus have tripled. More than half of all the synthetic nitrogen fertilizer, which was first manufactured in 1913, ever used on the planet has been used since 1985.
- Since 1750, the atmospheric concentration of carbon dioxide has increased by about 32 percent (from about 280 to 376 parts per million in 2003), primarily due to the combustion of fossil fuels and land use changes. Approximately 60 percent of that increase (60 parts per million) has taken place since 1959.

Humans are fundamentally, and to a significant extent irreversibly, changing the diversity of life on Earth, and most of these changes represent a loss ofbiodiversity.

- More than two thirds of the area of 2 of the world's 14 major terrestrial biomes and more than half of the area of 4 other biomes had been converted by 1990, primarily to agriculture.
- Across a range of taxonomic groups, either the population size or range or both of the majority of species is currently declining.
- The distribution of species on Earth is becoming more homogenous; in other words, the set of species in any one region of the world is becoming more similar to the set in other regions primarily as a result of introductions of species, both intentionally and inadvertently in association with increased travel and shipping.

- The number of species on the planet is declining. Over the past few hundred years, humans have increased the species extinction rate by as much as 1,000 times over background rates typical over the planet's history *(medium certainty).* Some 10–30 percent of mammal, bird, and amphibian species are currently threatened with extinction *(medium to high certainty).* Freshwater ecosystems tend to have the highest proportion of species threatened with extinction.
- Genetic diversity has declined globally, particularly among cultivated species.

Most changes to ecosystems have been made to meet a dramatic growth in the demand for food, water, timber, fiber, and fuel. Some ecosystem changes have been the inadvertent result of activities unrelated to the use of ecosystem services, such as the construction of roads, ports, and cities and the discharge of pollutants. But most ecosystem changes were the direct or indirect result of changes made to meet growing demands for ecosystem services, and in particular growing demands for food, water, timber, fiber, and fuel (fuelwood and hydropower).

Between 1960 and 2000, the demand for ecosystem services grew significantly as world population doubled to 6 billion people and the global economy increased more than sixfold. To meet this demand, food production increased by roughly two-and-a-half times, water use doubled, wood harvests for pulp and paper production tripled, installed hydropower capacity doubled, and timber production increased by more than half.

The growing demand for these ecosystem services was met both by consuming an increasing fraction of the available supply (for example, diverting more water for irrigation or capturing more fish from the sea) and by raising the production of some services, such as crops and livestock. The latter has been accomplished through the use of new technologies (such as new crop varieties, fertilization, and irrigation) as well as through increasing the area managed for the services in the case of crop and livestock production and aquaculture.

Finding #2: The changes that have been made to ecosystems have contributed to substantial netgains in human well-being and economic development, but these gains have been achieved, at growing costs in the form of the degradation of many ecosystem services, increased risks of nonlinear changes, and the exacerbation of poverty for some groups of people. These problems, unless addressed, will substantially diminish the benefits that future generations obtain from ecosystems.

In the aggregate, and for most countries, changes made to the world's ecosystems in recent decades have provided substantial benefits for human well-being and national development. Many of the most significant changes to ecosystems have been essential to meet growing needs for food and water; these changes have helped reduce the proportion of malnourished people and improved human health. Agriculture, including fisheries and forestry, has been the mainstay of strategies for the development of countries for centuries, providing revenues that have enabled investments in industrialization and poverty alleviation. Although the value of food production in 2000 was only about 3 percent of gross world product, the agricultural labor force accounts for approximately 22 percent of the world's population, half the world's total labor force, and 24 percent of GDP in countries with per capita incomes of less than $765 (the low-income developing countries, as defined by the World Bank).

These gains have been achieved, however, at growing costs in the form of the degradation of many ecosystem services, increased risks of nonlinear changes in ecosystems, the exacerbation of poverty for some people, and growing inequities and disparities across groups of people.

Degradation and Unsustainable Use of Ecosystem Services

Approximately 60 percent (15 out of 24) of the ecosystem services evaluated in this assessment (including 70 percent of regulating and cultural services) are being degraded or used unsustainably. Ecosystem services that have been degraded over the past 50 years include capture fisheries, water supply, waste treatment and detoxification, water purification, natural hazard protection, regulation of air quality, regulation of regional and local climate, regulation of erosion, spiritual fulfillment, and aesthetic enjoyment. The use of two ecosystem services—capture fisheries and fresh water—is now well beyond levels that can be sustained even at current demands, much less future ones. At least one quarter of important commercial fish stocks are overharvested *(high certainty).* From 5 percent to possibly 25 percent of global freshwater use exceeds long-term accessible supplies and is now met either through engineered water transfers or overdraft of groundwater supplies *(low to medium certainty).* Some 15–35 percent of irrigation withdrawals exceed supply rates and are therefore unsustainable *(low to medium certainty).* While 15 services have been degraded, only 4 have been enhanced in the past 50 years, three of which involve food production: crops, livestock, and aquaculture. Terrestrial ecosystems were on average a net source of CO_2 emissions during the nineteenth and early twentieth centuries, but became a net sink around the middle of the last century, and thus in the last 50 years the role of ecosystems in regulating global climate through carbon sequestration has also been enhanced.

Actions to increase one ecosystem service often cause the degradation of other services. For example, because actions to increase food production typically involve increased use of water and fertilizers or expansion of the area of cultivated land, these same actions often degrade other ecosystem services, including reducing the availability of water for other uses, degrading water quality, reducing biodiversity, and decreasing forest cover (which in turn may lead to the loss of forest products and the release of greenhouse gasses). Similarly, the conversion of forest to agriculture can significantly change the frequency and magnitude of floods, although the nature of this impact depends on the characteristics of the local ecosystem and the type of land cover change.

The degradation of ecosystem services often causes significant harm to human well-being. The information available to assess the consequences of changes in ecosystem services for human well-being is relatively limited. Many ecosystem services have not been monitored, and it is also difficult to estimate the influence of changes in ecosystem services relative to other social, cultural, and economic factors that also affect human well-being. Nevertheless, the following types of evidence demonstrate that the harmful effects of the degradation of ecosystem services on livelihoods, health, and local and national economies are substantial.

- *Most resource management decisions are most strongly influenced by ecosystem services entering markets; as a result, the nonmarketed benefits are often lost or degraded. These nonmarketed benefits are often high and sometimes more valuable than the marketed ones.* For example, one of the most comprehensive studies to date, which examined the marketed and nonmarketed economic values associated with forests in eight Mediterranean countries, found that timber and fuelwood generally accounted for less than a third of total economic value of forests in each country. Values associated with non-wood forest products, recreation, hunting, watershed protection, carbon sequestration, and passive use (values independent of direct uses) accounted for between 25 percent and 96 percent of the total economic value of the forests.
- *The total economic value associated, with managing ecosystems more sustainably is often higher than the value associated with the conversion of the ecosystem through farming, clear-cutlogging, or other intensive uses.* Relatively few studies have compared the total economic value (including values of both marketed and nonmarketed ecosystem services) of ecosystems under alternate management regimes, but some of the studies that do exist have found that the benefit of managing the ecosystem more sustainably exceeded that of converting the ecosystem.
- *The economic andpublic health costs associated with damage to ecosystem services can be substantial.*
    - The early 1990s collapse of the Newfoundland cod fishery due to overfishing resulted in the loss of tens of thousands ofjobs and cost at least $2 billion in income support and retraining.
    - In 1996, the cost of U.K. agriculture resulting from the damage that agricultural practices cause to water (pollution and eutrophication, a process whereby excessive plant growth depletes oxygen in the water), air (emissions of greenhouse gases), soil (off-site erosion damage, emissions of greenhouse gases), and biodiversity was $2.6 billion, or 9 percent of average yearly gross farm receipts for the 1990s. Similarly, the damage costs of freshwater eutrophication alone in England and Wales (involving factors including reduced value of waterfront dwellings, water treatment costs, reduced recreational value of water bodies, and tourism losses) was estimated to be $105–160 million per year in the 1990s, with an additional $77 million a year being spent to address those damages.
    - The incidence of diseases of marine organisms and the emergence of new pathogens is increasing, and some of these, such as ciguatera, harm human health. Episodes of harmful (including toxic) algal blooms in coastal waters are increasing in frequency and intensity, harming other marine resources such as fisheries as well as human health. In a particularly severe outbreak in Italy in 1989, harmful algal blooms cost the coastal aquaculture industry $10 million and the Italian tourism industry $11.4 million.
    - The frequency and impact of floods and fires has increased significantly in the past 50 years, in part due to ecosystem changes. Examples are the increased susceptibility of coastal populations to tropical storms when mangrove forests are cleared and the increase in downstream flooding that followed land use changes in the upper Yangtze River. Annual economic losses from extreme events increased tenfold from the 1950s to approximately $70 billion in 2003, of which natural catastrophes (floods, fires, storms, drought, earthquakes) accounted for 84 percent of insured losses.
- *The impact of the loss of cultural services is particularly difficult to measure, but it is especially important for many people.* Human cultures, knowledge systems, religions, and social interactions have been strongly influenced by ecosystems. A number of the MA sub-global assessments found that spiritual and cultural values of ecosystems were as important as other services for many local communities, both in developing countries (the importance of sacred groves of forest in India, for example) and industrial ones (the importance of urban parks, for instance).

The degradation of ecosystem services represents loss of a capital asset. Both renewable resources such as ecosystem services and nonrenewable resources such as mineral deposits, some soil nutrients, and fossil fuels are capital assets. Yet traditional national accounts do not include measures of resource depletion or of the degradation of these resources. As a result, a country could cut its forests and deplete its fisheries, and this would show only as a positive gain in GDP (a measure of current economic well-being) without registering the corresponding decline in assets (wealth) that is the more appropriate measure of future economic well-being. Moreover, many ecosystem services (such as fresh water in aquifers and the use of the atmosphere as a sink for pollutants) are available freely to those who use them, and so again their degradation is not reflected in standard economic measures.

When estimates of the economic losses associated with the depletion of natural assets are factored into measurements of the total wealth of nations, they significantly change the balance sheet of countries with economies significantly dependent on natural resources. For example, countries such as Ecuador, Ethiopia, Kazakhstan, Democratic Republic of Congo, Trinidad and Tobago, Uzbekistan, and Venezuela that had positive growth in net savings in 2001, reflecting a growth in the net wealth of the country, actually experienced a loss in net savings when depletion of natural resources (energy and forests) and estimated damages from carbon emissions (associated with contributions to climate change) were factored into the accounts.

While degradation of some services may sometimes be warranted to produce a greater gain in other services, often more degradation of ecosystem services takes place than is in society's interests because many of the services degraded are "public goods." Although people benefit from ecosystem services such as the regulation of air and water quality or the presence of an aesthetically pleasing landscape, there is no market for these services and no one person has an incentive to pay to maintain the good. And when an action results in the degradation of a service that harms other individuals, no market mechanism exists (nor, in many cases, could it exist) to ensure that the individuals harmed are compensated for the damages they suffer.

Wealthy populations cannot be insulated from the degradation of ecosystem services. Agriculture, fisheries, and forestry once formed the bulk of national economies, and the control of natural resources dominated policy agendas. But while these natural resource industries are often still important, the relative economic and political significance of other industries in industrial countries has grown over the past century as a result of the ongoing transition from agricultural to industrial and service economies, urbanization, and the development of new technologies to increase the production of some services and provide substitutes for others. Nevertheless, the degradation of ecosystem services influences human well-being in industrial regions and among wealthy populations in developing countries in many ways:

- The physical, economic, or social impacts of ecosystem service degradation may cross boundaries. For example, land degradation and associated dust storms or fires in one country can degrade air quality in other countries nearby.
- Degradation of ecosystem services exacerbates poverty in developing countries, which can affect neighboring industrial countries by slowing regional economic growth and contributing to the outbreak of conflicts or the migration of refugees.
- Changes in ecosystems that contribute to greenhouse gas emissions contribute to global climate changes that affect all countries.
- Many industries still depend directly on ecosystem services. The collapse of fisheries, for example, has harmed many communities in industrial countries. Prospects for the forest, agriculture, fishing, and ecotourism industries are all directly tied to ecosystem services, while other sectors such as insurance, banking, and health are strongly, if less directly, influenced by changes in ecosystem services.
- Wealthy populations of people are insulated from the harmful effects of some aspects of ecosystem degradation, but not all. For example, substitutes are typically not available when cultural services are lost.
- Even though the relative economic importance of agriculture, fisheries, and forestry is declining in industrial countries, the importance of other ecosystem services such as aesthetic enjoyment and recreational options is growing.

It is difficult to assess the implications of ecosystem changes and to manage ecosystems effectively because many of the effects are slow to become apparent, because they may be expressed primarily at some distance from where the ecosystem was changed, and because the costs and benefits of changes often accrue to different sets of stakeholders. Substantial inertia (delay in the response of a system to a disturbance) exists in ecological systems. As a result, long time lags often occur between a change in a driver and the time when the full consequences of that change become apparent. For example, phosphorus is accumulating in large quantities in many agricultural soils, threatening rivers, lakes, and coastal oceans with increased eutrophication. But it may take years or decades for the full impact of the phosphorus to become apparent through erosion and other processes. Similarly, it will take centuries for global temperatures to reach equilibrium with changed concentrations of greenhouse gases in the atmosphere and even more time for biological systems to respond to the changes in climate.

Moreover, some of the impacts of ecosystem changes may be experienced only at some distance from where the change occurred. For example, changes in upstream catchments affect water flow and water quality in downstream regions; similarly, the loss of an important fish nursery area in a coastal wetland may diminish fish catch some distance away. Both the inertia in ecological systems and the temporal and spatial separation of costs and benefits of ecosystem changes often result in situations where the individuals experiencing harm from ecosystem changes (future generations, say, or downstream landowners) are not the same as the individuals gaining the benefits. These temporal and spatial patterns make it extremely difficult to fully assess costs and benefits associated with ecosystem changes or to attribute costs and benefits to different stakeholders. Moreover, the institutional arrangements now in place to manage ecosystems are poorly designed to cope with these challenges.

Increased Likelihood of Nonlinear (Stepped) and Potentially Abrupt Changes in Ecosystems

There is *established but incomplete* evidence that changes being made in ecosystems are increasing the likelihood of nonlinear changes in ecosystems (including accelerating, abrupt, and potentially irreversible changes), with important consequences for human well-being. Changes in ecosystems generally take place gradually. Some changes are nonlinear, however: once a threshold is crossed, the system changes to a very different state. And these nonlinear changes are sometimes abrupt; they can also be large in magnitude and difficult, expensive, or impossible to reverse. Capabilities for predicting some nonlinear changes are improving, but for most ecosystems and for most potential nonlinear changes, while science can often warn of increased risks of change it cannot predict the thresholds at which the change will be encountered. Examples of large-magnitude nonlinear changes include:

- *Disease emergence.* If, on average, each infected person infects at least one other person, then an epidemic spreads, while if the infection is transferred on average to less than one person, the epidemic dies out. During the 1997-98 El Nino, excessive flooding caused cholera epidemics in Djibouti, Somalia, Kenya, Tanzania, and Mozambique. Warming of the African Great Lakes due to climate change may create conditions that increase the risk of cholera transmission in the surrounding countries.
- *Eutrophication and hypoxia.* Once a threshold of nutrient loading is achieved, changes in freshwater and coastal ecosystems can be abrupt and extensive, creating harmful algal blooms (including blooms of toxic species) and sometimes leading to the formation of oxygen-depleted zones, killing most animal life.

- *Fisheries collapse.* For example, the Atlantic cod stocks off the east coast of Newfoundland collapsed in 1992, forcing the closure of the fishery after hundreds of years of exploitation. Most important, depleted stocks may take years to recover, or not recover at all, even if harvesting is significantly reduced or eliminated entirely.
- *Species introductions and losses.* The introduction of the zebra mussel into aquatic systems in the United States, for instance, resulted in the extirpation of native clams in Lake St. Clair and annual costs of $100 million to the power industry and other users.
- *Regional climate change.* Deforestation generally leads to decreased rainfall. Since forest existence crucially depends on rainfall, the relationship between forest loss and precipitation decrease can form a positive feedback, which, under certain conditions, can lead to a nonlinear change in forest cover.

The growing bushmeat trade poses particularly significant threats associated with nonlinear changes, in this case accelerating rates of change. Growth in the use and trade of bushmeat is placing increasing pressure on many species, especially in Africa and Asia. While the population size of harvested species may decline gradually with increasing harvest for some time, once the harvest exceeds sustainable levels, the rate of decline of populations of the harvested species will tend to accelerate. This could place them at risk of extinction and also reduce the food supply of people dependent on these resources in the longer term. At the same time, the bushmeat trade involves relatively high levels of interaction between humans and some relatively closely related wild animals that are eaten. Again, this increases the risk of a nonlinear change, in this case the emergence of new and serious pathogens. Given the speed and magnitude of international travel today, new pathogens could spread rapidly around the world.

The increased likelihood of these nonlinear changes stems from the loss of biodiversity and growing pressures from multiple direct drivers of ecosystem change. The loss of species and genetic diversity decreases the resilience of ecosystems, which is the level of disturbance that an ecosystem can undergo without crossing a threshold to a different structure or functioning. In addition, growing pressures from drivers such as overharvesting, climate change, invasive species, and nutrient loading push ecosystems toward thresholds that they might otherwise not encounter.

Exacerbation of Poverty for Some Individuals and Groups of People and Contribution to Growing Inequities and Disparities across Groups of People

Despite the progress achieved in increasing the production and use of some ecosystem services, levels of poverty remain high, inequities are growing, and many people still do not have a sufficient supply of or access to ecosystem services.

- In 2001, 1.1 billion people survived on less than $1 per day of income, with roughly 70 percent of them in rural areas where they are highly dependent on agriculture, grazing, and hunting for subsistence.
- Inequality in income and other measures of human well-being has increased over the past decade. A child born in sub-Saharan Africa is 20 times more likely to die before age 5 than a child born in an industrial country, and this disparity is higher than it was a decade ago. During the 1990s, 21 countries experienced declines in their rankings in the Human Development Index (an aggregate measure of economic well-being, health, and education); 14 of them were in sub-Saharan Africa.
- Despite the growth in per capita food production in the past four decades, an estimated 852 million people were undernourished in 2000–02, up 37 million from the period 1997–99. South Asia and sub-Saharan Africa, the regions with the largest numbers of undernourished people, are also the regions where growth in per capita food production has lagged the most. Most notably, per capita food production has declined in sub-Saharan Africa.
- Some 1.1 billion people still lack access to improved water supply, and more than 2.6 billion lack access to improved sanitation. Water scarcity affects roughly 1—2 billion people worldwide. Since 1960, the ratio of water use to accessible supply has grown by 20 percent per decade.

The degradation of ecosystem services is harming many of the world's poorest people and is sometimes the principal factor causing poverty.

- Half the urban population in Africa, Asia, Latin America, and the Caribbean suffers from one or more diseases associated with inadequate water and sanitation. Worldwide, approximately 1.7 million people die annually as a result of inadequate water, sanitation, and hygiene.
- The declining state of capture fisheries is reducing an inexpensive source of protein in developing countries. Per capita fish consumption in developing countries, excluding China, declined between 1985 and 1997.
- Desertification affects the livelihoods of millions of people, including a large portion of the poor in drylands.

The pattern of "winners" and "losers" associated with ecosystem changes—and in particular the impact of ecosystem changes on poor people, women, and indigenous peoples—has not been adequately taken into account in management decisions. Changes in ecosystems typically yield benefits for some people and exact costs on others who may either lose access to resources or livelihoods or be affected by externalities associated with the change. For several reasons, groups such as the poor, women, and indigenous communities have tended to be harmed by these changes.

- Many changes in ecosystem management have involved the privatization of what were formerly common pool resources. Individuals who depended on those resources (such as indigenous peoples, forest-dependent communities, and other groups relatively marginalized from political and economic sources of power) have often lost rights to the resources.
- Some of the people and places affected by changes in ecosystems and ecosystem services are highly vulnerable and poorly equipped to cope with the major changes in

ecosystems that may occur. Highly vulnerable groups include those whose needs for ecosystem services already exceed the supply, such as people lacking adequate clean water supplies, and people living in areas with declining per capita agricultural production.

- Significant differences between the roles and rights of men and women in many societies lead to increased vulnerability of women to changes in ecosystem services.
- The reliance of the rural poor on ecosystem services is rarely measured and thus typically overlooked in national statistics and poverty assessments, resulting in inappropriate strategies that do not take into account the role of the environment in poverty reduction. For example, a recent study that synthesized data from 17 countries found that 22 percent ofhousehold income for rural communities in forested regions comes from sources typically not included in national statistics, such as harvesting wild food, fuel-wood, fodder, medicinal plants, and timber. These activities generated a much higher proportion of poorer families' total income than of wealthy families', and this income was of particular significance in periods of both predictable and unpredictable shortfalls in other livelihood sources.

Development prospects in dryland regions of developing countries are especially dependent on actions to avoid the degradation of ecosystems and slow or reverse degradation where it is occurring. Dryland systems cover about 41 percent of Earth's land surface and more than 2 billion people inhabit them, more than 90 percent of whom are in developing countries. Dryland ecosystems (encompassing both rural and urban regions of drylands) experienced the highest population growth rate in the 1990s of any of the systems examined in the MA. Although drylands are home to about one third of the human population, they have only 8 percent of the world's renewable water supply. Given the low and variable rainfall, high temperatures, low soil organic matter, high costs of delivering services such as electricity or piped water, and limited investment in infrastructure due to the low population density, people living in drylands face many challenges. They also tend to have the lowest levels of human well-being, including the lowest per capita GDP and the highest infant mortality rates.

The combination of high variability in environmental conditions and relatively high levels of poverty leads to situations where people can be highly vulnerable to changes in ecosystems, although the presence of these conditions has led to the development of very resilient land management strategies. Pressures on dryland ecosystems already exceed sustainable levels for some ecosystem services, such as soil formation and water supply, and are growing. Per capita water availability is currently only two thirds of the level required for minimum levels ofhuman well-being. Approximately 10–20 percent of the world's drylands are degraded *(medium certainty)* directly harming the people living in these areas and indirectly harming a larger population through biophysical impacts (dust storms, greenhouse gas emissions, and regional climate change) and through socioeconomic impacts (human migration and deepening poverty sometimes contributing to conflict and instability). Despite these tremendous challenges, people living in drylands and their land management systems have a proven resilience and the capability of preventing land degradation, although this can be either undermined or enhanced by public policies and development strategies.

Finding #3: The degradation of ecosystem services could grow significantly worse during the first half of this century and is a barrier to achieving the Millennium Development Goals.

The MA developed four scenarios to explore plausible futures for ecosystems and human well-being. The scenarios explored two global development paths, one in which the world becomes increasingly globalized and the other in which it becomes increasingly regionalized, as well as two different approaches to ecosystem management, one in which actions are reactive and most problems are addressed only after they become obvious and the other in which ecosystem management is proactive and policies deliberately seek to maintain ecosystem services for the long term.

Most of the direct drivers of change in ecosystems currently remain constant or are growing in intensity in most ecosystems. In all four MA scenarios, the pressures on ecosystems are projected to continue to grow during the first half of this century. The most important direct drivers of change in ecosystems are habitat change (land use change and physical modification of rivers or water withdrawal from rivers), overexploitation, invasive alien species, pollution, and climate change. These direct drivers are often synergistic. For example, in some locations land use change can result in greater nutrient loading (if the land is converted to high-intensity agriculture), increased emissions of greenhouse gases (if forest is cleared), and increased numbers of invasive species (due to the disturbed habitat).

- *Habitat transformation, particularly from conversion to agriculture:* Under the MA scenarios, a further 10–20 percent of grassland and forestland is projected to be converted between 2000 and 2050 (primarily to agriculture). The projected land conversion is concentrated in low-income countries and dryland regions. Forest cover is projected to continue to increase within industrial countries.
- *Overexploitation, especially overfishing:* Over much of the world, the biomass of fish targeted in fisheries (including that of both the target species and those caught incidently) has been reduced by 90 percent relative to levels prior to the onset of industrial fishing, and the fish being harvested are increasingly coming from the less valuable lower trophic levels as populations of higher trophic level species are depleted. These pressures continue to grow in all the MA scenarios.
- *Invasive alien species:* The spread of invasive alien species and disease organisms continues to increase because of both deliberate translocations and accidental introductions related to growing trade and travel, with significant harmful consequences to native species and many ecosystem services.
- *Pollution, particularly nutrient loading:* Humans have already doubled the flow of reactive nitrogen on the continents, and some projections suggest that this may increase by roughly a further two thirds by 2050. Three out of four MA scenarios project that the global flux of nitrogen to coastal ecosystems will increase by a further

Box 1
MA Scenarios

The MA developed four scenarios to explore plausible futures for ecosystems and human well-being based on different assumptions about driving forces of change and their possible interactions:

Global Orchestration – This scenario depicts a globally connected society that focuses on global trade and economic liberalization and takes a reactive approach to ecosystem problems but that also takes strong steps to reduce poverty and inequality and to invest in public goods such as infrastructure and education. Economic growth in this scenario is the highest of the four scenarios, while it is assumed to have the lowest population in 2050.

Order from Strength – This scenario represents a regionalized and fragmented world, concerned with security and protection, emphasizing primarily regional markets, paying little attention to public goods, and taking a reactive approach to ecosystem problems. Economic growth rates are the lowest of the scenarios (particularly low in developing countries) and decrease with time, while population growth is the highest.

Adapting Mosaic – In this scenario, regional watershed-scale ecosystems are the focus of political and economic activity. Local institutions are strengthened and local ecosystem management strategies are common; societies develop a strongly proactive approach to the management of ecosystems. Economic growth rates are somewhat low initially but increase with time, and population in 2050 is nearly as high as in *Order from Strength.*

TechnoGarden – This scenario depicts a globally connected world relying strongly on environmentally sound technology, using highly managed, often engineered, ecosystems to deliver ecosystem services, and taking a proactive approach to the management of ecosystems in an effort to avoid problems. Economic growth is relatively high and accelerates, while population in 2050 is in the mid-range of the scenarios.

The scenarios are not predictions; instead they were developed to explore the unpredictable features of change in drivers and ecosystem services. No scenario represents business as usual, although all begin from current conditions and trends.

Both quantitative models and qualitative analyses were used to develop the scenarios. For some drivers (such as land use change and carbon emissions) and ecosystem services (water withdrawals, food production), quantitative projections were calculated using established, peer-reviewed global models. Other drivers (such as rates of technological change and economic growth), ecosystem services (particularly supporting and cultural services, such as soil formation and recreational opportunities), and human well-being indicators (such as human health and social relations) were estimated qualitatively. In general, the quantitative models used for these scenarios addressed incremental changes but failed to address thresholds, risk of extreme events, or impacts of large, extremely costly, or irreversible changes in ecosystem services. These phenomena were addressed qualitatively by considering the risks and impacts of large but unpredictable ecosystem changes in each scenario.

Three of the scenarios – *Global Orchestration, Adapting Mosaic,* and *TechnoGarden* incorporate significant changes in policies aimed at addressing sustainable development challenges. In *Global Orchestration* trade barriers are eliminated, distorting subsidies are removed, and a major emphasis is placed on eliminating poverty and hunger. In *Adapting Mosaic,* by 2010, most countries are spending close to 13 percent of their GDP on education (as compared to an average of 3.5 percent in 2000), and institutional arrangements to promote transfer of skills and knowledge among regional groups proliferate. In *TechnoGarden* policies are put in place to provide payment to individuals and companies that provide or maintain the provision of ecosystem services. For example, in this scenario, by 2015, roughly 50 percent of European agriculture, and 10 percent of North American agriculture is aimed at balancing the production of food with the production of other ecosystem services. Under this scenario, significant advances occur in the development of environmental technologies to increase production of services, create substitutes, and reduce harmful trade-offs.

10–20 percent by 2030 (medium certainty), with almost all of this increase occurring in developing countries. Excessive flows of nitrogen contribute to eutrophication of freshwater and coastal marine ecosystems and acidification of freshwater and terrestrial ecosystems (with implications for biodiversity in these ecosystems). To some degree, nitrogen also plays a role in creation of ground-level ozone (which leads to loss of agricultural and forest productivity), destruction of ozone in the stratosphere (which leads to depletion of the ozone layer and increased UV-B radiation on Earth, causing increased incidence of skin cancer), and climate change. The resulting health effects include the consequences of ozone pollution on asthma and respiratory function, increased allergies and asthma due to increased pollen production, the risk of blue-baby syndrome, increased risk of cancer and other chronic diseases from nitrates in drinking water, and increased risk of a variety of pulmonary and cardiac diseases from the production of fine particles in the atmosphere.

- *Anthropogenic Climate Change:* Observed recent changes in climate, especially warmer regional temperatures, have already had significant impacts on biodiversity and ecosystems, including causing changes in species distributions, population sizes, the timing of reproduction or migration events, and an increase in the frequency of pest and disease outbreaks. Many coral reefs have undergone major, although often partially reversible, bleaching episodes when local sea surface temperatures have increased during one month by 0.5 – 1° Celsius above the average of the hottest months

By the end of the century, climate change and its impacts may be the dominant direct driver of biodiversity loss and changes in ecosystem services globally. The scenarios developed by the Intergovernmental Panel on Climate Change project an increase in global mean surface temperature of 2.0 – 6.4° Celsius above preindustrial levels by 2100, increased incidence of floods and droughts, and a rise in sea level of an additional 8–88 centimeters between 1990 and 2100. Harm to biodiversity will grow worldwide with increasing rates of change in climate and increasing absolute amounts of change. In contrast, some ecosystem services in some regions may initially be enhanced by projected changes in climate (such as increases in temperature or precipitation), and thus these regions may experience net benefits at low levels of climate change. As climate change becomes more severe, however, the harmful impacts on ecosystem services outweigh the benefits in most regions of the world. The balance of scientific evidence suggests that there will be a significant net harmful impact on ecosystem services worldwide if global mean surface temperature increases more than 2° Celsius above preindustrial levels or at rates greater than 0.2° Celsius per decade *(medium certainty)*. There is a wide band of uncertainty in the amount of warming that would result from any stabilized greenhouse gas concentration, but based on IPCC projections this would require an eventual CO_2 stabilization level of less than 450 parts per million carbon dioxide *(medium certainty)*.

Under all four MA scenarios, the projected changes in drivers result in significant growth in consumption of ecosystem services, continued loss of biodiversity, and further degradation of some ecosystem services.

- During the next 50 years, demand for food crops is projected to grow by 70–85 percent under the MA scenarios, and demand for water by between 30 percent and 85 percent. Water withdrawals in developing countries are projected to increase significantly under the scenarios, although these are projected to decline in industrial countries *(medium certainty)*.
- Food security is not achieved under the MA scenarios by 2050, and child malnutrition is not eradicated (and is projected to increase in some regions in some MA scenarios) despite increasing food supply and more diversified diets *(medium certainty)*.
- A deterioration of the services provided by freshwater resources (such as aquatic habitat, fish production, and water supply for households, industry, and agriculture) is found in the scenarios, particularly in those that are reactive to environmental problems *(medium certainty)*.
- Habitat loss and other ecosystem changes are projected to lead to a decline in local diversity of native species in all four MA scenarios by 2050 *(high certainty)*. Globally, the equilibrium number of plant species is projected to be reduced by roughly 10–15 percent as the result of habitat loss alone over the period of 1970 to 2050 in the MA scenarios *(low certainty)*, and other factors such as overharvesting, invasive species, pollution, and climate change will further increase the rate of extinction.

The degradation of ecosystem services poses a significant barrier to the achievement of the Millennium Development Goals and the MDG targets for 2015. The eight Millennium Development Goals adopted by the United Nations in 2000 aim to improve human well-being by reducing poverty, hunger, child and maternal mortality, by ensuring education for all, by controlling and managing diseases, by tackling gender disparity, by ensuring environmental sustainability, and by pursuing global partnerships. Under each of the MDGs, countries have agreed to targets to be achieved by 2015. Many of the regions facing the greatest challenges in achieving these targets coincide with regions facing the greatest problems of ecosystem degradation.

Although socioeconomic policy changes will play a primary role in achieving most of the MDGs, many of the targets (and goals) are unlikely to be achieved without significant improvement in management of ecosystems. The role of ecosystem changes in exacerbating poverty (Goal 1, Target 1) for some groups of people has been described already, and the goal of environmental sustainability, including access to safe drinking water (Goal 7, Targets 9, 10, and 11), cannot be achieved as long as most ecosystem services are being degraded. Progress toward three other MDGs is particularly dependent on sound ecosystem management:

- Hunger (Goal 1, Target 2): All four MA scenarios project progress in the elimination of hunger but at rates far slower than needed to attain the internationally agreed target of halving, between 1990 and 2015, the share of people suffering from hunger. Moreover, the improvements are slowest in the regions in which the problems are greatest: South Asia and sub-Saharan Africa. Ecosystem condition, in particular climate, soil degradation, and water availability, influences progress toward this goal through its effect on crop yields as well as through impacts on the availability of wild sources of food.
- Child mortality (Goal 4): Undernutrition is the underlying cause of a substantial proportion of all child deaths. Three of the MA scenarios project reductions in child undernourishment by 2050 of between 10 percent and 60 percent but undernourishment increases by 10 percent in *Order from Strength (low certainty)*. Child mortality is also strongly influenced by diseases associated with water quality. Diarrhea is one of the predominant causes of infant deaths worldwide. In sub-Saharan Africa, malaria additionally plays an important part in child mortality in many countries of the region.
- Disease (Goal 6): In the more promising MA scenarios, progress toward Goal 6 is achieved, but under *Order from Strength* it is plausible that health and social conditions for the North and South could further diverge, exacerbating health problems in many low-income regions. Changes in ecosystems influence the abundance of human pathogens such as malaria and cholera as well as the risk of emergence of new diseases. Malaria is responsible for 11 percent of the disease burden in Africa, and it is estimated that Africa's GDP could have been $100 billion larger in 2000 (roughly a 25 percent increase) if malaria had been eliminated 35 years ago. The prevalence of the following infectious diseases is particularly strongly influenced by ecosystem change: malaria, schistosomiasis, lymphatic filariasis, Japanese encephalitis, dengue fever, leishmaniasis, Chagas disease, meningitis, cholera, West Nile virus, and Lyme disease.

Finding #4: The challenge of reversing the degradation of ecosystems while meeting increasing demands for their services can be partially met under some scenarios that the MA considered, but these involve significant changes inpolicies, institutions, and practices that are not currently under way. Many options exist to conserve or enhance specific ecosystem services in ways that reduce negative trade-offs or that provide positive synergies with other ecosystem services.

Three of the four MA scenarios show that significant changes in policies, institutions, and practices can mitigate many of the negative consequences of growing pressures on ecosystems, although the changes required are large and not currently under way. All provisioning, regulating, and cultural ecosystem services are projected to be in worse condition in 2050 than they are today in only one of the four MA scenarios *(Order from Strength).* At least one of the three categories of services is in better condition in 2050 than in 2000 in the other three scenarios. The scale of interventions that result in these positive outcomes are substantial and include significant investments in environmentally sound technology, active adaptive management, proactive action to address environmental problems before their full consequences are experienced, major investments in public goods (such as education and health), strong action to reduce socioeconomic disparities and eliminate poverty, and expanded capacity of people to manage ecosystems adaptively. However, even in scenarios where one or more categories of ecosystem services improve, biodiversity continues to be lost and thus the long-term sustainability of actions to mitigate degradation of ecosystem services is uncertain.

Past actions to slow or reverse the degradation of ecosystems have yielded significant benefits, but these improvements have generally not kept pace with growing pressures and demands. Although most ecosystem services assessed in the MA are being degraded, the extent of that degradation would have been much greater without responses implemented in past decades. For example, more than 100,000 protected areas (including strictly protected areas such as national parks as well as areas managed for the sustainable use of natural ecosystems, including timber or wildlife harvest) covering about 11.7 percent of the terrestrial surface have now been established, and these play an important role in the conservation ofbiodiversity and ecosystem services (although important gaps in the distribution of protected areas remain, particularly in marine and freshwater systems). Technological advances have also helped lessen the increase in pressure on ecosystems caused per unit increase in demand for ecosystem services.

Substitutes can be developed for some but not all ecosystem services, but the cost of substitutes is generally high, and substitutes may also have other negative environmental consequences. For example, the substitution of vinyl, plastics, and metal for wood has contributed to relatively slow growth in global timber consumption in recent years. But while the availability of substitutes can reduce pressure on specific ecosystem services, they may not always have positive net benefits on the environment. Substitution of fuelwood by fossil fuels, for example, reduces pressure on forests and lowers indoor air pollution but it also increases net greenhouse gas emissions. Substitutes are also often costlier to provide than the original ecosystem services.

Ecosystem degradation can rarely be reversed without actions that address the negative effects or enhance the positive effects of one or more of the five indirect drivers of change: population change (including growth and migration), change in economic activity (including economic growth, disparities in wealth, and trade patterns), sociopolitical factors (including factors ranging from the presence of conflict to public participation in decision-making), cultural factors, and technological change. Collectively these factors influence the level of production and consumption of ecosystem services and the sustainability of the production. Both economic growth and population growth lead to increased consumption of ecosystem services, although the harmful environmental impacts of any particular level of consumption depend on the efficiency of the technologies used to produce the service. Too often, actions to slow ecosystem degradation do not address these indirect drivers. For example, forest management is influenced more strongly by actions outside the forest sector, such as trade policies and institutions, macroeco-nomic policies, and policies in other sectors such as agriculture, infrastructure, energy, and mining, than by those within it.

An effective set of responses to ensure the sustainable management of ecosystems must address the indirect and drivers just described and must overcome barriers related to:

- Inappropriate institutional and governance arrangements, including the presence of corruption and weak systems of regulation and accountability.
- Market failures and the misalignment of economic incentives.
- Social and behavioral factors, including the lack of political and economic power of some groups (such as poor people, women, and indigenous peoples) that are particularly dependent on ecosystem services or harmed by their degradation.
- Underinvestment in the development and diffusion of technologies that could increase the efficiency of use of ecosystem services and could reduce the harmful impacts of various drivers of ecosystem change.
- Insufficient knowledge (as well as the poor use of existing knowledge) concerning ecosystem services and management, policy, technological, behavioral, and institutional responses that could enhance benefits from these services while conserving resources.

All these barriers are further compounded by weak human and institutional capacity related to the assessment and management of ecosystem services, underinvestment in the regulation and management of their use, lack of public awareness, and lack of awareness among decision-makers of both the threats posed by the degradation of ecosystem services and the opportunities that more sustainable management of ecosystems could provide.

Box 2
Examples of Promising and Effective Responses for Specific Sectors

Illustrative examples of response options specific to particular sectors judged to be promising or effective are listed below. (See Appendix B.) A response is considered effective when it enhances the target ecosystem services and contributes to human well-being without significant harm to other services or harmful impacts on other groups of people. A response is considered promising if it does not have a long track record to assess but appears likely to succeed or if there are known ways of modifying the response so that it can become effective.

Agriculture

- Removal of production subsidies that have adverse economic, social, and environmental effects.
- Investment in, and diffusion of, agricultural science and technology that can sustain the necessary increase of food supply without harmful tradeoffs involving excessive use of water, nutrients, or pesticides.
- Use of response polices that recognize the role of women in the production and use of food and that are designed to empower women and ensure access to and control of resources necessary for food security.
- Application of a mix of regulatory and incentive- and market-based mechanisms to reduce overuse of nutrients.

Fisheries and Aquaculture

- Reduction of marine fishing capacity.
- Strict regulation of marine fisheries both regarding the establishment and implementation of quotas and steps to address unreported and unregulated harvest. Individual transferable quotas may be appropriate in some cases, particularly for cold water, single species fisheries.
- Establishment of appropriate regulatory systems to reduce the detrimental environmental impacts of aquaculture.
- Establishment of marine protected areas including flexible no-take zones.

Water

- Payments for ecosystem services provided by watersheds.
- Improved allocation of rights to freshwater resources to align incentives with conservation needs.
- Increased transparency of information regarding water management and improved representation of marginalized stakeholders.
- Development of water markets.
- Increased emphasis on the use of the natural environment and measures other than dams and levees for flood control.
- Investment in science and technology to increase the efficiency of water use in agriculture.

Forestry

- Integration of agreed sustainable forest management practices in financial institutions, trade rules, global environment programs, and global security decision-making.
- Empowerment of local communities in support of initiatives for sustainable use of forest products; these initiatives are collectively more significant than efforts led by governments or international processes but require their support to spread.
- Reform of forest governance and development of country-led, strategically focused national forest programs negotiated by stakeholders.

The MA assessed 74 response options for ecosystem services, integrated ecosystem management, conservation and sustainable use of biodiversity, and climate change. Many of these options hold significant promise for overcoming these barriers and conserving or sustainably enhancing the supply of ecosystem services. Promising options for specific sectors are shown in Box 2, while cross-cutting responses addressing key obstacles are described in the remainder of this section.

Institutions and Governance

Changes in institutional and environmental governance frameworks are sometimes required to create the enabling conditions for effective management of ecosystems, while in other cases existing institutions could meet these needs but face significant barriers. Many existing institutions at both the global and the national level have the mandate to address the degradation of ecosystem services but face a variety of challenges in doing so related in part to the need for greater cooperation across sectors and the need for coordinated responses at multiple scales.However, since a number of the issues identified in this assessment are recent concerns and were not specifically taken into account in the design of today's institutions, changes in existing institutions and the development of new ones may sometimes be needed, particularly at the national scale.

In particular, existing national and global institutions are not well designed to deal with the management of common pool resources, a characteristic of many ecosystem services. Issues of ownership and access to resources, rights to participation in decision-making, and regulation of particular types of resource use or discharge of wastes can strongly influence the sustainability of ecosystem management and are fundamental determinants of who wins and loses from changes in ecosystems. Corruption, a major obstacle to effective management of ecosystems, also stems from weak systems of regulation and accountability.

Promising interventions include:

- *Integration of ecosystem management goals within other sectors and within broader development planning frameworks.* The most important public policy decisions affecting ecosystems are often made by agencies and in policy arenas other than those charged with protecting ecosystems. For example, the Poverty Reduction Strategies prepared by developing-country governments for the World Bank and other institutions strongly shape national development priorities, but in general these have not taken into account the importance of ecosystems to improving the basic human capabilities of the poorest.
- *Increased coordination among multilateral environmental agreements and between environmental agreements and other international economic and social institutions.* International agreements are indispensable for addressing ecosystem-related concerns that span national boundaries, but numerous obstacles weaken their current effectiveness. Steps are now being taken to increase the coordination among these mechanisms, and this could help to broaden the focus of the array of instruments. However, coordination is also needed between the multilateral environmental agreements and more politically powerful international institutions, such as economic and trade agreements, to ensure that they are not acting at cross-purposes. And implementation of these agreements needs to be coordinated among relevant institutions and sectors at the national level.
- *Increased transparency and accountability of government and private-sector performance on decisions that have an impact on ecosystems, including through greater involvement ofconcerned stakeholders in decision-making.* Laws, policies, institutions, and markets that have been shaped through public participation in decision-making are more likely to be effective and perceived as just. Stakeholder participation also contributes to the decision-making process because it allows a better understanding of impacts and vulnerability, the distribution of costs and benefits associated with trade-offs, and the identification of a broader range of response options that are available in a specific context. And stakeholder involvement and transparency of decision-making can increase accountability and reduce corruption.

Economics and Incentives

Economic and financial interventions provide powerful instruments to regulate the use of ecosystem goods and services. Because many ecosystem services are not traded in markets, markets fail to provide appropriate signals that might otherwise contribute to the efficient allocation and sustainable use of the services. A wide range of opportunities exists to influence human behavior to address this challenge in the form of economic and financial instruments. However, market mechanisms and most economic instruments can only work effectively if supporting institutions are in place, and thus there is a need to build institutional capacity to enable more widespread use of these mechanisms.

Promising interventions include:

- *Elimination of subsidies that promote excessive use of ecosystem services (and, where possible, transfer of these subsidies to payments for non-marketed ecosystem services).* Government subsidies paid to the agricultural sectors of OECD countries between 2001 and 2003 averaged over $324 billion annually, or one third the global value of agricultural products in 2000. A significant proportion of this total involved production subsidies that led to greater food production in industrial countries than the global market conditions warranted, promoted overuse of fertilizers and pesticides in those countries, and reduced the profitability of agriculture in developing countries. Many countries outside the OECD also have inappropriate input and production subsidies, and inappropriate subsidies are common in other sectors such as water, fisheries, and forestry. Although removal of perverse subsidies will produce net benefits, it will not be without costs. Compensatory mechanisms may be needed for poor people who are adversely affected by the removal of subsidies, and removal of agricultural subsidies within the OECD would need to be accompanied by actions designed to minimize adverse impacts on ecosystem services in developing countries.
- *Greater use of economic instru*ments *and market-based approaches in the management of ecosystem services.* These include:
  - Taxes or user fees for activities with "external" costs (tradeoffs not accounted for in the market). Examples include taxes on excessive application of nutrients or ecotourism user fees.
  - Creation of markets, including through cap-and-trade systems. One of the most rapidly growing markets related to ecosystem services is the carbon market. Approximately 64 million tons of carbon dioxide equivalent were exchanged through projects from January to May 2004, nearly as much as during all of 2003. The value of carbon trades in 2003 was approximately $300 million. About one quarter of the trades involved investment in ecosystem services (hydropower or biomass). It is ***speculated*** that this market may grow to $10 billion to $44 billion by 2010. The creation of a market in the form of a nutrient trading system may also be a low-cost way to reduce excessive nutrient loading in the United States.
  - Payment for ecosystem services. For example, in 1996 Costa Rica established a nationwide system of conservation payments to induce landowners to provide ecosystem services. Under this program, Costa Rica brokers contracts between international and domestic "buyers" and local "sellers" of sequestered carbon, biodiversity, watershed services, and scenic beauty. Another innovative conservation financing mechanism is "biodiversity offsets," whereby developers pay for conservation activities as compensation for unavoidable harm that a project causes to biodiversity.

- Mechanisms to enable consumer preferences to be expressed through markets. For example, current certification schemes for sustainable fisheries and forest practices provide people with the opportunity to promote sustainability through their consumer choices.

Social and Behavioral Responses

Social and behavioral responses—including population policy, public education, civil society actions, and empowerment of communities, women, and youth—can be instrumental in responding to the problem of ecosystem degradation. These are generally interventions that stakeholders initiate and execute through exercising their procedural or democratic rights in efforts to improve ecosystems and human well-being.

Promising interventions include:

- *Measures to reduce aggregate consumption of unsustainably managed ecosystem services.* The choices about what individuals consume and how much are influenced not just by considerations of price but also by behavioral factors related to culture, ethics, and values. Behavioral changes that could reduce demand for degraded ecosystem services can be encouraged through actions by governments (such as education and public awareness programs or the promotion of demand-side management), industry (commitments to use raw materials that are from sources certified as being sustainable, for example, or improved product labeling), and civil society (through raising public awareness). Efforts to reduce aggregate consumption, however, must sometimes incorporate measures to increase the access to and consumption of those same ecosystem services by specific groups such as poor people.
- *Communication and education.* Improved communication and education are essential to achieve the objectives of environmental conventions and the Johannesburg Plan of Implementation as well as the sustainable management of natural resources more generally. Both the public and decision-makers can benefit from education concerning ecosystems and human well-being, but education more generally provides tremendous social benefits that can help address many drivers of ecosystem degradation. While the importance of communication and education is well recognized, providing the human and financial resources to undertake effective work is a continuing problem.
- *Empowerment of groups particularly dependent on ecosystem services or affected by their degradation, including women, indigenous peoples, and young people.* Despite women's knowledge about the environment and the potential they possess, their participation in decision-making has often been restricted by economic, social, and cultural structures. Young people are also key stakeholders in that they will experience the longer-term consequences of decisions made today concerning ecosystem services. Indigenous control of traditional homelands can sometimes have environmental benefits, although the primary justification continues to be based on human and cultural rights.

Technological Responses

Given the growing demands for ecosystem services and other increased pressures on ecosystems, the development and diffusion of technologies designed to increase the efficiency of resource use or reduce the impacts of drivers such as climate change and nutrient loading are essential. Technological change has been essential for meeting growing demands for some ecosystem services, and technology holds considerable promise to help meet future growth in demand. Technologies already exist for reduction of nutrient pollution at reasonable costs—including technologies to reduce point source emissions, changes in crop management practices, and precision farming techniques to help control the application of fertilizers to a field, for example—but new policies are needed for these tools to be applied on a sufficient scale to slow and ultimately reverse the increase in nutrient loading (even while increasing nutrient application in regions such as sub-Saharan Africa where too little fertilizer is being applied). However, negative impacts on ecosystems and human well-being have sometimes resulted from new technologies, and thus careful assessment is needed prior to their introduction. Promising interventions include:

- *Promotion of technologies that enable increased crop yields without harmful impacts related to water, nutrient, and pesticide use.* Agricultural expansion will continue to be one of the major drivers of biodiversity loss well into the twenty-first century. Development, assessment, and diffusion of technologies that could increase the production of food per unit area sustainably without harmful trade-offs related to excessive consumption of water or use of nutrients or pesticides would significantly lessen pressure on other ecosystem services.
- *Restoration of ecosystem services.* Ecosystem restoration activities are now common in many countries. Ecosystems with some features of the ones that were present before conversion can often be established and can provide some of the original ecosystem services. However, the cost of restoration is generally extremely high compared with the cost of preventing the degradation of the ecosystem. Not all services can be restored, and heavily degraded services may require considerable time for restoration.
- *Promotion of technologies to increase energy efficiency and reduce greenhouse gas emissions.* Significant reductions in net greenhouse gas emissions are technically feasible due to an extensive array of technologies in the energy supply, energy demand, and waste management sectors. Reducing projected emissions will require a portfolio of energy production technologies ranging from fuel switching (coal/oil to gas) and increased power plant efficiency to increased use of renewable energy technologies, complemented by more efficient use of energy in the transportation, buildings, and industry sectors. It will also involve the development and implementation of supporting institutions and policies to overcome barriers to the diffusion of these technologies into the market-place, increased public and private-sector funding for research and development, and effective technology transfer.

Knowledge Responses

Effective management of ecosystems is constrained both by the lack of knowledge and information about different aspects of ecosystems and by the failure to use adequately the information that does exist in support of management decisions. In most regions, for example, relatively limited information exists about the status and economic value of most ecosystem services, and their depletion is rarely tracked in national economic accounts. Basic global data on the extent and trend in different types of ecosystems and land use are surprisingly scarce. Models used to project future environmental and economic conditions have limited capability of incorporating ecological "feedbacks," including nonlinear changes in ecosystems, as well as behavioral feedbacks such as learning that may take place through adaptive management of ecosystems.

At the same time, decision-makers do not use all of the relevant information that is available. This is due in part to institutional failures that prevent existing policy-relevant scientific information from being made available to decision-makers and in part to the failure to incorporate other forms of knowledge and information (such as traditional knowledge and practitioners' knowledge) that are often of considerable value for ecosystem management.

Promising interventions include:

- *Incorporation of nonmarket values of ecosystems in resource management and investment decisions.* Most resource management and investment decisions are strongly influenced by considerations of the monetary costs and benefits of alternative policy choices. Decisions can be improved if they are informed by the total economic value of alternative management options and involve deliberative mechanisms that bring to bear noneconomic considerations as well.
- *Use of all relevant forms of knowledge and information in assessments and decision-making, including traditional and practitioners' knowledge.* Effective management of ecosystems typically requires "place-based" knowledge—that is, information about the specific characteristics and history of an ecosystem. Traditional knowledge or practitioners' knowledge held by local resource managers can often be of considerable value in resource management, but it is too rarely incorporated into decision-making processes and indeed is often inappropriately dismissed.
- *Enhancing and sustaining human and institutional capacity for assessing the consequences of ecosystem change for human well-being and acting on such assessments.* Greater technical capacity is needed for agriculture, forest, and fisheries management. But the capacity that exists for these sectors, as limited as it is in many countries, is still vastly greater than the capacity for effective management of other ecosystem services.

A variety of frameworks and methods can be used to make better decisions in the face of uncertainties in data, prediction, context, and scale. Active adaptive management can be a particularly valuable tool for reducing uncertainty about ecosystem management decisions. Commonly used decision-support methods include cost-benefit analysis, risk assessment, multicriteria analysis, the precautionary principle, and vulnerability analysis. Scenarios also provide one means to cope with many aspects of uncertainty, but our limited understanding of ecological systems and human responses shrouds any individual scenario in its own characteristic uncertainty. Active adaptive management is a tool that can be particularly valuable given the high levels of uncertainty surrounding coupled socioecological systems. This involves the design of management programs to test hypotheses about how components of an ecosystem function and interact, thereby reducing uncertainty about the system more rapidly than would otherwise occur.

Sufficient information exists concerning the drivers of change in ecosystems, the consequences of changes in ecosystem services for human well-being, and the merits of various response options to enhance decision-making in support of sustainable development at all scales. However, many research needs and information gaps were identified in this assessment, and actions to address those needs could yield substantial benefits in the form of improved information for policy and action. Due to gaps in data and knowledge, this assessment was unable to answer fully a number of questions posed by its users. Some of these gaps resulted from weaknesses in monitoring systems related to ecosystem services and their linkages with human well-being. In other cases, the assessment revealed significant needs for further research, such the need to improve understanding of nonlinear changes in ecosystems and of the economic value of alternative management options. Investments in improved monitoring and research, combined with additional assessments of ecosystem services in different nations and regions, would significantly enhance the utility of any future global assessment of the consequences of ecosystem change for human well-being.

Critical Thinking

1. Does everyone depend equally on ecosystems and the services they provide?
2. The report refers to "significant changes in policies, institutions, and practices that are not currently underway." Given that it is now more than five years since this report was published, are they underway now?
3. The report suggests "the harmful consequences of [environmental] degradation could grow significantly worse in the next 50 years." This will likely be during the prime of your life. How does this make you feel?

The Anthropocene: Are Humans Now Overwhelming the Great Forces of Nature?

We explore the development of the Anthropocene, the current epoch in which humans and our societies have become a global geophysical force. The Anthropocene began around 1800 with the onset of industrialization, the central feature of which was the enormous expansion in the use of fossil fuels. We use atmospheric carbon dioxide concentration as a single, simple indicator to track the progression of the Anthropocene. From a preindustrial value of 270–275 ppm, atmospheric carbon dioxide had risen to about 310 ppm by 1950. Since then the human enterprise has experienced a remarkable explosion, the Great Acceleration, with significant consequences for Earth System functioning. Atmospheric CO_2 concentration has risen from 310 to 380 ppm since 1950, with about half of the total rise since the preindustrial era occurring in just the last 30 years. The Great Acceleration is reaching criticality. Whatever unfolds, the next few decades will surely be a tipping point in the evolution of the Anthropocene.

WILL STEFFEN, PAUL J. CRUTZEN, AND JOHN R. MCNEILL

Introduction

Global warming and many other human-driven changes to the environment are raising concerns about the future of Earth's environment and its ability to provide the services required to maintain viable human civilizations. The consequences of this unintended experiment of humankind on its own life support system are hotly debated, but worst-case scenarios paint a gloomy picture for the future of contemporary societies.

Underlying global change (Box 1)are human-driven alterations of *i)* the biological fabric of the Earth; *ii)* the stocks and flows of major elements in the planetary machinery such as nitrogen, carbon, phosphorus, and silicon; and *iii)* the energy balance at the Earth's surface (2). The term *Anthropocene* (Box 2) suggests that the Earth has now left its natural geological epoch, the present interglacial state called the Holocene. Human activities have become so pervasive and profound that they rival the great forces of Nature and are pushing the Earth into planetary *terra incognita.* The Earth is rapidly moving into a less biologically diverse, less forested, much warmer, and probably wetter and stormier state.

The phenomenon of global change represents a profound shift in the relationship between humans and the rest of nature. Interest in this fundamental issue has escalated rapidly in the international research community, leading to innovative new research projects like Integrated History and future of People on Earth (IHOPE) (8). The objective of this paper is to explore one aspect of the IHOPE research agenda—the evolution of humans and our societies from hunter-gatherers to a global geophysical force.

To address this objective, we examine the trajectory of the human enterprise through time, from the arrival of humans on Earth through the present and into the next centuries. Our analysis is based on a few critical questions:

- Is the imprint of human activity on the environment discernible at the global scale? How has this imprint evolved through time?
- How does the magnitude and rate of human impact compare with the natural variability of the Earth's environment? Are human effects similar to or greater than the great forces of nature in terms of their influence on Earth System functioning?
- What are the socioeconomic, cultural, political, and technological developments that change the relationship between human societies and the rest of nature and lead to accelerating impacts on the Earth System?

Pre-Anthropocene Events

Before the advent of agriculture about 10000–12000 years ago, humans lived in small groups as hunter-gatherers. In recent centuries, under the influence of noble savage myths, it was often

Box 1
Global Change and the Earth System

The term *Earth System* refers to the suite of interacting physical, chemical and biological global-scale cycles and energy fluxes that provide the life-support system for life at the surface of the planet (1). This definition of the Earth System goes well beyond the notion that the geophysical processes encompassing the Earth's two great fluids—the ocean and the atmosphere—generate the planetary life-support system on their own. In our definition biological/ecological processes are an integral part of the functioning of the Earth System and not merely the recipient of changes in the coupled ocean-atmosphere part of the system. A second critical feature is that forcings and feedbacks *within* the Earth System are as important as external drivers of change, such as the flux of energy from the sun. Finally, the Earth System includes humans, our societies, and our activities; thus, humans are not an outside force perturbing an otherwise natural system but rather an integral and interacting part of the Earth System itself.

We use the term *global change* to mean both the biophysical and the socioeconomic changes that are altering the structure and the functioning of the Earth System. Global change includes alterations in a wide range of global-scale phenomena: land use and land cover, urbanisation, globalisation, coastal ecosystems, atmospheric composition, riverine flow, nitrogen cycle, carbon cycle, physical climate, marine food chains, biological diversity, population, economy, resource use, energy, transport, communication, and so on. Interactions and linkages between the various changes listed above are also part of global change and are just as important as the individual changes themselves. Many components of global change do not occur in linear fashion but rather show strong nonlinearities.

Box 2
The Anthropocene

Holocene ("Recent Whole") is the name given to the postglacial geological epoch of the past ten to twelve thousand years as agreed upon by the International Geological Congress in Bologna in 1885 (3). During the Holocene, accelerating in the industrial period, humankind's activities became growing geological and morphological force, as recognised early by number of scientists. Thus, in 1864, Marsh published book with the title "Man and Nature," more recently reprinted as "The Earth as Modified by Human Action" (4). Stoppani in 1873 rated human activities as "new telluric force which in power and universality may be compared to the greater forces of earth" (quoted from Clark [5]). Stoppani already spoke of the anthropozoic era. Humankind has now inhabited or visited all places on Earth; he has even set foot on the moon. The great Russian geologist and biologist Vernadsky (6) in 1926 recognized the increasing power of humankind in the environment with the following excerpt ". . .the direction in which the processes of evolution must proceed, namely towards increasing consciousness and thought, and forms having greater and greater influence on their surroundings." He, the French Jesuit priest P. Teilhard de Chardin and E. Le Roy in 1924 coined the term "noösphere," the world of thought, knowledge society, to mark the growing role played by humankind's brainpower and technological talents in shaping its own future and environment. A few years ago the term "Anthropocene" has been introduced by one of the authors (P.J.C.) (7) for the current geological epoch to emphasize the central role of humankind in geology and ecology. The impact of current human activities is projected to last over very long periods. For example, because of past and future anthropogenic emissions of CO_2, climate may depart significantly from natural behaviour over the next 50 000 years.

thought that preagricultural humans lived in idyllic harmony with their environment. Recent research has painted a rather different picture, producing evidence of widespread human impact on the environment through predation and the modification of landscapes, often through use of fire (9). However, as the examples below show, the human imprint on environment may have been discernible at local, regional, and even continental scales, but preindustrial humans did not have the technological or organizational capability to match or dominate the great forces of nature.

The mastery of fire by our ancestors provided humankind with a powerful monopolistic tool unavailable to other species, that put us firmly on the long path towards the Anthropocene. Remnants of charcoal from human hearths indicate that the first use of fire by our bipedal ancestors, belonging to the genus *Homo erectus,* occurred a couple of million years ago. Use of fire followed the earlier development of stone tool and weapon making, another major step in the trajectory of the human enterprise.

Early humans used the considerable power of fire to their advantage (9). Fire kept dangerous animals at a respectful distance, especially during the night, and helped in hunting protein-rich, more easily digestible food. The diet of our ancestors changed from mainly vegetarian to omnivorous, a shift that led to enhanced physical and mental capabilities. Hominid brain size nearly tripled up to an average volume of about 1300 cm^3, and gave humans the largest ratio between brain and body size of any species (10). As a consequence, spoken and then, about 10 000 years ago, written language could begin to develop, promoting communication and transfer of knowledge within and between generations of humans, efficient accumulation of knowledge, and social learning over many thousands of years in an impressive catalytic process, involving many human brains and their discoveries and innovations. This power is minimal in other species.

Among the earliest impacts of humans on the Earth's biota are the late Pleistocene megafauna extinctions, a wave of extinctions during the last ice age extending from the woolly mammoth in northern Eurasia to giant wombats in Australia (11–13). A similar wave of extinctions was observed later in the Americas. Although there has been vigorous debate about the relative roles of climate variability and human predation in driving these extinctions, there is little doubt that humans played a significant role, given the strong correlation between the extinction events and human migration patterns. A later but even more profound impact of humans on fauna was the domestication of animals, beginning with the dog up to 100 000 years ago (14) and continuing into the Holocene with horses, sheep, cattle, goats, and the other familiar farm animals. The concomitant domestication of plants during the early to mid-Holocene led to agriculture, which initially also developed through the use of fire for forest clearing and, somewhat later, irrigation (15).

According to one hypothesis, early agricultural development, around the mid-Holocene, affected Earth System functioning so fundamentally that it prevented the onset of the next ice age (16). The argument proposes that clearing of forests for agriculture about 8000 years ago and irrigation of rice about 5000 years ago led to increases in atmospheric carbon dioxide (CO_2) and methane (CH_4) concentrations, reversing trends of concentration decreases established in the early Holocene. These rates of forest clearing, however, were small compared with the massive amount of land transformation that has taken place in the last 300 years (17). Nevertheless, deforestation and agricultural development in the 8000 to 5000 BP period may have led to small increases in CO_2 and CH_4 concentrations (maybe about 5–10 parts per million for CO_2) but increases that were perhaps large enough to stop the onset of glaciation in northeast Canada thousands of years ago. However, recent analyses of solar forcing in the late Quaternary (18) and of natural carbon cycle dynamics (19, 20) argue that natural processes can explain the observed pattern of atmospheric CO_2 variation through the Holocene. Thus, the hypothesis that the advent of agriculture thousands of years ago changed the course of glacial-interglacial dynamics remains an intriguing but unproven beginning of the Anthropocene.

The first significant use of fossil fuels in human history came in China during the Song Dynasty (960–1279) (21, 22). Coal mines in the north, notably Shanxi province, provided abundant coal for use in China's growing iron industry. At its height, in the late 11th century, China's coal production reached levels equal to all of Europe (not including Russia) in 1700. But China suffered many setbacks, such as epidemics and invasions, and the coal industry apparently went into a long decline. Meanwhile in England coal mines provided fuel for home heating, notably in London, from at least the 13th century (23, 24). The first commission charged to investigate the evils of coal smoke began work in 1285 (24). But as a concentrated fuel, coal had its advantages, especially when wood and charcoal grew dear, so by the late 1600s London depended heavily upon it and burned some 360 000 tons annually. The iron forges of Song China and the furnaces of medieval London were regional exceptions, however; most of the world burned wood or charcoal rather than resorting to fuel subsidies from the Carboniferous.

Preindustrial human societies indeed influenced their environment in many ways, from local to continental scales. Most of the changes they wrought were based on knowledge, probably gained from observation and trial-and-error, of natural ecosystem dynamics and its modification to ease the tasks of hunting, gathering, and eventually of farming. Preindustrial societies could and did modify coastal and terrestrial ecosystems but they did not have the numbers, social and economic organisation, or technologies needed to equal or dominate the great forces of Nature in magnitude or rate. Their impacts remained largely local and transitory, well within the bounds of the natural variability of the environment.

The Industrial Era (ca. 1800–1945): Stage 1 of the Anthropocene

One of the three or four most decisive transitions in the history of humankind, potentially of similar importance in the history of the Earth itself, was the onset of industrialization. In the footsteps of the Enlightenment, the transition began in the 1700s in England and the Low Countries for reasons that remain in dispute among historians (25). Some emphasize material factors such as wood shortages and abundant water power and coal in England, while others point to social and political structures that rewarded risk-taking and innovation, matters connected to legal regimes, a nascent banking system, and a market culture. Whatever its origins, the transition took off quickly and by 1850 had transformed England and was beginning to transform much of the rest of the world.

What made industrialization central for the Earth System was the enormous expansion in the use of fossil fuels, first coal and then oil and gas as well. Hitherto humankind had relied on energy captured from ongoing flows in the form of wind, water, plants, and animals, and from the 100-or 200-year stocks held in trees. Fossil fuel use offered access to carbon stored from millions of years of photosynthesis: a massive energy subsidy from the deep past to modern society, upon which a great deal of our modern wealth depends.

Industrial societies as a rule use four or five times as much energy as did agrarian ones, which in turn used three or four times as much as did hunting and gathering societies (26). Without this transition to a high-energy society it is inconceivable that global population could have risen from a billion around 1820 to more than six billion today, or that perhaps one billion of the more fortunate among us could lead lives of comfort unknown to any but kings and courtiers in centuries past.

Prior to the widespread use of fossil fuels, the energy harvest available to humankind was tightly constrained. Water and wind power were available only in favoured locations, and only in societies where the relevant technologies of watermills, sailing ships, and windmills had been developed or imported. Muscular energy derived from animals, and through them from plants, was limited by the area of suitable land for crops and forage, in many places by shortages of water, and everywhere by inescapable biological inefficiencies: plants photosynthesize less than a percent of the solar energy that falls on the Earth, and animals eating those plants retain only a tenth of the chemical energy stored in plants. All this amounted to a bottleneck upon human numbers, the global economy, and the ability of humankind to shape the rest of the biosphere and to influence the functioning of the Earth System.

The invention (some would say refinement) of the steam engine by James Watt in the 1770s and 1780s and the turn to fossil fuels shattered this bottleneck, opening an era of far looser constraints upon energy supply, upon human numbers, and upon the global economy. Between 1800 and 2000 population grew more than six-fold, the global economy about 50-fold, and energy use about 40-fold (27). It also opened an era of intensified and ever-mounting human influence upon the Earth System.

Fossil fuels and their associated technologies—steam engines, internal combustion engines—made many new activities possible and old ones more efficient. For example, with abundant energy it proved possible to synthesize ammonia from atmospheric nitrogen, in effect to make fertilizer out of air, a process pioneered by the German chemist Fritz Haber early in the 20th century. The Haber-Bosch synthesis, as it would become known (Carl Bosch was an industrialist) revolutionized agriculture and sharply increased crop yields all over the world, which, together with vastly improved medical provisions, made possible the surge in human population growth.

The imprint on the global environment of the industrial era was, in retrospect, clearly evident by the early to mid 20th century (28). Deforestation and conversion to agriculture were extensive in the midlatitudes, particularly in the northern hemisphere. Only about 10 percent of the global terrestrial surface had been "domesticated" at the beginning of the industrial era around 1800, but this figure rose significantly to about 25–30 percent by 1950 (17). Human transformation of the hydrological cycle was also evident in the accelerating number of large dams, particularly in Europe and North America (29). The flux of nitrogen compounds through the coastal zone had increased over 10-fold since 1800 (30).

The global-scale transformation of the environment by industrialization was, however, nowhere more evident than in the atmosphere. The concentrations of CH_4 and nitrous oxide (N_2O) had risen by 1950 to about 1250 and 288 ppbv, respectively, noticeably above their preindustrial values of about 850 and 272 ppbv (31, 32). By 1950 the atmospheric CO_2 concentration had pushed above 300 ppmv, above its preindustrial value of 270–275 ppmv, and was beginning to accelerate sharply (33).

Quantification of the human imprint on the Earth System can be most directly related to the advent and spread of fossil fuel-based energy systems, the signature of which is the accumulation of CO_2 in the atmosphere roughly in proportion to the amount of fossil fuels that have been consumed. We propose that atmospheric CO_2 concentration can be used as a single, simple indicator to track the progression of the Anthropocene, to define its stages quantitatively, and to compare the human imprint on the Earth System with natural variability.

Around 1850, near the beginning of Anthropocene Stage 1, the atmospheric CO_2 concentration was 285 ppm, within the range of natural variability for interglacial periods during the late Quaternary period. During the course of Stage 1 from 1800/50 to 1945, the CO_2 concentration rose by about 25 ppm, enough to surpass the upper limit of natural variation through the Holocene and thus provide the first indisputable evidence that human activities were affecting the environment at the global scale. We therefore assign the beginning of the Anthropocene to coincide with the beginning of the industrial era, in the 1800–1850 period. This first stage of the Anthropocene ended abruptly around 1945, when the most rapid and pervasive shift in the human-environment relationship began.

The Great Acceleration (1945–ca. 2015): Stage 2 of the Anthropocene

The human enterprise suddenly accelerated after the end of the Second World War (27) Population doubled in just 50 years, to over 6 billion by the end of the 20th century, but the global economy increased by more than 15-fold. Petroleum consumption has grown by a factor of 3.5 since 1960, and the number of motor vehicles increased dramatically from about 40 million at the end of the War to nearly 700 million by 1996. From 1950 to 2000 the percentage of the world's population living in urban areas grew from 30 to 50 percent and continues to grow strongly. The interconnectedness of cultures is increasing rapidly with the explosion in electronic communication, international travel and the globalization of economies.

The pressure on the global environment from this burgeoning human enterprise is intensifying sharply. Over the past 50 years, humans have changed the world's ecosystems more rapidly and extensively than in any other comparable period in human history (37). The Earth is in its sixth great extinction event, with rates of species loss growing rapidly for both terrestrial and marine ecosystems (38). The atmospheric concentrations of several important greenhouse gases have increased substantially, and the Earth is warming rapidly (39). More nitrogen is now converted from the atmosphere into reactive forms by fertilizer production and fossil fuel combustion than by all of the natural processes in terrestrial ecosystems put together (40).

The remarkable explosion of the human enterprise from the mid-20th century, and the associated global-scale impacts on many aspects of Earth System functioning, mark the second stage of the Anthropocene—the Great Acceleration (41). In many respects the stage had been set for the Great Acceleration by 1890 or 1910. Population growth was proceeding faster than at any previous time in human history, as well as economic growth. Industrialization had gathered irresistible momentum, and was spreading quickly in North America, Europe, Russia, and Japan. Automobiles and airplanes had appeared, and soon rapidly transformed mobility. The world economy was growing ever more tightly linked by mounting flows of migration, trade, and capital. The years 1870 to 1914 were, in fact, an age of globalization in the world economy. Mines and plantations in diverse lands such as Australia, South Africa, and Chile were opening or expanding in response to the emergence of growing markets for their products, especially in the cities of the industrialized world.

At the same time, cities burgeoned as public health efforts, such as checking waterborne disease through sanitation measures, for the first time in world history made it feasible for births consistently to outnumber deaths in urban environments. A major transition was underway in which the characteristic habitat of the human species, which for several millennia had been the village, now was becoming the city. (In 1890 perhaps 200 million people lived in cities worldwide, but by 2000 the figure had leapt to three billion, half of the human population). Cities had long been the seats of managerial and technological innovation and engines of economic growth, and in the Great Acceleration played that role with even greater effect.

However, the Great Acceleration truly began only after 1945. In the decades between 1914 and 1945 the Great Acceleration was stalled by changes in politics and the world economy. Three great wrenching events lay behind this: World War I, the Great Depression, and World War II. Taken together, they slowed population growth, checked—indeed temporarily reversed—the integration and growth of the world economy. They also briefly checked urbanization, as city populations led the way in reducing their birth rates. Some European cities in the 1930s in effect went on reproduction strikes, so that (had they maintained this reluctance) they would have disappeared within decades. Paradoxically, however, these events also helped to initiate the Great Acceleration.

The lessons absorbed about the disasters of world wars and depression inspired a new regime of international institutions after 1945 that helped create conditions for resumed economic growth. The United States in particular championed more open trade and capital flows, reintegrating much of the world economy and helping growth rates reach their highest ever levels in the period from 1950 to 1973. At the same time, the pace of technological change surged. Out of World War II came a number of new technologies—many of which represented new applications for fossil fuels—and a commitment to subsidized research and development, often in the form of alliances among government, industry, and universities. This proved enormously effective and, in a climate of renewed prosperity, ensured unprecedented funding for science and technology, unprecedented recruitment into these fields, and unprecedented advances as well.

The Great Acceleration took place in an intellectual, cultural, political, and legal context in which the growing impacts upon the Earth System counted for very little in the calculations and decisions made in the world's ministries, boardrooms, laboratories, farmhouses, village huts, and, for that matter, bedrooms. This context was not new, but it too was a necessary condition for the Great Acceleration.

The exponential character of the Great Acceleration is obvious from our quantification of the human imprint on the Earth System, using atmospheric CO_2 concentration as the indicator. Although by the Second World War the CO_2 concentration had clearly risen above the upper limit of the Holocene, its growth rate hit a take-off point around 1950. Nearly three-quarters of the anthropogenically driven rise in CO_2 concentration has occurred since 1950 (from about 310 to 380 ppm), and about half of the total rise (48 ppm) has occurred in just the last 30 years.

Stewards of the Earth System? (ca. 2015–?): Stage 3 of the Anthropocene

Humankind will remain a major geological force for many millennia, maybe millions of years, to come. To develop a universally accepted strategy to ensure the sustainability of Earth's life support system against human-induced stresses is one of the greatest research and policy challenges ever to confront humanity. Can humanity meet this challenge?

Signs abound to suggest that the intellectual, cultural, political and legal context that permitted the Great Acceleration after 1945 has shifted in ways that could curtail it (41). Not surprisingly, some reflective people noted human impact upon the environment centuries and even millennia ago. However, as a major societal concern it dates from the 1960s with the rise of modern environmentalism. Observations showed incontrovertibly that the concentration of CO_2 in the atmosphere was rising markedly (42). In the 1980s temperature measurements showed global warming was a reality, a fact that encountered political opposition because of its implications, but within 20 years was no longer in serious doubt (39). Scientific observations showing the erosion of the earth's stratospheric ozone layer led to international agreements reducing the production and use of CFCs (chlorofluorocarbons) (43). On numerous ecological issues local, national, and international environmental policies were devised, and the environment routinely became a consideration, although rarely a dominant one, in political and economic calculations.

This process represents the beginning of the third stage of the Anthropocene, in which the recognition that human activities are indeed affecting the structure and functioning of the Earth System as a whole (as opposed to local-and regional-scale environmental issues) is filtering through to decision-making at many levels. The growing awareness of human influence on the Earth System has been aided by *i)* rapid advances in research and understanding, the most innovative of which is interdisciplinary work on human-environment systems; *ii)* the enormous power of the internet as a global, self-organizing information system; *iii)* the spread of more free and open societies, supporting independent media; and *iv)* the growth of democratic political systems, narrowing the scope for the exercise of arbitrary state power and strengthening the role of civil society. Humanity is, in one way or another, becoming a self-conscious, active agent in the operation of its own life support system (44).

This process is still in train, and where it may lead remains quite uncertain. However, three broad philosophical approaches can be discerned in the growing debate about dealing with the changing global environment (28, 44).

Business-As-Usual

In this conceptualisation of the next stage of the Anthropocene, the institutions and economic system that have driven the Great Acceleration continue to dominate human affairs. This approach is based on several assumptions. First, global change will not be severe or rapid enough to cause major disruptions to the global economic system or to other important aspects of societies, such as human health. Second, the existing market-oriented economic system can deal autonomously with any adaptations that are required. This assumption is based on the fact that as societies have become wealthier, they have dealt effectively with some local and regional pollution problems (45). Examples include the clean-up of major European rivers and the amelioration of the acid rain problem in western Europe and eastern North America. Third, resources required to mitigate global change proactively would be better spent on more pressing human needs.

The business-as-usual approach appears, on the surface, to be a safe and conservative way forward. However, it entails considerable risks. As the Earth System changes in response to human activities, it operates at a time scale that is mis-matched with human decision-making or with the workings of the economic system. The long-term momentum built into the Earth System means that by the time humans realize that a business-as-usual approach may not work, the world will be committed to further decades or even centuries of environmental change. Collapse of modern, globalized society under uncontrollable environmental change is one possible outcome.

An example of this mis-match in time scales is the stability of the cryosphere, the ice on land and ocean and in the soil. Depending on the scenario and the model, the Intergovernmental Panel on Climate Change (IPCC) (39) projected a global average warming of 1.1–6.4°C for 2094–2099 relative to 1980–1999, accompanied by a projected sea-level rise of 0.18–0.59 m (excluding contributions from the dynamics of the large polar ice sheets). However, warming is projected to be more than twice as large as the global average in the polar regions, enhancing ice sheet instability and glacier melting. Recent observations of glacial dynamics suggest a higher degree of instability than estimated by current cryospheric models, which would lead to higher sea level rise through this century than estimated by the IPCC in 2001 (46). It is now conceivable that an irreversible threshold could be crossed in the next several decades, eventually (over centuries or a millennium) leading to the loss of the Greenland ice sheet and consequent sea-level rise of about 5 m.

Mitigation

An alternative pathway into the future is based on the recognition that the threat of further global change is serious enough that it must be dealt with proactively. The mitigation pathway attempts to take the human pressure off of the Earth System by vastly improved technology and management, wise use of Earth's resources, control of human and domestic animal population, and overall careful use and restoration of the natural environment. The ultimate goal is to reduce the human modification of the global environment to avoid dangerous or difficult-to-control levels and rates of change (47), and ultimately to allow the Earth System to function in a pre-Anthropocene way.

Technology must play a strong role in reducing the pressure on the Earth System (48). Over the past several decades rapid advances in transport, energy, agriculture, and other sectors have led to a trend of dematerialization in several advanced economies. The amount and value of economic activity continue to grow but the amount of physical material flowing through the economy does not.

There are further technological opportunities. Worldwide energy use is equivalent to only 0.05 percent of the solar radiation reaching the continents. Only 0.4 percent of the incoming solar radiation, 1 W m^2, is converted to chemical energy by photosynthesis on land. Human appropriation of net primary production is about 10 percent, including agriculture, fiber, and fisheries (49). In addition to the many opportunities for energy conservation, numerous technologies—from solar thermal and photovoltaic through nuclear fission and fusion to wind power and biofuels from forests and crops—are available now or under development to replace fossil fuels.

Although improved technology is essential for mitigating global change, it may not be enough on its own. Changes in societal values and individual behaviour will likely be necessary (50). Some signs of these changes are now evident, but the Great Acceleration has considerable momentum and appears to be intensifying (51). The critical question is whether the trends of dematerialization and shifting societal values become strong enough to trigger a transition of our globalizing society towards a much more sustainable one.

Geo-Engineering Options

The severity of global change, particularly changes to the climate system, may force societies to consider more drastic options. For example, the anthropogenic emission of aerosol particles (e.g., smoke, sulphate, dust, etc.) into the atmosphere leads to a net cooling effect because these particles and their influence on cloud properties enhance backscattering of incoming solar radiation. Thus, aerosols act in opposition to the greenhouse effect, masking some of the warming we would otherwise see now (52). Paradoxically, a clean-up of air pollution can thus increase greenhouse warming, perhaps leading to an additional 1°C of warming and bringing the Earth closer to "dangerous" levels of climate change. This and other amplifying effects, such as feedbacks from the carbon cycle as the Earth warms (53), could render mitigation efforts largely ineffectual. Just to stabilize the atmospheric concentration of CO_2, without taking into account these amplifying effects, requires a reduction in anthropogenic emissions by more than 60 percent—a herculean task considering that most people on Earth, in order to increase their standard of living, are in need of much additional energy. One engineering approach to reducing the amount of CO_2 in the atmosphere is its sequestration in underground reservoirs (54). This "geo-sequestration" would not only alleviate the pressures on climate, but would also lessen the expected acidification of the ocean surface waters, which leads to dissolution of calcareous marine organisms (55).

In this situation some argue for geo-engineering solutions, a highly controversial topic. Geo-engineering involves purposeful manipulation by humans of global-scale Earth System processes with the intention of counteracting anthropogenically driven environmental change such as greenhouse warming (56). One proposal is based on the cooling effect of aerosols noted in the previous paragraph (57). The idea is to artificially enhance the Earth's albedo by releasing sunlight-reflective material, such as sulphate particles, in the stratosphere, where they remain for

1–2 years before settling in the troposphere. The sulphate particles would be produced by the oxidation of SO_2, just as happens during volcanic eruptions. In order to compensate for a doubling of CO_2, if this were to happen, the input of sulphur would have to be about 1–2 Tg S y^1 (compared to an input of about 10 Tg S by Mount Pinatubo in 1991). The sulphur injections would have to occur for as long as CO_2 levels remain high.

Looking more deeply into the evolution of the Anthropocene, future generations of *H. sapiens* will likely do all they can to prevent a new ice-age by adding powerful artificial greenhouse gases to the atmosphere. Similarly, any drop in CO_2 levels to low concentrations, causing strong reductions in photosynthesis and agricultural productivity, might be combated by artificial releases of CO_2, maybe from earlier CO_2 sequestration. And likewise, far into the future, H. sapiens will deflect meteorites and asteroids before they could hit the Earth.

For the present, however, just the suggestion of geoengineering options can raise serious ethical questions and intense debate. In addition to fundamental ethical concerns, a critical issue is the possibility for unintended and unanticipated side effects that could have severe consequences. The cure could be worse than the disease. For the sulphate injection example described above, the residence time of the sulphate particles in the atmosphere is only a few years, so if serious side-effects occurred, the injections could be discontinued and the climate would relax to its former high CO_2 state within a decade.

The Great Acceleration is reaching criticality. Enormous, immediate challenges confront humanity over the next few decades as it attempts to pass through a bottleneck of continued population growth, excessive resource use and environmental deterioration. In most parts of the world the demand for fossil fuels overwhelms the desire to significantly reduce greenhouse gas emissions. About 60 percent of ecosystem services are already degraded and will continue to degrade further unless significant societal changes in values and management occur (37). There is also evidence for radically different directions built around innovative, knowledge-based solutions. Whatever unfolds, the next few decades will surely be a tipping point in the evolution of the Anthropocene.

References and Notes

1. Oldfield, F. and Steffen, W. 2004. The earth system. In: Global Change and the Earth System: A Planet Under Pressure. Steffen, W., Sanderson, A., Tyson, P., Jäger, J., Matson, P., Moore, B. III, Oldfield, F., Richardson, K., et al. (eds). The IGBP Global Change Series, Springer-Verlag, Berlin, Heidelburg, New York, p. 7.
2. Hansen, J., Nazarenko, L., Ruedy, R., Sato, M., Willis, J., Del Genio, A., Koch, D., Lacis, A., et al. 2005. Earth's energy imbalance: comfirmation and implications. *Science 308,* 1431–1435.
3. Encyclopaedia Britannica. 1976. Micropædia, IX. London.
4. Marsh, G.P. 1965. *The Earth as Modified by Human Action.* Belknap Press, Harvard University Press, Cambridge, MA, 504 pp.
5. Clark, W.C. 1986. Chapter 1. In: *Sustainable Development of the Biosphere.* Clark, W.C. and Munn, R.E. (eds). Cambridge University Press, Cambridge, UK, 491 pp.
6. Vernadski, V.I. 1998. *The Biosphere (translated and annotated version from the original of 1926).* Copernicus, Springer, New York, 192 pp.
7. Crutzen, P. J. 2002. Geology of mankind: the anthropocene. *Nature 415,* 23.
8. Costanza, R., Graumlich, L. and Steffen, W. (eds). 2006. *Integrated History and Future of People on Earth.* Dahlem Workshop Report 96, MIT Press, Cambridge, MA, 495 pp.
9. Pyne, S. 1997. *World Fire: The Culture of Fire on Earth.* University of Washington Press, Seattle, 379 pp.
10. Tobias, P.V. 1976. The brain in hominid evolution. In: *Encyclopaedia Britannica,* Macropaedia Volume 8. Encyclopedia Britannica, London, p. 1032.
11. Martin, P.S. and Klein, R.G. 1984. *Quaternary Extinctions: A Prehistoric Revolution.* University of Arizona Press, Tucson. 892 pp.
12. Alroy, J. 2001. A multispecies overkill simulation of the End-Pleistocene Megafaunal mass extinction. *Science 292,* 1893–1896.
13. Roberts, R.G., Flannery, T.F., Ayliffe, L.K., Yoshida, H., Olley, J.M., Prideaux, G.J., Laslett, G.M., Baynes, A., et al. 2001. New ages for the last Australian Megafauna: continent-wide extinction about 46,000 years ago. *Science 292,* 1888–1892.
14. Leach, H.M. 2003. Human domestication reconsidered. *Curr. Anthropol. 44,* 349–368.
15. Smith, B.D. 1995. *The Emergence of Agriculture.* Scientific American Library, New York, 231 pp.
16. Ruddiman, W.F. 2003. The anthropogenic greenhouse era began thousands of years ago. *Climat. Chang. 61,* 261–293.
17. Lambin, E.F. and Geist, H.J. (eds). 2006. *Land-Use and Land-Cover Change: Local Processes and Global Impacts.* The IGBP Global Change Series, Springer-Verlag, Berlin, Heidelberg, New York, 222 pp.
18. EPICA Community Members. 2004. Eight glacial cycles from an Antarctic ice core. *Nature 429,* 623–628.
19. Broecker, W.C. and Stocker, T.F. 2006. The Holocene CO_2 rise: anthropogenic or natural? *Eos 87,* (3), 27–29.
20. Joos, F., Gerber, S., Prentice, I.C., Otto-Bliesner, B.L. and Valdes, P.J. 2004. Transient simulations of Holocene atmospheric carbon dioxide and terrestrial carbon since the Last Glacial Maximum. *Glob.l Biogeochem.* Cycles 18, GB2002.
21. Hartwell, R. 1962. A revolution in the iron and coal industries during the Northern Sung. *J. Asian Stud. 21,* 153–162.
22. Hartwell, R. 1967. A cycle of economic change in Imperial China: coal and iron in northeast China, 750–1350. *J. Soc. and Econ. Hist. Orient 10,* 102–159.
23. TeBrake, W.H. 1975. Air pollution and fuel crisis in preindustrial London, 1250–1650. *Technol. Culture 16,* 337–359.
24. Brimblecombe, P. 1987. *The Big Smoke: A History of Air Pollution in London since Medieval Times.* Methuen, London, 185 pp.
25. Mokyr, J. (ed). 1999. *The British Industrial Revolution: An Economic Perspective.* Westview Press, Boulder, CO, 354 pp.
26. Sieferle, R.-P. 2001. Der Europäische Sonderweg: Ursachen und Factoren. Stuttgart, 53 pp. (In German).
27. McNeill, J.R. 2001. *Something New Under the Sun.* W.W. Norton, New York, London, 416 pp.
28. Steffen, W., Sanderson, A., Tyson, P.D., Jäger, J., Matson, P., Moore, B. III, Oldfield, F., Richardson, K., et al. 2004. *Global Change and the Earth System: A Planet Under Pressure.* The IGBP Global Change Series, Springer-Verlag, Berlin, Heidelberg, New York, 336 pp.
29. Vörösmarty, C.J., Sharma, K., Fekete, B., Copeland, A.H., Holden, J., Marble, J. and Lough, J.A. 1997. The storage and aging of continental runoff in large reservoir systems of the world. *Ambio 26,* 210–219.

30. Mackenzie, F.T., Ver, L.M. and Lerman, A. 2002. Century-scale nitrogen and phosphorus controls of the carbon cycle. *Chem. Geol. 190,* 13–32.
31. Blunier, T., Chappellaz, J., Schwander, J., Barnola, J.-M., Desperts, T., Stauffer, B. and Raynaud, D. 1993. Atmospheric methane record from a Greenland ice core over the last 1000 years. *J. Geophys. Res. 20,* 2219–2222.
32. Machida, T., Nakazawa, T., Fujii, Y., Aoki, S. and Watanabe, O. 1995. Increase in the atmospheric nitrous oxide concentration during the last 250 years. *Geophys. Res. Lett. 22,* 2921–2924.
33. Etheridge, D.M., Steele, L.P., Langenfelds, R.L., Francey, R.J., Barnola, J.-M. and Morgan, V.I. 1996. Natural and anthropogenic changes in atmospheric CO_2 over the last 1000 years from air in Antarctic ice and firn. *J. Geophys. Res. 101,* 4115–4128.
34. Barnola, J.-M., Raynaud, D., Lorius, C. and Barkov, N.I. 2003 Historical CO_2 record from the Vostok ice core. In: *Trends: A Compendium of Data on Global Change.* Carbon Dioxide Information Analysis Cener, Oak Ridge National Laboratory, U.S. Department of Energy, Oak Ridge, TN.
35. Etheridge, D.M., Steele, L.P., Langenfelds, R.L., Francey, R.J., Barnola, J.-M. and Morgan, V.I. 1998. Historical CO_2 records from the Law Dome DE08, DE08-2, and DSS ice cores. In: *Trends: A Compendium of Data on Global Change.* Carbon Dioxide Information Analysis Center, Oak Ridge National Laboratory, U.S. Department of Energy, Oak Ridge, TN.
36. Indermuhle, A., Stocker, T.F., Fischer, H., Smith, H.J., Joos, F., Wahlen, M., Deck, B., Mastroianni, D., et al. 1999. High-resolution Holocene CO_2-record from the Taylor Dame ice core (Antarctica). *Nature 398,* 121–126.
37. Millennium Ecosystem Assessment. 2005. *Ecosystems & Human Well-bing: Synthesis.* Island Press, Washington.
38. Pimm, S.L., Russell, G.J., Gittleman, J.L. and Brooks, T.M. 1995. The future of biodiversity. *Science 269,* 347–350.
39. Intergovernmental Panel on Climate Change (IPCC). 2007. *Climate Change 2007: The Physical Science Basis. Summary for Policymakers.* IPCC Secretariat, World Meteorological Organization, Geneva, Switzerland, 18 pp.
40. Galloway, J.N. and Cowling, E.B. 2002. Reactive nitrogen and the world: two hundred years of change. *Ambio 31,* 64–71.
41. Hibbard, K.A., Crutzen, P.J., Lambin, E.F., Liverman, D., Mantua, N.J., McNeill, J.R., Messerli, B. and Steffen, W. 2006. Decadal interactions of humans and the environment. In: Integrated History and Future of People on Earth. Costanza, R., Graumlich, L. and Steffen, W. (eds). Dahlem Workshop Report 96. MIT Press, Cambridge, MA, pp 341–375.
42. Keeling, C.D. and Whorf, T.P. 2005. Atmospheric CO_2 records from sites in the SIO air sampling network. In: *Trends: A Compendium of Data on Global Change.* Carbon Dioxide Information Analysis Center, Oak Ridge National Laboratory, U.S. Department of Energy, Oak Ridge, TN.
43. Crutzen, P. 1995. My life with O_3, NO_x and other YZO_xs. In: *Les Prix Nobel (The Nobel Prizes) 1995.* Almqvist & Wiksell International, Stockholm. pp. 123–157.
44. Schellnhuber, H.-J. 1998. Discourse: Earth System analysis: the scope of the challenge. In: *Earth System Analysis.* Schellnhuber, H.-J. and Wetzel, V. (eds). Springer-Verlag, Berlin, Heidelberg, New York, pp. 3–195.
45. Lomborg, B. 2001. *The Skeptical Environmentalist: Measuring the Real State of the World.* Cambridge University Press, Cambridge, UK, 548 pp.
46. Rahmstorf, S. 2007. A semi-empirical approach to projecting future sea-level rise. *Science 315,* 368–370.
47. Schellnhuber, H.J., Cramer, W., Nakicenovic, N., Wigley, T. and Yohe, G. (eds). 2006. *Avoiding Dangerous Climate Change.* Cambridge University Press, Cambridge, UK, 406 pp.
48. Steffen, W. 2002. Will technology spare the planet? In: *Challenges of a Changing Earth: Proceedings of the Global Change Open Science Conference. Amsterdam, The Netherlands, 10–13 July 2001.* Steffen, W., Jäger, J., Carson, D. and Bradshaw, C. (eds). The IGBP Global Change Series, Springer-Verlag, Berlin, Heidelberg, New York, pp 189–191.
49. Haberl, H. 2006. The energetic metabolism of the European Union and the United States, decadal energy inputs with an emphasis on biomass. *J. Ind. Ecol. 10,* 151–171.
50. Fischer, J., Manning, A.D., Steffen, W., Rose, D.B., Danielle, K., Felton, A., Garnett, S., Gilna, B., et al. 2007. Mind the sustainability gap. *Trends Ecol.* Evol. in press.
51. Rahmstorf, S., Cazenave, A., Church, J.A., Hansen, J.E., Keeling, R.F., Parker, D.E., Somerville, R.C.J., et al. 2007. Recent climate observations compared to projections. *Science, 316,* 709.
52. Andreae, M.O., Jones, C.D. and Cox, P.M. 2005. Strong present day aerosol cooling implies a hot future. *Nature 435,* 1187–1190.
53. Friedlingstein, P., Cox, P., Betts, R., Bopp, L., von Bloh, W., Brovkin, V., Doney, V.S., Eby, M.I., et al. 2006. Climate-carbon cycle feedback analysis, results from the C^4MIP model intercomparison. *J. Clim. 19,* 3337–3353.
54. Intergovernmental Panel on Climate Change (IPCC). 2005. *Carbon Dioxide Capture and Storage. A Special Report of Working Group III.* Intergovernmental Panel on Climate Change, Geneva, Switzerland, 430 pp.
55. The Royal Society. 2005. *Ocean Acidification Due to Increasing Atmospheric Carbon Dioxide.* Policy document 12/05. The Royal Society, UK, 68 pp.
56. Schneider, S.H. 2001. Earth systems engineering and management. *Nature 409,* 417–421.
57. Crutzen, P. J. 2006. Albedo enhancement by stratospheric sulfur injections: A contribution to resolve a policy dilemma. *Clim. Chang. 77,* 211–220.
58. Raupach, M.R., Marland, G., Ciais, P., Le Quere, C., Canadell, J.G., Klepper, G. and Field, C.B. 2007. Global and regional drivers of accelerating CO_2 emissions. *Proc. Nat. Acad. Sci. USA.* in press.
59. This paper grew out of discussions at the 96th Dahlem Conference ("Integrated History and future of People on Earth [IHOPE]"), held in Berlin in June 2005. We are grateful to the many colleagues at the Conference who contributed to the stimulating discussions, and to Dr Julia Lupp, the Dahlem Conference organizer, for permission to base this paper on these discussions.
60. First submitted 31 May 2007. Accepted for publication, October 2007.

Critical Thinking

1. Is it important to label a period on Earth's history as dominated by human actions?
2. If humans are truly a part of nature and not apart from it, then what's wrong with them altering the planet in a manner equal to or greater than the "great forces of nature"?
3. Is it reasonable to wait until 2015 to begin Stage 3?

The State of the Nation's Ecosystems 2008

What the Indicators Tell Us

THE H. JOHN HEINZ CENTER

Each of the indicators in *The State of the Nation's Ecosystems 2008* is important, as each contributes to a comprehensive picture of a specific ecosystem or the nation as a whole. However, this *Highlights* report focuses on a subset of the indicators that are relevant across multiple ecosystem types. These indicators were selected because they represent important features of the nation as a whole (the core national indicators) or because they report on features that are critical to more than one ecosystem (for example, nitrate in streams is important to farmlands, forests, and urban and suburban landscapes; fire is important to forests and grasslands and shrublands; and large-scale mortalities are important to coastal and freshwater ecosystems).

A Varied and Changing Landscape

America's ecosystems are vast and immensely varied, ranging from vibrant kelp forests to wide grassy plains, from Arctic tundra to subtropical cypress swamps. Grasslands, shrublands, and forests together cover much of the landscape of the lower 48 states, but there are many regional patterns of the mix of ecosystem types. Croplands—the main component of farmland ecosystems—make up much of the remaining area, followed by freshwater and coastal wetlands and ponds that dot the landscape. While accounting for no more than a few percent of total area, developed land is not only found in major urban centers but is also scattered across much of the remaining landscape. Alaska, which is about one-quarter the size of the lower 48 states, is mostly covered by forests, grasslands, and shrublands, including large expanses of tundra.

In the lower 48 states there are

- 694 million acres of grasslands and shrublands
- 621 million acres of forests
- 400 million acres of croplands
- 96 million acres of freshwater wetlands
- 45 million acres of urban and suburban landscapes
- 6 million acres of freshwater ponds
- 5 million acres of coastal wetlands on the Gulf and Atlantic Coast

In Alaska there are

- 205 million acres of grassland and shrubland (including 135 million acres of tundra)
- 127 million acres of forests
  [Sources: USDA Forest Service (forests), USDA Economic Research Service (croplands), Multi-Resolution Land Characterization (MRLC) Consortium, and ESRI (roadmap used in analysis of urban and suburban landscapes); analysis by the U.S. Environmental Protection Agency and the U.S. Forest Service (grasslands and shrublands, urban and suburban landscapes), and the U.S. Fish and Wildlife Service National Wetlands Inventory (wetlands and ponds).]

The U.S. landscape has changed substantially in the last half-century. There have been increases in the area of urban development and the area of ponds and declines in the area of croplands and freshwater and coastal wetlands. Forest area has not changed significantly nationwide.

- Between 1945 and 2002, the area of "developed land" (a proxy for the area of urban and suburban landscapes) increased from 15 million acres to 60 million acres. (Source: U.S. Census Bureau.)
- Cropland area has declined by 12 percent nationwide since 1982; however, the cropland acreage in the two river basins with the greatest agricultural acreage—the Missouri and the Souris-Red Rainy/Upper Mississippi—has remained relatively stable. (Sources: USDA Natural Resources Conservation Service, National Resources Inventory, and USDA Economic Research Service.)
- Since 1953, forest area for the nation as a whole has changed by less than 1 percent; however, forest area has increased in the North and decreased in the South and Pacific Coast. (Source: USDA Forest Service.)
- Since 1955, the freshwater wetland area has declined by 9 percent. Over the same period, the area of ponds has more than doubled. (Source: U.S. Fish and Wildlife Service.)
- From the mid-1950s to 2004, more than 400,000 acres of vegetated wetlands on the Gulf and Atlantic coasts were lost, a decline of about 9 percent. (Source: U.S. Fish and Wildlife Service.)

As land is converted from one ecosystem type to another, the *pattern* of ecosystems in the landscape often changes along with extent—suburban developments may be built in areas that were formerly forests or grasslands, or abandoned farms may become forest again. Changes in the proximity of ecosystems to one another and the way they are intermingled can affect how these ecosystems function and the good and services they provide. We describe the pattern of small parcels of "natural" land based on the mix of land cover ("natural," cropland, and development) in the 240 acres surrounding each parcel. Highly managed landscapes—be they croplands or developed areas—break up expanses of "natural" lands.

- About 68 percent of the land cover of the lower 48 states is "natural" (forest, grasslands, shrublands, wetlands, lakes, or coastal waters). (Source: Multi-Resolution Land Characterization Consortium and ESRI.)
- Twenty-three percent of the "natural" land cover is described as "core natural" (that is, it has only other "natural" land in the 240 acres surrounding it). The Rocky Mountain region has the highest percentage of "core natural" parcels, and the Midwest has the lowest. (Source: Multi-Resolution Land Characterization Consortium and ESRI).
- Patches of "core natural" parcels were most often 10–100 square miles in size, with 11 percent of these patches at least 1000 square miles—the Rocky Mountain region had the highest percentage of these large patches. (Source: Multi-Resolution Land Characterization Consortium and ESRI.)

Conversion of land from rural to urban or suburban is generally permanent, and this conversion profoundly changes the benefits and services the land provides. Development in the lower-density areas of the urban-suburban landscape and in rural areas can have more ecological consequences than development in areas with higher preexisting housing densities because it can deforest and fragment habitat, interfere with the movement of animals, and reduce stream quality, and it often leads to further development.

- The Eastern United States has a larger proportion of its total area (4 percent to 5 percent) in urban and suburban landscapes than other regions (0.5 percent to 3 percent). (Source: Multi-Resolution Land Characterization Consortium and ESRI.)
- Between 1990 and 2000, most new housing development in rural and suburban areas took place in areas with preexisting housing densities of between 1 and 40 housing units per acre. More housing units were built in the East than in the West. [Source: U.S. Census Bureau, analyzed by D.M. Theobald (Colorado State University).]

Alterations of "natural" stream banks or coastal shorelines can allow pollutants to enter streams more easily, reduce shading and thus increase water temperature, and reduce habitat quality for species that need both in-stream and shoreline habitat. "Armored" coastlines can help protect against erosion and storm damage but may isolate coastal wetlands from tidal influence and may ultimately result in unexpected erosion, either locally or in adjacent areas.

Six percent of the nation's coastline is armored with bulkheads or riprap, while 20 percent of the nation's stream and river banks are in urban or agricultural land use. (Sources: National Oceanic and Atmospheric Administration, Multi-Resolution Land Characterization Consortium and ESRI.)

Nutrients—On The Land and In The Water

Nitrogen is a vital nutrient for plants and animals, but one that can change the makeup of forests, contaminate groundwater wells, or trigger the growth of algae in coastal waters if present in excess. Nitrogen reaches fresh waters primarily through runoff from fertilized farms, lawns, and gardens, as wastewater treatment discharge, and in precipitation (from fossil fuel combustion). Elevated nitrogen in untreated drinking water can cause health problems, while nitrogen in streams or groundwater can eventually reach the coast as well, where it can contribute to water quality problems (see "Oxygen—the Lifeblood of Our Coasts," below). Nitrogen (and sulfate) deposited in rain or snow ("acid rain") can also harm lakes and streams by raising their acidity.

- In more than half of the areas monitored, more than 600 pounds per square mile of nitrogen are delivered each year to streams and rivers. (Source: USGS, National Water Quality Assessment and National Stream Quality Accounting Network.)
- Streams and groundwater in farmland areas have higher concentrations of nitrate—a common form of nitrogen—than streams in forested or urban and suburban areas, most likely from nitrogen fertilizer applied by farmers. Between 1992 and 2003, 20 percent of groundwater wells had nitrate concentrations that exceeded the federal drinking water standard, and between 1992 and 2001, 13 percent of streams in farmland areas had nitrate concentrations that exceeded the standard. (Source: USGS, National Water Quality Assessment.)
- Three rivers—the Mississippi, the Columbia, and the Susquehanna—together discharge approximately 1 million tons of nitrogen in the form of nitrate per year to coastal waters, with more than 90 percent of that nitrogen carried by the Mississippi. (Source: USGS, National Water Quality Assessment and National Stream Quality Accounting Network.)
- Discharge of nitrate from the Mississippi River rose substantially from the 1950s to the 1980s, but there has been no clear upward or downward trend since 1983; for the Susquehanna and Columbia Rivers, there has been no clear trend since the 1970s. (Source: USGS, National Water Quality Assessment and National Stream Quality Accounting Network.)
- Just over 2 percent of U.S. wadeable streams are considered "highly acidic," with almost twice that percentage in parts of the Appalachians. (Source: U.S.

Environmental Protection Agency, Wadeable Streams Assessment.)

Elevated levels of phosphorus in rivers, streams, and lakes can lead to the excessive growth of algae and other aquatic plants, which can be unsightly, interfere with recreation, clog industrial and municipal water intakes, and harm fish and other aquatic animals by causing dissolved oxygen levels to drop. The EPA has recommended 0.1 parts per million (ppm) as a goal for preventing excess algae growth in streams not draining directly into a lake or other impoundment.

- Between 1992 and 2001, streams in urban areas and farmlands had higher phosphorus concentrations than streams in forested areas. (Source: USGS, National Water Quality Assessment.)
- Half of major rivers sampled for phosphorus had concentrations of at least 100 ppb. (Source: USGS, National Water Quality Assessment and National Stream Quality Accounting Network.)

Oxygen—The Lifeblood of Coastal Waters

Nitrogen delivered to estuaries and other coastal areas can promote excessive growth of algae whose decay removes oxygen (phosphorus can cause the same phenomenon in rivers and lakes). Low oxygen levels can cause stress or death to fish, shellfish, and marine mammals. Prolonged periods of low oxygen levels can affect recreational and commercial fisheries and harm plant and animal communities.

- The area of the Gulf of Mexico with low oxygen levels (measured in July) has more than doubled over the past 22 years, from about 3800 square miles in 1985 to about 7900 square miles in 2007 (press reports indicate that in 2008 the zone will be yet larger). (Source: N. N. Rabalais and R. E. Turner.)
- There has been no clear upward or downward trend in the area of the hypoxic zone in Chesapeake Bay since 1985, but during that time the hypoxic zone has covered between 10 percent and 25 percent of the area of the bay. (Source: Chesapeake Bay Program.)

Carbon—On Land, In The Soil, and In The Air

Carbon in the form of organic matter is a key element of productive ecosystems. When stored ("sequestered") in ecological reservoirs such as soils and in durable plant materials like tree trunks and large roots, carbon serves to offset emissions of carbon dioxide and methane to the atmosphere, where they trap solar radiation and contribute to the greenhouse effect. National-scale estimates of carbon storage are unavailable for many terrestrial and aquatic ecosystem types, so it is not possible to provide a complete picture of carbon sequestration in the United States; however, where national-scale estimates have been made, carbon levels have increased.

- From 1995 to 2005, forests gained nearly 150 million metric tons annually in above- and below-ground plant materials; information on forest soils is not widely available. (Source: USDA Forest Service.)
- In the 1990s, cropland soils added 16.5 million metric tons of carbon per year, and private grasslands and shrubland soils gained 1.6 million metric tons per year. (Source: Natural Resource Ecology Laboratory, Colorado State University.)

A strong scientific consensus exists that additional increases in atmospheric greenhouse gas concentrations are very likely to alter climate patterns and have significant effects on people and ecosystems worldwide. Positive effects, such as longer growing seasons, may be offset by negative effects, such as water shortages.

- In 2006, the atmospheric concentration of carbon dioxide (381 parts per million) was 36 percent greater than the average concentration during pre-industrial times and has increased by 20 percent since the 1950s. [Source: Multiple data sources used by Working Group I of the Intergovernmental Panel on Climate Change (IPCC), 2007.]
- The atmospheric concentration of methane in 2005 (1805 parts per billion) was 160 percent higher than the average preindustrial concentration and has increased by 55 percent since the 1950s. [Source: Multiple data sources used by Working Group I of the Intergovernmental Panel on Climate Change (IPCC), 2007.]

Chemical Contamination—In Water, Sediments, Air, and Fish

Modern society produces a host of useful compounds, many of which are now present in the air, water, sediment, soil, and animal and human tissues. In sufficient quantities, these chemical contaminants can affect human health, restrict people's use of ecosystems, and harm plants and animals. Contaminants in drinking water affect human health, contaminants in fish trigger consumption advisories, and many wildlife species have been harmed by biological concentration of pesticides (as with bald eagles and DDT).

(Data below are from U.S. Geological Survey, National Water Quality Assessment Program and U.S. Environmental Protection Agency, National Coastal Assessment Program.) At least one contaminant was detected in

- Seventy-five percent of the groundwater wells tested
- Virtually all the streams and stream sediments tested
- About 80 percent of the estuarine sediments tested
- About 80 percent of the freshwater fish tested
- Nearly all of the saltwater fish tested

At least one contaminant was detected at levels above benchmarks set to protect aquatic life in more than 50 percent of the

- Stream water samples
- Stream sediment samples
- Estuarine sediment samples
- Freshwater fish tissue samples

At least one contaminant was detected at levels above benchmarks to protect human health in

- One-third of the groundwater samples
- One-third of the saltwater fish samples
- One-fifth of stream water samples

Certain contaminants are of particular interest or concern in specific ecosystems. In farmlands, pesticides may affect water quality; in urban and suburban landscapes, outdoor air quality is of particular concern because of its effects on human health, vegetation, animals, and the built environment.

- In farmlands, about 57 percent of streams had at least one pesticide at concentrations exceeding benchmarks for the protection of aquatic life; about 16 percent had at least one pesticide at levels exceeding benchmarks for protection of human health. (Source: U.S. Geological Survey, National Water Quality Assessment Program.)
- In 2005, ozone levels were above the level set for the national air quality standard at 30 percent of urban and suburban monitoring stations on four or more days; 61 percent had high levels on at least one day. (Source: U. S. Environmental Protection Agency.)
- In 2005, 28 percent of urban sites nationwide reported fine particulate matter at concentrations of 15 micrograms per cubic meter or above—a concentration comparable to EPA's national annual standard. (Source: U.S. Environmental Protection Agency.)

Amount, Timing, and Availability of Water

All ecosystems depend on water to support life. The amount available and the timing of its availability help shape ecosystems physically (such as through erosion), influence what species can live in an area, and determine how much water people can withdraw from ecosystems. While precipitation and resulting stream flows vary naturally over both years and decades (for example, the drought of the 1930s), there is substantial scientific evidence that a warming climate will be accompanied by significant shifts in the timing and amount of rainfall, with some areas becoming drier than they were historically, and others wetter. In addition, whether precipitation comes as rain or snow will have a major effect on stream flows, as some areas are reliant upon snowmelt for large parts of the year.

- Many U.S. streams have shown a change of more than 30 percent in the volume and variability of stream flow compared to a baseline period in the 1940s and 50s. Change has included both increases and decreases in high flow volume, low flow volume, and variability of volume. (Source: U.S. Geological Survey.)
- Nationwide, a growing proportion of streams have low flows with a substantially higher low flow rate than during the baseline period, and grassland-shrubland streams show fewer and shorter zero-flow incidents. (Source: U.S. Geological Survey.)
- A growing proportion of streams have shown a decrease in both high flow rates and in the variability of flow compared to the 1941–1960 baseline period. (Source: U.S. Geological Survey.)

These results do not yet provide evidence for major precipitation shifts, but this indicator—especially at the regional level—will be crucial to understanding how climate is affecting ecosystems.

Erosion and Sediments—Rivers and Streams, Farmlands and Coasts

Erosion in agricultural areas reduces soil quality and degrades water and habitat quality, while erosion in coastal areas can threaten developed areas, result in habitat loss or alteration, redistribute nutrients, and affect coastal recreation. While data are not adequate for reporting on coastal erosion, data are available on potential soil erosion in farmlands (erosion itself is not measured—this indicator measures soil conditions that promote erosion).

- From 1982 to 2003, the proportion of U.S. croplands with the greatest potential for wind erosion decreased by nearly a third.
- A similar decline was seen for cropland soils with the greatest potential for water erosion.
- The potential for wind erosion tends to be greater in the West, while the potential for water erosion is greater in the East. (Source: USDA Natural Resources Conservation Service.)

Change in the condition of streambed sediments—ranging from fine sediment to pebbles and cobbles—is a measure of alteration of stream features such as the amount and velocity of water and the amount of eroded sediment it receives. Degradation of sediment quality compared to reference streams implies a reduction in the quality of habitat for fish, other animals, and plants. For example, excess sediment from agricultural erosion or after wildfires can smother the eggs of fish.

- In fresh waters, about 25 percent of stream-miles in the lower 48 states have "degraded" sediments and about half have "natural" sediments. Sediments are in moderate condition for 20 percent of stream-miles. (Source: U.S. Environmental Protection Agency, Wadeable Streams Assessment.)

Changing Temperature—Oceans and Cities

In coastal ecosystems, water temperature directly affects the type of algae, seagrass, marsh plants, mangroves, fish, birds, mammals, and other plants and animals that live in a particular region. In addition, increases in temperature are thought to be associated with the degradation of coral reefs (bleaching) and may increase the frequency or extent of blooms of harmful algae.

- From 1985 to 2006, sea surface temperature increased significantly in U.S. coastal waters (within 200 miles of the coast) in three regions—Gulf of Alaska, Gulf of Mexico, and South Atlantic—and showed no observable trend for the North and Mid-Atlantic, Southern California, Bering Sea, Pacific Northwest or Hawaii regions. (Source: National Oceanic and Atmospheric Administration, National Aeronautics and Space Administration.)

Air temperatures in urban areas are often higher than in surrounding rural areas—the "urban heat island" effect. Heat waves are often responsible for the loss of human life, and they are considered likely to increase in intensity, duration, and geographical range as climate warms. The heat island effect may change the community of plants and animals that live in an area (including pathogens) and accelerate the formation of ground-level ozone and other pollutants that adversely affect human health. Some cities are taking steps, such as encouraging "green roofs," designed to keep cities cooler. Unfortunately, data are not adequate for national reporting on the urban heat island effect.

At-Risk Plants, Animals, and Communities

Ecosystems are defined in part by individual species and communities (groups of plants and animals that tend to occur in similar environmental conditions). These species and communities provide people with food, fiber, and a vast array of recreational opportunities. Species can also provide genetic materials that may have various industrial, agricultural, and medicinal uses—for example, the Pacific yew tree is the source of paclitaxel, a compound used in cancer treatment.

While some species are naturally rare, many have experienced historical or more recent declines and as a result many species and plant and animal communities are at risk of extinction (species) or elimination (communities). The loss of native species changes community composition and may affect the ability of the ecosystem to provide benefits or to respond to stresses, such as changing climate, particularly if there are few species with similar ecological roles.

- In 2006, one-third of native plant and animal species (excluding marine species) were at risk of extinction, with the highest incidence of at-risk species in Hawaii (81 percent), California (29 percent) and Nevada (16 percent). In contrast, the Midwest and Northeast/Mid-Atlantic had the lowest percentages—generally below 6 percent.
- The percentage of at-risk native animals is higher in fresh waters (37 percent) than in forests (19 percent) or grasslands and shrublands (18 percent).
- Nationally, about 28 percent of native vertebrate animal species at risk have declining populations, 23 percent have stable populations, and 1 percent have increasing populations. Population trends for the remaining native vertebrate animal species (48 percent) were unknown.
- In fresh waters, forests, and grasslands and shrublands, a large majority of native animal species with known population trends have populations that are either stable or declining, and fewer than 3 percent have populations that are increasing. (Source: NatureServe and its Natural Heritage member programs.)

Many freshwater plant communities are also at risk of elimination. In 2006

- Sixty-two percent of wetland and river- and stream-bank communities were at risk of elimination.
- In all states but West Virginia, Maine, New Hampshire, Rhode Island, and Vermont more than 20 percent of freshwater plant communities were at risk of elimination; in nine states, including several in the Southeast, more than 60 percent of freshwater communities are at risk. (Source: NatureServe and its Natural Heritage member programs.)

Non-Native Species—Changing The Native Landscape

Established non-native species may act as predators or parasites of native species, cause diseases, compete for food or habitat, or alter habitat. They may also provide ecosystem services such as soil stabilization or forage for grazing animals. Significant public and private funds are spent to control the most troublesome non-native species—often called invasive species—such as zebra mussels, cheatgrass, English ivy, and melaleuca.

While understanding the spread of these species is crucial to understanding ecological condition, data are currently not adequate to report on established non-native species on a national scale, with the exception of non-native fish.

- Fifty-eight percent of watersheds have more than 10 established non-native fish species. Only two watersheds in the lower 48 states have no established non-native fish species. Watersheds in the central United States generally have the fewest non-native fish species.

Condition of Biological Communities

Assessing the condition of a species is relatively straightforward—one measures population size and trends, area occupied (range), status of threats, and the like. Assessing the status of biological communities is more difficult. In fact, at a national scale, well-accepted methods (and data) are available for assessing communities only in estuaries and wadeable streams, using measures of the condition of bottom-dwelling animals. The condition of insects, worms, mollusks, and crustaceans in bottom sediments is of particular importance because these animals directly reflect changes in water quality and other disturbances and are a key part of the food chain. Changes in biological condition reflect the influence of contaminants, oxygen levels, physical changes in habitats (such as from trawl fishing in coastal areas or sediment deposition in streams), shifts in temperature or salinity, and the amount and timing of stream flows.

- In 1999–2002, from 60 percent to 90 percent of the estuarine area on the Atlantic and Pacific coasts had bottom-dwelling animals in "natural" condition; about one-third of the estuary area in the Gulf of Mexico and Puerto Rico had bottom-dwelling animals in "natural" condition. "Degraded" communities covered 44 percent of the estuarine area in the Gulf of Mexico.

(Source: U.S. Environmental Protection Agency, National Coastal Assessment.)

- Between 2000 and 2004, bottom-dwelling animals were in "natural" condition in 28 percent of wadeable streams in the lower 48 states; 42 percent of streams had bottom-dwelling animals that were in "degraded" condition; and 25 percent of wadeable streams had bottom-dwelling animals in "moderate" condition. The West had a higher proportion of stream-miles with "natural" bottom-dwelling communities than the Eastern Highlands and the Plains and Lowlands. (Source: U.S. Environmental Protection Agency, Wadeable Streams Assessment.)

Disturbance and Mortality—Forests, Oceans, and Fresh Waters

Periodic disturbances such as fire, floods, and insect outbreaks are "normal" in many ecosystems. In other ecosystems, "die-offs" of birds, whales, dolphins, or other species are believed to be a signal of ecosystem disruption. Assessing whether these disturbances are more or less frequent compared to long-term trends is a useful part of determining a system's condition.

While fires and insects are a natural part of forest life, the introduction of non-native pests such as gypsy moths or severe fire events following long periods of fire suppression can devastate large areas of forest. Large-scale fires can also increase erosion and sedimentation in streams and increase the likelihood of invasion by non-native species.

- Although there has been a significant decline in the acreage of forests and grasslands and shrublands burned since 1916, in recent years (1979–2006) this trend has reversed, with a total of 9.8 million acres burned in 2006. (Source: USDA Forest Service and National Interagency Fire Center.)
- Since 1997, the number of acres of tree mortality due to insect damage has also increased. (Source: USDA Forest Service.)

In coastal waters, harmful algal events can sicken people and cause mass mortalities of fish and wildlife. Increased nutrient loads, some aquaculture practices, ballast water discharge from ships, and overfishing may contribute to the frequency and severity of such events. Further indicator development is required before it is possible to report fully on harmful algal events.

Unusual marine mortalities may threaten sensitive marine populations and may indicate that stresses such as toxins, pollution, or changing weather are affecting marine ecosystems.

- Between 1990 and 2006, the number of whales, dolphins, porpoises, seals, sea lions, sea otters, and manatees dying each year in unusual marine mortality events fluctuated widely, from zero to several hundred animals, with one year (2002) having almost twenty times the average for other years. Most of these mortality events are believed to have been caused by infectious disease or by toxins produced by algae. (Source: National Marine Fisheries Service.)

At present, data are not adequate for reporting on the number of animal deaths and deformities in fresh waters.

Ecosystem Productivity

The ability of plants to use energy from the sun to build plant matter drives and sustains nearly all life. Therefore, changes in plant growth can signal alterations in how an ecosystem is functioning and can be related to increases or decreases in yields of timber and food crops and possibly to changes in the numbers and types of species that live in the region. Altered productivity may result from changing climate, exposure to ground-level ozone, as well as from changes in land use or farm or forest management. In marine ecosystems, conversion of sunlight to plant material is measured as the concentration of chlorophyll (from algae and similar marine plants) in the water.

- Nationwide, the plant growth index—a measure of plant growth or productivity—has shown little annual variability over the 1982–2003 time period.
- Cropland and grassland areas showed slight increases in the plant growth index, while forest and shrubland areas showed no clear up or down trend.
- Compared to other regions, the Southeast and portions of the Midwest had the most land with increases in the plant growth index. (Source: National Aeronautics and Space Administration.)
- Chlorophyll concentrations in coastal waters have increased in the Pacific Northwest, Southern California, and North Atlantic regions (1997–2006). (Source: National Oceanic and Atmospheric Administration, National Ocean Service; National Aeronautics and Space Administration.)

Food, Fiber, and Water Withdrawal—The Goods We Use

The United States relies heavily on domestic resources to meet its food, fiber, and water needs. We build homes with timber from U.S. forests; dine on fruits and vegetables from local farms as well as large-scale farming operations in distant states; eat meat from livestock grazed for part of the year on our grasslands and shrublands; and divert water from our rivers, lakes, and aquifers to drink, irrigate our crops, run our factories, and power our hydroelectric plants. Changes in the quantities of these extracted goods can affect both the economy and human well-being.

- Each year the United States harvests or withdraws
- 4.6 million tons of fish and shellfish from coastal waters (commercial landings only, 2005)
- 21.2 billion cubic feet of timber from forests (2005)
- Agricultural products valued at $239 billion from farmlands (2005)
- 126 trillion gallons of water from fresh waters (2000) (Sources: National Marine Fisheries Service, USDA Forest Service, USDA Economic Research Service, U.S. Geological Survey.)

- Nationally, the production of agricultural goods, the harvest of forest products, and our withdrawals of fresh water have all increased in the past half-century. However, only the production of agricultural products has grown at a rate exceeding population growth. (Sources: USDA Economic Research Service, USDA Forest Service, U.S. Geological Survey, U.S. Census Bureau.)

The nation's efficiency at producing farm and forest products has also increased. Agricultural output has increased by approximately 170 percent since 1948, and the amount of inputs (energy, fertilizer, and so on) needed to produce each unit of farm output has changed; timber harvest has also increased.

- The amount of land needed to produce each unit of agricultural output has dropped by 70 percent since 1948, and the amounts of purchased energy and durable goods like tractors has also declined (by 60 percent and 42 percent), accompanied by increases in pesticide and fertilizer use. Between 1948 and 2004, fertilizer inputs per unit output increased 46 percent and pesticide inputs doubled. (Source: USDA National Agricultural Statistics Service.)
- Yields per acre of five major crops—wheat, corn, soybeans, cotton, and hay—have increased since 1950, with corn yields alone increasing nearly fourfold. (Source: USDA National Agricultural Statistics Service.)
- While the area covered by forests in the United States has remained stable—dropping less than 1 percent over the past 50 years—over the same period harvest of forest products increased by 40 percent. Since 1952, timber growth on both public and private timberlands has increased. As of 2005, more than half of all U.S. timber was harvested from southern forests. Southern forests are predominantly privately owned (87 percent), are younger, more frequently harvested, and have a greater proportion of forested land in planted timberland (sometimes referred to as "plantations" or "tree farms"), compared to forest stands in the western United States. (Source: USDA Forest Service.)

In the past half-century, we have also obtained more goods from our freshwater and coastal ecosystems.

- Between 1960 and 2000, surface water and groundwater withdrawals combined increased by 46 percent. Municipal, rural and thermoelectric water uses increased during this period, while industrial withdrawals declined. (Source: U.S. Geological Survey.)
- Between 1950 and 2005, commercial fish and shellfish landings in the United States increased by almost 90 percent. Since 1990, Alaskan waters have accounted for the bulk of U.S. commercial landings. Alaska is the only region where landings have increased since 1978. Landings have decreased between 1978 and 2005 in the West Coast and Hawaii, the Gulf of Mexico, and the North, Mid-, and South Atlantic. From 1996 to 2005, with the exception of Alaskan and migratory stocks, a greater percentage of known fish stocks have increasing population trends than decreasing trends. (Source: National Marine Fisheries Service.)

Recreation

As the popularity of our national parks and other recreational areas attests, the U.S. public enjoys outdoor recreation. Our ecosystems offer a diversity of settings in which to engage in a wide range of activities—everything from whitewater rafting in the Rockies, deep-sea fishing off Florida, biking across the vast Midwest plains, or hunting in the Maine woods, to dog sledding in Alaska. Recreation provides enjoyment, health benefits, and even educational opportunities.

- Americans over the age of 16 participated in outdoor recreational activities 58 billion times per year, and almost half (45 percent) of total recreation occurs in forests. Walking is the most popular activity (23 billion times per year), followed by nature viewing (15 billion times per year) and all other land-based activities (15 billion times per year). Americans participate in water-based activities approximately 5 billion times a year.

In general, participation in outdoor recreation appears to be increasing over the three time periods—additional years of monitoring data will be needed to determine if observed increases are part of statistically significant trends. (Source: USDA Forest Service.)

Natural Ecosystem Services—The "Hidden" Services

Other services we receive from our nation's ecosystems are less familiar but no less important. They include such critical natural processes as purification of air and water, regulation of climate and floodwaters, erosion control, pollination, seed dispersal, carbon storage, and renewal of soil fertility. Changes in these natural ecosystem services can affect not only the condition of our environment, but also our ability to obtain more tangible goods and services from the nation's ecosystems on a sustainable basis. At present, the scientific community is wrestling with how best to describe the extent and value of these services and to detect and evaluate changes. Our indicators reflect this need for continued development.

Critical Thinking

1. Why are urban and suburban landscapes identified as a "proxy" for "developed land?" Are there alternatives to this perspective?
2. Are non-native species always harmful to ecosystems?
3. How do the findings of this study compare to the Millennium Ecosystem Assessment?

Global Biodiversity Outook 3

Executive Summary

SECRETARIAT OF THE CONVENTION ON BIOLOGICAL DIVERSITY

The target agreed by the world's Governments in 2002, "to achieve by 2010 a significant reduction of the current rate of biodiversity loss at the global, regional and national level as a contribution to poverty alleviation and to the benefit of all life on Earth", has not been met.

There are multiple indications of continuing decline in biodiversity in all three of its main components–genes, species and ecosystems–including:

- Species which have been assessed for extinction risk are on average moving closer to extinction. Amphibians face the greatest risk and coral species are deteriorating most rapidly in status. Nearly a quarter of plant species are estimated to be threatened with extinction.
- The abundance of vertebrate species, based on assessed populations, fell by nearly a third on average between 1970 and 2006, and continues to fall globally, with especially severe declines in the tropics and among freshwater species.
- Natural habitats in most parts of the world continue to decline in extent and integrity, although there has been significant progress in slowing the rate of loss for tropical forests and mangroves, in some regions. Freshwater wetlands, sea ice habitats, salt marshes, coral reefs, seagrass beds and shellfish reefs are all showing serious declines.
- Extensive fragmentation and degradation of forests, rivers and other ecosystems have also led to loss of biodiversity and ecosystem services.
- Crop and livestock genetic diversity continues to decline in agricultural systems.
- The five principal pressures directly driving biodiversity loss (habitat change, overexploitation, pollution, invasive alien species and climate change) are either constant or increasing in intensity.
- The ecological footprint of humanity exceeds the biological capacity of the Earth by a wider margin than at the time the 2010 target was agreed.

The loss of biodiversity is an issue of profound concern for its own sake. Biodiversity also underpins the functioning of ecosystems which provide a wide range of services to human societies. Its continued loss, therefore, has major implications for current and future human well-being. The provision of food, fibre, medicines and fresh water, pollination of crops, filtration of pollutants, and protection from natural disasters are among those ecosystem services potentially threatened by declines and changes in biodiversity. Cultural services such as spiritual and religious values, opportunities for knowledge and education, as well as recreational and aesthetic values, are also declining.

The existence of the 2010 biodiversity target has helped to stimulate important action to safeguard biodiversity, such as creating more protected areas (both on land and in coastal waters), the conservation of particular species, and initiatives to tackle some of the direct causes of ecosystem damage, such as pollution and alien species invasions. Some 170 countries now have national biodiversity strategies and action plans. At the international level, financial resources have been mobilized and progress has been made in developing mechanisms for research, monitoring and scientific assessment of biodiversity.

Many actions in support of biodiversity have had significant and measurable results in particular areas and amongst targeted species and ecosystems. This suggests that with adequate resources and political will, the tools exist for loss of biodiversity to be reduced at wider scales. For example, recent government policies to curb deforestation have been followed by declining rates of forest loss in some tropical countries. Measures to control alien invasive species have helped a number of species to move to a lower extinction risk category. It has been estimated that at least 31 bird species (out of 9,800) would have become extinct in the past century, in the absence of conservation measures.

However, action to implement the Convention on Biological Diversity has not been taken on a sufficient scale to address the pressures on biodiversity in most places. There has been insufficient integration of biodiversity issues into broader policies, strategies and programmes, and the underlying drivers of biodiversity loss have not been addressed significantly. Actions to promote the conservation and sustainable use of biodiversity receive a tiny fraction of funding compared to activities aimed at promoting infrastructure and industrial developments. Moreover, biodiversity considerations are often ignored when such developments are designed, and opportunities to plan in ways that minimize unnecessary negative impacts on biodiversity are missed. Actions to address the underlying drivers of biodiversity loss, including demographic, economic, technological, socio-political and cultural pressures, in meaningful ways, have also been limited.

Most future scenarios project continuing high levels of extinctions and loss of habitats throughout this century, with associated decline of some ecosystem services important to human well-being.

For example:

- Tropical forests would continue to be cleared in favour of crops and pastures, and potentially for biofuel production.
- Climate change, the introduction of invasive alien species, pollution and dam construction would put further pressure on freshwater biodiversity and the services it underpins.
- Overfishing would continue to damage marine ecosystems and cause the collapse of fish populations, leading to the failure of fisheries.

Changes in the abundance and distribution of species may have serious consequences for human societies. The geographical distribution of species and vegetation types is projected to shift radically due to climate change, with ranges moving from hundreds to thousands of kilometres towards the poles by the end of the 21st century. Migration of marine species to cooler waters could make tropical oceans less diverse, while both boreal and temperate forests face widespread dieback at the southern end of their existing ranges, with impacts on fisheries, wood harvests, recreation opportunities and other services.

There is a high risk of dramatic biodiversity loss and accompanying degradation of a broad range of ecosystem services if ecosystems are pushed beyond certain thresholds or tipping points. The poor would face the earliest and most severe impacts of such changes, but ultimately all societies and communities would suffer.

Examples include:

- The Amazon forest, due to the interaction of deforestation, fire and climate change, could undergo a widespread dieback, with parts of the forest moving into a self-perpetuating cycle of more frequent fires and intense droughts leading to a shift to savanna-like vegetation. While there are large uncertainties associated with these scenarios, it is known that such dieback becomes much more likely to occur if deforestation exceeds 20–30 percent (it is currently above 17 percent in the Brazilian Amazon). It would lead to regional rainfall reductions, compromising agricultural production. There would also be global impacts through increased carbon emissions, and massive loss of biodiversity.
- The build-up of phosphates and nitrates from agricultural fertilizers and sewage effluent can shift freshwater lakes and other inland water ecosystems into a long-term, algaedominated (eutrophic) state. This could lead to declining fish availability with implications for food security in many developing countries. There will also be loss of recreation opportunities and tourism income, and in some cases health risks for people and livestock from toxic algal blooms. Similar, nitrogen-induced eutrophication phenomena in coastal environments lead to more oxygen-starved dead zones, with major economic losses resulting from reduced productivity of fisheries and decreased tourism revenues.
- The combined impacts of ocean acidification, warmer sea temperatures and other human-induced stresses make tropical coral reef ecosystems vulnerable to collapse. More acidic water-brought about by higher carbon dioxide concentrations in the atmosphere-decreases the availability of the carbonate ions required to build coral skeletons. Together with the bleaching impact of warmer water, elevated nutrient levels from pollution, overfishing, sediment deposition arising from inland deforestation, and other pressures, reefs worldwide increasingly become algae-dominated with catastrophic loss of biodiversity and ecosystem functioning, threatening the livelihoods and food security of hundreds of millions of people.

There are greater opportunities than previously recognized to address the biodiversity crisis while contributing to other social objectives. For example, analyses conducted for this Outlook identified scenarios in which climate change is mitigated while maintaining and even expanding the current extent of forests and other natural ecosystems (avoiding additional habitat loss from the widespread deployment of biofuels). Other opportunities include "rewilding" abandoned farmland in some regions, and the restoration of river basins and other wetland ecosystems to enhance water supply, flood control and the removal of pollutants.

Well-targeted policies focusing on critical areas, species and ecosystem services are essential to avoid the most dangerous impacts on people and societies. Preventing further human-induced biodiversity loss for the nearterm future will be extremely challenging, but biodiversity loss may be halted and in some aspects reversed in the longer term, if urgent, concerted and effective action is initiated now in support of an agreed long-term vision. Such action to conserve biodiversity and use its components sustainably will reap rich rewards–through better health, greater food security, less poverty and a greater capacity to cope with, and adapt to, environmental change.

Placing greater priority on biodiversity is central to the success of development and poverty-alleviation measures. It is clear that continuing with "business as usual" will jeopardize the future of all human societies, and none more so than the poorest who depend directly on biodiversity for a particularly high proportion of their basic needs. The loss of biodiversity is frequently linked to the loss of cultural diversity, and has an especially high negative impact on indigenous communities.

The linked challenges of biodiversity loss and climate change must be addressed by policymakers with equal priority and in close co-ordination, if the most severe impacts of each are to be avoided. Reducing the further loss of carbonstoring ecosystems such as tropical forests, salt marshes and peatlands will be a crucial step in limiting the build-up of greenhouse gases in the atmosphere. At the same time, reducing other pressures on ecosystems can increase their resilience, make them less vulnerable to those impacts of climate change which are already unavoidable, and allow them to continue to provide services to support people's livelihoods and help them adapt to climate change.

Better protection of biodiversity should be seen as a prudent and cost-effective investment in risk-avoidance for the global community. The consequences of abrupt ecosystem changes on a large scale affect human security to such an extent, that it is rational to minimize the risk of triggering them–even if we are not clear about the precise probability that they will occur. Ecosystem degradation, and the consequent loss of ecosystem services, has been identified as one of the main sources of disaster risk. Investment in resilient and diverse ecosystems, able to withstand the multiple pressures they are subjected to, may be the best-value insurance policy yet devised.

Scientific uncertainty surrounding the precise connections between biodiversity and human well-being, and the functioning of ecosystems, should not be used as an excuse for inaction. No one can predict with accuracy how close we are to ecosystem tipping points, and how much additional pressure might bring them about. What is known from past examples, however, is that once an ecosystem shifts to another state, it can be difficult or impossible to return it to the former conditions on which economies and patterns of settlement have been built for generations.

Effective action to address biodiversity loss depends on addressing the underlying causes or indirect drivers of that decline.

This will mean:

- Much greater efficiency in the use of land, energy, fresh water and materials to meet growing demand.
- Use of market incentives, and avoidance of perverse subsidies to minimize unsustainable resource use and wasteful consumption.
- Strategic planning in the use of land, inland waters and marine resources to reconcile development with conservation of biodiversity and the maintenance of multiple ecosystem services. While some actions may entail moderate costs or tradeoffs, the gains for biodiversity can be large in comparison.
- Ensuring that the benefits arising from use of and access to genetic resources and associated traditional knowledge, for example through the development of drugs and cosmetics, are equitably shared with the countries and cultures from which they are obtained.
- Communication, education and awarenessraising to ensure that as far as possible, everyone understands the value of biodiversity and what steps they can take to protect it, including through changes in personal consumption and behaviour.

The real benefits of biodiversity, and the costs of its loss, need to be reflected within economic systems and markets. Perverse subsidies and the lack of economic value attached to the huge benefits provided by ecosystems have contributed to the loss of biodiversity. Through regulation and other measures, markets can and must be harnessed to create incentives to safeguard and strengthen, rather than to deplete, our natural infrastructure. The re-structuring of economies and financial systems following the global recession provides an opportunity for such changes to be made. Early action will be both more effective and less costly than inaction or delayed action.

Urgent action is needed to reduce the direct drivers of biodiversity loss. The application of best practices in agriculture, sustainable forest management and sustainable fisheries should become standard practice, and approaches aimed at optimizing multiple ecosystem services instead of maximizing a single one should be promoted. In many cases, multiple drivers are combining to cause biodiversity loss and degradation of ecosystems. Sometimes, it may be more effective to concentrate urgent action on reducing those drivers most responsive to policy changes. This will reduce the pressures on biodiversity and protect its value for human societies in the short to medium-term, while the more intractable drivers are addressed over a longer time-scale. For example the resilience of coral reefs–and their ability to withstand and adapt to coral bleaching and ocean acidification–can be enhanced by reducing overfishing, land-based pollution and physical damage.

Direct action to conserve biodiversity must be continued, targeting vulnerable as well as culturally-valued species and ecosystems, combined with steps to safeguard key ecosystem services, particularly those of importance to the poor. Activities could focus on the conservation of species threatened with extinction, those harvested for commercial purposes, or species of cultural significance. They should also ensure the protection of functional ecological groups–that is, groups of species that collectively perform particular, essential roles within ecosystems, such as pollination, control of herbivore numbers by top predators, cycling of nutrients and soil formation.

Increasingly, restoration of terrestrial, inland water and marine ecosystems will be needed to re-establish ecosystem functioning and the provision of valuable services. Economic analysis shows that ecosystem restoration can give good economic rates of return. However the biodiversity and associated services of restored ecosystems usually remain below the levels of natural ecosystems. This reinforces the argument that, where possible, avoiding degradation through conservation is preferable (and even more cost-effective) than restoration after the event.

Better decisions for biodiversity must be made at all levels and in all sectors, in particular the major economic sectors, and government has a key enabling role to play. National programmes or legislation can be crucial in creating a favourable environment to support effective "bottom-up" initiatives led by communities, local authorities, or businesses. This also includes empowering indigenous peoples and local communities to take responsibility for biodiversity management and decision-making; and developing systems to ensure that the benefits arising from access to genetic resources are equitably shared.

We can no longer see the continued loss of and changes to biodiversity as an issue separate from the core concerns of society: to tackle poverty, to improve the health, prosperity and security of our populations, and to deal with climate change. Each of those objectives is undermined by current trends in the state of our ecosystems, and each will be greatly strengthened if we correctly value the role of biodiversity in supporting the shared priorities of the international community. Achieving this will involve placing biodiversity in the mainstream of decision-making in government, the private sector, and other institutions from the local to international scales.

The action taken over the next decade or two, and the direction charted under the Convention on Biological Diversity, will determine whether the relatively stable environmental conditions on which human civilization has depended for the past 10,000 years will continue beyond this century. If we fail to use this opportunity, many ecosystems on the planet will move into new, unprecedented states in which the capacity to provide for the needs of present and future generations is highly uncertain.

Critical Thinking

1. What are the five main pressures that continue to affect biodiversity?
2. How is it that governments around the world agreed to achieve a particular target and by 2010 but did not do so?
3. Describe the three main components of biodiversity?

UNIT 2

Sustainability—A New Paradigm?

Unit Selections

Learning Outcomes

After reading this unit, you should be able to:

- Identify, by naming, at least 80 percent of the 10 myths regarding sustainability.
- Generate, by synthesizing relevant details, a paragraph describing the differences and similarities of four possible scenarios for sustainability in the 21st century.
- Discriminate by matching names, eras, or practices with key notions of sustainability.
- Identify, by drawing, the various visual representations of sustainability.
- Generate, by synthesizing provided information, a paragraph describing the strengths and weaknesses of the various interpretations of sustainability.
- Generate, by evaluating relevant scientific principles, a rule for the simultaneous development of humanity and the life support systems of Earth.
- Discriminate, by matching, the strategies used by Lincoln regarding slavery with the appropriate issues related to sustainability.
- Generate, by synthesizing provided information, an analysis of the status of sustainability in the United States.

Student Website

www.mhhe.com/cls

Internet References

International Union for the Conservation of Nature (IUCN)
www.iucn.org

United States Environmental Protection Agency (USEPA)
www.epa.gov/sustainability

United States National Park Service (NPS)
www.nps.gov/sustainability

World Resources Institute (WRI)
www.wri.org

In the first unit, the idea that the current issue of sustainability is really an indictment of human actions and behaviors was explored. The "World Scientists' Warning to Humanity" clearly identified what could be characterized as crimes against the life support systems of Earth by humans. Supporting evidence was provided by a variety of reports produced by a variety of different organizations. That humans have in the past 200 years become capable of so altering the planet that they now rival geological forces resulted in the creation of a name to identify this period in the history of Earth: the Anthropocene. Still, it is possible that human actions are simply part of the grand scheme of life on Earth and that sustainability is simply a new paradigm, or worldview, in which humans begin to live more benignly on the planet. It is possible that prior human actions, while perhaps unwise or apparently lacking of an understanding of the life support systems of the planet, are not indictable offenses, and the concept of sustainability is a recognition of the acquisition of wisdom that will result in changes in human actions.

© Cultura RF/Getty Images

Article 7 explores some common misconceptions about the concept of sustainability. As with many new concepts, there are conflicting ideas about the definition and objectives of sustainability. If sustainability is to be achieved, people must all be on the proverbial same page. If they are not, they may find strategies selected to help achieve it are actually counterproductive. By clearly identifying what sustainability is and what it isn't, such obstacles might be minimized, if not avoided.

Article 8 explores four different scenarios about what the world might be like as a result of the challenges associated with paradigm change. History is full of examples of how what seemed like a good idea turned out to be full of unanticipated consequences. So, trying to understand the long-term outcome of different strategies associated with sustainability is probably a good idea.

Article 9 explores the history of the concept of sustainability. It is interesting to note that the concept of sustainability appears to predate the actual birth of the world. So, the question could be asked, "Is sustainability really a new paradigm, or is it just the word used to describe the mindset new?" If it is the latter, then the question of why humanity is not living more sustainably becomes even more intriguing. If embedded in history is the notion of protecting resources for future generations, why are we not doing so?

Article 10 explores the issue of sustainability in the 21st century, the century some are calling the sustainability century. It might seem paradoxical that humans would have to reflect in an intellectual and scholarly manner about what sustainability is and how it is achieved. It might seem as though it would come "naturally." Apparently, this is not the case. So, many have tended to overanalyze the concept and attempt to configure it in such a way so as to be able to identify it in much the same manner as an architect would be able to identify the blueprints for a building. Contributing to the paradox is the observation that despite all of the attention being given to the idea of sustainability, there is evidence that human actions are becoming less, not more, sustainable.

Article 11 explores the curious relationship that exists between humanity and Earth. Humans are dependent upon the life support systems of the planet for their very survival and the health of the life support systems are dependent upon human actions. The challenges associated with sustainability provide the disconnect between the two. Humans are taking far more than they are giving. The only way sustainability is going to be achieved is if humanity becomes as concerned about the health of the life support systems of the planet as it is about the benefits it receives from the systems.

Article 12 makes an interesting comparison between how President Abraham Lincoln framed the concept of slavery and the way sustainability is framed today. The rationale for such a comparison is that slavery was the major issue during Lincoln's presidency and sustainability is the major issue today. If sustainability could be addressed as successfully as was slavery, then humanity will surely benefit. The strategies used by Lincoln may very well be applicable to sustainability.

Article 13 is a synthesis of the 1,000-plus-page tome, *Stumbling Toward Sustainability.* It explores the question, of How has the United States performed regarding a transition toward sustainability since the 1992 Earth Summit? The news is not always positive; in fact, is it more often negative. This is problematic given the standing the United States holds in global affairs. This situation is a classic illustration of Thomas Paine's advice to "lead, follow, or get out of the way."

Top 10 *Myths* About Sustainability

Even advocates for more responsible, environmentally benign ways of life harbor misunderstandings of what "sustainability" is all about

MICHAEL D. LEMONICK

When a word becomes so popular you begin hearing it everywhere, in all sorts of marginally related or even unrelated contexts, it means one of two things. Either the word has devolved into a meaningless cliché, or it has real conceptual heft. "Green" (or, even worse, "going green") falls squarely into the first category. But "sustainable," which at first conjures up a similarly vague sense of environmental virtue, actually belongs in the second. True, you hear it applied to everything from cars to agriculture to economics. But that's because the concept of sustainability is at its heart so simple that it legitimately applies to all these areas and more.

Despite its simplicity, however, sustainability is a concept people have a hard time wrapping their minds around. To help, Scientific *American Earth 3.0* has consulted with several experts on the topic to find out what kinds of misconceptions they most often encounter. The result is this take on the top 10 myths about sustainability. And after this introduction, it's clear which myth has to come first. . . .

Myth 1

Nobody Knows What Sustainability Really Means

That's not even close to being true. By all accounts, the modern sense of the word entered the lexicon in 1987 with the publication of *Our Common Future,* by the United Nations World Commission on Environment and Development (also known as the Brundtland commission after its chair, Norwegian diplomat Gro Harlem Brundtland). That report defined sustainable development as "development that meets the needs of the present without compromising the ability of future generations to meet their own needs." Or, in the words of countless kindergarten teachers, "Don't take more than your share."

Note that the definition says nothing about protecting the environment, even though the words "sustainable" and "sustainability" issue mostly from the mouths of environmentalists. That point leads to the second myth. . . .

Myth 2

Sustainability is all about the Environment

The sustainability movement itself—not just the word—also dates to the Brundtland commission report. Originally, its focus was on finding ways to let poor nations catch up to richer ones in terms of standard of living. That goal meant giving disadvantaged countries better access to natural resources, including water, energy and food—all of which come, one way or another, from the environment. "The economy," says Anthony Cortese, founder and president of the sustainability education organization Second Nature, "is a wholly owned subsidiary of the biosphere. The biosphere provides everything that makes life possible, assimilates our waste or converts it back into something we can use."

If too many of us use resources inefficiently or generate waste too quickly for the environment to absorb and process, future generations obviously won't be able to meet their needs. Says Paul Hawken, the author (his latest book is *Blessed Unrest: How the Largest Movement in the World Came into Being, and Why No One Saw it Coming)* and entrepreneur (he's a co-founder of the Smith & Hawken garden tools company) who helped to found the sustainability movement: "We have an economy where we steal the future, sell it in the present, and call it GDP [gross domestic product]."

If people continue to pour carbon dioxide (CO_2) into the air, for example, we won't necessarily exhaust resources (there's plenty of coal still in the ground), but we will change the climate in ways that could very likely impose huge burdens on future generations. The same, of course, goes for the poisonous by-products other than CO_2 from all kinds of human activity, from manufacturing to mining to energy generation to agriculture, that get dumped onto the land and into streams, oceans and the atmosphere.

The nonenvironmental rationales for sustainability get a little squishier when we talk about intangibles, such as the beauty of nature or the value of wilderness. "In wildness is the preservation of the world," wrote Henry David Thoreau; the national parks movement that began in the U.S. at the end of the 19th century and has since spread internationally springs from that idea. In modern terms, because humans evolved in a nontechnological world, we seem to need some connection to nature to be content. That concept is tough to prove scientifically. Nevertheless, says Nancy Gabriel, program director at the Sustainability Institute in Hartland, Vt., "If you look at Western society, you have huge rates of depression, isolation, [and] people who are disenfranchised. I think that reconnecting to the land is an important way of reestablishing a basic level of happiness." That kind of intangible connection has led towns, cities and states all over the U.S., but especially in built-up areas, to preserve land for open space.

A related but separate myth is. . . .

Myth 3

"Sustainable" is a Synonym for "Green"

Although there's a fair amount of overlap between the terms, "green" usually suggests a preference for the natural over the artificial. With some six billion people on the planet today, and another three billion expected by the middle of the century, society cannot hope to give them a comfortable standard of living without a heavy dependence on technology. Electric cars, wind turbines and solar cells are the antithesis of natural—but they allow people to get around, warm their houses and cook their food with renewable resources (or at least, a much smaller input of nonrenewables) while emitting fewer noxious chemicals.

It's probably more difficult to see nuclear power as sustainable. Unlike the other alternative energy sources, it has long been anathema to environmentalists, largely because of the problem of storing radioactive waste. But nuclear reactors are also a highly efficient source of power, emit no pollutant gases and—with some types, anyway—can be designed to generate minimal waste and to be essentially meltdown-proof. That's why Patrick Moore, a co-founder of Greenpeace, has become a nuclear booster and why many other environmentalists are beginning—sometimes grudgingly—to entertain the idea of embracing nuclear. Calling it green would be a stretch. Calling it sustainable is much less of one.

Myth 4

It's all about Recycling

"I get that a lot," says Shana Weber, the manager of sustainability at Princeton University. "For some reason, recycling was the enduring message that came out of the environmental movement in the early 1970s." And of course, recycling is important: reusing metals, paper, wood and plastics rather than tossing them reduces the need to extract raw materials from the ground, forests and fossil-fuel deposits. More efficient use of pretty much anything is a step in the direction of sustainability. But it is just a piece of the puzzle. "I deal with the people who run the recycling program here," Weber notes, "but also with purchasing, dining services, the people who clean the buildings. The most important areas by far in terms of sustainability are energy and transportation." If you think you are living sustainably because you recycle, she says, you need to think again.

Myth 5

Sustainability is too expensive

If there is an 800-pound gorilla in the room of sustainability, this myth is it. That's because, as Gabriel observes, "there's a grain of truth to it." But only a grain. "It's only true in the short term in certain circumstances," Cortese says, "but certainly not in the long term." The truth lies in the fact that if you already have an unsustainable system in place—a factory or a transportation system, for example, or a furnace in your house, an incandescent lightbulb in your lamp or a Hummer in your driveway—you have to spend some money up front to switch to a more sustainable technology.

In general, governments and companies can take that step more easily than individuals can. "Over the past seven years," Cortese explains, "DuPont has made investments that have reduced its greenhouse gas emissions by 72 percent over 1990 levels. They've saved $2 billion." The Pentagon is determined to cut its energy use by a third, both to save money and to reduce its dependence on risky foreign oil supplies.

Myth 6

Sustainability Means Lowering Our Standard of Living

Not at all true. It does mean that we have to do more with less, but as Hawken argues, "Once we start to organize ourselves and innovate within that mind-set, the breakthroughs are extraordinary. They will allow us to achieve greatly superior rates of resource productivity, which in turn allow us to be prosperous, fed, clad, secure." Moreover, he and others maintain that the innovation at the heart of sustainable living will be a powerful economic engine. "Addressing climate change," he says, "is the biggest job creation program there is."

Myth 7

Consumer Choices and Grassroots Activism, Not Government Intervention, Offer the Fastest, Most Efficient Routes to Sustainability

Popular grassroots actions are helpful and ultimately necessary. But progress on some reforms, such as curbing CO_2 emissions, can only happen quickly if central authorities commit to making it happen. That is why tax credits, mandatory fuel-efficiency standards and the like are pretty much inevitable. That conclusion drives free-market evangelists crazy, but they operate on the assumption that wasteful use of resources and the destruction of the environment is without cost, which is demonstrably untrue.

To cite just one example, economic devastation is very likely under even the mildest plausible climate change scenarios, in the form of disruptions to agriculture from shifts in rainfall patterns and growing zones; densely populated coastal areas will be rendered unlivable as sea level rises, and so on. Yet the price currently being charged to people who add greenhouse gases to the atmosphere is zero. Putting a per-ton tax on carbon emissions would be wildly unpopular, but it would for the first time account for the real costs of unsustainable energy use.

Free-market purists also argue that with respect to the depletion of natural resources, rising prices will automatically push people into more efficient behavior. True enough—but the transition can be painful and disruptive. The primary reason U.S. automakers are in such trouble is that they have been depending for years on high-profit gas-guzzling SUVs. When the price of oil shot up last year, the market for big cars plummeted (gas prices have only come down since then in the face of a worldwide recession, which hasn't helped the auto industry). So car buyers may have changed their behavior, but only at the cost of potential disaster for some of America's biggest companies and their employees.

Still, rising energy prices have had the effect of again galvanizing research into wind, solar and other alternatives—and if you leave economic disruption aside, we can at least count on car companies to make more efficient vehicles and on utilities to find more sustainable sources of energy. But that outcome may reflect another myth. . . .

Myth 8

New Technology is Always the Answer

Not necessarily. During his presidential campaign, Barack Obama made the tactical mistake of pointing out that proper tire inflation could save Americans millions of gallons of gasoline through better fuel economy. The Republicans ridiculed him, just as they did President Jimmy Carter for appearing on TV in a sweater during the energy crisis of the late 1970s. Both Carter and Obama were right, however (California's Republican governor Arnold Schwarzenegger has called for proper tire inflation as well).

In other words, sometimes existing technology can make a huge difference. Sometimes it takes a creative business model. Israeli entrepreneur Shai Agassi, for example, wants to electrify the world's car fleet—widely acknowledged as a big step toward cutting down carbon emissions—not by inventing a battery that gets 200 miles on a charge but by inventing a better system for letting drivers go as far as they want without recharging. His proposal, which has been adopted on a pilot basis by Israel and Denmark, would create battery exchange stations along highways, analogous to the gas canister exchanges that people now use for barbecue grills. What do you do if you are out on the road and your battery is running low? You pull into a station, your dead battery is swapped for a fully charged one and you're on the road again in a few minutes.

"He's delivering distance, not better batteries," says Mark Lee, CEO of the London consulting firm SustainAbility. "There's an Italian utility that's selling its customers hot water, not energy to heat water. It's a different way of measuring, and it gives the company an incentive to be more efficient so it can be more profitable."

Myth 9

Sustainability is Ultimately a Population Problem

This is not a myth, but it represents a false solution. Every environmental problem is ultimately a population problem. If the world's population were only 100 million people, we would be hard-pressed to generate enough waste to overwhelm nature's cleanup systems. We could dump all our trash in a landfill in some remote area, and nobody would notice.

Population experts agree that the best way to limit population is to educate women and raise the standard of living generally in developing countries. But that strategy cannot possibly happen quickly enough to put a dent in the population on any useful timescale. The U.N. projects that the planet will have to sustain another 2.6 billion people by 2050. But even at the current population level of 6.5 billion, we're using up resources at an unsustainable rate. There is no way to reduce the population significantly without trampling egregiously on individual rights (as China has done with its one-child policy), encouraging mass suicide or worse. None of those proposals seems preferable to focusing directly on less wasteful use of resources.

Myth 10:

Once you Understand the Concept, Living Sustainably is a Breeze to Figure Out

All too often, a choice that seems sustainable turns out on closer examination to be problematic. Probably the best current example is the rush to produce ethanol for fuel from corn. Corn is a renewable resource—you can harvest it and grow more, roughly indefinitely. So replacing gasoline with corn ethanol seems like a great idea. Until you do a thorough analysis, that is, and see how energy-intensive the cultivation and harvesting of corn and its conversion into ethanol really are.

One might get a bit more energy out of the ethanol than was sunk into making it, which could still make ethanol more sustainable than gasoline in principle, but that's not the end of the problem. Diverting corn to make ethanol means less corn is left to feed livestock and people, which drives up the cost of food. That consequence leads to turning formerly fallow land—including, in some cases, rain forest in places such as Brazil—into farmland, which in turn releases lots of carbon dioxide into the atmosphere. Eventually, over many decades, the energy benefit from burning ethanol would make up for that forest loss. But by then, climate change would have progressed so far that it might not help.

You cannot really declare any practice "sustainable" until you have done a complete life-cycle analysis of its environmental costs. Even then, technology and public policy keep evolving, and that evolution can lead to unforeseen and unintended consequences. The admirable goal of living sustainably requires plenty of thought on an ongoing basis.

Critical Thinking

1. If it is true that the "biosphere provides everything that makes life possible, assimilates our waste or converts it back into something we can use," then why do we have a problem with pollution and landfills?
2. Do you believe that nuclear reactors can be designed to generate minimal waste?
3. Explain the statement, if you think you are living sustainably because you recycle, you need to think again.

MICHAEL D. LEMONICK is a senior writer at Climate Central, a nonprofit climate change think tank in Princeton, N.J.

The Century Ahead: Searching for Sustainability

PAUL D. RASKIN, CHRISTI ELECTRIS, AND RICHARD A. ROSEN

Introduction

Concern about the sustainability of nature and society is rising, and with good reason. Scientists report with ever-greater urgency the need for action to avoid destabilizing climate change and widespread destruction of the world's ecosystems 1, 2. Parallel efforts are required to ease looming shortages of critical resources such as oil, water, and food. Meanwhile development specialists call for mitigating poverty, strengthening social justice, and enhancing human well-being. Other observers appeal for more effective transnational governance to regulate the growth and impact of globalizing capital, finance, and product markets that threaten the long-term stability and fairness of the world economy.

These concerns are central to the broad challenge of sustainable development, an international commitment assumed, at least rhetorically, nearly two decades ago at the 1992 Earth Summit in Rio de Janeiro. At the core of the concept of sustainability lies a moral imperative to pass on an undiminished world to future generations. This clarion call to take responsibility for the welfare of the unborn requires that, in making choices today, we weigh the consequences for the long-term tomorrow.

This paper explores the implications of this challenge by considering four contrasting global scenarios representing alternative worlds that might emerge from the turbulence and uncertainty of the present. Market Forces and Policy Reform are evolutionary futures that, despite episodic setbacks, emerge gradually from the dominant forces governing world development today. The other two envision a fundamental restructuring of the global order: fragmentation in Fortress World and positive transformation in Great Transition. Each scenario tells a different story of the twenty-first century with varying patterns of resource use, environmental impacts, and social conditions.

It is important to note the distinction between scenarios and forecasts. The interactions among co-evolving human and environmental systems are highly complex and inherently uncertain, rendering predictive forecasts impossible in any rigorous statistical sense 3, 4. Instead, scenarios are intended as renderings of plausible possibilities, designed to stretch the imagination, stimulate debate, and, by warning of pitfalls ahead, prompt corrective action. Of course, the plausibility, and even the internal consistency, of different visions is itself uncertain. How will the climate system respond to increased greenhouse gas concentrations from human activity? What geo-political formations will emerge? How will human values adjust? Indeed, limiting the ways surprises and feedback might knock a scenario off course is an illuminating and underappreciated aspect of scenario analysis.

We have examined our scenarios in great quantitative detail to the year 2100 for eleven world regions. The summary presented here focuses on selected global-scale results, painting broad-brush pictures of these contrasting futures, and revealing the fundamental forces driving development away from or toward sustainability. Cross-scenario comparisons offers lessons for policy strategies, institutional change, and, ultimately, for human values.

This research updates and enhances an earlier series of scenario assessments conducted by the Tellus Institute on behalf of the Global Scenario Group 5–7. The base year has been advanced from 1995 to 2005, adding ten additional years to the massive database on which the analysis rests. That data feeds the PoleStar System, a computational framework originally developed in the early 1990s by the Tellus Institute and the Stockholm Environment Institute. PoleStar is designed to explore a full spectrum of integrated long-range scenarios in quantitative detail, including unconventional pathways of structural discontinuity 8.

The global simulations are disaggregated by region, major sectors and subsectors of the economy, key social variables, and numerous aspects of the environment and natural resources (see Table 1). Assumptions and computations are documented in 9, with regional results reported online at www.tellus.org/result_tables/results.cgi.

The Scenarios: An Overview

The four scenarios are listed in Table 2. Conventional Worlds scenarios assume the persistence of many of the dominant forces driving development and globalization in recent decades. GDP growth remains the primary measure of successful economies

Table 1 Key Issues Simulated

Sector	Issue
Social	• Population • Gross Domestic Product (GDP) and value-added by sector • Income (GDP per capita) • Income distribution within and between regions • Poverty • Hunger line (income for adequate diet) • Employment (productivity and length of work week)
Household	• Energy use by fuel • Water use • Air pollution • Water pollution
Service	• Energy use by fuel • Water use • Air pollution • Water pollution
Transportation	• Passenger by mode: public road (buses, *etc.*), private road, rail, air • Freight transportation in following modes: road, rail, water, air • Energy use by mode and fuel • Air pollution
Agriculture	• Diet by crop and animal product categories • Livestock: animal type, seafood (wild, farmed), other products (milk, etc) • Crops: coarse grains, rice, other (fruits, vegetables, *etc.*), sugarcane, biofuels • Energy use by fuel • Irrigation • Fertilizer use • Air pollution • Water pollution
Industry	• Energy use by fuel and subsector: iron and steel, non-ferrous metals, stone, glass, and clay, paper and pulp, chemical, other • Energy feedstock by subsector. • Water use by subsector • Air pollution from both fuel combustion and process • Water and toxic pollution
Forestry	• Primary wood requirements • Secondary wood for final demand, and input to paper and pulp, lumber, biofuel
Land-Use	• Conversions between built environment, cropland, pasture, forest types (unexploitable, exploitable, plantation, and protected), other protected (marshes, bays, *etc.*), other • Each category broken down by arable and non-arable areas • Cropland disaggregated by crop type, and irrigated/non-irrigated
Energy Conversion	• Conversion from primary to secondary fuels (*i.e.,* electricity production and oil refining) • Requirements for coal, biomass, natural gas, renewable (wind, solar, geothermal, etc), crude oil, nuclear, hydropower • Air pollution
Water	• Freshwater resources • Desalinization and waste-water recycling for water resources • Use-to-resource ratios • Water stress
Solid Waste	• Generation from household and service sectors • Landfill, incineration, recycling and other disposal technologies • Energy generation from incineration

Table 2 The Scenarios

Type	Name	Description
Conventional Worlds	*Market Forces (MF)*	Market-centered growth-oriented globalization
	Policy Reform (PR)	Government-led redirection of growth toward sustainability goals
Alternative Visions	*Fortress World (FW)*	An authoritarian path in response to mounting crises
	Great Transition (GT)	A fundamental transformation

and poor countries gradually converge toward the consumption and production patterns of rich nations. The degree of plausibility of these evolutionary scenarios rests with the validity of a basic premise: Conventional Worlds strategies will have the resilience to tolerate and recover from socio-ecological crises as they appear and succeed in maintaining rapid economic expansion.

Market Forces: Market-Centered Development

Market Forces is constructed as a future in which free market optimism remains dominant and proves well-founded. As population expands by 40 percent by 2050 10 and free trade and deregulation drive growth, the global economy expands over three-fold by 2050, eightfold by 2100. All economic figures in this paper are expressed in purchasing power parity (PPP) dollars, which takes account of national differences in the cost of living when converting to a common currency. The availability of sufficient resources—raw materials, land, water, energy—and the means of maintaining ecological resilience in such a huge economy are critical uncertainties. The challenge of satisfying bio-physical sustainability constraints would be compounded by the challenge of maintaining social and economic sustainability in a world of profound inequalities between rich and poor countries, and within each country. Instability and conflict could undercut the evolutionary dynamics of the scenario, triggering a descent of civilization toward a Fortress World or more chaotic outcomes.

Policy Reform: Directing Growth

Policy Reform assumes the emergence of a massive government-led effort achieves sustainability without major changes in the state-centric international order, modern institutional structures, and consumerist values. Strong and harmonized policies are implemented that, by redirecting the world economy and promoting technological innovation, are able to achieve internationally recognized goals for poverty reduction, climate change stabilization, ecosystem preservation, freshwater protection, and pollution control. The scenario meets tough stabilization targets for carbon dioxide emissions and, in rough compatibility with United Nations Millennium Development Goals 11, halves world hunger between 2005 and 2025 (then halves it again by 2050).

Policy Reform is designed as a backcast constrained to meet the objectives shown in Table 5. As total greenhouse emissions decline, growth continues in developing countries for two decades as redistribution policies raise incomes of the poorest regions and most impoverished people. Although such transfers have been debated at climate negotiations, with little success to date, our analysis indicates that a Policy Reform approach will require a deep and widespread commitment to economic equity. As poorer countries converge toward the living standards of richer countries, they accelerate investment in environmentally sustainable practices.

Implementing this grand policy program in the context of Conventional Worlds values and institutions would not be easy: intergovernmental efforts to address sustainability challenges over the past two decades have not succeeded. The Policy Reform path would require unprecedented political will for establishing the necessary regulatory, economic, social, technological, and legal mechanisms.

Fortress World: An Authoritarian Path

If the market adaptations and policy reforms of Conventional Worlds were to prove insufficient for redirecting development away from destabilization, the global trajectory could veer in an unwelcome direction. Fortress World explores the possibility that powerful world forces, faced with a dire systemic crisis, impose an authoritarian order where elites retreat to protected enclaves, leaving impoverished masses outside. In our troubled times, Fortress World seems the true "business-as-usual" scenario to many. In this dark vision, the global archipelago of connected fortresses seeks to control a damaged environment and restive population. Authorities employ geo-engineering techniques to stabilize the global climate, while dispatching "peace-keeping" militia to multiple hotspots in an attempt to quell social conflict and mass migration. But the results are mixed: emergency measures and spotty infrastructure investment cannot keep pace with habitat loss and climate change, nor provide adequate food and water to desperate billions. In this kind of future, sustainable development is not in the cards, a half-remembered dream of a more hopeful time.

Great Transition: A Sustainable Civilization

In dramatic contrast, Great Transition envisions a values-led change in the guiding paradigm of global development. The transformation is catalyzed by the "push" of deepening crises and the "pull" of desire for a just, sustainable, and planetary civilization. A pluralistic transnational world order coalesces as a growing cultural and political movement of global citizens spurs the establishment of effective governance institutions 12. The new paradigm is rooted in a triad of ascendant values: human solidarity, ecological resilience, and quality of life. Less consumerist lifestyles moderate the growth thrust of Conventional Worlds scenarios, as notions of the "good life" turn toward qualitative dimensions of well-being: creativity, leisure, relationships, and community engagement.

Population stabilizes more rapidly than in other scenarios as more equal gender roles and universal access to education and health care services lower birth rates in developing countries. The world approaches a steady-state economy with incomes reaching about $30,000 per person by 2100, three times the current average. Although this figure is well below the $50,000 of Conventional Worlds, the egalitarian income distributions of Great Transition leave most people far better off, while the improved social cohesion reduces conflict. In this deeply sustainable vision, crises still linger, but the world is able to confront them with enhanced institutions for reconciliation and cooperation.

Guidelines for Sustainability

The notion of sustainable development, though evocative, lacks precision. We operationalize the concept with a set of environmental and social goals that, if realized, would correlate to a resilient, just, and desirable form of global development in the coming decades. These indicators serve as a lens through which we evaluate the compatibility of each scenario with sustainability broadly construed. The broad aim is to consider the *quality of development*—the degree of well-being in human lives, the strength of communities, and the resilience of the biosphere—rather than gross domestic product, the misleading conventional measure of "development".

The primary economic and social objectives of sustainability are, in a strict sense, the resilience and persistence of existing institutions. However, such structural stasis would fail to reflect the full normative content of sustainable development, the desire to reduce poverty and disparity, and enhance social cohesion. Proximate goals advancing these desiderata include universal access to clean water, adequate nutrition, stabilized populations, high quality of work, guaranteed basic rights, and adequate leisure for individual fulfillment. Prominent social dimensions of sustainability are listed in Table 3.

Environmental sustainability objectives include mitigating anthropogenic climate change, reducing pollution, preserving natural resources, and protecting ecosystems and habitats. Actions that would contribute to these ends are summarized in Table 4. Reducing human impacts will depend on the scale and composition of consumption and production, the technologies deployed, the degree of social equity and stability, and, ultimately, on the human values and institutions that underpin development.

Selected sustainability objectives, broadly compatible with social and environmental goals widely discussed in the international discourse, are shown in Table 5. The Policy Reform scenario is constrained to meet these targets, which set the scale and timing for strategic interventions to decarbonize the energy system, conserve resources, and reduce poverty. The diminished environmental pressures and heightened equity embodied in the Great Transition scenario allow for transcending these minimal objectives.

Table 3 Social Dimensions of Sustainability

Enhance social stability and resilience
- *Enhance social cohesion*
- *Democratize governance of key institutions*
- *Strengthen cultural diversity*

Reduce poverty and hunger
- *Decrease income and wealth disparities*
- *Raise income to a sufficient level for all*
- *Stabilize then reduce population*
- *Improve access to adequate nutrition, sanitation, and freshwater*

De-materialize lifestyles
- *Moderate materialistic values*
- *Reduce formal work time*
- *Promote quality of life activities*

Table 4 Environmental and Resource Dimensions of Sustainability

Mitigate greenhouse gas emissions
- *Reduce combustion of fossil-fuels and sequester CO_2 emissions*
- *Minimize then reverse emissions from land-use changes*
- *Reduce other greenhouse gas emissions*

Protect natural resources
- *Reduce air and water pollution*
- *Eliminate emissions of toxic chemicals*
- *Reduce mineral flows through economy, and recycle intensively*
- *Reduce water stress*

Preserve habitats
- *Reduce urban sprawl*
- *Protect forests and other ecosystems*
- *Fish sustainably*
- *Promote ecological agriculture*

Views of the Future

This section offers a bird's eye view of the quantitative patterns of the scenarios. We begin with the Quality of Development Index (QDI) that we created to provide an overarching measure of sustainability performance. The QDI combines sub-indices representing three key aspects of sustainable development: human well-being, community cohesion, and environmental

Table 5 Selected Sustainability Targets

Dimension	Indicator	2005	2025	2050	2100
Poverty	*Chronic hunger (millions of people)*	893	446	223	56
	Percentage of 2005 value	–	50 Percent	25 Percent	6 Percent
Climate	*CO_2 concentration*	380 ppm	Stabilize at ≤ 350 ppm by 2100		
	Warming	–	<2.0° C		
	Cumulative CO_2 emissions since 2005	–	≤ 265 GtC		
Freshwater	*Use-to-resource ratio*	Varies by basin	Decrease in areas of water stress		
	People in water stress (billions)	1.73	<2.0		
Ecosystem Pressure	*Deforestation*	Varies by region	Slow and reverse		
	Land degradation	Varies by region	Slow and reverse		
	Marine over-fishing	Pervasive	Slow and restore stocks		

protection. The material Well-being sub-index includes indicators for Time Affluence (essentially leisure time) and Prosperity (a logarithmic function of income to account for well-documented diminishing returns to well-being); the Community sub-index includes indicators for Poverty Reduction and Social Cohesion (correlated to income disparity); the Environment sub-index includes indicators for Climate and Habitat. For details, see pp. 321–329 of the technical documentation for these scenarios 9.

Global trends for each scenario are shown in Figure 1. Despite rapid and continuous economic growth, Market Forces shows no discernible improvement in QDI (similar results are found at the regional scale where the QDI falls in OECD regions and rises only modestly in non-OECD regions.) By contrast, Policy Reform, with its commitments to meeting environmental and poverty reduction targets, shows steady and significant improvement. The QDI rises still higher in Great Transition, where a strong emphasis on quality-of-life and social cohesion contributes additional human well-being and community cohesion. Not surprisingly, Fortress World experiences a decline in all dimensions of development.

Figure 2 contrasts trends in QDI with GDP/capita ("income"), the more conventional measure of development. In Conventional Worlds scenarios, GDP growth has pride of place, but we can see by comparing Figures 2 (a)–(d), rising income alone is a poor predictor of the quality of development: QDI languishes in Market Forces, but shows significant increase in Policy Reform. The quality of development is still higher in Great Transition despite its more modest average incomes in the long run. Not surprisingly, the Fortress World QDI falls continuously as all dimensions deteriorate: incomes are lower, communities are less cohesive, and the environment degrades.

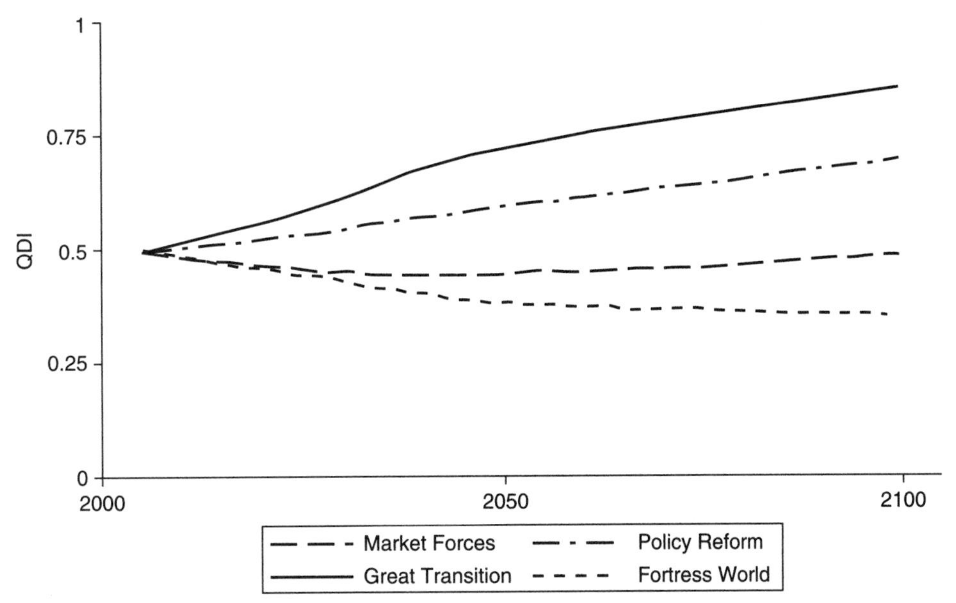

Figure 1 Global Quality of Development Index (QDI)

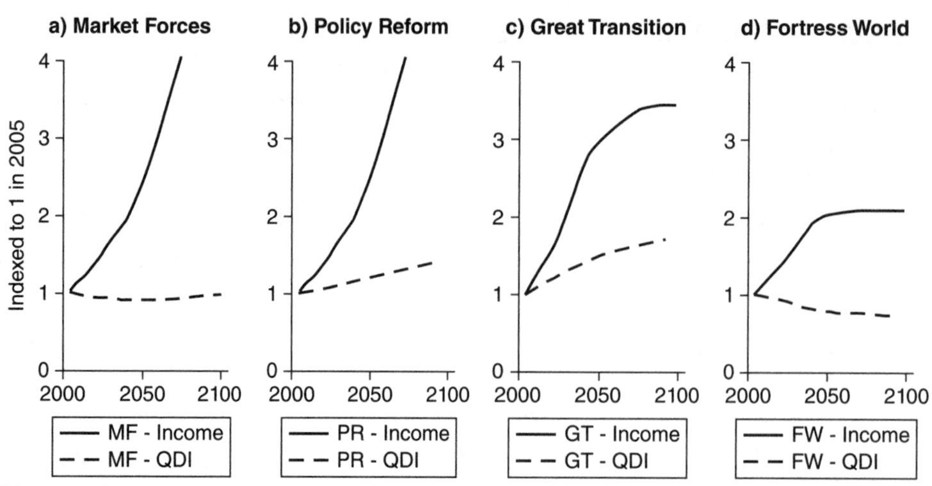

Figure 2 Global QDI *versus* Income

Economic and Social Patterns

Population

World population grows from 6.5 billion to between 7.2 and 10.2 billion depending on the scenario (Figure 3(a)), with most of the increase in developing countries. The variation in population trends is due primarily to differences in fertility rates (children per female), with lower rates correlating to higher levels of access to education (especially for girls) and family planning services, and the degree of poverty reduction. Correspondingly, the demographic transition to lower birth rates in the process of modernization is accelerated in Policy Reform and, especially, in Great Transition, but reversed in the de-developing Fortress World.

Income

As shown in Figure 3(b), average income per capita soars in both Policy Reform and Market Forces, and stagnates in Fortress World where most people become mired in poverty. In Great Transition, income rises rapidly before 2050, as strong commitments to equity spur rapid development in the global South, then moderates as high equity is achieved and the world economy approaches a steady state 12, 13. Note that, underlying these average patterns, Great Transition is simulated as a pluralistic future where social and cultural diversity flourishes across and within regions. Although regional incomes converge toward a similar level, Great Transition is envisioned as a pluralistic scenario with significant social and cultural diversity across and within regions.

International Equity

The current North-South disparity in economic development is extreme. International equity—defined here as the ratio of average income in non-OECD to OECD nations—was 0.13 in 2005, with markedly different trends in each scenario (Figure 4). International equity rises to 0.23 in Market Forces by 2100, a result of the somewhat higher growth in poorer countries found in standard economic projections (even so, the absolute difference in incomes between rich and poor regions increases). Ironically, international equity improves more in Fortress World than in Market Forces as regions become more equally poor. Through proactive pursuit of poverty reduction,

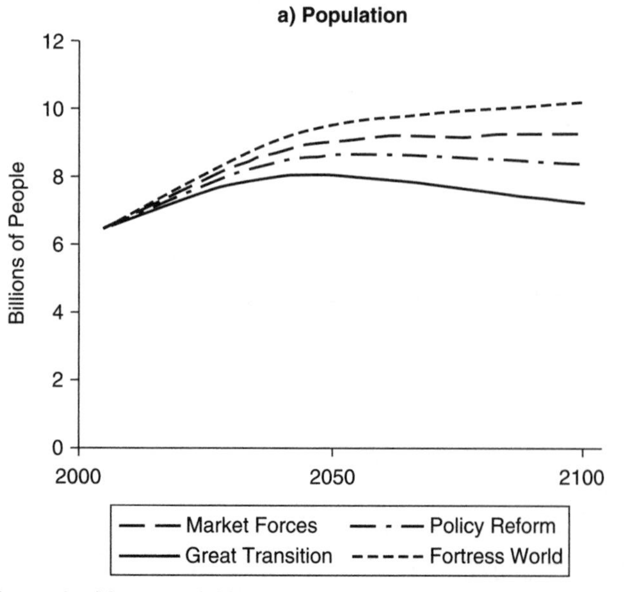

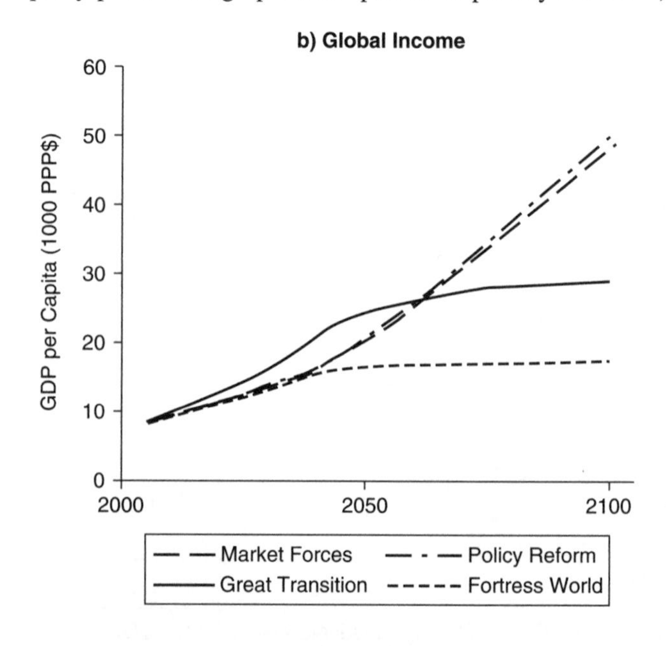

Figure 3 Macro-variables

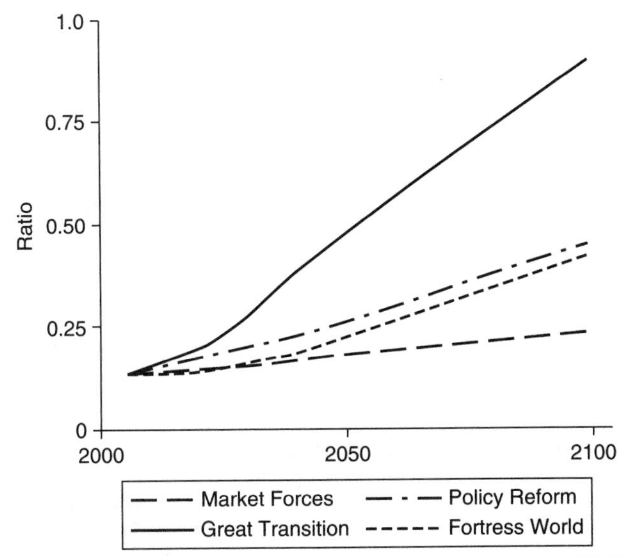

Figure 4 International Equity (Ratio of non-OECD to OECD regions)

Policy Reform drives international equity significantly higher, reaching 0.45 as a result of financial transfers and development aid from OECD to non-OECD regions. Great Transition, rooted in the core values of justice and solidarity, reaches an equitable world by 2100, where the gaps between regions have nearly vanished.

Intra-regional Equity

The distribution of income within regions also varies substantially across scenarios. Figure 5(a) illustrates the range of variation of intra-regional equity—defined as the ratio of the income of the poorest 20 percent to the richest 20 percent—for North America. The patterns for the highly equitable Great Transition are shown in Figure 5(b), where all regions approach a ratio of about 0.35, nearly 60 percent higher than Western Europe today.

Poverty and Hunger

The incidence of chronic hunger, our primary indicator of poverty, varies dramatically in the scenarios. The number of chronically undernourished people in a region depends on four key parameters: population, average income, income distribution, and the "hunger line," the income below which most people today are hungry. Hunger lines increase as countries get richer as traditional, non-monetary coping mechanisms lose efficacy (for details on the methods used for computing hunger, see 9, p. 26). Since these various factors evolve differently across the scenarios, hunger trends differ as well (Figure 6). The sustainability target for the Policy Reform scenario is to halve world hunger by 2025, and halve it again by 2050. The Great Transition scenario reduces hunger far more rapidly as incomes converge more quickly both between and within regions. In Market Forces, current levels of hunger persist through 2050 as population growth counterbalances the poverty reducing effects of income growth. In the polarized Fortress World, hunger rises persistently.

Work and Leisure Time

A key factor enhancing human well-being is the amount of time people have available for discretionary activities. Work time is defined here as the average number of hours worked per person across a whole population, including children, unemployed, and the elderly. Declining work time may be due to individuals working fewer hours and/or fewer people in the labor force due to, e.g., the elimination of child labor or earlier retirements. Work time is related to two other variables: GDP per capita and productivity (GDP per hour), since income equals productivity times work time. Market Forces and Policy Reform assume the maintenance of current work weeks in developed countries with gradual convergence to those levels in developing countries (leading to the decreases in global averages shown in Figure 7). Work time trends higher in Fortress World as the poor, who shoulder higher work loads, increase as a percentage

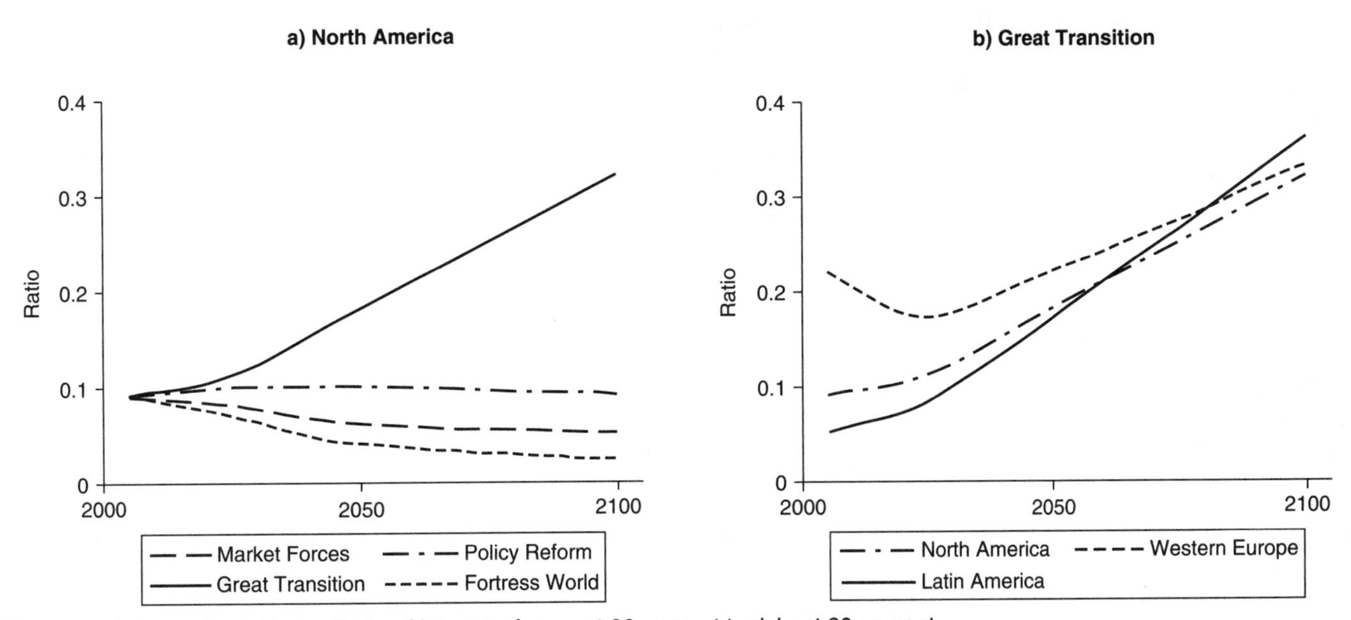

Figure 5 Intra-regional Equity Ratio of income of poorest 20 percent to richest 20 percent

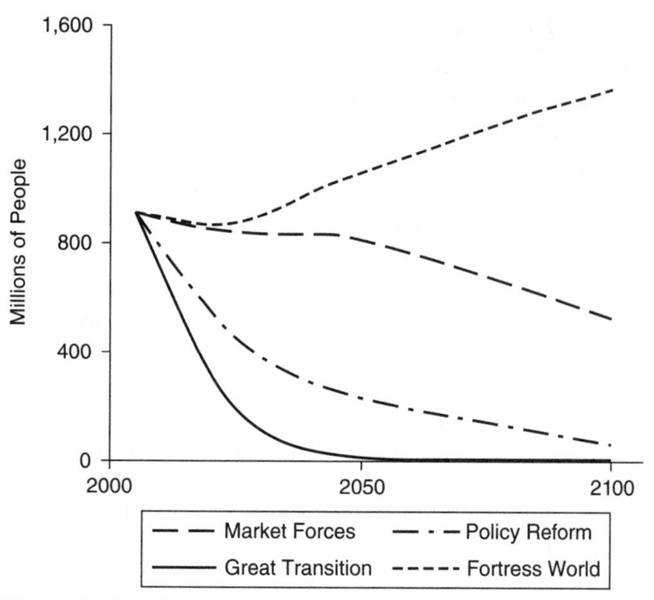

Figure 6 Incidence of Hunger

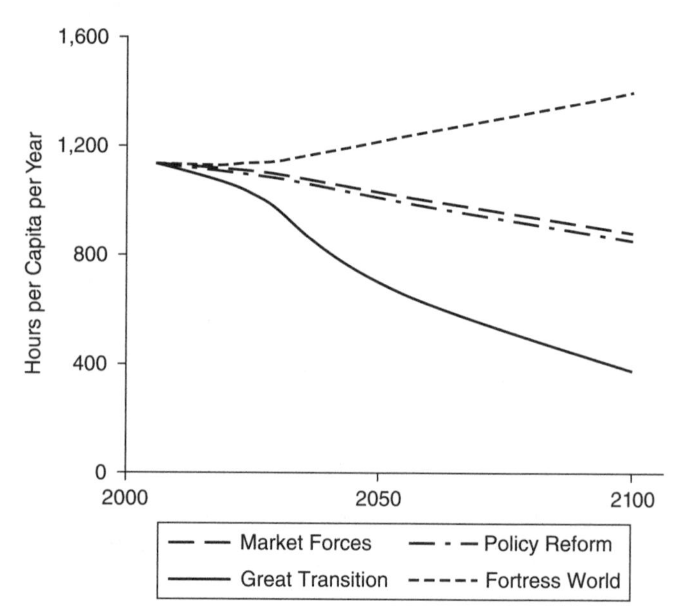

Figure 7 Work Time

of the population. In contrast, work time falls substantially in Great Transition as emphasis moves from production and consumption to quality of life. In this scenario, a worker in the United States might, for example, put in a 3-day work weeks at seven hours per day, with vacation time at current relatively high Western European levels.

Travel

An energy intensive activity, travel can be a positive or negative feature of life, a source of pleasure and cosmopolitan enrichment, on the one hand, or the drudgery of commuting and onerous business travel, on the other. In Market Forces and Policy Reform, past trends showing a strong correlation between travel levels and rising incomes persist (Figure 8). As large segments of the population become mired in poverty, travel declines in Fortress World. In Great Transition, rapid economic development in poorer regions drives average per capita distance traveled higher through mid-century, with slower growth thereafter with the approach of steady-state economies. Notably, the energy required per kilometer traveled becomes far lower in Great Transition due to greater dependence on environmentally-friendly modes: 26 percent of passenger travel is via public transportation by 2100 (*versus* 7 percent in Market Forces), and compact settlement patterns encourage more bicycling and walking.

Environment and Natural Resources

Climate Change

Carbon dioxide, the most significant anthropogenic greenhouse gas, is emitted from fossil fuel combustion, industrial processes, and land-use change. In Market Forces, where proactive energy efficiency and renewable energy measures remain weak, CO_2 emissions rise from 30 to 73 billion tonnes per year in 2100 as economic growth outpaces a 1.3 percent per annum decline in carbon intensity (CO_2 emissions per dollar of GDP). In Policy Reform and Great Transition, scenarios designed to prevent global warming greater than 2°C, CO_2 emissions decrease quickly and dramatically (Figure 9(a)). This daunting constraint requires deep improvements in energy efficiency, rapid uptake of renewable energy, and soil and forest conservation. Even with maximal effort, we see from the figure that emissions must become negative after 2075, accomplished by capturing and sequestering CO_2 from the waste stream of biomass-burning power plants. When biomass harvest is balanced by regeneration, burning biomass for electricity is carbon neutral. If, in addition, post-combustion CO_2 is captured and stored underground (say, in abandoned mines), the net effect is to remove that quantity of CO_2 from the atmosphere.

The long atmospheric residence time of CO_2 requires a focus on cumulative emissions, the sum of annual net

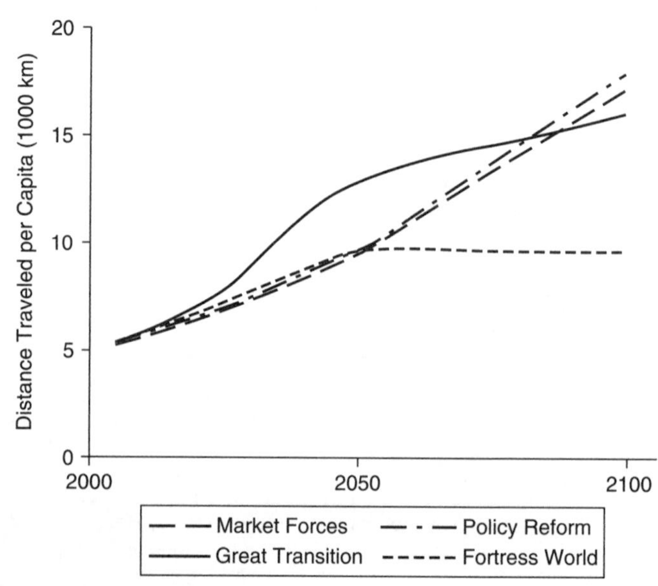

Figure 8 Travel Intensity

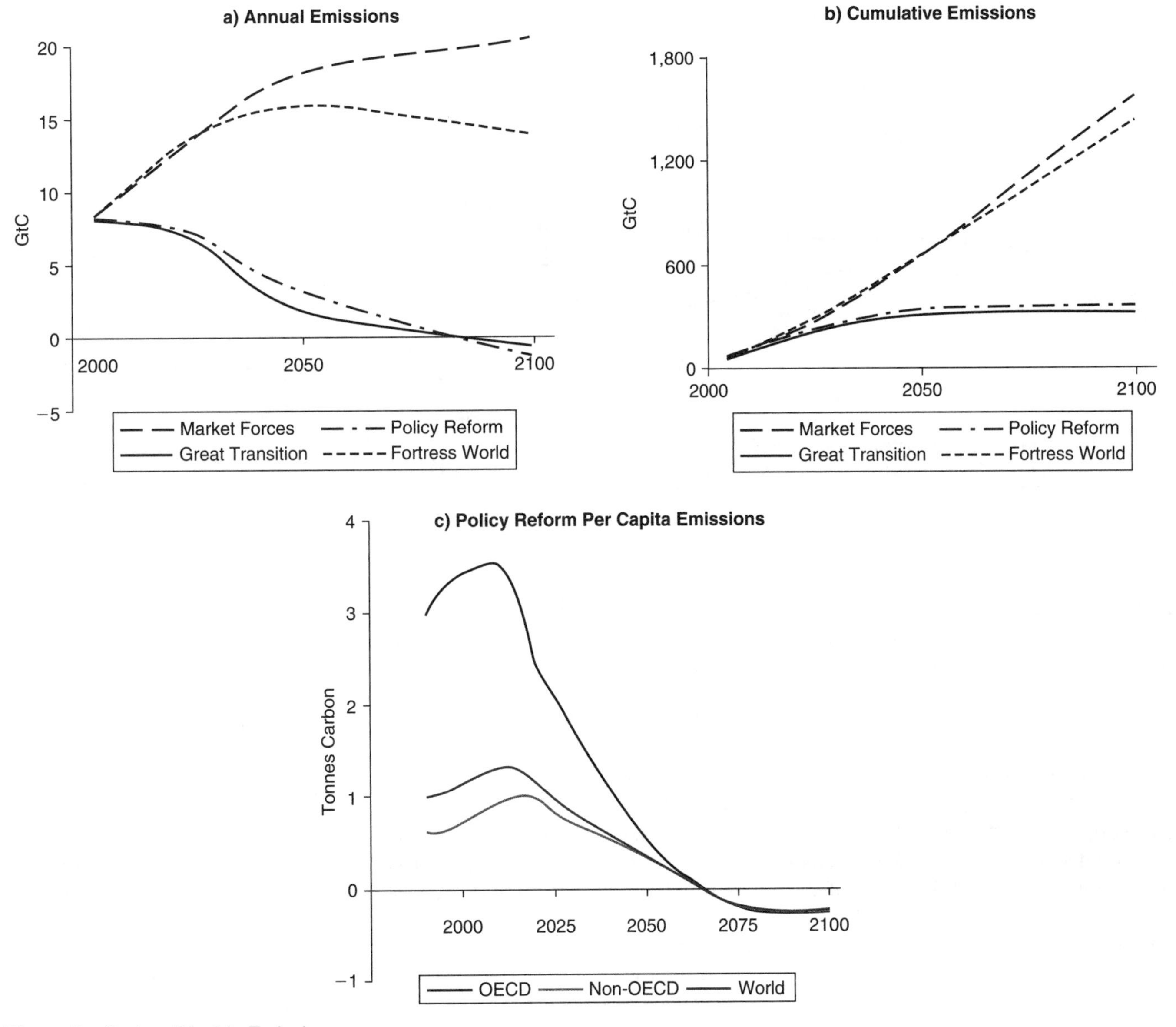

Figure 9 Carbon Dioxide Emissions

emissions (Figure 9(b)). Meeting the sustainability target of 350 parts per million (ppm) atmospheric CO_2 concentration allows cumulative emissions of about 265 GtC between 2005 and 2100, which implies a CO_2-equivalent concentration of approximately 450 ppm when other greenhouse gases are included 14, 15. This would give an approximately 70 percent chance of meeting the 2°C temperature change constraint 16; some stress that even lower concentrations levels may be necessary 17. In Policy Reform and Great Transition, global emission rights are distributed on the "contraction and convergence" equity principle wherein aggregate annual global emissions are capped and allocated to countries on the basis of population, approaching equal per capita emission rights by mid-century. This criterion allows temporary increases in emissions in poorer countries in the transition to a decarbonized global development path (Figure 9(c)). Emissions in OECD countries must fall some 85 percent by 2050.

Energy Demand

Achieving these climate goals requires strongly moderating energy demand and rapidly reducing the use of fossil fuels. We see from Figure 10(a) that energy demand in Policy Reform is far below Market Forces (a scenario that itself encompasses significant market-induced energy efficiency improvements) as a result of across-the-board efforts to promote highly efficient vehicles, green building regulations, and appliance and industrial process standards. Great Transition achieves additional reductions in energy demand, primarily through dematerialized lifestyles: moderated consumption, compact settlement patterns, reduced travel, and less meat-intensive diets. The lower energy demands in the Great Transition scenario substantially reduce the burdens of building the new post-fossil fuel energy infrastructure and depending on carbon sequestration to meet emissions constraints. Fortress World energy demand decreases eventually, but for the unwelcome reasons that development goes into reverse.

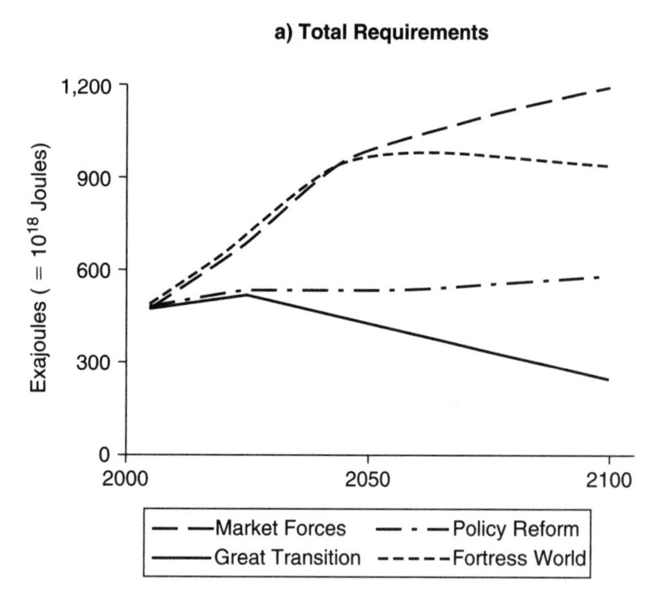

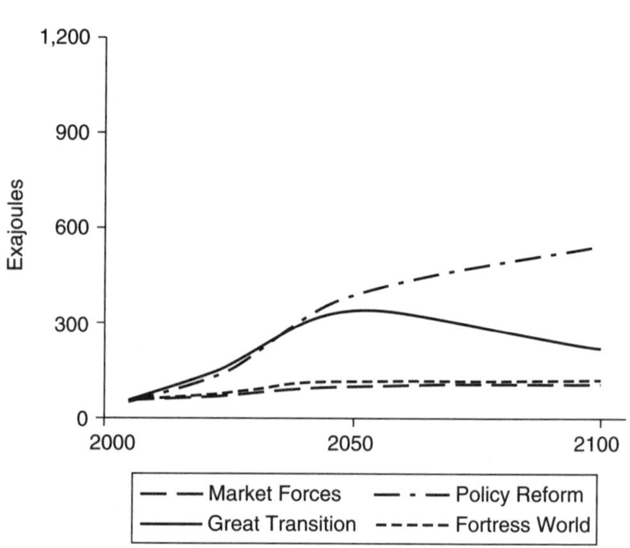

Figure 10 Energy

Energy Supply

Cost-effective conventional oil and gas resources are becoming scarce as requirements continue to rise. Any future gap between supply and demand will need to be filled by unconventional sources: shale oil, tar sands, biomass-based oils, and oil from coal. Table 6 shows estimates of the year of exhaustion of conventional resources in each scenario, suggesting that the onset of a widening shortfall is imminent in Market Forces and Fortress World. The evolutionary Market Forces vision assumes that a "peak oil" economic crisis can be avoided by bringing unconventional fossil fuel alternatives seamlessly to market in vast quantities, a Herculean technological task compounded by the heavier environmental burdens of the substitutes, which also require higher energy inputs in extraction and reformation stages. Policy Reform, by rapidly deploying additional energy efficiency technologies and renewable energy resources, postpones the exhaustion of conventional oil and gas for nearly a half century; Great Transition is able to avoid entirely the use of unconventional fossil fuels. Figure 10(b) shows trends in the use of renewable energy: gradual increase in Market Forces and Fortress World, and rapid increase in Policy Reform. Great Transition requirements decrease after 2050 in concert with decreasing total energy demand. These simulations assume nuclear-generated electricity remains a marginal energy source, limited by a host of risks: proliferation of nuclear weapons, storage of highly radioactive waste, safety, and high costs.

Biomass

Biomass is used as a fuel (firewood, bio-diesel, bio-kerosene, and ethanol) and a raw material for the paper and pulp, lumber, furniture, and construction industries. While lowering requirements for traditional firewood, rapid economic growth in Market Forces increases biomass demands in modern sectors (Figure 11). Despite strong waste recycling and efficiency measures, requirements are comparable in Policy Reform as biomass energy becomes a more significant element in the energy mix. However, land-intensive bio-fuel production competes with food production, putting additional pressure on agricultural innovations to increase yields and stabilize the prices of food staples. It takes the moderated energy demands of Great Transition for biomass demand to eventually decline from current levels, thereby sparing land for agriculture and nature.

Food and agriculture

Adequate nutrition and sound agriculture practices are central to human and environmental well-being. In recent decades, the conversion of land to crops and pastures has had major detrimental impacts on natural forests and other important ecosystems. By rapidly increasing crop yields, the Green Revolution of the past half century helped avoid the then looming food shortages, while moderating land requirements. But modern farming practices require high inputs of chemicals and irrigation water that pollute, stress water resources, and degrade soil. A more sustainable development pathway would abjure crop-intensive

Table 6 Year Conventional Fossil Fuels Exhausted

	Market Forces	*Policy Reform*	*Fortress World*	*Great Transition*
Crude Oil	2034	2071	2034	>2105
Natural Gas	2047	2088	2049	>2123

Note: Estimates of resources from 18, 19.

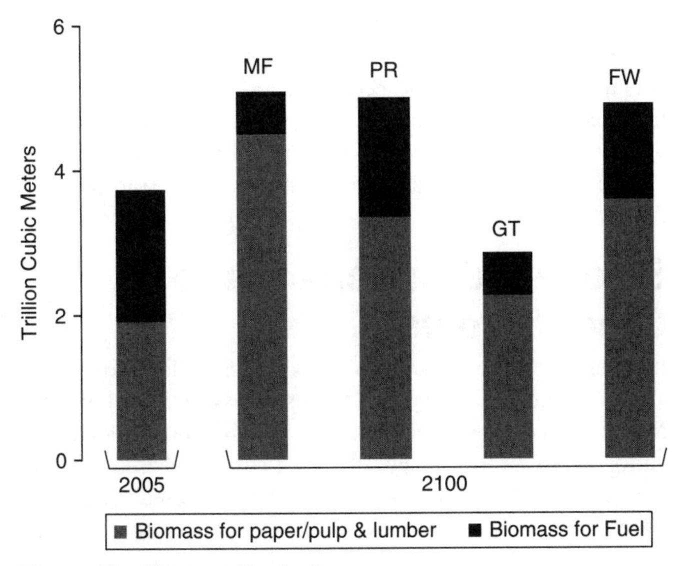

Figure 11 Biomass Production

and unhealthy high-meat diets, while adopting ecologically-sound farming practices. Global food requirements increase in all scenarios with rising incomes (Figure 12(a)), but decrease after 2050 in Great Transition as health and environmental motives foster diets high in nutrition and low in meat products. These diet adjustments, along with lower population, reduce aggregate requirements after 2050 (Figure 12(b)). On the production side, Market Forces and Fortress World rely heavily on chemical inputs and genetically modified organisms, while Policy Reform and Great Transition adopt organic, ecological agricultural methods.

Land

A major challenge in the transition to sustainability will be protecting ecosystems and habitats in the face of increasing pressures on land resources. Finding solutions that balance the needs of economies and nature is an urgent matter of human self-interest, since healthy ecosystems provide vital and valuable resources and services, though these are usually not monetized. For those who place inherent worth on the vitality of the natural world, preserving biodiversity and natural beauty is also an ethical concern. Rising populations and incomes drive the demand for land for agriculture, pasturing, human settlements, forest products, and bio-fuels. Reversing the momentum of habitat loss will take action on both the demand and supply side of the equation, moderating requirements and managing land resources sustainably. We illustrate scenario patterns for the case of forest land in Figure 13. The failure to align economies with environmental objectives in Market Forces leads to continued loss and degradation of land. Masses of land-hungry poor and technological stagnation in Fortress World further diminish ecosystems. Intensive governmental efforts to protect and restore ecosystems in Policy Reform slow, and then modestly reverse habitat loss. Lower population and economic growth, compact settlements, less land-intensive diets, and much lower use of biomass in Great Transition combine to enhance ecosystems, though even under these salutary conditions the protection of nature remains a long-term challenge.

Water

Providing adequate freshwater for the maintenance of human and natural systems will be another persistent challenge in this century. Today, 1.7 billion people live in areas of water stress, *i.e.*, where there is significant competition for water among agricultural, industrial, public, and environmental claims. In addition, several hundred million people endure "high water stress" of absolute and chronic shortages of freshwater resources. We deem a population in 'water stress' when water demand divided by renewable water resources (use-to-resource ratio) exceeds critical values and in 'high' stress at more stringent values (see 9, pp. 229–234). Future requirements in the scenarios, shown in Figure 14(a), exhibit a now familiar pattern. The decreases in Policy Reform and Great Transition are traced to vigorous

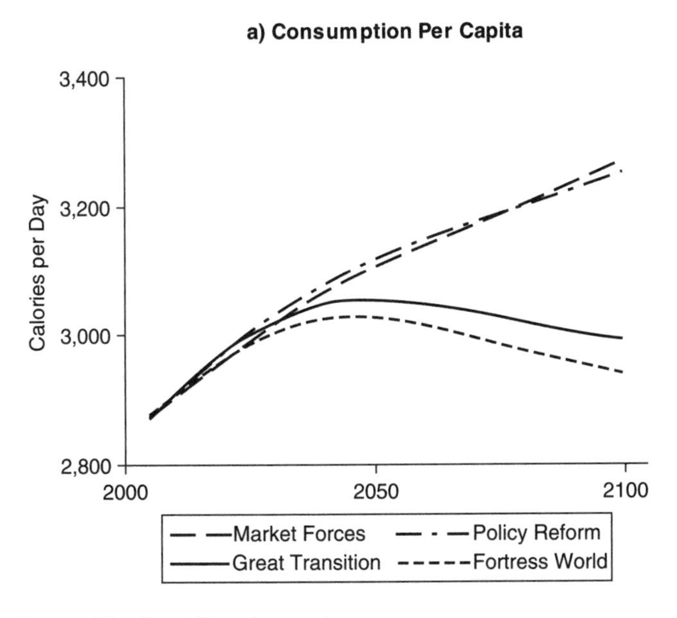

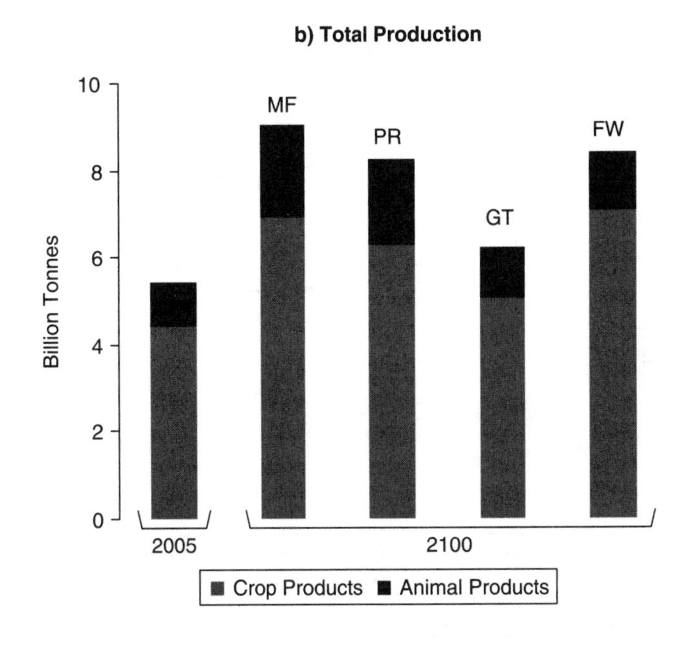

Figure 12 Food Requirements

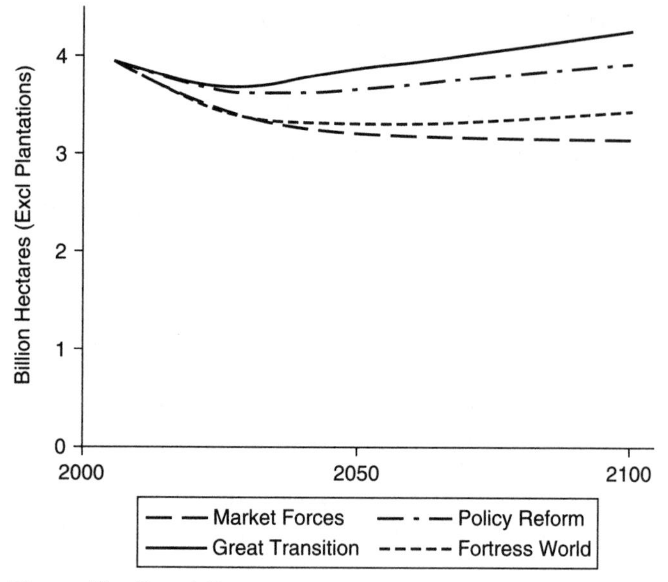

Figure 13 Forest Cover

efforts to deploy best-practice efficiency improvements for irrigation, which accounts for 70 percentage of requirements in 2005, and other end-uses. Figure 14(b) shows that these interventions, along with more sustainable water harvesting, meet our sustainability target (Table 5), despite high population growth in the Mideast, northern Africa, and swaths of many other regions short of water. In the absence of such intensive effort in Market Forces and Fortress World, water stress more than doubles. Moreover, climate change induced alteration of hydrological patterns, not accounted for in this analysis, could further exacerbate the problem of freshwater unsustainability.

Local Pollution

The scenario simulations track changes in representative air pollutants (e.g., sulfur oxides), water contaminants (nitrogen and biochemical oxygen demand), industrial toxics, and municipal solid waste. The broad patterns are illustrated in Figure 15 using the example of toxic chemical loads from industrial products and processes. Rapid economic growth and weak regulation lead to extreme contamination in Market Forces, while the deployment of clean technology and recycling substantially reduce pollution in Policy Reform and the Great Transition.

Selected Comparisons to Other Studies

In recent years, scenario analysis has been applied extensively at local, regional, and national scales across a staggering range of themes and issues. See 25 and 26 for histories of global scenarios, and 27 for a review of recent scenario-based environmental assessments. Global studies have focused on energy, water, environment, population, food and agriculture, and the economy 10, 28-33, but only a handful have undertaken a comprehensive and detailed socio-ecological assessment of the long term prospects for sustainability 34, 35. Significant efforts include the PoleStar-based work of the Global Scenario Group 9, 36, 37, upon which the current study builds, and applications of the IMAGE model 38 of the Netherlands Environmental Agency.

For the sake of brevity, we focus our comparisons of scenario analyses on a single aspect of the sustainability problématique: the climate and energy nexus. Figure 16 displays CO_2 emissions trajectories for our four scenarios along with others drawn from the work of the Intergovernmental Panel on Climate Change (IPCC). The shaded background shows the range of results of the Special Report on Emissions Scenarios (SRES) 39, relied on in IPCC's third and fourth assessments 40, 41. The solid lines are provisional "representative concentration pathways" slated for use in the upcoming fifth assessment 25, 42. This set is labeled "RCP X" in the figure with the X corresponding to the anthropogenic radiative forcing in 2100 expressed in watts/meter^2.

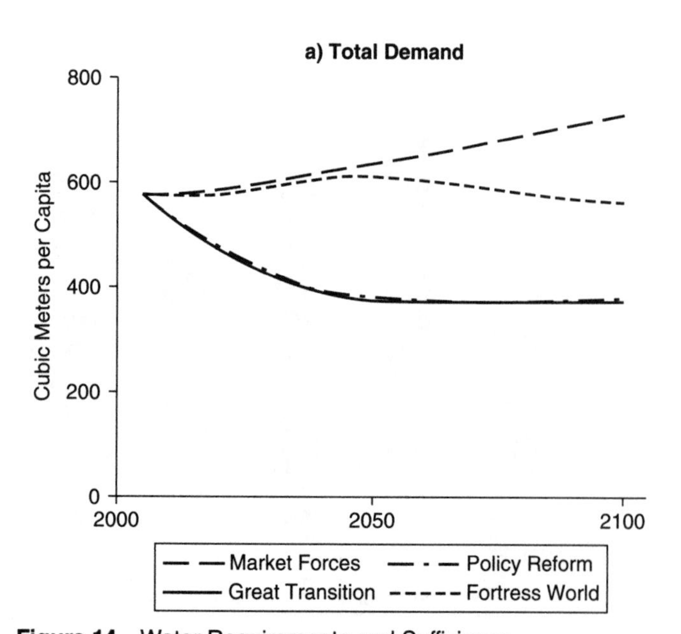

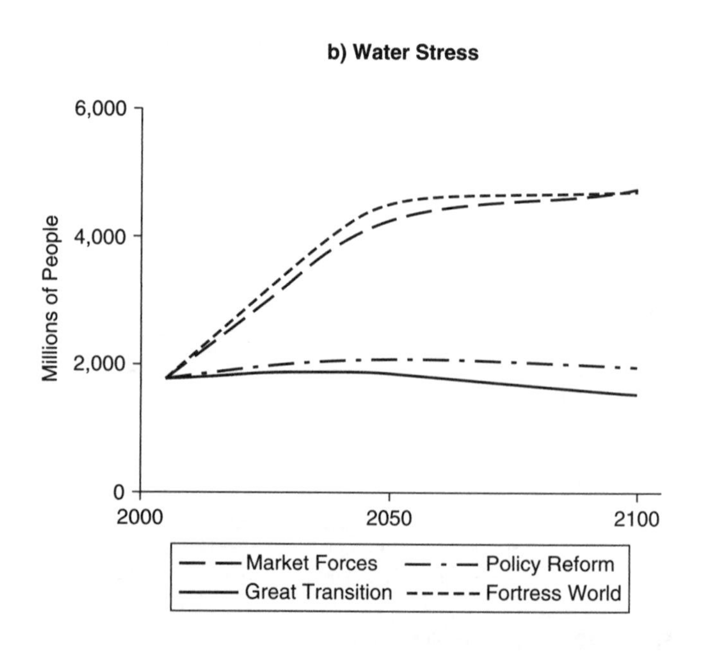

Figure 14 Water Requirements and Sufficiency

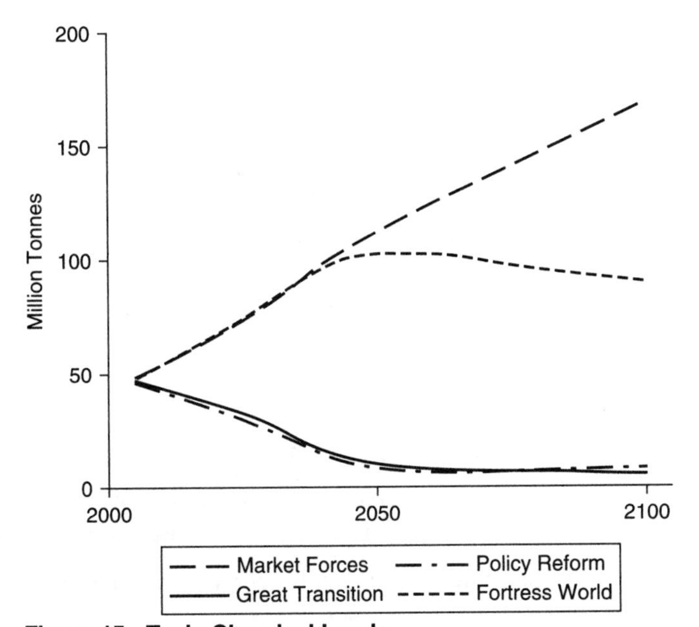

Figure 15 Toxic Chemical Loads

Carbon dioxide emissions in the Policy Reform and Great Transition scenarios fall below the lowest range of the IPCC scenarios. The RCP 2.6 trajectory, the most ambitious emissions reduction scenario currently being considered by IPCC, relies on massive deployment of carbon sequestration (capture of CO_2 from power plant waste streams with subsequent underground storage), though this remains an unproven technology at anything like the scales envisioned. By contrast, deeper and more rapid penetration of renewable energy and efficiency in Policy Reform reduces the need and delays the deployment of sequestration technology, while the dematerialized life-styles and moderated population growth in Great Transition reduces its role still further.

Figure 17 turns to a comparison of the energy requirements of our scenarios with the RCP set. The shaded area of the figure is bounded below by RCP 2.6, and above by the RCP "baseline" scenario 43, which has an emissions profile comparable to Market Forces. Although Figure 16 shows similar emissions trajectories for Policy Reform and RCP 2.6, energy requirements deviate radically. The far higher energy use in RCP 2.6 requires comparably heavier reliance on carbon sequestration with its inherent energy inefficiencies and uncertain prospects. The low energy requirements in Great Transition suggest the importance of transcending conventional development visions in designing global climate simulations to consider possibilities for more fundamental alterations in long-range social conditions. This, in turn, will encourage overcoming rigidities in standard integrated assessment models that might circumscribe the range of possible development paths for this century.

As a final comparison, Figure 18 displays the energy requirements of our four scenarios along with those from "GEO-4", the most recent Global Environmental Outlook (GEO) of the United Nations Environment Programme 44. The time horizon in the figure extends only to 2050, the last year reported in the GEO-4 report. The comparison is apt since the GEO scenarios were originally based on the Global Scenario Group scenarios herein updated and enhanced (the Global Scenario Group was originally established as the "scenario working group" for the GEO process) 45, 46. We see that the Policy Reform and Great Transition scenario follow trajectories below the GEO range as a result of more aggressive energy conservation policies and, in Great Transition, more basic adjustments in the scale and composition of energy demand. Where

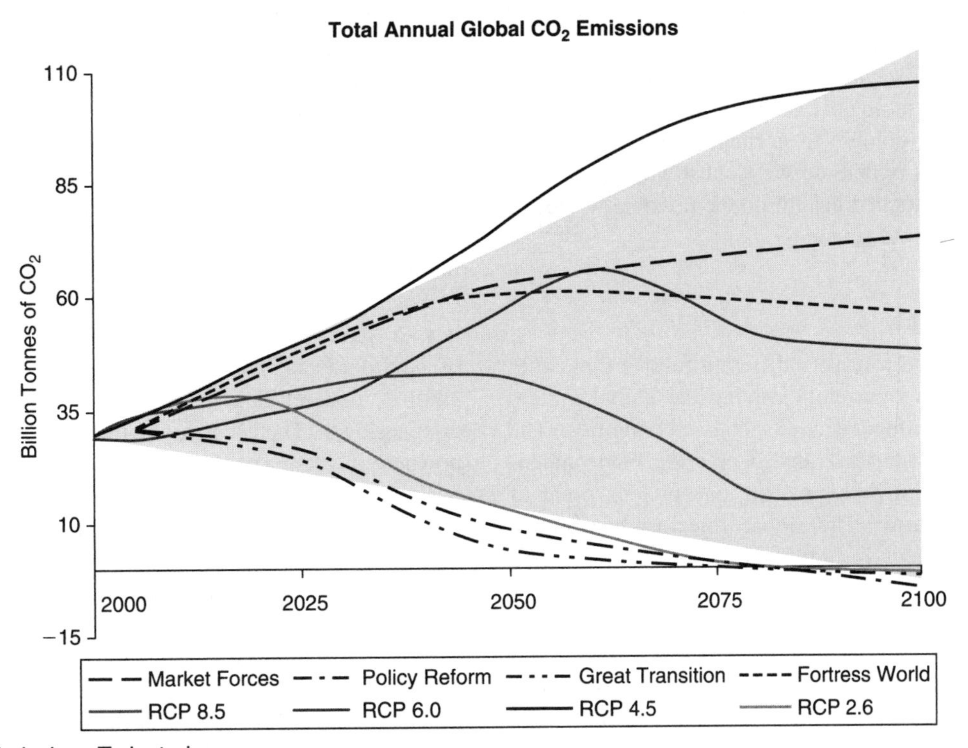

Figure 16 CO_2 Emissions Trajectories

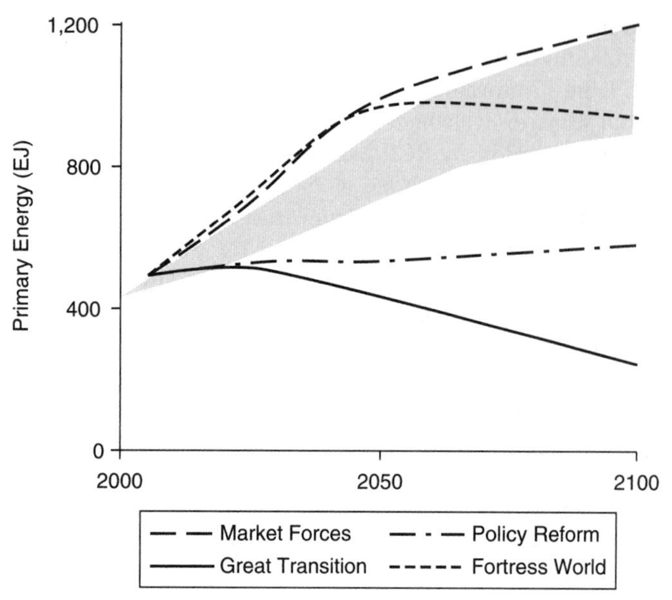

Figure 17 Comparison of Primary Energy Requirements to IPCC RCP Trajectories (shaded area)

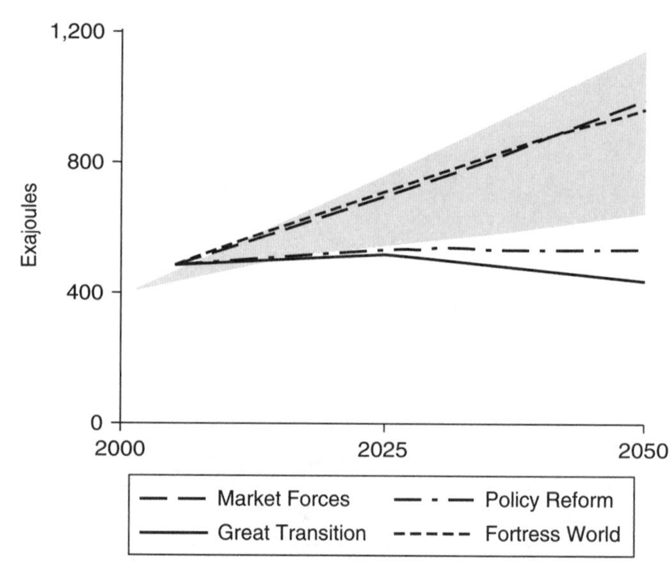

Figure 18 Comparison of Primary Energy Requirements to GEO-4 Trajectories (shaded area)

the GEO-4 scenarios rely heavily on nuclear power and carbon sequestration of fossil-fuel-generated energy, the Policy Reform and Great Transition scenarios require neither by 2050.

These comparisons illustrate the basic pattern that our results tend to span a wider range than other scenarios studies. In particular, the Great Transition and Fortress World, by introducing unconventional narratives of global development, expand the aperture for scanning possible futures. Standard modeling tools, built on algorithms specified on the basis of historic development patterns and structural relationships, have limited utility for illuminating scenarios in which the global system undergoes such fundamental restructuring 26. The PoleStar system relied on here casts a wider net by allowing flexible specification of novel outcomes and images of the future that serve as constraints on scenario trajectories. This backcasting technique shifts epistemological emphasis from passive projection (where are we going?) to normative consideration of alternative destinations (where do we want to go? how do we get there?), offering enriched insight on the possible worlds of tomorrow, and how best to act today.

Conclusions

This paper has sketched four very different futures that could emerge from the forces currently driving the global system toward critical environmental and social uncertainties. The world now faces multifaceted and interacting environmental, resource, and social problems, an inauspicious point of departure for all scenarios. The global trajectory can branch in alternative directions in the coming decades, depending on how bio-physical and cultural stresses manifest themselves and how society responds. The destiny of both people and planet rest ultimately with human choice as we anticipate and respond to crises and seize opportunities for positive transformation. Will our actions be tardy and tepid? Or timely and consequential?

If we muddle forward in a complacent Market Forces mode, the risks rise of deterioration of life-support ecosystems and civilized norms. By contrast, a long and tenacious process of proactive adjustments in policy and technology—as embodied in the Policy Reform scenario—could, in principle, redirect world development toward sustainability. However, this approach confronts the daunting challenge of marshalling the massive globally-coordinated interventions at the pace and magnitude required. If the strategy of incremental change fails, and crises mount, the global trajectory could swerve toward the authoritarian order of Fortress World, or even the collapse of organized institutions.

Yet, the sustainability challenge presents, as well, the prospect of transcending technological solutions with a transformation in human values and restructuring of economic and governance institutions. If humanity musters the will for a Great Transition, a new and vital phase of human history could open in this century: a planetary civilization that pursues peace and justice, delivers material sufficiency and rich lives, and understands humanity as a respectful member of a wider community of life.

Each scenario raises critical questions of feasibility. The laissez-faire optimism of Market Forces would invite a host of environmental and social crises that could feedback and amplify, undercutting its rosy assumption of perpetual economic growth. The necessary political will for Policy Reform is nowhere in sight. The organized cooperation of the global elite in Fortress World, in the face of unfolding crises and the resistance, no doubt, of the excluded masses, would be extremely difficult.

The Great Transition can only emerge as a collective cultural and political project of global citizens, a development that is far from guaranteed. We can only hope that our scan of the global future, suggesting the desirability—even necessity—of such a deep change, will help spur action to achieve sustainability.

References

Richardson, K.; Steffen, W.; Schellnhuber, H.J.; Alcamo, J.; Barker, T.; Kammen, D.M.; Leemans, R; Liverman, D.; Munasinghe, M.; Osman-Elasha, B.; Stern, N.; Wæver, O. Synthesis Report, Climate Change: Global Risks, Challenges & Decisions. In *Proceedings of the Climate Change Congress,* University of Copenhagen, Copenhagen, Denmark, 10–12 March 2009.

Millennium Ecosystem Assessment. *Ecosystems and Human Well-Being: Current State and Trends*; Island Press: Washington, DC, USA, 2005; Volume 1.

Raskin, P. World lines: A framework for exploring global pathways. *Ecol. Econ.* 2008, *65,* 461–470.

Swart, R.; Raskin, P.; Robinson, J. The problem of the future: Sustainability science and scenario analysis. *Glob. Environ. Change* 2004, *14,* 137–146.

Gallopín, G.; Raskin, P. *Global Sustainability: Bending the Curve*; Routledge: St Paul, MN, USA, 2002.

Raskin, P.; Banuri, T.; Gallopín, G.; Gutman, P.; Hammond, A.; Kates, R.; Swart, R. *Great Transition: The Promise and Lure of the Times Ahead;* Stockholm Environment Institute: Boston, MA, USA, 2002.

Global Scenario Group Homepage. www.gsg.org (accessed on 1 June 2010).

The PoleStar Project Homepage. www.polestarproject.org (accessed on 1 June 2010).

Electris, C.; Raskin, P.; Rosen, R.; Stutz, J. *The Century Ahead: Four Global Scenarios, Technical Documentation;* Tellus Institute: Boston, MA, USA, 2009; Available online: www.tellus.org/publications/files/TheCenturyAhead_TechDoc.pdf (accessed on 1 June 2010).

United Nations. *Population, Resources, Environment and Development: The 2005 Revision.* Available online: www.un.org/esa/population/unpop.htm (accessed on 1 June 2010).

The Millennium Development Goals Report; United Nations: New York, NY, USA, 2009.

Raskin, P. *The Great Transition Today: A Report from the Future;* Tellus Institute: Boston, MA, USA, 2006; Available online: www.gtinitiative.org/resources/paperseries.html (accessed on 1 June 2010).

Rosen, R.; Schweichart, D. *Visions of Regional Economies in a Great Transition World;* Tellus Institute: Boston, MA, USA, 2006; Available online: www.gtinitiative.org/resources/paperseries.html (accessed on 1 June 2010).

Nakicenovic, N. *World Energy Outlook 2007: CO_2 Emissions Pathways Compared to Long-Term CO_2 Stabilization Scenarios in the Literature and IPCC AR4;* Organisation for Economic Co-operation and Development (OECD): Paris, France, 2007; Available online: www.iea.org/weo/docs/weo2007/CO2_Scenarios.pdf (accessed on 1 June 2010).

International Institute for Applied Systems Analysis (IIASA). *Greenhouse Gas Initiative Scenario Database;* Available online: www.iiasa.ac.at/Research/GGI/DB/ (accessed on 1 June 2010).

Meinshausen, M.; Meinshausen, N.; Hare, W.; Raper, S.C.B.; Frieler, K.; Knutti, R.; Frame, D.J.; Allen, M.R. Greenhouse gas emission targets for limiting global warming to 2 °C. *Nature* 2009, *458,* 1158–1162.

Hansen, J.; Sato, M.; Kharecha, P.; Beerling, D.; Berner, R.; Masson-Delmotte, V.; Pagani, M.; Raymo, M.; Royer, D.L.; Zachos, J.C. Target atmospheric CO_2: Where should humanity aim? *Open Atmos. Sci. J.* 2008, *2,* 217–231.

BP. *BP Statistical Review of World Energy;* Available online: www.bp.com/liveassets/bp_internet/globalbp/globalbp_uk_english/reports_and_publications/statistical_energy_review_2007/STAGING/local_assets/downloads/pdf/statistical_review_of_world_energy_full_report_2007.pdf (accessed on 1 June 2010).

World Energy Council. *Survey of Energy Resources 2007;* Available online: www.worldenergy.org/publications/survey_of_energy_resources_2007/default.asp (accessed on 1 June 2010).

International Institute for Applied Systems Analysis (IIASA). *Global Energy Assessment (GEA).* Available online: www.iiasa.ac.at/Research/ENE/GEA/index.html (accessed on 1 June 2010).

Cosgrove, W.; Rijsberman, F. *World Water Vision: Making Water Everybody's Business;* Earthscan: London, UK, 2000.

Environmental Outlook to 2030; Organisation for Economic Co-operation and Development (OECD): Washington, DC, USA, 2008.

World Energy Outlook 2009; International Energy Agency (IEA): Paris, France, 2009.

Rosengrant, M.W.; Paisner, M.S.; Meijer, S.; Witcover, J. *2020 Global Food Outlook;* International Food Policy Research Institute: Washington, DC, USA, 2001.

Moss, R.H.; Edmonds, J.A.; Hibbard, K.A.; Manning, M.R.; Rose, S.K.; van Vuuren, D.P.; Carter, T.R.; Emori, S.; Kainuma, M.; Kram, T.; Meehl, G.; Mitchell, J.F.B.; Nakicenovic, N.; Riahi, K.; Smith, S.J.; Stouffer, R.J.; Thomson, A.M.; Weyant, J.P.; Wilbanks, T.J. The next generation of scenarios for climate change research and assessment. *Nature* 2010, *463,* 747–756.

Raskin, P. Worldlines: A Framework for Exploring Global Pathways. *Ecol. Econ.* 2008, *65,* 461–470.

Kok, M.T.J.; Bakkes, J.A.; Eickhout, B.; Manders, A.J.G.; van Oorschot, M.M.P.; van Vuuren, D.P.; van Wees, M.; Westhoek, H.J. *Lessons from Global Environmental Assessments;* Netherlands Environmental Assessment Agency: Bilthoven, The Netherlands, 2008.

World Agriculture: Toward 2030/2050; Global Perspective Studies Unit, Food and Agriculture Organization of the United Nations (FAO): Rome, Italy, 2006.

OECD-FAO Agricultural Outlook 2009–2018; Organisation for Economic Co-operation and Development (OECD): Paris, France, 2009.

Prinn, R.; Paltsev, S.; Sokolov, A.; Sarofim, M.; Reilly, J.; Jacoby, H. The Influence on Climate Change of Differing Scenarios for Future Development Analyzed Using the MIT Integrated Global System Model. In *Join Program Report Series;* MIT Joint Program on the Science and Policy of Global Change: Cambridge, MA, USA, 2008.

Wise, M.A.; Calvin, K.V.; Thomson, A.M.; Clarke, L.E.; Bond-Lamberty, B.; Sands, R.D.; Smith, S.J.; Janetos, A.C.; Edmonds, J.A. *The Implications of Limiting CO_2 Concentrations for Agriculture, Land Use, Land-use Change Emissions and Bioenergy;* Pacific Northwest National Laboratory: Richland, WA, USA, 2009.

Barker, T.; Foxon, T.; Scrieciu, S. Achieving the G8 50 percentage target: Modelling induced and accelerated technological change using the macro-econometric model E3MG. *Clim. Policy* 2008, *8,* S30–S45.

Stern, N. *The Economics of Climate Change: The Stern Review;* Cambridge University Press: Cambridge, UK, 2007.

Millennium Ecosystem Assessment; Island Press: Washington, DC, USA, 2005.

Raskin, P. Global scenarios: Background review for the millennium ecosystem assessment. *Ecosystems* 2005, *8,* 133–142.

Gallopín, G.; Hammond, A.; Raskin P.; Swart, R. *Branch Points: Global Scenarios and Human Choice;* Stockholm Environment Institute: Stockholm, Sweden, 1997.

Raskin, P.; Gallopín, G.; Gutman, P.; Hammond, A.; Swart, R. *Bending the Curve: Toward Global Sustainability;* Stockholm Environment Institute: Stockholm, Sweden, 1998.

Integrated Model to Assess the Global Environment (IMAGE). *IMAGE Model Site;* Available online: www.pbl.nl/en/themasites/image/index.html (accessed on 1 June 2010).

Intergovernmental Panel on Climate Change (IPCC). *IPCC Special Report on Emissions Scenarios;* Nakicenovic, N., Swart, R., Eds.; Cambridge University Press: Cambridge, UK, 2000.

Intergovernmental Panel on Climate Change (IPCC). *Third Assessment Report: Climate Change 2001;* Cambridge University Press: Cambridge, UK, 2001.

Intergovernmental Panel on Climate Change (IPCC). *Fourth Assessment Report: Climate Change 2007;* Cambridge University Press: Cambridge, UK, 2007.

Weyant, J.; Azar, C.; Kainuma, M.; Kejun, J.; Nakicenovic, N.; Shukla, P.R.; La Rovere, E.; Yohe, G. *Report of 2.6* Versus *2.9 Watts/m2 RCPP Evaluation Panel;* Intergovernmental Panel on Climate Change (IPCC): Geneva, Switzerland, 2009.

Rao, S.; Riahi, K.; Stehfest, E.; van Vuuren, D.; Cho, C.; den Elzen, M.; Isaac, M.; van Vliet, J. *IMAGE and MESSAGE Scenarios Limiting GHG Concentration to Low Levels;* Interim Report, IR-08-020; International Institute for Applied Systems Analysis (IIASA): Laxenburg, Austria, 2008.

Global Environment Outlook GEO-4: Environment for Development; United Nations Environment Programme: Washington, DC, USA, 2007.

Raskin, P.D.; Kemp-Benedict, E. *Global Environment Outlook Scenario Framework: Background Paper for UNEP's Third global Environmental Outlook Report (GEO-3);* UNEP: Nairobi, Kenya, 2004.

The GEO-3 Scenarios 2002–2032: Quantification and Analysis of Environmental Impacts; Potting, J., Bakkes, J., Eds.; Netherland Environmental Assessment Agency: Bilthoven, The Netherlands, 2004; p. 25.

Critical Thinking

1. How are the four scenarios different? How are they similar?
2. How might the scenarios differ if some other measure besides GDP was used in the analysis?
3. Is it reasonable to have a target of 56 million people living in poverty in 2100?

Acknowledgments—This study carries forward a two-decade program of exploration of alternative global futures conducted by the Tellus Institute, the Global Scenario Group and the Stockholm Environment Institute. We are grateful to each of the scores of colleagues who have contributed along the way, with special mention due to Gordon Goodman for originally conceiving of the PoleStar Project, Gilberto Gallopin for intellectual leadership in scenario formulation and Eric Kemp-Benedict for technical innovations and support. Of course, the authors alone are responsibility for any errors of fact or inadequacies of judgment in the current document.

The Invention of Sustainability

PAUL WARDE

This essay attempts something a little peculiar: a study of the genesis of a concept within discourses which did not, in fact, use the word. This is at least true of "sustainability" in English. The emergence of the German equivalent, *Nachhaltigkeit,* which might also be expressed by the idea of "lasting-ness", is, however, usually dated to the use of the word *nachhalthende* by Hanns Carl von Carlowitz in his *Sylvicultura oeconomica* of 1713, the first great forestry manual of the eighteenth century. In fact, the term can be found in the 1650s.[1]

The most familiar modern definition of the "sustainable" comes from the Brundtland Commission's report of 1987, where the term "sustainable development" was defined thus: "development that meets the needs of the present generation without compromising the ability of future generations to meet their own needs".[2] Formulations of sustainability are frequently rather vague, but they generally address the sense that humankind must ensure its *material* reproduction in a way that does not diminish the fortunes of future generations. This issue is, of course, a very current preoccupation for us, but preindustrial societies that laboured under the exigencies of the "organic economy" have frequently been assumed by historians to be effective ecological "optimizers", and to have developed institutional structures and economic practices to ensure their sustainability. In contrast, I will suggest that the modern notion of sustainability largely draws on ideas developed in the late eighteenth and early nineteenth centuries when new understandings of soil science and agricultural practice combined to develop the idea of a *circulation* of essential nutrients within ecologies, and hence allow the perception that disruption to circulatory processes could lead to permanent degradation. Whether agricultural practice was in most cases sustainable or not is, of course, a separate issue.

From the sixteenth century, handbooks of advice and estate management were published in northern Europe, mostly in imitation of classical predecessors such as Virgil and Xenophon, of which the first English translations emerged in the 1530s.[3] When Conrad von Heresbach came to write his *Foure bookes of husbandry* (originally published in German in 1570), he cited some fifty-seven ancient authors, but only eighteen of his near contemporaries.[4] Early, and enduringly popular, forms of the genre were calendars advising on what agricultural tasks to perform throughout the year. While authors of what was known in Germany as the *Hausväterliteratur* (writings for the "father of the house", or the "patriarch") drew liberally on the classics, they nevertheless also felt compelled to assert their own experience in farming, and they by no means absorbed other thinking uncritically.[5] The *Hausväterliteratur* also provided a slow diffusion of ideas from south to north, sometimes, of course, bringing specific recommendations to inappropriate climes, and provided a kind of guidebook for the management of a country estate.

The bottom line in these handbooks was the maintenance of household income. Thus in Gervase Markham's 1616 English edition of Claude Estienne's *La maison rustique,* a work that originally appeared in 1554, he proclaims,

> it is my purpose . . . to lay out unto you the waies, so to dwell upon, order, and maintaine a Farme, Meese or Inheritance in the Fields (name it as you please), as that it may keepe and maintaine with the profit and increase thereof, a painefull and skilfull Husbandman, and all his Familie.[6]

There was little, if any, attention given to either the capital stock, or the land itself as a source of revenue that could be alienated or used to secure debt. The focus was squarely on year-to-year production (and to a large extent autoconsumption) on an estate that was expected to endure into an unforeclosed future. The moral centre of this idealized household was the virtue of thrift, as Thomas Tusser put it, or to "Eate within thy tedure", as John Fitzherbert put it in his *Boke of Husbandry* of 1523. By "eating within thy tedure . . . thou nedest to bege nor borowe of no man, so longe shalte thou ecncrease and growe in richesse"; if needs to beg or borrow, "that wylle not long endure, but thou shallt fall in to poverty".[7] Behind these encomia to making shift by thrift was the assumption that it was not overworking of the land that could prove one's undoing, but the overexpenditure of its products. These "how-to" books provided commentary on best practice for particular tasks, often organized according to the layout of the estate. This made them practical and easy to navigate, but also presented a conceptual obstacle that they could not convey a sense of how the various elements of farming were interconnected. The possibility of certain agricultural practice compounding difficulties or advantages in other areas of farming would be enduringly difficult to formulate effectively.

A Theory of the Soil

Any sensible explanation of how agriculture could endure had to address the soil and the weather. The able husbandman had to be able to read both, a capacity that might require "long and assured experience" to nurture.[8] It was not long before the international range of the *Hausväterliteratur* confronted readers with regional differences in soil, clime and practice; some of the latter having no obvious explanation except for customary practice.[9] Such problems came to the fore in agronomic literature with an outpouring of works in the 1610s and 1620s from Gervase Markham, who exemplified a great agronomic tradition of publishing numerous versions of the same book under a slightly different title with minor emendations. Markham made soil type the organizing principle of his treatises, and this would exercise a profound influence thereafter: local environmental variations became the foremost theme for agronomy.[10]

What was the soil? According to Estienne, soils are either simple or compound, and loose or binding.[11] In an English context, "loose" or "binding" was understood as a sandy or a clayey soil. These represented the two simple kinds of earth, and all compounds were a mix of these. In turn, all soils displayed qualities derived from the four humors: clays were cold and moist, sands hot and dry, and most soils some blend of this palette. The farmer had to start, stated Markham, from a "true knowledge of the Nature and Condition of your Ground" that was to be combined with "the Clyme and Continent wherein they lye" and inferred by their "outward faces and charracters". Colour, texture and taste were the indicators on which the husbandman could develop such knowledge, largely derived from long observation: "every man in his owne workes knows the alteration of climates".[12]

Farmers knew very well that soil could become exhausted, fields worn out, or the land out of "heart". But as this focus on the "heart" of the particular field was preoccupied with balance, or at least an appropriate matching up of soil, husbandry, crop and climate, as yet it precluded a broader sense of the possibility of permanent degradation. Land was "out of heart" simply because the soil embodied an imbalance of the humors, or was being put to an inappropriate use. There was no essential element that could be, or come to be, deficient: all imbalances could be meliorated by the requisite lashings of some substance which provided the appropriate quality, or as Sir Hugh Plat wrote, "contraries are remedied by their contraries".[13] The prevailing theory of the soil thus helps explain the surprising optimism of early modern improvers, and equally their complete lack of sense that there could be a long-term trend in soil quality. Durability, or as we might say sustainability, was conceived as a question of *balance,* not, as would later be the case, of *flows.*

Agronomists would hold true to humoral thinking about the soil long into the eighteenth century. But the sixteenth century also provides us with an inkling of what was to come, from the pen of Sir Hugh Plat. His theory of "vegetative salts" as the source of generation was the first trial of a theory of soil nutrients that offered the possibility of circulation-based thinking about agriculture and ecology. His 1594 work on salts drew very heavily on translating the work of Bernard Palissy, another Renaissance polymath who wrote on fossils, hydrology and chemistry and practised ceramics, painting and surveying. It was a kind of salt (and for Plat there were many, including copper, nitre and sugar),

> that maketh all seedes to flourish, and growe, and although the number of those men is verie small, which can giue anie true reason whie dungue shoulde doe anie good in arable groundes, but are ledde thereto more by custome than anie Philosophicall reason, neuerthlesse it is apparaunt, that no dungue, which is layde vppon barraine groundes, coulde anie way enrich the same, if it were not for the salt which the straw and hay left behinde . . . it is not the dung itselfe which causeth fruitfulnes: but the salt which the seed hath sucked out of the ground.[14]

Plat prefigured later writers in seeking to identify more exactly the *active agent* that brought herbaceous growth.

Whatever that agent might be, everyone agreed that the soil could be subjected to amelioration to improve its fertility. In 1523 Fitzherbert had dwelt on the matter fairly briefly, drawing on Virgil: the dung of doves was best, that of all animals that chew the cud good, and the horse worst.[15] As fields also seemed to recover heart of themselves, in part because of inundations from the atmosphere, but also because fallowing permitted the removal of weeds, it was important to give many grounds their "rest" or "recreation". A field had to be kept in balance in a manner not unlike a person: "as a field starveth, if it not be dunged at all, so it burneth if it be over-dunged."[16]

Providence and Particularity

By the middle of the seventeenth century, however, the English focus had shifted from the maintenance to the *increase* of profit. Writing was still a mix of reportage on best local practice (with considerable respect for the Dutch and Flemish), sometimes sceptical reference to the ancients, and an increasing zeal for Baconian experimentation. Two phases of more coordinated networks of correspondents emerged: one associated with the Commonwealth and the circle around the Baltic immigrant and reformer Samuel Hartlib, the second connected with the Royal Society.[17] This era was in some ways the culmination of an accelerating transfer of information into and around England, but also reflected the special circumstances of displacement by war and the development of wide networks of letter writers who were confronted, like Sir Richard Weston in Flanders in the 1640s, with agricultural and botanical knowledge that appeared developed far beyond that of Britain. Weston's observations facilitated a shift from a preoccupation with the maintenance to the *increase* of profit: "But I advize you to make Trial your selv's of all these several *Husbandries,* and then to follow that which you finde cheapest and best."[18] Markham had rather tentatively and apologetically made some estimates of the costs of several inputs into agricultural production; Weston, developing the emerging art of the estate and farm account, provided a balance sheet of the benefits of new techniques calculated over several years.[19] Authors remained, however, very particular in their observations, stressing the different practices to be employed for the melioration of particular soils and aspects of the land. Nevertheless, their confidence in their transformatory power was high; witness Walter Blith's assertion of 1649:

> All sorts of lands, of what nature or quality soever they be, under what Climate soever, of what constitution of condition soever, of what face or character soever they be (unless it be such as Naturally participates of so much fatnesse, which Artificially it may be raised unto) will admit of a very large Improvement.[20]

Thus the literature of improvement was not at all concerned with potential failings in production, but with the level of yields and profit attained. However, this was not a discourse of "mastery" over nature. The mid- and late seventeenth-century authors still operated with a strong sense of providentialism. The great synthesis of agronomic works provided by John Worlidge in 1669 consistently gave space to the "uncertain Dispositions of an Over-ruling Providence".[21] Some thirty years later, Timothy Nourse would agree: the farmer, "after he has cast his Business to the best Method his Reason can propose, must still depend upon Providence, as to the event, here being so many Accidents which may traverse his Designs, and such as can be never provided against, nor foreseen."[22] Indeed, part of the explanation for this general confidence in the responsiveness of the soil to human endeavour may have arisen from the increasingly widespread view that God rewarded industry and that godly virtues were most manifest in the industrious and sober husbandman. It was a highly disturbing and bewildering aspect of enclosure for its opponent Henry Hallhead that it could lead to higher yields apparently being achieved by less labour.[23] There was no space for the idea that misplaced labour could actually diminish the fertility of God's earth. Neglect could allow "barreness both by little and little [to] increase", but "nature is no niggard, but giveth riches to all that are industrious".[24] And in the end, fertility was in God's gift: in the collections of letters and discourses edited by Samuel Hartlib in the 1650s, "It is the Lord that maketh barren places fruitfull", so a husbandman must "walk as becommeth a Christian, in all Sobriety, Righteousnesse and Godlinesse: not to trust his confidence in his own labours, and good Husbandry; but on the Lord that hath made all things."[25]

Agronomy engaged with a broader debate about nature in the seventeenth century that nevertheless only tangentially touched upon agriculture: was Nature subject to a long-term process of degeneration over time, requiring the active and continuing intervention of God in the workings of the world? Or did Nature operate according to a set of constant laws—the more mechanistic view—that the virtuous could come to understand and use to restore the fertility of the past? Either way, whether farming practices were unsustainable or not was not a matter for concern, and those writers who on a more epochal scale suggested that tillage over the centuries could cause soil erosion simply saw this as a displacement elsewhere—partly an explanation of why soils were thin on mountains.[26] Success or failure in agriculture was determined by localized virtuosity subject to the long- or short-term course of providence.

Forestry and Sustainability

Forestry, however, presented a rather different case to farming. Here we find from an early date the linkage of concerns for the durability of local wood supplies with the fortunes of the state itself. In part this rested simply upon the biological properties of wood: any concern with it operates on a different time horizon to crops and animals.

Nevertheless, early writing about wood still related it to the household economy: did woodland management ensure that you obtained an affordable supply? Regulations and advice sought to demarcate space dedicated to wood production, protecting the wood from grazing and hunting, rather than mastery of the processes that brought good wood yields. Hence much earlier writing was about jurisdiction and access. This was the key preoccupation of the first two books devoted to forest matters published in Europe in 1560 and 1576 by the German jurist and bureaucrat Noé Meurer.[27] Coppicing, the systematic cutting back of trees to the rootstock and reharvesting after a set period of years, had been widely practiced in Europe for very many centuries. During the sixteenth and seventeenth centuries, governments began to try and set the periods by which this cyclical harvest should occur, although this often related to a desire to calculate revenue. In the case of Colbert's famous Forest Ordinance of 1669, for example, the rather arbitrary ten-year cycle seems to have been set to make it easier to draw up leases of Crown estates.[28]

From the late fifteenth century, governments began to pass state-wide legislation relating to the supply of wood and the condition of the forests, pioneered in Italy and particularly in Germany. This legislation was largely concerned with demarcating forest space, preventing waste, and subjecting felling to the approval of a new forestry administration: a power that was largely negative and juridical. Everywhere—whether true or not—the word was of impending shortages and scarcity. In Bavaria the looming wood shortage would imminently lead to men leaving "their goods, homes and sustenance including even their wives and children and go from the same because of its lack." The fate of stands of trees was clearly linked to the general welfare.[29]

An English act of 1544, much like German contemporaries, spoke of "the great decaye of Tymber and Woodes" meaning "a great and manifest likelihood of scarcity". By 1577 William Harrison was also concerned about the possibility of general shortage: "it is to be feared that brome, turfe, . . . heth, firze, brakes, whinnes, ling, . . . straw, sedge, reede, rush, & seacole will be good marchadze even in the citie of London". And by 1611, after a period when woodland had become a prominent concern of commissions examining Crown land revenues in the first Jacobean years, Arthur Standish produced his *Commons Complaint:* his first grievance was "the generall destruction and waste of woods in this kingdome", there being "too many destroyers, but few or none at all doth plant or preserve". The consequences could be dramatic: "so it may be conceived, no wood, no kingdome".[30]

Standish presents something of a watershed in that he sets out his work as a systematic attempt, along with plenty of rough calculations, to lay out a plan for a secure national wood supply. What increasingly marked writing about wood was its concern for, as we might say, intergenerational justice: a new conception of struggles over allocation. As Noé Meurer put it in 1576 (echoing in fact language to be found in court cases from the 1550s), forestry officials were to prevent anyone overcutting wood, so that "not they alone, but also their descendants, heirs and children, will always have from their woods what they need [*die notdurft*] for building and burning".[31] Standish condemned men who overexploited woods, "desiring to become heyres of their owne time, without respect had to such heyres as shall succeed them". He thought that recent history contained a salutary lesson: "forty years ago . . . the poorer sort scorned to eate a piece of meate roasted with sea-coles, which now the best Magistrates are constrained to do".[32] The most famous of seventeenth-century works on timber is John Evelyn's *Sylva,* a compendium of discussions among various members of the Royal Society in the early 1660s with a particular concern for the shortage of timber for shipbuilding (in fact, a rather exaggerated fear). As Evelyn put it, each generation was not born for itself, but for "posterity". The same word justifying action, *posterité,* was employed in Colbert's great forest ordinance of 1669. Defending "posterity" had long been a theme asserted by those defending customary rights, such as to commons. But to assert that that the resources themselves, rather than the right to utilize them, should belong to posterity represented a new sensibility that arose in the context of wood.[33]

The proposals in Evelyn's *Sylva,* which went through many subsequent expanded editions and became the standard text on arboriculture, were in fact rather more modest than those of Samuel Hartlib, who had wanted a Crown Officer of the Woods on the Continental model. Evelyn provided what could be an extended chapter in the *Hausväterliteratur* tradition: short essays on how best to propagate particular trees, rather than—as Standish did—a project of national regeneration. The estate owner's handling of his plantations was linked to the fate of the nation, most explicitly in having a national store of shipbuilding timber, but there was no advance on the notion propagated by Standish that a proper and systematic balance in wood management could be developed to ensure supplies in perpetuity.[34] In the end, the English Crown's interest in wood supplies was weak; unlike most parts of Europe, they no longer controlled many woodland assets, having leased Crown woodlands out as part of the desperate efforts to shore up state finances under James I and Charles I. In the same period, coal had rapidly overtaken firewood as the primary source of heat.[35]

Further steps were taken in Germany. It was there that the particular preoccupation with the durability of wood supplies came together with a tradition of state-wide resource regulation and a new science of government in the shape of "cameralism". Since the mid-seventeenth century, German political theory had displayed a preoccupation derived from the *Hausväterliteratur* with fostering good agricultural practice, including the notion that "each region can properly maintain only so many people from its own resources as can get their means of support from its yield".[36] Potential problems, as with the farm or estate, were conceived as arising for the hungry populace, not in the environment. Of course, the idea that if you continually extract more wood from a given area than grows back in the same time period then your wood will disappear was well known. Only gradually, however, was the notion implanted that this process could be controlled artificially, with conifers as well as deciduous trees, in such a way that the yield itself could be predicted even after long periods between harvests. This was the particular contribution of German forestry, especially in areas of high industrial demand—such as the Saxon mining districts overseen by von Carlowitz, who used the term *nachhaltende* in 1713.[37]

The eighteenth century saw the development of "sustained-yield" theory, the cornerstone of modern forestry. This came to rely on a limited number of reliable conifer species (in fact the systematic cultivation of conifers could be found around Nuremberg as early as 1368), the surveying and maintenance of fixed areas of growth divided into "age classes" that could be calculated from the moment of planting, or measuring the basal circumference of the tree.[38] This in turn required an expansion of the supply of professional foresters trained in surveying and geometry. One step forward was taken by Carl Christoph Oettel in the 1760s with the recognition that tree trunks should be treated as cones, not cylinders. Oettel and his successors realized that by accurate calculation of growth volumes one could accurately calculate area yields by the multiplication of individual trees of a specified age. The logical consequence of this was that forests were most economically managed where standard trees could be grown on points where their growth rates could be predicted. Surveying the forest became a search for districts of homogeneity whose yields could be planned long into the future. The concept of the "normal tree", the *Normalbaum,* was born; the forest became the aggregate of the individual tree; and the best forest was that where what the famed "classic" forester Gottlieb Hartig called the "arbitrary" deviations of nature could be eliminated. The role of quantification shifted from being descriptive to being prescriptive.[39]

Not just studies by foresters, but general cameralist works on the fiscal state also could contain large sections on forest management, notably in the work of Johann Heinrich Gottlieb von Justi in the 1750s and 1760s. The purpose of the management was straightforwardly to work out how much "wood can be annually felled sustainably, economically and without ruin to the woodlands". Von Justi noted that the forester must adapt his methods to the what "was possible given the nature of the matter and the qualities of the ground"; wood was an important source of revenue but more important was an "indispensable necessity for the maintenance of the Inhabitants".[40] Government should also seek to record demand and keep that, too, in balance. All these were tasks that proved beyond the capacity of eighteenth-century administrations, but they set out a clear framework for the ambitions of scientific forestry.

Management and Improvement in the Eighteenth Century

In the early eighteenth century, the number of publications on agricultural improvement grew rapidly. Renowned authors could become consultants to estate managers, whether directly on the ground, as with the Hertfordshire farmer William Ellis, or via networks of correspondents, as with Richard Bradley, who probably conned his way to the chair of botany in Cambridge in 1720. In the case of lawyer and landowner Jethro Tull fame came by word of mouth, but he was eventually persuaded into publication and eager antagonism from 1732.[41] As the number of publications expanded, so disputes became sharper, and authors became more insistent that their recommendations were based on observation, practice and success.

Most eighteenth-century works began with a detailed discussion of the properties of soil, and the wide range of substances that might be used to manure them: other soils, marl, lime, dung, rags, soot, ashes, animal remains. Agriculture remained a localized and particular enterprise; as Ellis wrote,

> every Farmer ought to make it his primary Study to inform himself of the several Sorts of Ground that often belong

> to his Farm, and that besides his own Judgment to consult his Neighbours, who as Natives on the Place may be able to let him know more than the Dictates of his own Reason, that formerly were more remote from the same.

The development of careful comparative analyses of soil types and local climates also gave a clearer sense of the limits of fertility, as Ellis again observed: "Manures to the Earth are in some degree as Food to Animals . . . as they exhaust, the other feeds and supplies, or else the tone of the Ground's strength will soon be debilitated."[42] This perception may have arisen as a result of the clearly increasing application of all kinds of manure, the most prominent of which were marl, lime and animal dung, that had effects of varying duration.

Yet this remained a question of balance, and one that could be remedied by not persistently sowing the same crop, although the reason why different crops seemed to extract different things from the soil was not clear; or mixing soils with dressings and manures from elsewhere. Thus for Bradley, "the earth can never be rendered unprolific, unless she is constantly constrained to feed one kind of herb or plant". The mix was key, because in line with humoral theory (and in this regard Aristotle was still held in high regard), excess of any one element was the source of problems: "As the barrenness of most soils depends on the abundance of some one ingredient, there is scarce any one kind that may not serve as manure for some other."[43] All who wrote on the subject could agree that plant nutrition came from some kind of salt, as Hugh Plat had averred, and generally the consensus viewed that salt as being derived from the atmosphere in combination with sunshine and rain, a viewpoint that in part went back to experiments by Van Helmont indicating that plant growth did not diminish the mass of the soil. The significant properties of different soils were in the manner they absorbed salts from the atmosphere.

The focus on atmospheric salt infusions meant that little of nutritional value for plants was thought to reside in the soil itself, although writers were inconsistent on this point. Many authors recognized that the quality of the tilth was an important determinant for plant growth, above all its degree of fragmentation. The finer the tilth, the greater the surface area of soil exposed and greater its ability to absorb "air salts" and promote the growth of plants. The explanation for the action of animal dung was thus that, like yeast in dough, its fermentation produced air pockets that helped break down the soil. Jethro Tull could, controversially, go as far as arguing that dung was unnecessary if the soil was properly worked in his "horse-hoeing husbandry", an idea that was partly justified by the theory that matter was infinitely divisible.[44] But the tide of competitive experimentation made it clear, as well as necessary to argue, that, as Hale put it, "A soil may be render'd worse by bad management; as certainly as improved by good." A providential view had been supplanted by one that insisted that a full understanding of nature's secrets, rather than simple labour, was the key to raising yield.[45]

The "New Husbandry"

It was only later in the eighteenth century that the discourse of agronomy shifted towards the pattern already established in forestry, to become a generalized theory of the management of agrarian resources. Major steps in this direction were taken in England, but systematic refinement was done in Germany. None of these later agronomists waxed larger in their self-importance, or international influence, than the prolific Arthur Young.

Young claimed in his *Rural oeconomy* of 1770 that not a single previous author had written one page of use to practical men armed with his new system of "general management".[46] Young's German equivalent and follower was Albrecht Thaer, a Hanoverian doctor turned farmer who founded the Prussian agricultural academy. According to him, "the science of agriculture rests on experience", as opposed to "simple tradition", although this was hackneyed language in both Britain and Germany. Yet it was important to Thaer's reputation to insist that the combination of "accident and necessity" in his own *experimental* conduct had brought him success in discovering efficacious new crop rotations, and not reading English authors such as Young, of whom he was in fact a conscious imitator. In fact the virtues of experimentation had long been exhorted, but the new vogue made them central to a process of theoretical refinement. Both the prestige of experiment and confidence in the rewards of agronomic investigation went hand in hand with the rapid advances in chemistry.[47]

"Experimental" agriculture was built on a new theory of the soil, where the farmer had the power to set the quality of the soil rather than having to merely adapt to local edaphic conditions. This relied on maintaining a balance between the extent of meadow and pasture, and the extent of arable. Each harvest of cereal crops reduced soil fertility as the crop removed nutrients. Continued high yields thus rested on replacing these nutrients. Some manure could come from feeding livestock the straw of the previous harvest, but obviously this still involved net loss of "succulent juices" in the recycling process. The answer was to pasture livestock on meadows, transferring biomass onto the arable fields, and logically to optimize output one needed the correct ratio of meadow to arable. Establishing this ratio, and all the "proportions" derived from this, was the stated aim of Young at the very beginning of his *Rural oeconomy* of 1770, and "if any of the proportions . . . are broken, the whole chain is affected . . . so much does one part of a well managed farm depend on the other".[48] Getting it wrong would eventually cause soil exhaustion. There was nothing new in the idea of balancing livestock numbers and tillage to produce an optimum supply of manure, but Young was right to insist on the novelty of understanding cultivation as a closed system where manure was *the* critical vehicle for recycling nutrients.

Thus what really marked Young and his successors out from predecessors was neither their practicality nor their experimentation, but a *theory of the soil* that viewed fertility as inhering in a substance that was transmitted through feed and animal dung. Other dressings of the soil, such as marl or lime, simply sought to optimize conditions for the uptake of this substance. This was a systematic *model* that, through experiment and calculation, could guarantee success. As much as experiment, farming was to become a matter of accounting, a training in which Young had probably acquired in a short and unhappy apprenticeship to a merchant in King's Lynn: a farmer should be "very ready at figures", and have some knowledge of mechanics, geometry and "the application of mathematical studies".[49]

Albrecht Thaer drew extensively on Young in the conduct of his "experiments", and his hugely influential classic *The Principles of Agriculture* (1809–12). Thaer argued that "the produce generally depends more on the quantity of the manure than on the nature of the soil", and poor land was commensurately a product not of poor soil but of "want of manure". Naturally this required a correct proportioning of tillage and livestock, or, more precisely, tillage and fodder. Working from the principle that it was manure that recycled the necessary "nutritive juices" into the soil for crop growth, Thaer developed a form of accounting for this process based on "degrees" of fertility and advanced a model of the ideal proportions to be established between fodder input, animal numbers and agricultural output.[50] Young and Thaer, the two most influential writers of their age (among very many) on agriculture in Britain and Germany thus established early in their careers a relatively "closed" system-like model of nutrient flows within the farm, mediated via animal dung. While they remained keen observers of wider farming practice and the importance of producing a good tilth and combating weeds, they argued that only in this way could yields be improved. Sustainability consisted of effective management of this cycle.

The End of the Isolated State: von Thünen and von Liebig

It had long been argued that the state of tillage in a country was important for welfare, especially in the short run in preventing dearth and disorder. Yet unlike the case with trees, where from an early date wood supplies were perceived to be dwindling and where the resource suffered from an appreciable finitude, shortfalls in agricultural production were not linked to a deficient resource base. Only very gradually did "improvers" perceive that yield improvement could be seen as akin to expanding the national territory, while energy supplies or timber had long been linked to the fate of the state, and coal seams could be viewed as the equivalent of colonies ("England's Peru", or a "subterranean forest".)[51] Attempts to relate agricultural production to national capacity emerged most prominently in France with the Physiocrats, but it was in Germany that the "closed system" of farming as conceived from the 1760s was allied in a systematic way with the wider world in which farms were embedded. This development took two very different courses, one a theory of the market, the other a theory of national ecology.

Johann Gottlieb von Thünen managed a large estate in eastern Germany, and his classic work of the 1820s, "The Isolated State", was above all an exercise in accounting and determining what kind of farming would turn a profit. He followed Thaer in developing crop and manure accounting, and absorbed critically the work of some of the new soil chemists, although this still formalized biological relationships into equations with rather vague elements such as the "Richness", and "Quality" of the land. His major contribution was to analyse how agriculture in a market economy had to adapt to what the market could bear. Imagining a land that was an "isolated state" with invariant environmental conditions and one urban major centre, von Thünen laid the foundations of economic geography. What could be profitably produced at any given point in space was a function of price, transport costs and costs of production (wages, capital and rents), producing an idealized landscape of a set of concentric circles displaying different land uses laid outwards from the urban centre.[52]

The system of the farm remained ecologically closed and thus dependent on management of the nutrient cycle, but the balance of products that cycle was to generate was determined by the market. This model of sustainability closely reflected the arguments aired in the brief florescence of liberal forestry in Germany in the early nineteenth century, when figures such as Pfeil argued that the cutting cycle of stands of trees should be determined by a comparison between their growth rates, the value of the stock of trees and rental of the land, and prevailing interest rates.[53] In practice, forestry practice could not be adjusted to the high variance in market indicators and the principles of sustainability in forestry remained vested in wood production, not revenue. For von Thünen, maintenance of household income was achieved by balancing local ecological with wider market considerations.

Young and Thaer provided models of material flows based on the farm that could provide a template for both success and stability, but the flip side was that mismanagement could lead into a spiral of degradation. This abstracted formal model would be both dramatically expanded in scope by Justus von Liebig, chemist and admirer and follower of Alexander von Humboldt, the naturalist who himself had been influenced as a young man by the prevalent cameralism and forest science.[54] Liebig gave ecological substance and historical traction to the emerging model of sustainability, bringing a novel understanding of soil dynamics derived from agricultural studies together with the concern for the polity as a whole that had defined the wood-shortage debates.

Early nineteenth-century chemists had developed an "organicist" theory of plant growth, arguing that it was living matter in the humus of the topsoil that was in turn absorbed by and fed growing plants, a kind of recycling of vitalism, a life force inhering in organic matter. Manure, of course, was the agent of transmission in this recycling.[55] It was this that gave real substance to the ideas of Young. Liebig published the first edition of his *Chemistry* in 1840, responding to a request to write up an address to the British Association for the Advancement of Science in Liverpool in 1837.[56] In this and subsequent editions Liebig criticized the notion that the results from one farm could be universalized, as did critics of the "new husbandry", a controversy that continued over the real value of "experimental farms". Liebig argued in opposition to the "organicist" perspective that the fundamental roots of plant physiology were not some vitalist force inhering in organic matter, but the complex interactions of inorganic trace elements contained in the soil and atmosphere. This understanding of the chemical underpinnings of agricultural production also led Liebig to the conclusion that simple recycling of organic matter was insufficient to maintain soil fertility because each stage of processing involved irretrievable loss of elements, and thus led to long-term decline in yields.[57]

Liebig did not dismiss the recycling of manure, indeed he insisted on making it as efficient as possible. But he argued that the particular contribution this could make would vary greatly according to soil type; that inevitable wastage meant that additional supplies of elements had to be obtained from the atmosphere and processes of weathering of the land; and that in the face of a rising European population, alternative sources of nutrients had to be found if food supplies were to be kept in step with demand. In other words, he placed concerns of sustainability in a *dynamic* setting. In conditions of rapid urbanization this huge proportion of necessary elements was not being recycled within the system, but due to the construction of the water-closet and sewers, was instead discharged uselessly into the sea. Young had already lamented the loss of London's "night soil" in 1799.[58] Liebig saw this as no less than a process of "self-annihilation" (*Selbstvernichtung*), where the future of civilization rested upon the resolution of this "sewer question" (*Kloakenfrage*) and the return of urbanites' ordure to the farmers' fields.[59]

Partly due to some wildly inaccurate population estimates, Liebig interpreted the path of civilizations ever since the Greeks as being a story of slow, relentless overexploitation of the earth and consequent decadence and decline. "A people arises and develops in proportion to the fertility of the land, and with its exhaustion they disappear." Only China and Japan represented an exception of steady growth because of their extreme success in recycling, through the removal of livestock husbandry and directly applying human faecal matter to the land. Europe, and especially England, were only able to sustain population expansion through a vampiric dependency on imports, especially of finite stocks of guano, but including the gruesome excavation of bones from battlefields and ossories.[60]

Thus in Liebig we find something like the modern conception of sustainability: that a society's development is beholden to fundamental biological and chemical processes, but also that this was a complex dynamic system with feedback effects. He had turned the argument of an earlier English treatise on agricultural chemistry, that "Nature has fix'd bounds to fertility beyond which we cannot proceed, however prompted by avarice", into a general developmental model. Indeed, that 1760 treatise was also prescient in its argument "chemistry may contribute more to our knowledge of soils, and their productions, by its several operations, than all the attempts hitherto made by practice and observation".[61] Taken up by Liebig, a new ethic emerged that knowledge of those fundamental biological and chemical processes, courtesy of the newly minted scientist, would dictate the ability of *societies* to endure.

Endnotes

1. H. C. von Carlowitz, *Sylvicultura oeconimica* (Leipzig, 1713); P.-M. Steinsiek, *Nachhaltigkeit auf Zeit. Waldschutz im Westharz vor 1800* (Münster, 1999), 78.
2. World Commission on Environment and Development, *Our Common Future* (Oxford, 1987). See www.un-documents.net/ocf-02.htm.
3. See discussions in J. Sieglerschmidt, "Die virtuelle Landwirtschaft der Hausväterliteratur", in R. P. Sieferle and H. Brueninger, eds., *Natur-Bilder. Wahrnehmungen von Natur und Umwelt in der Geschichte* (Frankfurt a.M, 1999), 223–54; A. McRae, *God Speed the Plough. The Representation of Agrarian England, 1500–1660* (Cambridge, 1996); G. E. Fussell, *The Old English Farming Books from Fitzherbert to Tull 1523–1730* (London, 1949); M. Ambrosoli, *The Wild and the Sown: Botany and Agriculture in Western Europe, 1350–1850* (Cambridge, 1997).
4. C. von Heresbach, *Foure bookes of husbandry,* trans. B. Googe (London, 1577), iv–v.
5. See Sieglerschmidt, "Die virtuelle Landwirtschaft".
6. C. Estienne, *Maison rustique, or The countrey farme,* trans R. Surflet, revised G. Markham (London, 1616), 1–2.
7. J. Fitzherbert, *The boke of husbandry* (1533), 71; T. Tusser, *Fiue hundred pointes of good husbandrie* (London, 1573), 7v.–12v.
8. Estienne, *Maison rustique,* 24.
9. T. Tusser, *Fiue hundred pointes of good husbandrie* (London, 1580), 41v, 43.
10. G. Markham, *Markhams farwell to husbandry* (London, 1620).
11. Estienne, *Maison rustique,* 528.
12. Markham, *Markhams farwell,* 1620, 7–8; *idem, The English husbandman* (London, 1635), 95.
13. Sir H. Plat, *The iewell house of art and nature* (London, 1594), 3.
14. Plat, *The iewell house,* 14–5.
15. Fitzherbert, *Boke of husbandry,* 23; Ambrosoli, *The Wild and the Sown,* 231.
16. Estienne, *Maison rustique,* 536.
17. C. Webster, *The Great Instauration: Science, Medicine and Reform, 1626–1660* (London, 1975).
18. Sir R. Weston, *A Discours of Husbandrie Used in Brabant and Flanders* (London, 1650), 15.
19. Weston, *Discours of husbandrie,* 16–9.
20. W. Blith, *The English improver improved* (London, 1652), 17.
21. J. Worlidge, *Systema agriculturae* (London, 1669), 179.
22. T. Nourse, *Campania Foelix* (London, 1706), 37.
23. J. O. Appleby, *Economic Thought and Ideology in Seventeenth-Century England* (Princeton, 1978), 69–70.
24. G. Plattes cited in Webster, *The Great Instauration,* 356–7.
25. S. Hartlib, *The compleat husband-man* (London, 1659), 80–81.
26. M. H. Nicolson, *Mountain Gloom and Mountain Glory. The Development of the Aesthetics of the Infinite* (Ithaca, 1959), 233–70; S. Schaffer, "The Earth's Fertility as a Social Fact in Early Modern Britain", in M. Teich, R. Porter and B. Gustafsson, eds., *Nature and Society in Historical Context* (Cambridge, 1997), 131–4; C. Glacken, *Traces on the Rhodian Shore* (Berkeley, 1967), 408–11.
27. N. Meurer, *Vom forstlicher Oberherrligkeit und Gerechtigkeit* (Pforzheim, 1560); N. Meurer, *Jag und Forstrecht* (Frankfurt a.M, 1576).
28. See, for example, P. Warde, "Fear of Wood Shortage and the Reality of the Woodland in Europe, c.1450–1850", *History Workshop Journal* 62 (2006), 28–57; A. Corvol, "La décadence des forêts. Leitmotiv", in A. Corvol, *La forêt malade. Debats anciens et phénomènes nouveaux XVIIe—XXe siècles* (Paris, 1994), 3–17.
29. Warde, "Fear of Wood Shortage", 42.
30. 35 Henry VIII c.17; W. Harrison, *An historical description of the Island of Britain* (London, 1577), 91; A. Standish, *The commons complaint* (London, 1611), 1. See also *idem, New directions of experience* (London, 1614).

31. Meurer, *Jag und Forstrecht,* 5; P. Warde, *Ecology, Economy and State Formation in Early Modern Germany* (Cambridge, 2006), 325.
32. Standish, *New directions,* 2.
33. J. Evelyn, *Sylva, or, A discourse of forest-trees* (London, 1664), 111. For a German example see K. Mantel, *Forstgeschichte des 16. Jahrhunderts unter dem Einfluß der Forstordnungen und Noe Meurers* (Hambrurg, 1980), 70.
34. Evelyn, *Sylva;* Worlidge, *Systema,* f.4v.
35. On the leasing of the Crown estates see R. Hoyle, ed., *The Estates of the English Crown, 1558–1640* (Cambridge, 1992); on coal use see P. Warde, *Energy Consumption in England and Wales, 1560–2000* (Naples, 2007).
36. From Seckendorff's *Der Christen Staat* of 1685, cited in A. W. Small, *The Cameralists: The Pioneers of German Social Policy* (New York, 1909), 48; see also K. Tribe, *Governing Economy: The Reformation of German Economic Discourse 1750–1840* (Cambridge, 1989).
37. See note 1.
38. There is a voluminous literature on these matters. See for example W. Schenk, *Waldnuztung, Waldzustand und regionale Entwicklung in vorindustrieller Zeit im mittleren Deutschland* (Stuttgart, 1996); U. E. Schmidt, *Der Wald in Deutschland im 18. und 19. Jahrhundert* (Saarbrücken, 2002); L. Sporhan and W. Stromer, "Die Nadelholzsaat in den Nürnberger Reichswäldern zwischen 1469 und 1600", *Zeitschrift für Agrargeschichte und Agrarsoziologie* 17 (1969), 79.
39. H. E. Lowood, "The Calculating Forester: Quantification, Cameral Science, and the Emergence of Scientific Forestry Management in Germany", in T. Frangsmäyer, J. L. Heilbron and R. R. Rider, eds., *The Quantifying Spirit in the Eighteenth Century* (Oxford, 1990), 315–42.
40. J. H. G. von Justi, *Politische und Finanzschriften über wichtige Gegenstände der Stattskunst, der Kriegswissenschaften und des Cameral- und Finanzwesens* (Kopenhagen und Leipzig, 1761), 440–44.
41. Fussell, *The Old English Farming Books,* 82–111; *idem, More Old English Farming Books from Tull to the Board of Agriculture, 1731 to 1793* (London, 1950), 1–12; D. E. Allen, *The Naturalist in Britain: A Social History* (London, 1976), 16. See Bradley's *Dictionary of National Biography* entry at www.oxforddnb.com/view/article/3189.
42. W. Ellis, *Chiltern and vale farming explained* (London, 1733), 48, 372.
43. R. Bradley, *A General Treatise of Agriculture* (London, 1757), 97; T. Hale, *A compleat body of husbandry* (London, 1758), 107.
44. J. Tull, *Horse-hoeing husbandry* (London, 1732)
45. Hale, *A compleat body of husbandry,* 84; see Schaffer, "Earth's Fertility", 133.
46. A. D. Young, *Rural oeconomy* (London, 1770), 2–4. Young was an extraordinarily prolific author and by no means trod the same narrow line in all of his works. See G. E. Mingay, *Arthur Young and His Times* (London, 1975); J. G. Gazeley, *The Life of Arthur Young 1741–1820* (Philadelphia, 1973).
47. Albrecht Thaer, *The Principles of Agriculture,* trans. W. Shaw (London, 1844), 3, 232–3. For an account of the interaction of chemistry and agriculture, see A. Clow and N. L. Clow, *The Chemical Revolution: A Contribution to Social Technology* (New York, 1952), 458–502.
48. Young, *Rural oeconomy,* 12.
49. Mingay, *Arthur Young,* 13.
50. Thaer, *Principles of Agriculture,* 24, 130–33, 138–9, 141.
51. R.-P. Sieferle, *The Subterranean Forest: Energy Systems and the Industrial Revolution* (Cambridge, 2001).
52. J. G. von Thünen, *Der isolierte Staat in Beziehung auf Landwirtschaft und Nationalökonmie* (Rostock, 1842), 243, 49–56, 83–7, 123, 56, 71–80. J. von Liebig, *Die Chemie in ihrer Anwendung auf Agricultur und Physiologie* (Braunschweig, 1862), 135.
53. H. Rubner, *Forstgeschichte im Zeitalter der industriellen Revolution* (Berlin, 1967), 126, 142.
54. W. H. Brock, *Justus von Liebig: The Chemical Gatekeeper* (Cambridge, 2002); also see dedication in Liebig, *Die Chemie in ihrer Anwendung.*
55. Thaer, *Principles of Agriculture,* 336; Liebig, *Die Chemie in ihrer Anwendung,* 13–14, 137.
56. Liebig, *Die Chemie in ihrer Anwendung,* vii–viii.
57. Ibid., 1–7, 13–5.
58. D. Woodward, "'Gooding the Earth': Manuring Practices in Britain 1500–1800", in S. Foster and T. C. Smout, *The History of Soils and Field Systems* (Aberdeen, 1994), 106.
59. Liebig, *Die Chemie in ihrer Anwendung,* 128–9, 153.
60. Ibid., 96–7, 108, 110–11, 120–29.
61. Farmer, *An essay on the theory of agriculture* (London, 1760), 5, 47.

Critical Thinking

1. If sustainability is such a old concept, why have we not achieved it?
2. How does the statement, "the principles of sustainability in forestry remained vested in wood production, not revenue" differ from current practices?
3. It has been suggested that many of the current unsustainable practices have their origins in Western philosophy. After reading this article, what do you think?

The Future of Sustainability: Re-thinking Environment and Development in the Twenty-First Century

W. M. Adams

1. Background

IUCN convened a meeting at the end of January 2006, to discuss the issue of sustainability in the twenty-first century[1]. The meeting considered the progress made towards global sustainability, the opportunities and the constraints facing the world and the World Conservation Union in attempting to meet the challenge of sustainability. This paper has been written to develop further key arguments explored at the meeting, and to provide a basis for discussion by IUCN Council of next steps in the 'rethinking sustainability' process[2].

2. The Idea of Sustainable Development

At the start of the twenty-first century, the problem of global sustainability is widely recognised by world leaders, and a common topic of discussion by journalists, scientists, teachers, students and citizens in many parts of the world. The World Summit on Sustainable Development (WSSD, 2002) confirmed that the first decade of the new century, at least, would be one of reflection about the demands placed by humankind on the biosphere.

The idea of sustainability dates back more than 30 years, to the new mandate adopted by IUCN in 1969[3]. It was a key theme of the United Nations Conference on the Human Environment in Stockholm in 1972[4]. The concept was coined explicitly to suggest that it was possible to achieve economic growth and industrialization without environmental damage. In the ensuing decades, mainstream sustainable development thinking was progressively developed through the World Conservation Strategy (1980)[5], the Brundtland Report (1987)[6], and the United Nations Conference on Environment and Development in Rio (1992), as well as in national government planning and wider engagement from business leaders and non-governmental organisations of all kinds.

Over these decades, the definition of sustainable development evolved. The Brundtland Report defined sustainable as 'development that meets the needs of the present without compromising the ability of future generations to meet their own needs'[6]. This definition was vague[7], but it cleverly captured two fundamental issues, the problem of the environmental degradation that so commonly accompanies economic growth, and yet the need for such growth to alleviate poverty.

The core of mainstream sustainability thinking has become the idea of three dimensions, environmental, social and economic sustainability. These have been drawn in a variety of ways, as 'pillars', as concentric circles, or as interlocking circles (Figure 1). The IUCN Programme 2005-8, adopted in 2005, used the interlocking circles model to demonstrate that the three objectives need to be better integrated, with action to redress the balance between dimensions of sustainability (Figure 1 c).

Governments, communities and businesses have all responded to the challenge of sustainability to some extent.

Almost every national government in the United Nations now has a minister and a department tasked with policy on the environment, and many regional and local governments have also developed this capacity. Since 1992 the volume and quality of environmental legislation (international, national and local) has expanded hugely, and international agreements (such as the Kyoto protocol) have not only raised the profile of environmental change but also begun to drive global policy change.

Public awareness of environmental and social issues in development are in many cases now well developed. Citizens in almost all countries not only know the issues, but tend to feel that the quality of the environment is important both to their own wellbeing and to the common good.

The 'greening' of business has grown to be a central issue in corporate social responsibility for many global companies, although for many it is still a boutique concern within wider

A. Pillars

The three pillars
of sustainable development

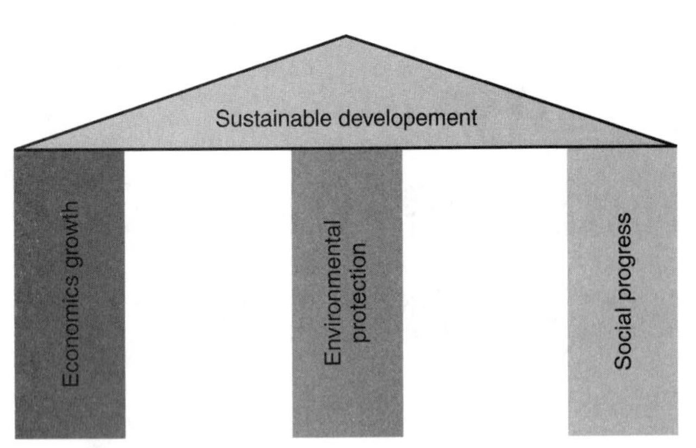

www.vda.de/en/service/jahresbericht/auto2002/auto+umwelt/u_3.html

B. Concentric Circles

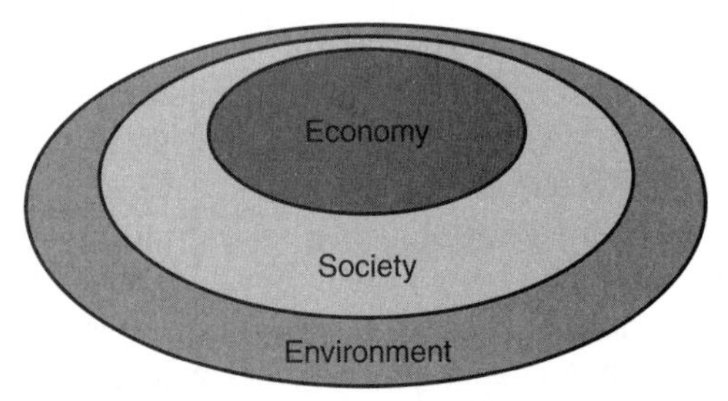

www.sustainablecampus.cornell.edu/sustainability-intro.htm

C. Overlapping Circles

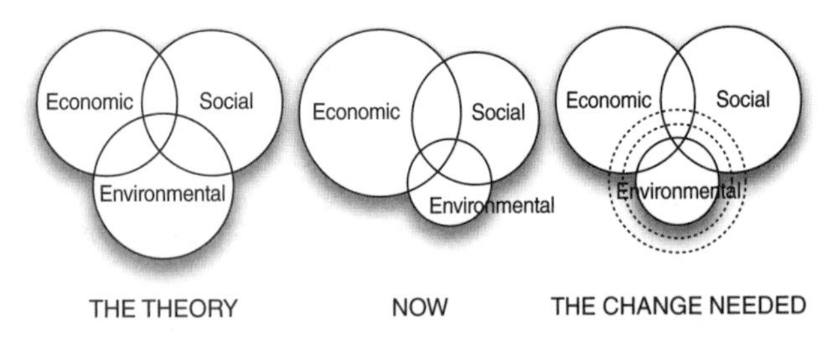

The three pillars of sustainable develoment, from left to right, the theory, the realty and the change needed to better balance the model

www.iucn.org/programme/

Figure 1 Three Visual Representations of Sustainable Development: Pillars, Circles, Interlocking Circles

relationship management, rather than something that drives structural change in the nature or scale of core business.

There is a profound paradox here. On the one hand, the twenty-first century is widely heralded as the era of sustainability, with a rainbow alliance of government, civil society and business devising novel strategies for increasing human welfare within planetary limits. On the other hand, the evidence is that the global human enterprise rapidly becoming *less* sustainable and not more. Much has been achieved - but is it enough? Are global trends towards sustainability or away from it? Have the concepts of sustainability and sustainable development offered a coherent basis for change?

2. Critiques of Sustainable Development

2.1 Is It Clear What Sustainable Development Means?

The phrase sustainable development covers a complex range of ideas and meanings[8]. *Our Common Future* located environmental issues within an economic and political frame, moving sustainability to the core of international development debate. Rio emphasised global environmental change, and the problems of biodiversity and resource depletion and climate change. The World Summit on Sustainable Development returned poverty to the top of the agenda, reflecting the Millennium Development Goals agreed at the United Nations Millennium Summit in September 2000[9]. Sustainability was one of eight Goals, associated with 18 targets and 48 indicators intended to be yardsticks for measuring improvements in people's lives[10].

Analysts agree that one reason for the widespread acceptance of the idea of sustainable development is precisely this looseness. It can be used to cover very divergent ideas[11]. Environmentalists, governments, economic and political planners and business people use 'sustainability' or 'sustainable development' to express sometimes very diverse visions of how economy and environment should be managed. The Brundtland definition was neat but inexact. The concept is holistic, attractive, elastic but imprecise. The idea of sustainable development may bring people together but it does not necessarily help them to agree goals. In implying everything sustainable development arguably ends up meaning nothing.

2.2 The Problem of Trade-Offs

The conventional understanding of sustainable development, based on the 'three pillars' model is flawed because it implies that trade-offs can always be made between environmental, social and economic dimensions of sustainability. In response to this, a distinction is often drawn between 'strong' sustainability (where such trade-offs are not allowed or are restricted) and 'weak' sustainability (where they are permissible). The concept of 'critical natural capital' is also used to describe elements of the biosphere that cannot be traded off (e.g. critical ecosystems or species). However, in practice, development decisions by governments, businesses and other actors do allow trade-offs and put greatest emphasis on the economy above other dimensions of sustainability. This is a major reason why the environment continues to be degraded and development does not achieve desirable equity goals.

The three 'pillars' cannot be treated as if equivalent. First, the economy is an institution that emerges from society: these are in many ways the same, the one a mechanism or set of rules created by society to mediate the exchange of economic goods or value. The environment is different, since it is not created by society. Thinking about trade-offs rarely acknowledges this. Second, the environment underpins both society and economy.

The resources available on earth and the solar system effectively present a finite limit on human activity. Effective limits are often much more specific and framing, in that the capacity of the biosphere to absorb pollutants, provide resources and services is clearly limited in space and time. In many areas (e.g. warm shallow coastal waters adjacent to industrialised regions) that capacity is close to its limits.

2.3 The Problem of Metrics

There is no agreed way of defining the extent to which sustainability is being achieved in any policy programme. Sustainability and sustainable development are effectively ethical concepts, expressing desirable outcomes from economic and social decisions. The term 'sustainable' is therefore applied loosely to policies to express this aspiration, or to imply that the policy choice is greener than it might otherwise be (e.g. the idea of a 'sustainable road building programme'). Everywhere the rhetoric of sustainable development is ignored in practical decisions. Often sustainable development ends up being development as usual, with a brief embarrassed genuflection towards the desirability of sustainability. The important matter of principle therefore becomes a victim of the desire to set targets and measure progress.

3. Is There a Problem with the State of the World?

The issue of environmental limits to the human project on earth was brought to international attention in the early 1970s, particularly by the Club of Rome's precocious computer modelling in *Limits to Growth*[12]. The World Conservation Strategy, published in 1980, offered the first coherent a analysis of environmental sustainability. It emphasised the need to maintain essential ecological processes and life support systems, to preserve genetic diversity, and to ensure the sustainable utilization of species and ecosystems.

In 2005, exactly a quarter of a century later, the findings of the Millennium Ecosystem Assessment offered a stark commentary on the state of the earth and the sustainability of humankind's management.

The significance and scale of the global human footprint is not in doubt. Consumption of living resources as raw material and sinks for waste materials is high and growing[13]. In 1997, Peter Vitousek and colleagues noted in *Science* that the rate and scale of change in the biosphere as well as the kinds and combinations of change were fundamentally different from those at any other time in planetary history[14]. The results of these transformations are almost universally negative in their impacts on the biosphere. In 1992, Edward Wilson noted that human activities have increased 'background' extinction rates by between 100 and 10,000 times. 'We are', he said, 'in the midst of one of the great extinction spasms of geological history'.[15]

The message is no better on poverty. The Millennium Assessment makes quite clear that not only does the level of poverty remain high, but inequality is growing.

Problems of environment and development are closely linked; degradation of ecosystem services harms poor people. Half the urban population in Africa, Asia, Latin America, and the Caribbean suffers from one or more diseases associated with inadequate water and sanitation. The declining state of capture fisheries is reducing an inexpensive source of protein in developing countries. Per capita fish consumption in developing countries, excluding China, declined between 1985 and 1997. Desertification affects the livelihoods of millions of people, including a large portion of the poor in drylands.

Since the Millennium Summit in 2000 (at which world leaders agreed the Millennium Development Goals), and the World Summit on Sustainable Development in 2002, there has been a renewed energy to policy debate about poverty and environment. The concept of sustainable development precisely embraces this challenge.

Yet despite over three decades of explicit concern about sustainability, a concern increasingly part of the mainstream of international debate, the human claim on nature is increasing almost everywhere unchecked, and the problem of poverty is deeply persistent. The implications for the poor of the current generation and for future generations is extremely serious.

The velocity of environmental change is fast, and increasing. As Peter Vitousek and colleagues comment, tellingly, 'we are changing the earth more rapidly than we are understanding it'[21]. Rates of human transformation of the earth are increasing, particularly in countries undergoing rapid industrialization or de-industrialisation. The human capacity to destroy life-support systems (ecosystem services) is new. Humanity is burning through natural assets and their capacity to support life and quality of human life without thought to the future and the rights and needs of today's people.

The current relationship between humans and biosphere is novel, outside all human historical experience (and therefore learned adaptive responses), and arguably outside the envelope of evolution adaptation of higher mammals.

4. Urgency, Risk and Opportunity

Although the issue of sustainability has been recognised explicitly since the 1970s, there is an acute urgency to the global problematique at the start of the twenty first century. However, at the same time the first decade of this century offers a unique opportunity to re-think the dominant patterns of global development.

The twentieth century was dominated by debates about 'development', how to promote Western models of economic growth, urbanisation and industrialisation globally. Environmentalist critique of development in the last 30 years argued that the conventional development model was unsustainable.

Several factors now offer a unique window for demonstrating that fact, and for convening a new discussion about human and environmental futures.

Today, at the start of the twenty first century, some developing countries had begun to achieve sustained economic growth and industrialisation on this model, first the 'Asian Tigers', then China and India. The success of development on the standard 'fossil fuel automobile-based throwaway consumer economy' in China and India offers a unique opportunity to assess its limitations. China's success, for example, is bringing massive increases in consumption (grain, meat, steel oil, timber)[22]. China's revolutionary economic growth demonstrates the flaws with the conventional growth model. It shows the need for systemic change in the way development is understood and brought about globally: in the west as much as elsewhere. The earth is at a tipping point: business as usual is no longer an option.

The present global dilemma offers huge risks, but also outstanding opportunities. The need to create a 'sustainable post-fossil-fuel society and economy'[23] has never been more widely recognised, although the challenges on the road to achieving it remain breathtaking.

The dominant development model based on the unlimited meeting of consumer wants leads inexorably to over-consumption. Yet continued physical expansion in the global reach of commodity supply systems means that consumers in developed countries continue to perceive resource flows as bountiful, and develop no sense of limits to consumption[24]. Whether as consumers or citizens, people in industrialised economies show no awareness that production systems are ecologically flawed or constrained. Yet this model is itself disseminated internationally by global media and advertising as unproblematic, uniformly good and desirable. Belief in the opportunity to consume without limits in an ecologically limited world is a powerful driving force increasing global risk.

Interestingly, the unsustainability of the present global development model is probably better understood in China than in the conventional industrial heartlands of Europe and North America. There, politicians fear backlash from citizens reacting as consumers to anything that alters their lifestyle in ways they perceive as deleterious. This results in demands for low fuel prices, profligate material and energy consumption, and persistent ignorance of the social and environmental conditions under which global products are created. Environmentalist challenges to business as usual remain outside the mainstream, and the unsustainable patterns of production and consumption of the developed world persist.

The global integration of once semi-independent national economies is advancing rapidly, eroding the capacity of the nation state to balance economic, social and environmental choices.

Social and cultural globalization is also rapid, creating both dizzying opportunities for information and cultural exchange, but also unprecedented challenges to the post-second world war institutions of international integration and governance. Disabling fears about security, cultural change and political threat are an issue in many countries.

Human influences on natural patterns of climatic variability undermine the comfortable assumption dominating the twentieth century that global climate would persist within known historical bounds. Scientific understanding, although growing, is still limited. However, it is clear that the ocean-atmosphere envelope demonstrates non-linear dynamics, making relatively rapid changes in climatic patterns a likely feature of the future earth; human forcing of the parameters of that change (through the greenhouse effect and other processes) will increase the speed and unpredictability of such changes. Climate change has immediate implications for other phenomena such as sea level and extreme events. The coastal location of the world's largest cities exposes huge numbers of people to potential future risk.

The growth of global human populations brings exciting benefits in terms of cultural achievements and creativity, and the generation of new ideas. However, the rate of growth of human populations and the rate of growth of the services needed to meet growing human need present huge challenges. The chronic nature of the poverty into which many children are born presents significant and rapidly advancing risks.

Technology also offers opportunities and risks. The novelty of some new technologies and the speed of technological innovation and adoption brings the potential for unforeseen social, environmental, economic or health consequences, (e.g. the adoption of new technologies or novel compounds by untrained users). Some technologies bring significant political and governance challenges (e.g. nuclear fission).

Developments in ecological restoration offer novel and inspiriting opportunities to enhance and reinstate biodiversity and ecosystem services, yet human skills in ecosystem assembly remain limited. For this reason, any argument for a strategy of 'develop now and restore damaged ecosystems later', based on extrapolation of the logic of the 'environmental Kutznets curve' is fundamentally flawed. 'Critical natural capital' cannot be replaced within realistic timeframes.

The concurrence of disasters in 2005 and 2006 (numerous hurricanes and tropical storms, earthquakes, flooding, famine) has concentrated the minds of Western media pundits on the shared fate of humanity. Some of these disasters (especially storminess and flooding) are connected in popular accounts to issues such as climate change. The parallel nature of environmental and humanitarian issues is thus clear to many people.

There is therefore, in the first decades of the twenty-first century, a powerful opportunity to start a new debate about development, economy, equity and environment. This must address both the human needs and aspirations of the poor of developing world, and the over-consumption in the industrialised world.

5. A New Challenge

5.1 The Need for a New Approach

Despite the achievements of the last three decades, the present concepts of sustainability and sustainable development are clearly inadequate to drive the transitions necessary to adapt human relations with the rest of the biosphere for the future. Something new is needed.

The problem with sustainability and sustainable development is not that the aspirational values they represent are wrong, but that they are over-worked and tired. As currently formulated they are too loose to drive effective change on the scale required.

The need at the start of the twenty-first century is clearly for systemic change. The experience of the last 30 years shows that this cannot be brought about using the metaphors, slogans and ideas that are currently available. The scale of transformation needed demands new concepts, new ideas, new ways of engaging citizens and opinion leaders in the search for solutions.

However, as an idea sustainability has been, and continues to be, powerful. While the concept is clearly burdened with a great deal of excess weight, and many potentially conflicting ideas have become attached to it like barnacles on a ship's hull, it still has considerable power. The concept of sustainability is widely recognised and discussed. It has taken a decade and a half's effort to build the concept into the thinking of local and national governments, business and schools and universities. To use a business analogy, sustainability is an established 'brand' that has wide recognition and still expresses core values to a wide audience. For a business with an established brand that has become tired, abandonment and re-launch of a replacement could bring just huge costs and confusion and lost public engagement.

Hypothesis 1: That the most effective strategy is to adopt an incremental or evolutionary approach, re-orientating the concept of sustainability, reemphasising what it means and moving forwards; a strategy of 'keep it but fix it'.

5.2 Timing

The manifold challenges to the world community first decade of the twenty-first century present a turbulent moment within which to push for a new engagement with the idea of sustainability. However, it also offers a window of opportunity for the development of a new approach to planetary management.

By 2020 responses to issues like climate change and 'peak oil' will be more obvious, but the room for manoeuvre will be much less. Moreover, the political stresses that result for these challenges will not necessarily be conducive to calm collaborative action. Change, particularly significant change, in 'business as usual, needs time, but the environment is the timekeeper. Human misuse of environmental assets is driving environmental change, and this demands action now.

Hypothesis 2: That the timing is right to develop a new strategic approach to global sustainability

5.3 The Role of The World Conservation Union - IUCN

IUCN has a unique constitution (incorporating government and non-governmental organisations) and unique convening power. IUCN therefore is therefore in a position to start to broker new forms of coalition, alliances and see if we can create innovation. If IUCN's membership can be mobilised, then it could provide the basis for a catalytic effect on current debate. IUCN can do little alone, but it can empower and mobilise others.

Hypothesis 3: That IUCN should take a lead in developing new thinking about sustainability

6. New Concepts, New Thinking

6.1 Sustainability and Resilience

The uncomfortable bottom line of sustainability is the insight that the biosphere is limited. In its crude form, the idea of 'limits to growth' dominated 1970s environmentalism. Evidence of resource substitution (fibre optics for copper cables, light plastics for steel) and improved resource use technologies (e.g. improved technologies for the discovery and exploitation of oil reserves) have allowed this view to be pilloried as unrealistic 'flat-earthism'. On the other hand, the spread of persistent organic pollutants, the ozone hole and the growing certainly of anthropogenic climate change caused by CO_2 and other greenhouse gases demonstrate that the fundamental point is perfectly valid. The earth's capacity to yield products for human consumption, to absorb or sequestrate human wastes (especially novel compounds), and to yield ecosystem services are all of them limited. The idea that that there is always somewhere to absorb externalities is flawed, and it is a myth of progress that living systems will always recover from human demands.

Moreover, as environmental capacity is reached, institutions for sharing the earth are placed under intolerable strain.

The science of resilience is central to an understanding of the planetary future, and the metaphor of resilience (and its limits) is valuable for its contribution to more general debate. For decades, message taken from the science of ecology by society more generally was that ecosystems were homeostatic—that once a stress was removed, they would bounce back to their former state. This comforting metaphor implied that there was no reason to fear that human misuse of the global environment would lead to irretrievable breakdown. The bleak message of the Gaia hypothesis, that the biosphere could be understood as a self-regulating system, was reinterpreted with shocking anthropocentric complacency to imply that it would therefore always support human life. The earth may function to maintain life, but not necessarily life in the stunning biodiversity we know today, and certainly not human life.

Ecology has moved on. Non-linear dynamics are accepted as an inherent element in ecosystem function. Polluted lakes do not necessarily return to their former state when pollution stops; climate can not be expected to vary around some mean approximating to the conditions of the last 30 years; it is highly likely that extinction of certain species will change the amplitude and frequency of ecosystem change in ways that constrain human opportunities; novel compounds and broad-taxon genetic manipulation may well generate shifts in ecosystem form and function.

The biosphere is not infinite. As Edward Wilson observes, 'the biosphere, all organisms combined, makes up only one part in ten billion of the earth's mass. It is sparsely distributed through a kilometre-thick layer of soil, water and air stretched over a half billion square kilometres of the surface'[25].

The capacity of nature to meet human needs depends on both its internal dynamics and its dynamic responses to human stresses. The resilience of the biosphere is critical to the sustainability of human enterprise on earth.

6.2. Sustainability and Human Wellbeing

The diversity of life is fundamental to human wellbeing[26]. The concept of nature has great strength, because it combines both a conventional conservation concern for species and ecosystems (biodiversity) and the diverse ways in which species and ecosystems have value (aesthetic, cultural and spiritual values as well as more directly material values, and the Millennium Ecological Assessment recognised).

Under the conventional development model, the 'good life' is defined in narrow economistic terms, in terms of access to good and services. This formulation is inadequate. Just as Amartya Sen's concept of 'development as freedom' (the expansion of the real freedoms that people enjoy) transforms understanding of attempts to achieve development, so too there is a need to concentrate not on the means to achieve sustainability, but on ends[27].

Sustainability needs to be made the basis of a new understanding of human aspiration and achievement. The relevant metric of sustainability is 'the production of human wellbeing (not necessarily material goods) per unit of extraction from or imposition upon nature'[28].

A key element here is the linkage between human wellbeing and security. The quality, diversity and functions of the environment underpin human health, solidarity and security. This is not currently central to thinking about social and economic development choices, which separate political and economic risk into the mainstream of debate, and sidelines environmental quality and risk wither to the arena of scientific disagreement or some secondary concern about 'quality of life'. Material consumption and political security are therefore treated as if they were separate from, more important than, issues of quality of life.

In fact, security between people depends fundamentally on issues of equity, within and between generations. David Orr suggests the principle that 'no human being has the right to diminish the life and well-being of another and no generation has the right to inflict harm on generations to come'[29]. Security and wellbeing are both rooted in issues of justice at global scale. Sustainability is the path that allows humanity as a whole to maintain and extend quality of life through diversity of life.

The importance of future generations are a central core concept of sustainability. Intra-generational equity (meeting human needs now) needs to be directly linked to the fulfilment of basic needs of all global citizens in the future (inter-generational equity). At present we lack political mechanisms to achieve the former, and we allow development only loosely tied to this goal to undermine capacity to achieve the latter

Justice is of fundamental importance to the planetary future: equity in the enjoyment of the benefits from the use of the earth's resources between and within generations.

6.3 A New Economy

The market is a human institution of unique power and efficiency. It is capable of driving massive changes in environment and human opportunity on a scale and at a speed that dwarfs the regulatory powers of citizen, state or global organisation. Human aspirations, and subsistence, are inextricably linked to the performance of that economy. The twentieth century was the first where the state of the environment became an issue for legislators. Environmentalists have long argued for tighter regulation of markets, but have only recently shown much sophistication in imaging how to engage the power of markets to secure environmental services and biological diversity. This will be vital if we are to map a transition pathway to low-carbon economy that works for both industrialised and non-industrialised economies, for rich and poor countries and for rich and poor within those countries.

We need to devise metrics to make the economy 'tell the economic truth', especially about the externalities of industrial, economic and social processes. This needs new metrics, arising from a new consensus about aims and means and new debates about human goals.

The market is central to the way the world works, but sustainability needs to be understood as a fundamental cultural idea: we need to plant a culture of sustainability. The planetary future depends on what kind of culture of consumerism we build. We need to redesign and engineer the global economy so that people can get more yet consume less. One aspect of this is an economy of services rather than objects, that generates value without generating waste or unnecessary physical or energetic throughput.

To deal with inequity between rich and poor within a finite world, we need to devise processes that allow gear-down in industrialised economies (in terms of energy and material throughput) as well as necessary gear-up in less industrialised economies.

6.4 Presenting New Thinking

The existing language of sustainability has become a prison for the imagination. It limits the capacity of partners to respond to the challenge of planetary future (e.g. language of choices, trade-offs). The elements needed for the future are easily stated, although very challenging to work through. They include imagination, vision, passion and emotion.

The issue of emotion is probably central to success. Existing approaches to sustainability have depended heavily on natural science (from which the concept came), and economics. 'Dismal science' in all forms remains essential to charting a course to the future, but it is not enough to drive changes needed. The world is not run by technocrats (even economists), but politicians and the citizens they represent or govern. In the past sustainability has engaged the mind, but the future demands an engagement with the hearts as well.

7. Managing Change

7.1 Beyond the Usual

The solution to unsustainable planetary management demands a move beyond both 'business as usual' and 'politics as usual'. There is nothing usual about the situation humankind is in: nobody has ever been here before.

The search for sustainability can be understood as a social trajectory, a choice of paths. This choice has to be offered in terms of a framework of choices. The challenge is to rationalise and reconcile the contradictory achievements of human progress, and provide choices that allow people to separate ends (happiness, freedom, fulfilment, a diversity of options) and means (jobs, income, wealth, possessions, consumption, power).

The language of 'environmental limits' is in many ways a political non-starter. However, it is also central to the challenge of sustainability. Failure to understand and live within limits is the main reason why current patterns of development are not sustainable.

A core challenge therefore is how to 'sell' structural change against the immediate short-term interests of non-destitute citizens, businesses locked into current markets, financial institutions that believe they have no role beyond maintaining shareholder value, and timid politicians. The policy conservatism and self-interest of wealthy consumers and citizens, the deadening effects of 'affluenza', and of narrow self-interest of the solvent, are key constraints on novel structural change. The parish pump political rhetoric that 'we will not negotiate our way of life' is an understandable position for wealthy countries to take, but it is a deeply negative in its implications. Those with a vested interest oppose change more strongly than those with a vision for change.

The solution to the dilemma of creating change which the rich and powerful mistrust has to be in terms of presenting opportunities and not threats. Consumption has to be made be a driver of positive change, not a driver of global degradation. The language of future possibilities is likely to be more effective than the language of risk. Environmentalism's traditional capacity to speak like the prophet Jeremiah, promising hell to come, does not promote creative thinking and openness to change. The path-dependence of environmentalist rhetoric in the twentieth century has become disfunctional.

Technology is critical to the transition from the 'old economy (fossil fuel, automobile throw-away) to the new economy (reuse, recycle, new energy)[30]. New technologies may be the key to substantial improvements in material and energy intensity. They may also pose risks to health, welfare and environment. New institutions may be needed to manage transitions to new technologies.

We are on the cusp of non-media mass communication (citizen-to-citizen learning, using the web). This has implications for the way information is stored and exchanged (search engines versus libraries), how information becomes knowledge and how opinion gains authority. These offer both opportunities and risks to the formulation and dissemination of new paradigms for imaging the planetary future.

7.2 Alliances for Change

To have credibility and success, environmentalists need to move beyond the comfort zone of their established professional rituals and partnerships. The changes needed cannot be brought about by environmentalists alone, let alone by IUCN. It will require numerous alliances with a diverse range of actors, big and small, including businesses, governments, development and environmental-developmental organisations and other civil society organisations such as religious groups. Capacity building will be critical to the ability of some partners to support and bring about change.

Businesses are an important part of the solution. A key dimension of an approach offering choices must be the effective combination of enterprise, market and regulation. The market is hugely powerful as a force, for good or bad. It is highly efficient, but needs regulation if it is to 'tell the ecological truth'. Taxation (with taxes restructured to reflect indirect costs of resource use, for example carbon throughput) is necessary if creative structural change is to be brought about. Relevant businesses are not necessarily large.

Conservation and environmentalism in the past have placed excessive emphasis on government and regulation: but why try to drive or coerce change by regulation if you can use the market to change behaviour? As the Grameen businesses demonstrate, social enterprise can be a powerful; force for positive change, far outstripping the capacity of government because of its capacity to harness individual human enterprise and self-interest. Such viral, bottom-of-pyramid solutions to sustainability challenges are in their infancy.

Businesses cannot bring about the needed changes alone. They need governments to regulate, and financiers to reward moves towards sustainability. Ultimately, citizens need to provide the driving forces for new economies through their decisions as consumers. Their ability to balance long term human interests as citizens, parents and neighbours in making short-term consumer choices will have a significant impact on the feasibility of a transition to a new sustainable global economy.

It is unlikely that an attempt to draw up a holistic 'plan for the future' will be effective. The economic, cultural and political changes needed are too complex to map out in detail. A more effective strategy would be based on evolving braided channels of change that different actors can own and drive forwards.

Different strategies will be needed in different contexts: no holistic 'one size fits all' plan will be effective. Los Angeles and Liberia are different places, with different challenges.

7.3 Vision and Expectations

The challenges ahead demand vision and boldness. Popular support for the complex and difficult transitions ahead demand popular support. This will only be realised if ideas connect with heart and emotion. The choices ahead are essentially political, and engagement in debate must centre on central questions of ethics.

At the same time, proposal must be realistic. Win-win solutions are rare. We need to understand how to make trade-offs between goals (between the interests of different people, between different environmental outcomes) better.

The next six decades are crucial. Sixty years is three human generations. Young people can imagine their grandchildren. What world will today's teenagers see their children and grandchildren try to live in?

Notes

1. 'The Future of Environmentalism: Re-thinking sustainability for the twenty-first century' 29–31 January, Hotel Uto Kulm in Zurich, attended by 20 people, including the President and Director General. It was facilitated by Angela Cropper, and attended by William M. Adams, Rubens Harry Born, Lester R. Brown, Sylvia Earle, ,Javed Jabbar, Bill Jackson, Sally Jeanrenaud, David Kaimowitz, Ashok Khosla, Lu Zhi, Gabriel Lopez, Christine Milne, Mark Moody-Stuart, Valli Moosa, Manfred Niekisch, Carlos Manuel Rodriguez, Achim Steiner,

Alexei Yablokov, Muhammad Yunus. This meeting was part of a process begun by a decision of the 63rd Meeting of the World Conservation Union Council (14–16 February 2005), which called upon the Director General to 'develop a statement of Council which would capture the conceptualization of conservation as it stands today'. This statement was intended 'to reflect the key conclusions from the 3rd IUCN World Conservation Congress, which sought to link the human and environmental agendas more effectively, and set out the direction for the future evolution of conservation. In addition, the value of ecosystems should be explored as a key concept. It could serve as a clarion call to the Union's members and Commissions, to the environmental movement as a whole and society at large'.

2. This paper has been drafted by W.M. Adams, University of Cambridge, Downing Place, Cambridge CB2 3EN, email: wa12@cam.ac.uk. It draws directly on the insights and suggestions of all those at the Uto Kulm meeting, but does not necessarily reflect their views.
3. The new IUCN mandate in 1969 spoke of 'the perpetuation and enhancement of the living world—man's natural environment—and the natural resources on which all living things depend', which referred to management of 'air, water, soils, minerals and living species including man, so as to achieve the highest sustainable quality of life'
4. McCormick, J.S. (*The Global Environmental Movement: reclaiming Paradise,* (London: Belhaven, 1992).
5. IUCN, *The World Conservation Strategy,* (Geneva: International Union for Conservation of Nature and Natural Resources, United Nations Environment Programme, World Wildlife Fund, 1980). WWF is now the Worldwide Fund for Nature, IUCN now the World Conservation Union - IUCN.
6. B Brundtland, H. *Our Common Future,* (Oxford: Oxford University Press, for the World Commission on Environment and Development, 1987), (p. 43).
7. S.M. Lélé, "Sustainable development: a critical review," *World Development* 19 (1991): 607–621.
8. S.M. Lélé, (1991) 'Sustainable development: a critical review', *World Development* 19: 607–621.
9. A.L. Mabogunje, (2002) 'Poverty and environmental degradation: challenges within the global economy', *Environment* 44 (1): 10–18.
10. www.developmentgoals.org/
11. W.M. Adams, *Green Development: environment and sustainability in the Third World* (London: Routledge, 2001).
12. Meadows, D., Randers, J. and Behrens, W.W. (1972) *The Limits to Growth.,* Universe Books, New York.
13. Wackernagel, M. and Rees, W. (1996) *Our Ecological Footprint: Reducing Human Impact on the Earth*
14. Vitousek P M , Mooney H A, Lubchenco J, Melillo J M (1997) 'Human domination of Earth's ecosystems', *Science* 277 (25 July): 494–499.
15. Edward Wilson (1992) *The Diversity of Life,* Harvard University Press.
16. Vitousek P M, Ehrlich P R, Ehrlich A H and Matson P A (1986) 'Human appropriation of the products of photosynthesis', *BioScience* 36: 368–373
17. Hannah, L *et al,* (1994) 'A preliminary inventory of human disturbance of world ecosystems', *Ambio* 23: 246–250.
18. Myers, R., and Worm, B. 2003, 'Rapid worldwide depletion of predatory fish communities', *Nature,* vol. 423, pp. 280–283.
19. Millennium Ecosystem Assessment
20. Millennium Ecosystem Assessment
21. Vitousek P M , Mooney H A, Lubchenco J, Melillo J M (1997) 'Human domination of Earth's ecosystems', *Science* 277 (25 July): 494–499
22. Lester R. Brown (2006) *Plan B. 2.0: rescuing a planet under stress and a civilization in trouble,* W.W. Norton, New York, for the Earth Policy Institute.
23. Paelke, R. (2005) 'Sustainability as a bridging concept', *Conservation Biology* 19: 36–8.
24. Newton, J.L. and Freyfogle, E.T. (2004) 'Sustainability: a dissent', *Conservation Biology* 19: 23–32.
25. Edward Wilson (1992) *The Diversity of Life,* Harvard University Press, p. 33.
26. Environmental health and human wellbeing are core concepts in the IUCN Programme 2005–8.
27. Sen, A. (2001) *Development as Freedom,* Oxford University Press.
28. Paehlke, R. (2005) 'Sustainability as a bridging concept', *Conservation Biology* 19: 36–8, p. 36.
29. Orr, D. (2006) 'Framing sustainability', *Conservation Biology* 20: 265–6, p. 266.
30. Lester R. Brown (2006) *Plan B. 2.0: rescuing a planet under stress and a civilization in trouble,* W.W. Norton, New York, for the Earth Policy Institute.

Critical Thinking

1. Do you agree with the statement: "the idea of sustainability dates back more than 30 years?"
2. Explain why the so-called three pillars of sustainability cannot be treated as equivalent.
3. Do you agree that "the human capacity to destroy life-support systems (ecosystem services) is new?" Why or why not?

Sustainable Co-evolution

JOHN CAIRNS, JR.

Summary

Humankind is dependent upon Earth's ecological life support system, whose well-being, in turn, depends upon the practices of human society. The health of both systems requires harmonious, mutualistic interactions between them. Because of its population size and demographic distribution (increasingly urbanized), humankind is also dependent upon its technological life support system, which, as currently managed, threatens the ecological life support system. A fundamental difference exists between the two systems—humankind is capable of using intelligence and reason to regulate its activities but the 30 + million other life forms that comprise the ecological life support system cannot. As a consequence, empathy for the other system is the responsibility of human society. Sustainable co-evolution requires that human society have a high level of ecological literacy and act in a nurturing, compassionate way toward the other system. Only then will sustainable co-evolution be possible since both systems are dynamic and continually changing.

> *The human race has only one or perhaps two generations to rescue itself . . . The longer that no remedial action is taken, the greater the degree of misery and biological impoverishment that mankind must be prepared to accept . . .*
>
> Paul Brown, The Guardian Weekly

Basic Definition

The basic definition of co-evolution is "the simultaneous development of adaptations in two or more populations, species *or other categories* [italics mine] that interact so closely that each is a strong selective force on the other" (Raven and Johnson, 1986). The concept of co-evolution has been used most commonly to describe paired changes in species such as butterflies and the flowers they feed on (e.g. Ehrlich and Raven, 1964), hosts and parasites (e.g. Pimental *et al.,* 1978), and predator/prey relationships (e.g. Thompson, 1986). The term co-evolution has been used to describe changes in more than species pairs, such as the reciprocal changes in agricultural practices and weeds (e.g. Ghersa *et al.,* 1994). In cultural anthropology, the concept has been used to describe paired changes in human culture and human genetics (Durham, 1991). Janzen (1988) and Cairns (1994) have used the concept to describe the relationship between human society and natural systems. Cairns (1994) also points out that the relationship can be either mutualistic or hostile.

Selective Forces

The key concept of the mutualistic co-evolution definition is that interacting entities must serve as selective forces on each other so that the changes enhance the survival of each partner or system, i.e. Earth's ecological life support system and human society should interact so that changes enhance the survival of each component. Human society has been a strong selective force on the global environment (Myers, 1979; Wilson, 1988; Ehrlich and Ehrlich, 1991; National Research Council, 1992; Brown *et al.,* 1992; Tilman and Lehman, 2001; Gardner *et al.,* 2003). Understanding the mechanisms of mutualistic co-evolution should enhance the debate on global environmental issues.

Interaction between the ecological life support system and humankind takes various forms. For example, submergence of islands due to sea level rise is a major cause of habitat loss (e.g. Taylor, 2003). Brown (2001a) describes a variety of stresses that damage the biota and their habitat. In addition to physical destruction of habitat (e.g. deforestation) and more destructive storms (Stevens, 1997), chemical stressors such as pollution and climate change (e.g. greenhouse gases) are also present.

Of course, other species can act as selective forces on human society. Best known are the effects of disease organisms and agricultural pests or other interactions that affect industrial systems in a variety of ways (e.g. Cairns and Bidwell, 1996). At present, it might appear that natural systems do not exert a strong selective force on humankind nor are they capable of doing so. A pandemic disease would quickly alter this misapprehension, and numerous other scenarios are known to the environmentally literate. A technological society can do much ecological damage quickly, but natural systems often take decades or more to react; however, eventually they do. Nature can and will overcome the damage caused by humankind. Unless humankind develops a more harmonious lifestyle, the relationship of natural systems and humankind will not be a mutualistic one.

Future Evolutionary Trends

For most people, a future Earth without humans is unthinkable, yet Dixon and Adams (2003) have explored this possibility using fundamental biological and evolutionary principles. They postulate both a human era and a post-human era. All species eventually become extinct; why should humans be an exception? The quest for sustainable use of the planet assumes that humankind will quickly learn enough about how Earth and its natural systems work to make continuity possible. Sustainability hypothesizes a harmonious, mutualistic relationship between humankind and the ecological life support system that is sustainable, i.e. capable of lasting indefinitely. Wilson (2002) believes that, in the end, success or failure (in humankind's relationship with nature) will depend on an ethical decision on which those now living will be defined and judged for all generations. Obviously, if I did not believe in the possibility of sustainable use of the planet, I would not, at 80 years of age, be spending time writing about it. However, excessive optimism is not justified since complex civilizations do collapse (e.g., Tainter, 1988)

Sustainability faces daunting issues. Myers and Norman (2001) state that the present biotic crisis will surely disrupt and deplete certain basic processes of evolution, with consequences that will persist for millions of years. Ehrlich (2000) has discussed the magnitude of the crisis, but believes, as I do, that there is still time for remedial action that is within the capabilities of humankind. However, evolution is not predictable, so adaptability must be continuous, not just a matter of a few generations.

The quest for sustainable use of the planet is based on the assumption that human births can continue for an indefinite period in order for the species to survive. Death of individuals is distressing, especially in very large numbers; however, if there are enough normal births, *Homo sapiens* will persist. However, dinosaurs did not recover from environmental change, although life continued despite their loss and the loss of a number of other species. To achieve sustainability for *Homo sapiens,* it is essential to protect both natural systems and evolutionary processes. One important task is to end the ecological and genetic isolation of populations that have been fragmented by human activities. Nigh *et al.* (1992) recommend that humankind preserve the *processes* underlying a dynamic biodiversity at all levels. Tilman and Lehman (2001) assert that human-caused environmental changes are creating regional combinations of environmental conditions that, within the 21st century, may fall outside the envelope within which many of the terrestrial plants of a region evolved. Clearly, animals, including humans, will be affected by these changes should they occur.

Making the Connections

Havel (1990) notes: 'Education is the ability to perceive the hidden connections between phenomena.' The age of specialization has produced much useful knowledge, but the connections between components have been badly neglected. As Morowitz (1992) remarks: 'Sustained life is a property of an ecological system rather than a single organism or species.' No species, including humans, can exist in isolation from the ecological life support system. This crucial connection of humankind with the ecological life support system has not received the attention it deserves and is an essential component in achieving sustainability. Since the connection is between two dynamic systems, mutualistic co-evolution is the only path to success. The transition to a sustainable future is partly a technical and scientific problem, but is primarily a matter of ecological and sustainability ethics (Cairns, in press).

A key concept is that humankind is a part, but only a part, of a complex network of species called the interdependent web of life or, from an anthropocentric point of view, the ecological life support system. Especially in this information age, humankind should be able to coevolve with the larger ecological life support system of which it is a part. In short, humankind must design practices that are compatible with the design of nature to reach the ultimate goal of sustainable use of the planet. A condition of reaching this goal is maintaining the integrity and health of ecosystems, so that these dynamic systems can function within normal variability and not be forced into disequilibrium. Natural laws cannot be ignored without severe penalties. A 'partner' unable to coevolve with the other partner is in serious, probably fatal, trouble. At present, human society is diverging markedly from a sustainable relationship with natural systems by damaging their integrity, health, and component species. Since these systems collectively represent natural capital, which provides ecosystem services (e.g., Daily and Ellison, 2002), human society will suffer more and more as ecosystems collapse and the course of biological evolution is altered.

Reducing Points of Instability

Havel (1990) remarked that hope 'is not the conviction that something will turn out well, but the certainty that something makes sense, regardless of how it turns out.' A satisfactory outcome for the quest for sustainable use of the planet will require much hard work and major reduction of and quick elimination of unsustainable practices. If the entire human population consumed like the population of the United States, at least four *more* planet Earths would be required. If Earth's entire population did so at the rate of people in high income countries, 2.4 *more* Earths would be required (Mastny, 2003).

An equally sobering point is that, *at present,* the average human already uses resources at a rate higher than the planet's biological capacity to replace them (based on calculations on the biologically productive area needed to produce the resources used and absorb the waste generated by the human population [*Ecological Footprint Accounts: Moving Sustainability from Concept to Measurable Goal, Redefining Progress,* Oakland, California, 2002]). Mastny (2003) gives numerous alternatives to present practices, i.e. sustainable practices are available. Human society cannot negotiate with nature or ask for forgiveness for past environmental damage. Failure of the ecological life support system will be the ultimate consequence of ignoring natural law.

Each day, humankind makes choices both individually and collectively. Some examples related to sustainability follow.

1. Economic growth cannot continue indefinitely at the expense of natural capital and ecosystem services. Humankind is now choosing economic growth.
2. Humankind, especially in the United States, has chosen material consumption over ecological and sustainability ethics.
3. The wealthiest 20 percent of the world's population possesses 85 percent of all automobiles, consumes 84 percent of all paper, uses 65 percent of all electricity, and consumes 45 percent of all meat and fish (United Nations Inter-Regional Expert Group Meeting on Consumer Protection and Sustainable Consumption: New Guidelines for the Global Consumer, Sao Paulo, Brazil, 28–30 January, 1998). Humankind in the wealthy nations chooses to pay only minor attention to this disparity.
4. Wealthy consumers choose not to make major alterations in their buying practices to favor less wasteful and environmentally sound production of material goods.
5. Humankind chooses to ignore the adverse effects of biotic impoverishment, which consists of both extinction of species and a mega-mass extinction of populations.
6. Humankind chooses to worship technology to a degree that humankind believes it can manipulate a biological future.
7. Humankind chooses to minimize problems of carrying capacity. Hardin (1993) discusses the challenge of limits and the inescapable conclusion that per capita share of environmental riches must decrease as population increases.
8. Arguably, humankind's worst choice has been the failure to show empathy for its descendents and those of the 30 + million other life forms sharing the planet. Leaving a habitable planet instead of a damaged one is the essence of sustainability.
9. Humankind chooses to ignore unsustainable practices, while politics and many corporations hinder attempts for public debate. Since corporations control the news media either directly or through advertising, the debate is finding life on the internet. For example, how will the 2 billion people, who are projected to be added to the world's population between 2000 and 2030, mostly in poor nations, be able to lead a quality life?

Environmental Economics

In order to promote the co-evolution of human society with Earth's ecological life support system, the accurate environmental costs of humankind's practices must be expressly included in all economic analyses. At present, valid assessments are not the norm, although some fine examples of "green" economics exist. Instead, environmental costs have been relegated to the status of externalities; futures are discounted; natural capital is not depreciated; and the environmental costs of waste products are not assessed. Fortunately, there is an environmentally friendly means of preserving natural capital and the services it provides (Hawken *et al.,* 1999). However, these attractive alternatives are diminished in effectiveness because the feedback loops, both economic and environmental, have markedly reduced effects due to government subsidies (Myers and Kent, 2001).

Even when economic and environmental feedback loops are not rendered less effective by subsidies, humans and the environment are still exposed to untested chemical substances for economic and political reasons. Scientific, social, economic, and political systems often unintentionally cooperate in this human experiment (e.g. Schettler *et al.,* 1999). On the other hand, increasing evidence suggests that present economic systems are beginning to recognize the value of natural capital and ecosystem services.

The US National Academy of Engineering (1996) states that a primary challenge for the future is to maximize the benefits of technological innovation and use while minimizing undesirable environmental effects. However, Brown (2001b) states that the issue is not whether humankind knows what needs to be done or whether technologies are available, but whether social institutions are capable of bringing about the change in the time remaining. Earlier, Wells (1920) wrote in *The Outline of History:* 'Human history becomes more and more a race between education and catastrophe.' Lovelock (1988) added a cautionary note that humankind is in a new world that is harder to make sense of and riskier to speculate about—not just more to be learned but everything must be learned. Lovelock (1988) further states that emphasis should be shifted to a concern for the planet rather than a concern for humans. Living sustainably is the *sine qua non* for this shift. Sustainable use of the planet is essential to a mutualistic, harmonious relationship with natural systems of which humankind is a part.

Conclusions

Even discussing the collapse of complex civilizations is likely to generate criticism, as is the case for any imminent disaster, despite persuasive evidence that this has occurred throughout human history. Even more apocalyptic is discussing the possibility of a post-human world, even though there is abundant evidence that most species have a finite period on the ecological stage in the evolutionary theater. However, discussing these unpleasant outcomes is the best way to determine how to avoid them.

Developing sustainable co-evolutionary policies is based on assumptions: (1) humankind is dependent upon Earth's ecological life support system, (2) the best way to avoid redirecting evolutionary processes in ways unfavorable to humankind is to develop a mutualistic relationship with natural systems, (3) since, at present, humankind is also dependent on its technological life support system, it should be managed so that it is not a threat to the ecological life support system, (4) despite overwhelming evidence that most species become extinct, *Homo sapiens* might be an exception if it uses reason coupled with a vastly improved level of environmental literacy; and (5) although science and technology are essential to achieving sustainability, they must be guided by ecological and sustainability ethics.

Achieving sustainable use of the planet is the "acid test" of human intelligence, reasoning, adaptability, and wisdom. Humans must demonstrate that they are, as a species, fit to meet these enormous challenges.

References

Brown LR. Signs of stress: the biological basis. In *Eco-Economy.* New York: W.W. Norton; 2001a: 49–76

Brown LR. *Eco-Economy.* New York: W.W. Norton; 2001b

Brown LR, Brough H, Durning A, Flavin C, French H, Jacobson J, Lenssen N, Lowe M, Postel S, Renner M, Ryan J, Starke L and Young J. *State of the World 1992.* New York: W.W. Norton; 1992

Cairns J Jr. *Ecological and Sustainability Ethics.* Eco-Ethics International Union, Inter-Research, www.esep.de/journals/esep/esepbooks/CairnsEsepBook2.pdf; 2004

Cairns J Jr. Eco-societal restoration: re-examining human society's relationship with natural systems. Abel Wolman Distinguished Lecture. Washington, DC: National Academy of Sciences; 1994

Cairns J Jr. and Bidwell JR. Discontinuities in technological and natural systems caused by exotic species. *Biodiversity Conservation* 1996;5: 1085–94

Daily GC and Ellison K. *The New Economy of Nature.* Washington, DC: Island Press; 2002

Dixon D and Adams J. *The Future is Wild.* Buffalo, NY: Firefly Books; 2003

Durham WH. *Co-evolution: Genes, Culture, and Human Diversity.* Stanford, CA: Stanford University Press; 1991

Erhlich PR. *Human Natures: Genes, Cultures and the Human Prospect.* Washington, DC: Island Press; 2000

Ehrlich PR and Ehrlich A. *Healing the Planet: Strategies for Solving the Environmental Crisis.* New York: Addison-Wesley; 1991

Ehrlich PR and Raven PH. Butterflies and plants: a study in co-evolution. *Evolution* 1964;18:96–608

Gardner TA, Cote IM, Gill JA, Grant A, Watkinson AR. Long-term region-wide declines in Caribbean corals. *Science* 2002;301(5635):958–60

Ghersa CM, Roush ML, Radosevich SR and Cordray SM. Co-evolution of agro-ecosystems and weed management. *BioScience* 1994;44:85–94

Hardin G. *Living within Limits: Ecology, Economics, and Population Taboos.* Oxford, UK: Oxford University Press; 1993

Harvel V. *Disturbing the Peace.* London, UK: Faber and Faber; 1990

Hawken P, Lovins A and Lovins H. *Natural Capitalism.* New York: Little, Brown and Company; 1999

Janzen DH, Gunacaste National Park: tropical ecological and biocultural restoration. In Cairns J Jr. (ed.), *Rehabilitating Damaged Ecosystems.* Boca Raton, FL: CRC Press;1998: Vol. II, 143–92

Lovelock J. *The Ages of Gaia.* New York: W.W. Norton;1988

Mastny L. *Purchasing Power: Harnessing Institutional Procurement for People and the Planet.* Washington, DC: Worldwatch Paper 166; 2003

Morowitz H. *Beginnings of Cellular Life.* New Haven, CT: Yale University Press; 1992

Myers N. *The Sinking Ark.* Oxford, UK: Pergamon Press; 1979

Myers N and Kent J. *Perverse Subsidies.* Washington, DC: Island Press; 2001

Myers N and Norman AH. The biotic crisis and the future of evolution. *Proceedings of the National Academy of Sciences USA,* 2001;98:5389–92

National Research Council. *Global Environmental Change.* Washington, DC: National Academy Press; 1992

Nigh TA, Pflieger WL, Redfearn PL Jr, Scroeder WA, Templeton AR and Thompson FR III. *The Biodiversity of Missouri.* Jefferson City, MO: Missouri Department of Conservation; 1992

Pimental D, Levins SA and Olson D. Co-evolution and the stability of exploiter-victim systems. *American Naturalist* 1978;112:19–125

Raven PH and Johnson GB. *Biology.* St. Louis, MO: Times Mirror/Mosby College Publishing; 1986

Schettler T, Solomon G, Valenti M and Huddle A. *Generations at Risk.* Cambridge, MA: The MIT Press; 1999

Stevens WK. Storm warning: bigger hurricanes and more of them. *New York Times,* 1997;3 June

Tainter J. *The Collapse of Complex Civilizations.* Cambridge, UK: Cambridge University Press; 1988

Taylor D. Small islands threatened by sea level rise. In *Vital Signs.* pp. 84–85. New York: W.W. Norton;2003

Thompson JN. Patterns in co-evolution. In Stone AR and Hawksworth DL (eds.), *Co-evolution and Systematics.* Oxford: Clarendon Press; 1986;119–43

Tilman D and Lehmann C. Human-caused environmental change: impacts on plant diversity and evolution. *Proceedings National Academy of Sciences, USA* 2001;98(10):5433–40

US National Academy of Engineering. *Engineering within Ecological Constraints.* Washington, DC: National Academy Press; 1996

Wells HG. *The Outline of History.* Classic Textbooks;1920

Wilson EO (ed.) *Biodiversity.* Washington, DC: National Academy Press; 1988

Wilson EO. *The Future of Life.* New York: Alfred E. Knopf; 2002

Critical Thinking

1. Explain the statement: "Human society cannot negotiate with nature or ask for forgiveness for past environmental damage."
2. If indeed "there is still time for remedial action that is within the capabilities of humankind" regarding sustainability, what are the top three most urgent actions we must take?
3. Does humankind really still need to "learn. . .about how Earth and its natural systems work?"

Acknowledgments—I am indebted to Karen Cairns for transferring the handwritten draft of this manuscript to typed format and to Darla Donald for valuable editorial assistance.

Framing Sustainability

David W. Orr

In June of 1858 Abraham Lincoln began his address at Springfield, Illinois by saying "If we could first know where we are, and whither we are tending, we could then better judge what to do and how, to do it." He spoke on the issue of slavery that day with a directness that other politicians were loath to practice. At Springfield he asserted that "A house divided against itself cannot stand . . . this government cannot endure, permanently half slave and half free." His immediate targets were the evasions and complications of the Kansas-Nebraska Act of 1854 and the Supreme Court ruling handed down in the Dred Scott decision as well as those whom he accused of conspiring to spread slavery to states where it did not already exist. In his speech Lincoln accused Senator Stephen Douglas, President Franklin Pierce, Supreme Court Justice Roger Taney and President James Buchanan of a conspiracy to spread slavery supported by circumstantial evidence such that it was "impossible to not believe that Stephen and Franklin and Roger and James all understood one another from the beginning, and all worked upon a common plan or draft drawn up before the first lick was struck." His opponent in the upcoming Senatorial election, Stephen Douglas, he described as a "caged and toothless" lion.

Lincoln had begun the process of "framing" the major issue of the day without equivocation but in a way that would build electoral support based on logic, evidence, and eloquence. On February 27, 1860, Lincoln's address at the Cooper Institute in New York extended and deepened the argument. He began with words from Stephen Douglas "Our fathers, when they framed the Government under which we live, understood this question just as well, and even better, than we do now," and proceeded to analyze the historical record to infer what the "fathers" actually believed. Lincoln in a masterful and lawyerly way identified thirty-nine of the founders who had "acted on the question" of slavery in decisions voted on in 1784, 1787, 1789, 1798, 1803, and 1820. In contrast to the position held by Douglas, Lincoln showed that twenty-one of the thirty-nine had acted in ways that clearly indicated their belief that the Federal government had the power to rule on the issue of slavery and the other sixteen who'd not been called upon to act on the issue had in various ways taken positions that suggested that they would have concurred with the majority.

Having destroyed Douglas' position, Lincoln proceeded to address the "the Southern people . . . if they would listen." He began with the assertion that every man has a right to his opinion, but "no right to mislead others, who have less access to history and less leisure to study it" and proceeded down the list of charges and counter charges in the overheated politics of 1860. His aim was to join the Republican cause with the Constitutional power to restrain the extension of slavery and not to assert the power of the Federal government to abolish it, while also saying bluntly that slavery was wrong. He admonished his followers to "calmly consider [the] demands" of the Southern people and "yield to them if, in our deliberate view of our duty, we possibly can." And then he closed by saying "LET US HAVE FAITH THAT RIGHT MAKES MIGHT, AND IN THAT FAITH, LET US TO THE END, DARE TO DO OUR DUTY AS WE UNDERSTAND IT." (emphasis is in Lincoln's text).

The Cooper Institute address was instrumental in Lincoln's election to the Presidency and also in framing the Constitutional issues over slavery and state's rights that had smoldered for seventy-four years before bursting into the conflagration of Civil War. As President, Lincoln further refined the issues of slavery, states' rights, and Constitutional law. His first inaugural address in 1861 Lincoln attempted to reach out to the "people of the Southern states," assuring them that he neither claimed nor would assert a right as President to "interfere with the institution of slavery in the States where it exists." The address is an extended description of the Constitutional realities as Lincoln saw them in which a national government could not be dissolved by the actions of the constituent states. The point was that the Union remained unbroken and that he'd sworn only to defend the Constitution and the union that it had created, not to abolish slavery. He regarded himself still as the President of the southern states and the conflict as a rebellion, not a war between independent countries. Lincoln admonished his "countrymen" to "think calmly and well" on the issues at hand and then closed with the words:

> "I am loth [sic] to close. We are not enemies, but friends. We must not be enemies. Though passion may have strained, it must not break our bonds of affection. The mystic chords of memory, stretching from every battle-field, and patriot grave to every living heart and hearthstone, all over this broad land, will yet swell the chorus of the Union, when again touched, as surely they will be, by the better angels of our nature."

The call went unheeded and war came.

Through the next four years Lincoln continued to frame the meaning of the Constitution relative to the issues of the Civil War, but always in measured strokes looking to a horizon that most did not see. The Emancipation Proclamation carefully calibrated to the war situation and the nuances of keeping the loyal slave states neutral proclaimed only a partial emancipation, drawing the ire of the impatient. At Gettysburg, Lincoln in a masterpiece of eloquent concision based on years of arguing the principle "If men are created equal, they cannot be property" corrected the Constitution, in Garry Wills' (1992) view, without overthrowing it. The "unfinished work" he described was that of restoring the Union and, in effect, taking a country of states to "a new birth of freedom" as a nation with a government "of the people, by the people, for the people."

Lincoln's second Inaugural Address is the capstone of his efforts to frame slavery, the Constitution, and describe a nation dedicated to the proposition that all men are created equal. The setting was the final months of the Civil War with Confederate armies on the threshold of defeat. Lincoln's tone is somber, not triumphal. While both sides in the war prayed to the same God, the prayers of neither were answered in full. "The Almighty," Lincoln reminds the nation, "has His own purposes" which transcend those of either side in the war. Drawing from Matthew 7:1 Lincoln reminds the victorious not to judge former slaveholders "that we be not judged." Lincoln closes by saying "With malice toward none; with charity for all . . . let us . . . bind up the nation's wounds; to care for him who shall have borne the battle, and for his widow, and his orphan—to do all which may achieve and cherish a just and lasting peace."

From his earliest utterances on slavery to "charity for all" Lincoln progressively framed the issues of slavery in ways that left no doubt that he thought it a great wrong but that preservation of the Constitution was the prior consideration. When war came, Lincoln's first aim was to maintain the Union but he then used the occasion to enlarge the concept of a "nation conceived in liberty and dedicated to the proposition that all men are created equal."

We are now engaged in a worldwide conversation about the issues of human longevity on Earth, but no national leader has yet framed a satisfactory vision of sustainability. It is still commonly regarded as one of many issues on a long and growing list, not as the linchpin that connects all of the other issues. Relative to the large issues of sustainability, we are virtually everywhere roughly where the United States was, say, in the year 1850 on the matter of slavery. On the art of framing political and moral issues, much of late has been written (Lakoff, 2004). What can be learned from how Lincoln cast the problem of slavery?

First, Lincoln did not equivocate or agonize about the essential nature of slavery. He did not over think the subject; he regarded slavery as a great wrong and said so plainly and often. "If slavery is not wrong," he wrote in 1864, "nothing is wrong." Moreover, he saw the centrality of the issue to other issues on the national agenda such as the tariff, sectionalism, and national growth. Second, more clearly than any other political figure of his time, he understood the priority of keeping the Constitutional foundation of the nation intact and dealing with slavery within the existing framework of law and philosophy. He did not set out to create something from whole cloth but built a case from sources ready at hand: the Bible, the Declaration of Independence, and the Constitution. Third, he used language and logic with a mastery superior to that of any President before or since. Lincoln was a relentless logician but always spoke with vernacular eloquence in words that could be plainly understood by everyone. Fourth, while the issue of slavery was a great moral wrong, Lincoln did not abuse religion to describe it. While his language was full of Biblical metaphors and allusions, he avoided the temptation to demonize the South and to make the war a religious crusade. Throughout the seven years from the House Divided speech to his assassination in 1865, Lincoln's is one of the masterpieces of political framing that combined shrewdness and sagacity with moral clarity. The result was a progressive clarification of the issues leading to a larger concept of nationhood.

From Lincoln's example we might learn, first, to avoid unnecessary complication and contentiousness. The issues of sustainability are primarily ones of fairness and intergenerational rights not ones of technology or economics, as important as these may be. Lincoln regarded slavery as wrong because no human had the right to hold property in another human being period, not because it was economically inefficient. This was magnetic North by which he oriented his politics. By a similar logic, ours is in the principle that no human has the right to diminish the life and well-being of another and no generation has the right to inflict harm on generations to come. Lincoln did not equivocate on the issue of slavery, nor should we on the tyranny one generation can now impose on another by leaving it ecologically impoverished. Climate change and biotic impoverishment are prime examples of intergenerational remote tyranny and as such constitute a great and permanent wrong and we should say so. Each generation ought to serve as a Trustee for posterity, a bridge of obligation stretching from the distant past to the far future. In that role each generation is required to act cautiously, carefully, and wisely (Brown, 1994). In Wendell Berry's words this "is a burden that falls with greatest weight on us of the industrial age who have been and are, by any measure, the humans most guilty of desecrating the world and of destroying creation" (Berry, 2005).

Second, Lincoln built his case from sources—the Declaration of Independence, the Constitution, and the Bible—familiar to his audience. In doing so, he took Jefferson's views on equality to their logical conclusion and recast the Constitution as the foundation for a truly more perfect union that could protect the dignity of all human beings. In our time we can draw on similar sources but now much enhanced by other constitutions and laws and proclamations of the world community. The Universal Declaration of Human Rights and Earth Charter, for example, describe an inclusive political universe that extends a moral covenant to all the people of Earth and all those yet to be born. It is reasonable to expand this covenant to include the wider community of life, as Aldo Leopold once proposed.

Third, Lincoln's use of religion is instructive both for its depth and for his restraint. He used Biblical imagery and language frequently, but did not do so to castigate Southerners or to inflate Northern pretensions. His use of religion was cautionary, aimed to heal, not divide. Lincoln oriented the struggle over slavery in a larger vision of an imperfect nation striving to fulfill God's justice on earth. The message for us is to ground the issue of sustainability in higher purposes resonant with what is best in the world's great religions but owned by no one creed.

Fourth, Lincoln understood the power of language to clarify, motivate, and to ennoble. Few, if any leaders have ever used words more powerfully or to better effect in a good cause than did Lincoln. This wasn't what we now call "spin" or manipulation of the gullible, but the art of persuasion at its best. Lincoln did not have speech writers to create his message and calibrate it to the latest polls. He wrote his own addresses and letters and is reported to have agonized sometimes for hours and days to find the right words. He spoke directly, often bluntly but softened by humor and the adroit use of metaphor and home spun stories. The result was to place the horrors of combat and the bitterness of sectional strife into a larger context that motivated many to make heroic sacrifices and a legacy of thought and words that "remade America," as Garry Wills (1992) puts it. Now perhaps more than ever we turn to Lincoln for perspective and inspiration.

The tragedy of the U.S. civil war originated in the evasions of the generations prior to 1861. The founders chose not to deal with the problem of slavery in 1787 and it subsequently grew into tragedy the effects of which are still evident. Similarly, without our foresight and action, future generations will attribute the tragedies of climate change and biotic impoverishment to our lies, evasions, and derelictions. But the issue of slavery and that of sustainability also differ in important respects. Slavery was practiced only in a few places and it could be ended by one means or another. The issues comprising the challenge of sustainability, on the other hand, affect everyone on the Earth and for as far into the future as one cares to imagine and they are once and for all time. Never again can we take for granted that the planet will recover from human abuse and insult. For all of its complications, slavery was a relatively simple issue compared to the complexities of sustainability. Progress toward sustainability, however defined, will require more complicated judgments involving intergenerational ethics, science, economics, politics, and much else as applied to problems of energy, agriculture, forestry, shelter, urban planning, health, livelihood, security and the distribution of wealth within and between generations.

Differences notwithstanding, Lincoln's example is instructive. He understood that the deeper problems of race had not been solved by war which had decided only the Constitutional issues about the right of states to secede. It did nothing to resolve the more volatile problems that created the conflict in the first place. He had the faith that they might someday be solved, but only in a nation in which strife and bitterness were set aside by the better angels of our nature. His aim was to create the framework, including the 13 amendment to the Constitution that prohibited slavery, in which healing and charity might take root. Lincoln continues to inspire in our time because he framed the legalities of Constitution and war in a larger context of history, obligation, human dignity, and fundamental rights.

The multiple problems of sustainability will not be solved by this generation or the next. Our role, however, is to frame them in such a way as to create the possibility that they might someday be resolved. Lincoln's example is instructive to us because he understood the importance of preserving the larger framework in which the lesser art of defining particular issues might proceed with adequate deliberation which is to say that he understood that the art of framing issues is a means to reach larger ends. In our time many things that ought to be and must be sustained are in jeopardy, the most important of which are those qualities Lincoln used in defining the specific issue of slavery: clarity, courage, generosity, kindness, wisdom, and humor.

Literature Cited

Berry, W. 2005. Blessed are the Peacemakers. Shoemaker & Hoard, Washington.

Briggs, J. 2005. Lincoln's speeches reconsidered. John Hopkins University Press, Baltimore, Maryland.

Brown, P. 1994. Restoring the public trust. Beacon Press, Boston.

Don, F., editor. 1989. Abraham Lincoln: speeches and writings, 1859–1865. The Library of America, New York.

Guelzo, A. 2004. Lincoln's emancipation proclamation. Simon & Schuster, New York.

Holzer, H. 2004. Lincoln at Cooper Union. Simon & Schuster, New York.

Lakoff, G. 2004. Don't think of an elephant. Chelsea Green, White River Junction, Vermont.

White, R. 2002. Lincoln's greatest speech. Simon & Schuster, New York.

White, R. 2005. The eloquent president. Random House, New York.

Wills, G. 1992. Lincoln at Gettysburg. Simon & Schuster, New York.

Critical Thinking

1. Is it logical to compare how an elected official addressed a societal issue to how society might address an ecological issue?
2. Describe how do you think sustainability has been framed?
3. What do you think of the statement: "The multiple problems of sustainability will not be solved by this generation or the next?"

Synthesis

JOHN C. DERNBACH

In June 1992, at the United Nations Conference on Environment and Development (UNCED, or Earth Summit) in Rio de Janeiro, the nations of the world agreed to implement an ambitious plan for sustainable development. The United States was one of those countries. Has the United States moved toward or away from sustainable development in the 10-year period since Rio? What should the country do next? The book has sought to answer both questions.

Sustainable development is ecologically sustainable human development; it includes but is not limited to economic development. Sustainable development affirms the basic goals of development since the end of World War II, but changes them in one key way. Development is based on peace, economic development, social betterment, and effective national governance. Its goals are human freedom, opportunity, and quality of life, and it has succeeded in many ways.

Unfortunately, we now face growing environmental degradation around the world, and a growing gap between rich and poor. Increasingly, these problems undermine and hinder traditional methods of economic and social development. Deforestation and overfishing mean that many people and businesses can no longer earn a livelihood. Pollution impairs human health and thus human betterment. Conflicts over water and other resources lead to violence and civil strife. These and other problems are profoundly destabilizing because they mean less freedom and opportunity and lower quality of life.

Sustainable development responds to these problems by adding environmental protection to the goals of traditional development. Instead of development at the environment's expense, or environmental protection at the expense of development, sustainable development would achieve both traditional development and environmental protection or restoration at the same time. Sustainable development affirms the importance of freedom, opportunity, and quality of life, for both present and future generations.

Sustainable development should matter to the United States because freedom, opportunity, and quality of life are among our core goals as a nation. Providing a better life for those who come after us is also a basic American value. Sustainable development would lead to a stronger, more efficient, and more productive America, because this country's economic, environmental, social, and security goals would support each other in greater and greater degrees over time, rather than undermine one another. Sustainable development would also both require and promote effective governance and legal systems, which Americans also value. By addressing the destabilizing effects of poverty and environmental degradation around the world, the United States could help make the world more secure. In addition, U.S. economic and military power, as well as the ethical and religious foundations for sustainability, suggest a special obligation to work for sustainable development.

The United States has, unquestionably, begun to take some steps toward sustainable development. In fact, those who see sustainable development as including prior and ongoing efforts, such as conservation and pollution control, could rightly say that the 1990s saw a continuation of activities that began before the Earth Summit. Yet, on balance, the United States is now far from being a sustainable society, and in many respects is farther away than it was in 1992.

While there is "good news" and "bad news" to report, the bad news is told in general trends, broad studies, and for entire economic sectors or program areas. All too frequently, the good news is limited to specific examples and particular programs. The United States has not responded in a way that corresponds to the seriousness of the problems we face or to the opportunities provided by sustainable development. Nevertheless, legal and policy tools are available to put the United States on a direct path to sustainability, to our great advantage and without major dislocations—if we can muster the will and the vision to use them.

This synthesis begins with an overview of the book's findings and recommendations, followed by an explanation of sustainable development and its importance to the United States. It then summarizes each of the book's major sections, which concern consumption and population; international trade, finance, and development assistance; conservation and management of natural resources; waste and toxic chemicals; education; institutions and infrastructure; and governance. Throughout, the synthesis summarizes and often excerpts from individual chapters.

Overview

A Little Good News

In virtually every area of American life, a few people and organizations are exercising leadership for sustainability. A small number of federal agencies, state governments, local governments, corporations, universities, and others have taken a leadership role in moving toward sustainable development over the past decade. Nearly all of these efforts contain room for improvement. Still, they demonstrate that it is both possible and desirable to reconcile environmental, social, and economic goals. For instance:

> The federal government greatly expanded its use of habitat conservation plans in the past decade to reconcile conflicts between economic development and endangered species protection. A few states have begun to implement strategies for sustainable development and use indicators for sustainability.
>
> At the community level, some sustainability initiatives have been undertaken, and are yielding some positive results.
>
> A handful of major corporations are seriously embracing the "triple bottom line" of environment, economy, and society or equity as a way of setting and achieving goals.
>
> A small minority of primary schools, high schools, and higher education institutions are teaching students to perform the kind of integrated and interdisciplinary analysis needed to make decisions that simultaneously further social, economic, and environmental goals.

In a few areas, the United States has played a significant and constructive international leadership role. These include the protection of high seas fisheries, the prevention of lead poisoning, integration of environmental considerations into trade agreements, and incorporation of environmental impact reviews and public participation in World Bank projects.

The President's Council on Sustainable Development (PCSD), an advisory council that existed between 1993 and 1999, developed hundreds of recommendations that would foster national security, economic development, job creation, and environmental protection at the same time. The PCSD and others outlined a policy framework showing that the United States actually could make significant progress toward sustainable development.

There is much better information about many environmental problems now than there was 10 years ago, and generally greater access to it. We also have a much better idea of the steps needed to achieve sustainable development, and have made significant progress in creating the policy and legal tools necessary to do so.

A Lot of Bad News

Energy and materials consumption grew substantially in the past decade, and reduced or outweighed many specific environmental achievements. With 5 percent of the world's population, the United States was at the time of the Earth Summit responsible for about 24 percent of the world's energy consumption and almost 30 percent of the world's raw materials consumption. Since the Earth Summit, materials use has increased 10 percent, primary energy consumption has increased 21 percent, and energy-related carbon dioxide (CO2) emissions have increased by 13 percent. Over and over, increases in materials and energy efficiency, and in the effectiveness of pollution controls for individual sources, are outweighed by increases in consumption. Despite a significant increase in municipal waste recycling in the past decade, for example, the U.S. generation and disposal of municipal solid waste per capita have been growing since 1996. U.S. population—the number of people consuming resources and energy—grew by 32.7 million, or 13.2 percent, from 1990 to 2000, the largest single decade of growth in the nation's history.

Moreover, the United States has not exercised the kind of international leadership necessary to encourage or support sustainable development around the world. The United States is not a Party to many treaties and international agreements that are intended to foster sustainable development in specific contexts, including the Convention on Biological Diversity and the Kyoto Protocol. Current patterns of international trade cause environmental harm and impair sustainable development in part because U.S. trade policy tends to put short-term domestic economic goals ahead of sustainable development. U.S. official development assistance has declined since Rio. Although the United States was the second largest provider of official development assistance in 2000, its contribution was the lowest of all industrialized countries, measured as a percentage of gross domestic income.

U.S. law and policy continue to encourage unsustainable development in a variety of ways. These include subsidies, "grandfather" provisions for existing and more-polluting facilities and activities in pollution control laws, and fragmented local decisionmaking that encourages sprawl. Such laws and policies mean that individuals and corporations have fewer choices, and less sustainable choices, than they would otherwise.

The United States has no national strategy for achieving sustainable development, and no generally accepted indicators to mark progress along the way. Nor does the United States have a meaningful or effective strategy to address climate change, biodiversity, and many other issues. Neither the executive branch nor the U.S. Congress systematically analyze proposed activities to find ways to make significant progress on economic, environmental, social, and security goals at the same time.

As a whole, the condition of America's natural resources and ecosystems has not improved, and appears to have deteriorated slightly, over the past decade. There was no discernible improvement in our rivers, streams, and lakes, and the quality of our ocean coastal waters appears to have deteriorated. Greenhouse gas (GHG) emissions increased, and a large number of plant and animal species continue to be at risk of extinction. U.S. agriculture is less sustainable, and urban sprawl continues relatively unabated. Air quality improved slightly, but not enough to fully protect human health.

The social and institutional infrastructure and supports needed for sustainable development continue to cause environmental degradation and underserve the poor. The negative environmental impacts of transportation increased during the past decade, despite significant legislative changes. The U.S. sanitation system remains vulnerable to breakdowns, the level of communicable diseases is high when compared to other developed countries, and there has been no discernible progress in improving access to medical care.

Recommendations for the Next Decade

The path to sustainability is not an easy one, but it is marked by basic American values. These include freedom, opportunity, and quality of life; greater efficiency; more effective and responsive governance; a desire to make a better world for those who follow us; a willingness to find and exploit opportunities; a quest for a safer world; and a sense of calling to play a constructive role in international affairs. All of these are underscored by our ethical and even religious obligations toward each other and the environment.

The United States would take a large and decisive step toward sustainability if individuals, businesses, educational institutions, local and state governments, federal agencies and others would simply adopt and build on the leading sustainability practices of their counterparts here and in other nations.

A national strategy for sustainable development, with specified goals and priorities, would harness all sectors of society to achieve our economic, social, environmental, and security goals. The strategy could be modeled on that of the European Union (EU) and states such as New Jersey, and specifically address climate change, biodiversity, and other major issues. An executive-level entity would be needed to coordinate and assist in the implementation of the strategy. A counterpart entity in Congress would also be helpful. The strategy would more likely be effective if there were a set of indicators to measure progress in achieving its goals. Comparable state and local strategies and indicators are also needed.

The United States needs to recognize that its substantial consumption levels, coupled with domestic population growth, have serious environmental, social, and economic impacts. Americans also need to understand that human well-being can be decoupled from high consumption of materials and energy. A shift in taxes from labor and income, on one hand, to materials and energy consumption, on the other, would encourage both greater efficiency and reduced negative environmental impacts.

Congress should repeal or modify laws, policies, and subsidies that encourage unsustainable development. The elimination of subsidies would also have positive budgetary impacts. The repeal or modification of such laws would provide more and better opportunities for individuals and corporations to act in a more sustainable manner, and would remove an important set of barriers to sustainability.

Protection of natural resources and the environment must focus more holistically on the resources to be protected, and on understanding those resources. Congress and the states need to assure that these resources are protected from all significant threats, and are protected from those threats to the same degree. In addition, the type of substantive goals that exist in the air and water pollution control programs, as well as supportive implementing mechanisms, should be applied to biodiversity, climate change, oceans under U.S. jurisdiction, forests, and other natural resources. The United States also needs to fund or support the development of more complete and reliable information about ecosystems as well as about the connections among its economic, environmental, social, and security goals.

Social infrastructure, institutions, and laws should be designed and operated to further economic, environmental, and social goals at the same time. Public health services and, at a minimum, basic medical services should be available to all. Transportation infrastructure should be more efficient and diverse, and provide people with more choices.

The United States needs to take a stronger and more constructive leadership role internationally, not only on terrorism but on the broad range of issues related to sustainable development. The United States should further increase its official development assistance, while taking measures to ensure that the money is spent effectively and for sustainable development. More broadly, U.S. foreign policy, including trade policy, needs to be more supportive of the development aspect of sustainable development. The United States should also become a Party to many of the international treaties that would foster sustainable development, including the Convention on Biological Diversity, the Cartagena Protocol on Biosafety, the Aarhus Convention on Access to Information, the Rotterdam Convention on Prior Informed Consent, the Stockholm Convention on Persistent Organic Pollutants, and the Basel Convention on the Control of Transboundary Movements of Hazardous Wastes.

Some longer term changes are also needed if the United States is to achieve sustainable development. They include the evolution of judicial understanding of property to update expectations about the productive value of ecosystems and the establishment of more inviting avenues for public participation in and challenge to decisions affecting sustainability.

What Is Sustainable Development?

Sustainable development is human development that is ecologically sustainable. Its aims are human freedom, opportunity, and higher quality of life. It is not another name for economic development, although it includes economic development.

Because "sustainable" modifies "development," it is first important to understand what development means. Although Americans understand development to mean the transformation of a field or woodlot into housing or a mall, development has a different meaning at the international level. Since the end of World War II, the United States and most of the world community have successfully sought greater peace and security, economic development, and social development or human rights. They have also sought national governance that supports these goals, even though they recognize that international efforts are also needed. As understood internationally, these are the four elements of development. This understanding of development grew out of the experiences of the last world war and the great depression that preceded and contributed to it, and a firm desire to ensure that the conditions that led to them would not occur again. More positively, development is intended to foster human freedom, opportunity, and quality of life.

For more than half a century, we have measured progress by the extent to which we have realized these goals. And there has been a great deal of progress. The world is more free, there is more opportunity, and most humans have a higher quality of life now than they did in 1945.

But until recently, protecting and restoring the environment was not among these goals. Indeed, progress in achieving these other goals was considered to outweigh or even justify any environmental degradation that may have occurred.

As the World Commission on Environment and Development concluded in 1987, progress in the past half century has come with a price we cannot ignore and can no longer afford—massive and growing environmental degradation, and a growing number of people in poverty. The commission concluded that countries should seek sustainable development—"development that meets the needs of the present without compromising the ability of future generations to meet their own needs." Sustainable development would thus meet human needs over the long term; the present generation would not benefit at the expense of future generations. When nations of the world endorsed sustainable development at the Earth Summit in 1992, they redefined progress to include environmental protection and restoration.

Sustainable development is based on a sober and realistic appraisal of how humans need to approach the problems of the next half century or more. Like traditional development, it is premised on a recognition of what can happen when freedom, opportunity, and quality of life are inequitably realized or are diminishing.

Every major international and regional report on the condition of the environment shows continuing and deteriorating environmental conditions. The gap between the rich and poor continues to grow. Poverty and environmental degradation are mutually reinforcing; poor people live in the most polluted or degraded environments, and this contributes to their poverty. Although poverty and environmental degradation are important in their own right, they also can cause or contribute to wars, starvation, ethnic tensions, and terrorism, which are more likely to get headlines than their underlying causes. Like terrorism, poverty and environmental degradation are destabilizing. The pressures caused by poverty and environmental degradation are likely to increase in the next half century. Global population is expected to grow from roughly six to nine billion, or 50 percent, by 2050. The global economy is likely to grow by a factor of three to five in the same period. As difficult and challenging as things now appear, they are likely to become much more difficult and challenging in the decades ahead.

Sustainable development also has deep ethical and religious roots. Sustainable development leads to two major shifts in ethical thinking and action. It recognizes the connections between humanity's social, ecological, and economic obligations, and it recognizes responsibility for future as well as present generations. Agenda 21, the blueprint for sustainable development adopted at the Earth Summit, thus calls for distributive justice, or a fair sharing of environmental resources by humans. The distributive justice theme was in response to demands by developing countries that they have the same right to use natural resources as developed countries. Agenda 21 also suggests that humans have a moral responsibility to limit activities that, if not curtailed or redirected, will severely degrade or even destroy ecosystems. Because human damage to the environment also hurts other humans, sustainable development recognizes the relationship between environmental protection and social justice.

The sacred texts and beliefs underlying the world's religions also support sustainable development, even if that has not been true of their practices. These religious traditions support appreciation for all life; human stewardship of creation; harmony among humans, their communities, and their environment; and a caring for place. They also indicate that the natural world is valuable in itself, not simply insofar as humans may value it. They articulate the importance of deep respect for creation, both human and nonhuman, and living in a manner that is ecologically sustainable. These texts and beliefs also indicate the importance of fair and equitable sharing of resources, which would mean both ceilings and floors for consumption. Finally, they suggest that people be given an opportunity to participate in decisions that will affect their lives and their communities.

To achieve sustainable development, nations at the Earth Summit endorsed two important but nonbinding texts, Agenda 21 and the Rio Declaration. (They also agreed to a separate set of principles for forestry.) As a global plan of action for sustainable development, Agenda 21 is intended to be carried out primarily, but not exclusively, by countries within their own borders. Agenda 21, which contains 40 separate chapters, runs several hundred pages regardless of how it is printed. These chapters focus on the social and economic dimensions of sustainable development, e.g., poverty, human health, and population; conservation and management of natural resources, e.g., atmosphere, forests, biological diversity, and various wastes and toxic chemicals; the role of major groups, e.g., children and youth, women, farmers, workers, and business and industry, in attaining sustainable development; and means of implementation, e.g., financial resources, technology transfer, science, education, and public information. Each chapter identifies specific actions to be taken, explains generally why these actions are necessary, identifies the persons or institutions who are to take action, and describes specific means of implementation.

The Rio Declaration is a set of 27 principles for sustainable development. Key principles include the integration of environment and development in decisionmaking, sustainable patterns of resource production and consumption, the polluter-pays principle, the precautionary approach or principle, developed country leadership, intergenerational equity, and public participation. The polluter-pays principle would have polluters bear the costs of preventing and cleaning up environmental problems rather than impose the costs of those problems on others. According to the precautionary principle, the absence of complete scientific certainty about serious problems is not an excuse for refusing to take action. These principles also are woven into Agenda 21.

In Rio, the international community also established a process for reviewing national and international progress toward sustainable development. Agenda 21 has been, and continues to be, the focal point of that process.

When countries agreed to Agenda 21 and the Rio Declaration, they agreed to implement these agreements, both at home and in their foreign policy. The United States, under the leadership of President George H.W. Bush, was one of those countries.

Why Should Sustainable Development Matter to the United States?

Americans should care about sustainable development because its goals—human freedom, opportunity, and quality of life—are also our goals. We sought independence for these purposes, established a legal and economic system premised on their importance, endured a civil war to protect that system and expand its opportunities to others, and fought two world wars and numerous other conflicts to protect ourselves and help make those same opportunities available to others.

Sustainable development, moreover, is not just about *us,* the current generation of Americans. It is, in the U.S. Constitution's words, about "ourselves *and* our posterity," our children, grandchildren, nieces, nephews, and others not yet born who will someday inhabit this country. We pride ourselves on providing our descendants greater opportunities and a better quality of life. Sustainable development would do precisely that. Without it, we cannot assure our children and grandchildren a better life, and are likely to leave them a poorer one.

Sustainable development would lead to a stronger and more efficient America because we would be pursuing social, economic, environmental, and security goals in ways that are more mutually reinforcing or supportive over time, not contradictory or antagonistic. The result would be a stronger, more efficient country that provides its citizens and their descendants increasingly more opportunities in a quality natural environment. Increased energy efficiency would reduce energy costs for manufacturers and consumers, and would also mean reduced pollution. In addition to securing an ongoing supply of timber and paper products, sustainable forestry matters because we rely on forests for watershed maintenance, pollution abatement, climate control, jobs, and recreation. Similarly, a sustainable transportation system would make it easier, less expensive, and less environmentally damaging for people of all incomes to travel from home to work and other destinations. Cleaner production is likely to be less costly and more efficient, reduce the economic and social burdens created by human exposure to hazardous wastes and substances, and improve the occupational health and safety of workers.

Sustainable development would also lead to better and more responsive governance, which is another basic American value. Ensuring that our economic, social, environmental, and security goals are mutually supportive would require that the government does not subsidize with one hand what it controls on the other. It would also require more public involvement in many decision-making processes because public input is more likely to ensure that these goals are harmonized.

Sustainable development would also lead to a safer, more stable and secure world outside American borders. That would have important and positive consequences for both ourselves and others, particularly after September 11, 2001. The world is deeply divided between haves and have-nots, and the risk of evolution toward an unstable, two-class world, with a huge global underclass, is quite real. Americans have a large stake in the prevention or avoidance of humanitarian emergencies, national and regional conflicts, environmental deterioration, terrorism, illicit drugs, the spread of diseases, illegal migration, and other disasters. These threats to our security do not need passports to cross borders. None of the goals that this country has pursued around the world—peace and stability, human rights and democratization, expansion of trade and markets, environmental protection, or putting an end to hunger and extreme deprivation—can be accomplished effectively except in the context of sustainable development. Thus, while sustainable development assistance in developing countries can be justified on humanitarian grounds, it is also consistent with the strategic interests of the United States.

Americans have a special role to play in sustainable development. We have the largest economy and the most powerful military in the world. Not only do we have enormous capability to bring to bear in the pursuit of sustainable development, we also bear a significant share of the responsibility for the global environmental problems that sustainable development is intended to address. The United States is the world's largest producer and consumer of materials and energy. Since the U.S. model of production and consumption is widely emulated throughout the world, U.S. domestic actions could also have a major international effect.

It is often said that nations or individuals can lead, follow, or get out of the way. The United States is in an unparalleled position to play a key international leadership role on sustainable development. The United States could instead permit the EU, Japan, and other developed countries to play the leadership role, and follow their lead. That would be unpalatable to many, but it would be better than doing nothing. Because of its dominant role in international affairs, however, the United States cannot simply get out of the way. If the United States does not lead or follow, it will be an obstacle to international efforts to achieve sustainable development.

The ethical and religious justifications for sustainable development also provide a reason that Americans should care. U.S. actions do not simply affect us; they affect others as well. Historic and continuing U.S. emissions of GHGs are likely to adversely affect others by contributing to rising sea levels and higher temperatures around the world, for example. Moreover, the texts and beliefs of each of the world's major religions teach responsibility toward other humans as well as the environment. Because Americans see themselves as a religious people, they should respond accordingly.

Finally, our government agreed to Agenda 21 and the Rio Declaration at the Earth Summit. These texts are not legally binding, but a nation's political commitment is not a trivial thing. Indeed, it is in the national interest to honor international political commitments.

The decisions we make about sustainable development are defining decisions for the United States. They will define the values for which our country stands.

Critical Thinking

1. Is it accurate to suggest, "the path to sustainability is not an easy one?" Is it productive to do so?
2. Why do you think the United States "has not exercised the kind of international leadership necessary to encourage or support" sustainability?
3. What do you think are the three most important actions or steps the United States should, perhaps must, take to provide leadership for sustainability?

UNIT 3

Earth's Life Support Systems and Ecosystem Services

Unit Selections

Learning Outcomes

After reading this unit, you should be able to:

- Identify, by listing, the ecosystem services used in the determination of their value.
- Demonstrate, by using Internet inflation calculators, the ability to determine the current value of the world's ecosystem services.
- Identify, by listing, the major threats to ecosystem services.
- Identify, by naming, the most significant change to ecosystems.
- Identify, by matching, the dominant specific factor for alteration with the carbon, nitrogen, and phosphorus cycles.
- Classify, by using definitions, the modern categories of ecosystem services.
- Identify, by listing, at least four specific services provided by each of the modern categories of ecosystem services.
- Generate, by synthesizing relevant concepts, a paragraph describing the relationship between the competitive exclusion principle and sustainable stewardship of ecosystems and ecosystem services.

Student Website

www.mhhe.com/cls

Internet References

Ecosystem Marketplace (EM)
www.ecosystemmarketplace.com

International Union for the Conservation of Nature (IUCN), Commission on Environmental Management (CEM)
www.iucn.org/about/union/commissions/cem/cem_work/cem_services

United States Environmental Protection Agency (USEPA)
www.epa.gov/ecology

United States Forest Service (USFS)
www.fs.fed.us/ecosystemservices

World Resources Institute (WRI)
www.wri.org/project/mainstreaming-ecosystem-services

If the issue of sustainability concerns what has been described as "maladaptive human behavior" regarding the life support systems of Earth, then it is important for those interested in sustainability to have a detailed understanding not only of the life support systems but also of how human actions have modified them. These systems have evolved over billions of years. The human presence on Earth is comparatively extremely brief when compared to the age of the planet. Sustainability studies concern four basic questions: (1) What are the unique characteristics of Earth that have allowed for life to appear and to evolve in the manner that it has; (2) How are human actions compromising those unique characteristics; (3) Why do humans behave in the manner that we do; that is, it seems utterly illogical that a species would so degrade it's habitat that it's own survival is questioned, so if that is indeed happening, there must be a logical explanation for such behavior; and (4) Dependent upon the answers one identifies for questions 1–3, what actions are necessary to achieve sustainability? This unit focuses primarily on the first question.

Article 14 is an explanation of many of the life support systems of the planet without the economic valuation of their individual worth. This type of information is vital because "historically, the nature and value of Earth's life support systems have largely been ignored until their disruption or loss highlighted their importance."

Article 15 presents the most current, scientifically comprehensive answer to the question, How have ecosystems changed as a result of human actions? Simply stated, humans have caused more changes to ecosystems in the past 50 years than at any other comparable period in all of recorded human history! The question related to sustainability is: Can this level of planetary remodeling be sustained?

Article 16 provides an update of the examination of ecosystem services provided nearly 20 years earlier by the authors of articles 1 and 2. A substantial contribution is the identification of categories of ecosystems services. This makes analysis of the issue easier than trying to account for nearly a score of different services. By clustering the services based on the categories, it also makes it easier to identify trends related to the categories rather than those associated with a single service.

Article 17 should be considered as much a Hardin classic as "The Tragedy of the Commons," which appears in Unit 4, but it is not. One of the fundamental issues surrounding the concept of sustainability is that humans are partitioning the resources of Earth (ecosystems and ecosystem services) in such a manner that we are competitively excluding other species from benefiting from them. Hardin explained all of this one-half of a century ago. Perhaps, if more people understood what he explained in this article, we would not be in the environmental predicament in which we find ourselves now. That is why it is imperative that students and practitioners of sustainability understand this root issue associated with the concept.

Ecosystem Services: Benefits Supplied to Human Societies by Natural Ecosystems

GRETCHEN C. DAILY ET AL.

Introduction

Many societies today have technological capabilities undreamed of in centuries past. Their citizens have such a global command of resources that even foods flown in fresh from all over the planet are taken for granted, and daily menus are decoupled from the limitations of regional growing seasons and soils. These developments have focused so much attention upon human-engineered and exotic sources of fulfillment that they divert attention from the local biological underpinnings that remain essential to economic prosperity and other aspects of our well-being.

These biological underpinnings are encompassed in the phrase ecosystem services, which refers to a wide range of conditions and processes through which natural ecosystems, and the species that are part of them, help sustain and fulfill human life. These services maintain biodiversity and the production of ecosystem goods, such as seafood, wild game, forage, timber, biomass fuels, natural fibers, and many pharmaceuticals, industrial products, and their precursors. The harvest and trade of these goods represent important and familiar parts of the human economy. In addition to the production of goods, ecosystem services support life through (Holdren and Ehrlich 1974; Ehrlich and Ehrlich 1981):

- purification of air and water.
- mitigation of droughts and floods.
- generation and preservation of soils and renewal of their fertility.
- detoxification and decomposition of wastes.
- pollination of crops and natural vegetation.
- dispersal of seeds.
- cycling and movement of nutrients.
- control of the vast majority of potential agricultural pests.
- maintenance of biodiversity.
- protection of coastal shores from erosion by waves.
- protection from the sun's harmful ultraviolet rays.
- partial stabilization of climate.
- moderation of weather extremes and their impacts.
- provision of aesthetic beauty and intellectual stimulation that lift the human spirit.

Although the distinction between "natural" and "human-dominated" ecosystems is becoming increasingly blurred, we emphasize the natural end of the spectrum, for three related reasons. First, the services flowing from natural ecosystems are greatly undervalued by society. For the most part, they are not traded in formal markets and so do not send price signals that warn of changes in their supply or condition. Furthermore, few people are conscious of the role natural ecosystem services play in generating those ecosystem goods that are traded in the marketplace. As a result, this lack of awareness helps drive the conversion of natural ecosystems to human-dominated systems (e.g., wheatlands or oil palm fields), whose economic value can be expressed, at least in part, in standard currency. The second reason to focus on natural ecosystems is that many human-initiated disruptions of these systems—such as introductions of exotic species, extinctions of native species, and alteration of the gaseous composition of the atmosphere through fossil fuel burning—are difficult or impossible to reverse on any time scale relevant to society. Third, if awareness is not increased and current trends continue, humanity will dramatically alter Earth's remaining natural ecosystems within a few decades (Daily 1997a, b).

The lack of attention to the vital role of natural ecosystem services is easy to understand. Humanity came into being after most ecosystem services had been in operation for hundreds of millions to billions of years. These services are so fundamental to life that they are easy to take for granted, and so large in scale that it is hard to imagine that human activities could irreparably disrupt them. Perhaps a thought experiment that removes these services from the familiar backdrop of the Earth is the best way to illustrate both the importance and complexity of ecosystem services, as well as how ill-equipped humans are to recreate them. Imagine, for example, human beings trying to colonize the moon. Assume for the sake of argument that the moon had already miraculously acquired some of the basic conditions for supporting human life, such as an atmosphere, a climate, and a physical soil structure similar to those on Earth. The big question facing human colonists would then be, which of Earth's millions of species would need to be transported to the moon to make that sterile surface habitable?

One could tackle that question systematically by first choosing from among all the species exploited directly for food, drink, spices, fiber, timber, pharmaceuticals, and industrial products such as waxes, rubber, and oils. Even if one were highly selective, the list could amount to hundreds or even thousands of species. And that would only be a start, since one would then need to consider which species are crucial to supporting those used directly: the bacteria, fungi, and invertebrates that help make soil fertile and break down wastes and organic matter; the insects, bats, and birds that pollinate flowers; and the grasses, herbs, and trees that hold soil in place, regulate the water cycle, and supply food for animals. The clear message of this exercise is that no one knows which combinations of species—or even approximately how many—are required to sustain human life.

Rather than selecting species directly, one might try another approach: Listing the ecosystem services needed by a lunar colony and then guessing at the types and numbers of species required to perform each. Yet determining which species are critical to the functioning of a particular ecosystem service is no simple task. Let us take soil fertility as an example. Soil organisms are crucial to the chemical conversion and physical transfer of essential nutrients to higher plants. But the abundance of soil organisms is absolutely staggering. Under a square-yard of pasture in Denmark, for instance, the soil is inhabited by roughly 50,000 small earthworms and their relatives, 50,000 insects and mites, and nearly 12 million roundworms. And that tally is only the beginning. The number of soil animals is tiny compared to the number of soil microorganisms: a pinch of fertile soil may contain over 30,000 protozoa, 50,000 algae, 400,000 fungi, and billions of individual bacteria (Overgaard-Nielsen 1955; Rouatt and Katznelson 1961; Chanway 1993). Which must colonists bring to the moon to assure lush and continuing plant growth, soil renewal, waste disposal, and so on? Most of these soil-dwelling species have never been subjected to even cursory inspection: no human eye has ever blinked at them through a microscope, no human hand has ever typed out a name or description of them, and most human minds have never spent a moment reflecting on them. Yet the sobering fact is, as E. O. Wilson put it: they don't need us, but we need them (Wilson 1987).

The Character of Ecosystem Services

Moving our attention from the moon back to Earth, let us look more closely at the services nature performs on the only planet we know that is habitable. Ecosystem services and the systems that supply them are so interconnected that any classification of them is necessarily rather arbitrary. Here we briefly explore a suite of overarching services that operate in ecosystems worldwide.

Production of Ecosystem Goods

Humanity obtains from natural ecosystems an array of ecosystem goods—organisms and their parts and products that grow in the wild and that are used directly for human benefit. Many of these, such as fishes and animal products, are commonly traded in economic markets. The annual world fish catch, for example, amounts to about 100 million metric tons and is valued at between $50 billion and $100 billion; it is the leading source of animal protein, with over 20 percent of the population in Africa and Asia dependent on fish as their primary source of protein (UNFAO 1993). The commercial harvest of freshwater fish worldwide in 1990 totaled approximately 14 million tons and was valued at about $8.2 billion (UNFAO 1994). Interestingly, the value of the freshwater sport fishery in the U.S. alone greatly exceeds that of the global commercial harvest, with direct expenditures in 1991 totaling about $16 billion. When this is added to the value of the employment generated by sport fishing activities, it raises the total to $46 billion (Felder and Nickum 1992, cited in Postel and Carpenter 1997). The future of these fisheries is in question, however, because fish harvests have approached or exceeded sustainable levels virtually everywhere. Nine of the world's major marine fishing areas are in decline due to overfishing, pollution, and habitat destruction. (UNFAO 1993; Kaufman and Dayton 1997).

Turning our attention to the land, grasslands are an important source of marketable goods, including animals used for labor (horses, mules, asses, camels, bullocks, etc.) and those whose parts or products are consumed (as meat, milk, wool, and leather). Grasslands were also important as the original source habitat for most domestic animals such as cattle, goats, sheep, and horses, as well as many crops, such as wheat, barley, rye, oats, and other grasses (Sala and Paruelo 1997). In a wide variety of terrestrial habitats, people hunt game animals such as waterfowl, deer, moose, elk, fox, boar and other wild pigs, rabbits, and even snakes and monkeys. In many countries, game meat forms an important part of local diets and, in many places, hunting is an economically and culturally important sport.

Natural ecosystems also produce vegetation used directly by humans as food, timber, fuelwood, fiber, pharmaceuticals and industrial products. Fruits, nuts, mushrooms, honey, other foods, and spices are extracted from many forest species. Wood and other plant materials are used in the construction of homes and other buildings, as well as for the manufacture of furniture, farming implements, paper, cloth, thatching, rope, and so on. About 15 percent of the world's energy consumption is supplied by fuelwood and other plant material; in developing countries, such "biomass" supplies nearly 40 percent of energy consumption (Hall et al. 1993), although the portion of this derived from natural rather than human-dominated ecosystems is undocumented. In addition, natural products extracted from many hundreds of species contribute diverse inputs to industry: gums and exudates, essential oils and flavorings, resins and oleo-resins, dyes, tannins, vegetable fats and waxes, insecticides, and multitudes of other compounds (Myers 1983; Leung and Foster 1996). The availability of most of these natural products is in decline due to ongoing habitat conversion.

Generation and Maintenance of Biodiversity

Biological diversity, or biodiversity for short, refers to the variety of life forms at all levels of organization, from the molecular to the landscape level. Biodiversity is generated and maintained in natural ecosystems, where organisms encounter a wide variety of living conditions and chance events that shape their evolution in unique ways. Out of convenience or necessity, biodiversity is usually quantified in terms of numbers of species, and this perspective has greatly influenced conservation goals. It is important to remember, however, that the benefits that biodiversity supplies to humanity are delivered through populations of species residing in living communities within specific physical settings— in other words, through complex ecological systems, or ecosystems (Daily and Ehrlich 1995). For human beings to realize most of the aesthetic, spiritual, and economic benefits of biodiversity, natural ecosystems must therefore be accessible. The continued existence of coniferous tree species somewhere in the world would not help the inhabitants of a town inundated by flooding because of the clearing of a pine forest upstream. Generally, the flow of ecosystem goods and services in a region is determined by the type, spatial layout, extent, and proximity of the ecosystems supplying them. Because of this, the preservation of only one minimum viable population of each non-human species on Earth in zoos, botanical gardens, and the world's legally protected areas would not sustain life as we know it. Indeed, such a strategy, taken to extreme, would lead to collapse of the biosphere, along with its life support services.

As described in the previous section, biodiversity is a direct source of ecosystem goods. It also supplies the genetic and biochemical resources that underpin our current agricultural and pharmaceutical enterprises and may allow us to adapt these vital enterprises to global change. Our ability to increase crop productivity in the face of new pests, diseases, and other stresses has depended heavily upon the transfer to our crops of genes from wild crop relatives that confer resistance to these challenges. Such extractions from biodiversity's "genetic library" account for annual increases in crop productivity of about 1 percent, currently valued at $ 1 billion (NRC 1992). Biotechnology now makes possible even greater use of this natural storehouse of genetic diversity via the transfer to crops of genes from any kind of organism— not simply crop relatives—and it promises to play a major role in future yield increases. By the turn of the century, farm-level sales of the products of agricultural biotechnology, just now entering the marketplace, are expected to reach at least $ 10 billion per year (World Bank 1991, cited in Reid et al. 1996). In addition to sustaining the production of conventional crops, the biodiversity in natural ecosystems may include many potential new foods. Human beings have utilized around 7,000 plant species for food over the course of history and another 70,000 plants are known to have edible parts (Wilson 1989). Only about 150 food plants have ever been cultivated on a large scale, however. Currently, 82 plant species contribute 90 percent of national per-capita supplies of food plants (Prescott-Allen and Prescott-Allen 1990), although a much smaller number of these supply the bulk of the calories humans consume. Many other species, however, appear more nutritious or better suited to the growing conditions that prevail in important regions than the standard crops that dominate world food supply today. Because of increasing salinization of irrigated croplands and the potential for rapid climate change, for instance, future food security may come to depend on drought- and salt-tolerant varieties that now play comparatively minor roles in agriculture.

Turning to medicinal resources, a recent survey showed that of the top 150 prescription drugs used in the United States, 118 are based on natural sources: 74 percent on plants, 18 percent on fungi, 5 percent on bacteria, and 3 percent on one vertebrate (snake) species. Nine of the top ten drugs in this list are based on natural plant products (Grifo and Rosenthal, in press, as cited in Dobson 1995). The commercial value of pharmaceuticals in the developed nations exceeds $40 billion per year (Principe 1989). Looking at the global picture, approximately 80 percent of the human population relies on traditional medical systems, and about 85 percent of traditional medicine involves the use of plant extracts (Farnsworth et al. 1985).

Saving only a single population of each species could have another cost. Different populations of the same species may produce different types or quantities of defensive chemicals that have potential use as pharmaceuticals or pesticides (McCormick et al. 1993); and they may exhibit different tolerances to environmental stresses such as drought or soil salinity. For example, the development of penicillin as a therapeutic antibiotic took a full 15 years after Alexander Fleming's famous discovery of it in common bread mold. In part, this was because scientists had great difficulty producing, extracting, and purifying the substance in needed quantities. One key to obtaining such quantities was the discovery, after a worldwide search, of a population of Fleming's mold that produced more penicillin than the original (Dowling 1977). Similarly, plant populations vary in their ability to resist pests and disease, traits important in agriculture. Many thousands of varieties of rice from different locations were screened to find one with resistance to grassy stunt virus, a disease that posed a serious threat to the world's rice crop (Myers 1983). Despite numerous examples like these, many of the localities that harbor wild relatives of crops remain unprotected and heavily threatened.

Climate and Life

Earth's climate has fluctuated tremendously since humanity came into being. At the peak of the last ice age 20,000 years ago, for example, much of Europe and North America were covered by mile-thick ice sheets. While the global climate has been relatively stable since the invention of agriculture around 10,000 years ago, periodic shifts in climate have affected human activities and settlement patterns. Even relatively recently, from 1550-1850, Europe was significantly cooler during a period known as the Little Ice Age. Many of these changes in climate are thought to be caused by alterations in Earth's orbital rotation or in the energy output of the sun, or even by events on the Earth itself—sudden perturbations such as violent volcanic eruptions and asteroid impacts or more gradual tectonic events such as the uplift of the Himalayas. Remarkably, climate has been buffered enough through all these changes to sustain life for at least 3.5 billion years (Schneider and Londer 1984). And life itself has played a role in this buffering.

Climate, of course, plays a major role in the evolution and distribution of life over the planet. Yet most scientists would agree that life itself is a principal factor in the regulation of global climate, helping to offset the effects of episodic climate oscillations by responding in ways that alter the greenhouse gas concentrations in the atmosphere. For instance, natural ecosystems may have helped to stabilize climate and prevent overheating of the Earth by removing more of the greenhouse gas carbon dioxide from the atmosphere as the sun grew brighter over millions of years (Alexander et al. 1997). Life may also exert a destabilizing or positive feedback that reinforces climate change, particularly during transitions between interglacial periods and ice ages. One example: When climatic cooling leads to drops in sea level, continental shelves are exposed to wind and rain, causing greater nutrient runoff to the oceans. These nutrients may fertilize the growth of phytoplankton, many of which form calcium carbonate shells. Increasing their populations would remove more carbon dioxide from the oceans and the atmosphere, a mechanism that should further cool the planet. Living things may also enhance warming trends through such activities as speeding up microbial decomposition of dead organic matter, thus releasing carbon dioxide to the atmosphere (Schneider and Boston 1991; Allegre and Schneider 1994). The relative influence of life's stabilizing and destabilizing feedbacks remains uncertain; what is clear is that climate and natural ecosystems are tightly coupled, and the stability of that coupled system is an important ecosystemservice.

Besides their impact on the atmosphere, ecosystems also exert direct physical influences that help to moderate regional and local weather. For instance, transpiration (release of water vapor from the leaves) of plants in the morning causes thunderstorms in the afternoon, limiting both moisture loss from the region and the rise in surface temperature. In the Amazon, for example, 50 pecent of the mean annual rainfall is recycled by the forest itself via evapotranspiration—that is, evaporation from wet leaves and soil combined with transpiration (Salati 1987). Amazon deforestation could so dramatically reduce total precipitation that the forest might be unable to reestablish itself following complete destruction (Shukla et al. 1990). Temperature extremes are also moderated by forests, which provide shade and surface cooling and also act as insulators, blocking searing winds and trapping warmth by acting as a local greenhouse agent.

Mitigation of Floods and Droughts

An enormous amount of water, about 1,19,000 cubic kilometers, is rained annually onto the Earth's land surface—enough to cover the land to an average depth of one meter (Shiklomanov 1993). Much of this water is soaked up by soils and gradually meted out to plant roots or into aquifers and surface streams. Thus, the soil itself slows the rush of water off the land in flash floods. Yet bare soil is vulnerable. Plants and plant litter shield the soil from the full, destructive force of raindrops and hold it in place. When landscapes are denuded, rain compacts the surface and rapidly turns soil to mud (especially if it has been loosened by tillage); mud clogs surface cavities in the soil, reduces infiltration of water, increases runoff, and further enhances clogging. Detached soil particles are splashed downslope and carried off by running water (Hillel 1991).

Erosion causes costs not only at the site where soil is lost but also in aquatic systems, natural and human-made, where the material accumulates. Local costs of erosion include losses of production potential, diminished infiltration and water availability, and losses of nutrients. Downstream costs may include disrupted or lower quality water supplies; siltation that impairs drainage and maintenance of navigable river channels, harbors, and irrigation systems; increased frequency and severity of floods; and decreased potential for hydroelectric power as reservoirs fill with silt (Pimentel et al. 1995). Worldwide, the replacement cost of reservoir capacity lost to siltation is estimated at $6 billion per year.

In addition to protecting soil from erosion, living vegetation—with its deep roots and above-ground evaporating surface—also serves as a giant pump, returning water from the ground into the atmosphere. Clearing of plant cover disrupts this link in the water cycle and leads to potentially large increases in surface runoff, along with nutrient and soil loss. A classic example comes from the experimental clearing of a New Hampshire forest, where herbicide was applied to prevent regrowth for a three-year period after the clearing. The result was a 40 percent increase in average stream flow. During one four-month period of the experiment, runoff was more than five times greater than before the clearing (Bormann 1968). On a much larger scale, extensive deforestation in the Himalayan highlands appears to have exacerbated recent flooding in Bangladesh, although the relative roles of human and natural forces remain debatable (Ives and Messerli 1989). In addition, some regions of the world, such as parts of Africa, are experiencing an increased frequency and severity of drought, possibly associated with extensive deforestation.

Wetlands are particularly well-known for their role in flood control and can often reduce the need to construct flood control structures. Floodplain forests and high salt marshes, for example, slow the flow of floodwaters and allow sediments to be deposited within the flood-plain rather than washed into downstream bays or oceans. In addition, isolated wetlands such as prairie potholes in the Midwest and cypress ponds in the Southeast, serve as detention areas during times of high rainfall, delaying saturation of upland soils and overland flows into rivers and thereby damping peak flows. Retaining the integrity of these wetlands by leaving vegetation, soils, and natural water regimes intact can reduce the severity and duration of flooding along rivers (Ewel 1997). A relatively small area of retained wetland, for example, could have largely prevented the severe flooding along the Mississippi River in 1993.

Services Supplied by Soil

Soil represents an important component of a nation's assets, one that takes hundreds to hundreds of thousands of years to build up and yet very few years to be lost. Some civilizations have drawn great strength from fertile soil; conversely, the loss of productivity through mismanagement is thought to have ushered many once flourishing societies to their ruin (Adams 1981). Today, soil degradation induced by human activities afflicts nearly 20 percent of the Earth's vegetated land surface (Oldeman et al. 1990).

In addition to moderating the water cycle, as described above, soil provides five other interrelated services (Daily et al. 1997). First, soil shelters seeds and provides physical support as they sprout and mature into adult plants. The cost of packaging and storing seeds and of anchoring plant roots would be enormous without soil. Human-engineered hydroponic systems can grow plants in the absence of soil, and their cost provides a lower bound to help assess the value of this service. The costs of physical support trays and stands used in such operations total about US$55,000 per hectare (for the Nutrient Film Technique Systems; FAO 1990).

Second, soil retains and delivers nutrients to plants. Tiny soil particles (less than two microns in diameter), which are primarily bits of humus and clays, carry a surface electrical charge that is generally negative. This property holds positively charged nutrients—cations such as calcium and magnesium—near the surface, in proximity to plant roots, allowing them to be taken up gradually. Otherwise, these nutrients would quickly be leached away. Soil also acts as a buffer in the application of fertilizers, holding onto the fertilizer ions until they are required by plants. Hydroponic systems supply water and nutrients to plants without need of soil, but the margin for error is much smaller—even small excesses of nutrients applied hydroponically can be lethal to plants. Indeed, it is a complex undertaking to regulate the nutrient concentrations, pH, and salinity of the nutrient solution in hydroponic systems, as well as the air and solution temperature, humidity, light, pests, and plant diseases. Worldwide, the area under hydroponic culture is only a few thousand hectares and is unlikely to grow significantly in the foreseeable future; by contrast, global cropped area is about 1.4 billion hectares (USDA 1993).

Third, soil plays a central role in the decomposition of dead organic matter and wastes, and this decomposition process also renders harmless many potential human pathogens. People generate a tremendous amount of waste, including household garbage, industrial waste, crop and forestry residues, and sewage from their own populations and their billions of domesticated animals. A rough approximation of the amount of dead organic matter and waste (mostly agricultural residues) processed each year is 130 billion metric tons, about 30 percent of which is associated with human activities (derived from Vitousek et al. 1986). Fortunately, there is a wide array of decomposing organisms—ranging from vultures to tiny bacteria—that extract energy from the large, complex organic molecules found in many types of waste. Like assembly-line workers, diverse microbial species process the particular compounds whose chemical bonds they can cleave and pass along to other species the end products of their specialized reactions. Many industrial wastes, including soaps, detergents, pesticides, oil, acids, and paper, are detoxified and decomposed by organisms in natural ecosystems if the concentration of waste does not exceed the system's capacity to transform it. Some modern wastes, however, are virtually indestructible, such as some plastics and the breakdown products of the pesticide DDT.

The simple inorganic chemicals that result from natural decomposition are eventually returned to plants as nutrients. Thus, the decomposition of wastes and the recycling of nutrients—the fourth service soils provide— are two aspects of the same process. The fertility of soils—that is, their ability to supply nutrients to plants— is largely the result of the activities of diverse species of bacteria, fungi, algae, crustacea, mites, termites, spring-tails, millipedes, and worms, all of which, as groups, play important roles. Some bacteria are responsible for "fixing" nitrogen, a key element in proteins, by drawing it out of the atmosphere and converting it to forms usable by plants and, ultimately, human beings and other animals. Certain types of fungi play extremely important roles in supplying nutrients to many kinds of trees. Earthworms and ants act as "mechanical blenders," breaking up and mixing plant and microbial material and other matter (Jenny 1980). For example, as much as 10 metric tonnes of material may pass through the bodies of earthworms on a hectare of land each year, resulting in nutrient rich "casts" that enhance soil stability, aeration, and drainage (Lee 1985).

Finally, soils are a key factor in regulating the Earth's major element cycles—those of carbon, nitrogen, and sulfur. The amount of carbon and nitrogen stored in soils dwarfs that in vegetation, for example. Carbon in soils is nearly double (1.8 times) that in plant matter, and nitrogen in soils is about 18 times greater (Schlesinger 1991). Alterations in the carbon and nitrogen cycles may be costly over the long term, and in many cases, irreversible on a time scale of interest to society. Increased fluxes of carbon to the atmosphere, such as occur when land is converted to agriculture or when wetlands are drained, contribute to the buildup of key greenhouse gases, namely carbon dioxide and methane, in the atmosphere (Schlesinger 1991). Changes in nitrogen fluxes caused by production and use of fertilizer, burning of wood and other biomass fuels, and clearing of tropical land lead to increasing atmospheric concentrations of nitrous oxide, another potent greenhouse gas that is also involved in the destruction of the stratospheric ozone shield. These and other changes in the nitrogen cycle also result in acid rain and excess nutrient inputs to freshwater systems, estuaries, and coastal marine waters. This nutrient influx causes eutrophication of aquatic ecosystems and contamination of drinking water sources—both surface and ground water—by high levels of nitrate-nitrogen (Vitousek et al. 1997).

Pollination

Animal pollination is required for the successful reproduction of most flowering plants. About 220,000 out of an estimated 240,000 species of plants for which the mode of pollination has been recorded require an animal such as a bee or hummingbird to accomplish this vital task. This includes both wild plants and about 70 percent of the agricultural crop species that feed the world. Over 100,000 different animal species—including bats, bees, beetles, birds, butterflies, and flies—are known to provide these free pollination services that assure the perpetuation of plants in our croplands, backyard gardens, rangelands, meadows and forests. In turn, the continued availability of these pollinators depends on the existence of a wide variety of habitat types needed for their feeding, successful breeding, and completion of their life cycles (Nabhan and Buchmann 1997).

One third of human food is derived from plants pollinated by wild pollinators. Without natural pollination services, yields of important crops would decline precipitously and many wild plant species would become extinct. In the United States alone, the agricultural value of wild, native pollinators—those sustained by natural habitats adjacent to farmlands— is estimated in the billions of dollars per year. Pollination by honey bees, originally imported from Europe, is extremely important as well, but these bees are presently in decline, enhancing the importance of pollinators from natural ecosystems. Management of the honey bee in the New World is currently threatened by the movement of, and hybridization with, an aggressive African strain of honey bee that was accidentally released in Brazil in 1956. Diseases of honey bee colonies are also causing a marked decline in the number of managed colonies. Meanwhile, the diversity of natural pollinators available to both wild and domesticated plants is diminishing: more than 60 genera of pollinators include species now considered to be threatened, endangered or extinct (Buchmann and Nabhan 1996).

Natural Pest Control Services

Humanity's competitors for food, timber, cotton, and other fibers are called pests, and they include numerous herbivorous insects, rodents, fungi, snails, nematodes, and viruses. These pests destroy an estimated 25 to 50 percent of the world's crops, either before or after harvest (Pimentel et al. 1989). In addition, numerous weeds compete directly with crops for water, light, and soil nutrients, further limiting yields.

Chemical pesticides, and the strategies by which they are applied to fight crop pests, can have harmful unintended consequences. First, pests can develop resistance, which means that higher and higher doses of pesticides must be applied or new chemicals developed periodically to achieve the same level of control. Resistance is now found in more than 500 insect and mite pests, over 100 weeds, and in about 150 plant pathogens (WRI 1994). Second, populations of the natural enemies of pests are decimated by heavy pesticide use. Natural predators are often more susceptible to synthetic poisons than are the pests because they have not had the same evolutionary experience with overcoming plant chemicals that the pests themselves have had. And natural predators also typically have much smaller population sizes than those of their prey. Destruction of predator populations leads to explosions in prey numbers, not only freeing target pests from natural controls but often "promoting" other non-pest species to pest status. In California in the 1970s, for instance, 24 of the 25 most important agricultural pests had been elevated to that status by the overuse of pesticides (NRC 1989). Third, exposure to pesticides and herbicides may pose serious health risks to humans and many other types of organisms; the recently discovered declines in human sperm counts may be attributable in part to such exposure (Colborn et al. 1996).

Fortunately, an estimated 99 percent of potential crop pests are controlled by natural enemies, including many birds, spiders, parasitic wasps and flies, lady bugs, fungi, viral diseases, and numerous other types of organisms (DeBach 1974). These natural biological control agents save farmers billions of dollars annually by protecting crops and reducing the need for chemical control (Naylor and Ehrlich 1997).

Seed Dispersal

Once a seed germinates, the resulting plant is usually rooted in place for the rest of its life. For plants, then, movement to new sites beyond the shadow of the parent is usually achieved through seed dispersal. Many seeds, such as those of the dandelion, are dispersed by wind. Some are dispersed by water, the most famous being the seafaring coconut. Many other seeds have evolved ways of getting around by using animals as their dispersal agents. These seeds may be packaged in sweet fruit to reward an animal for its dispersal services; some of these seeds even require passage through the gut of a bird or mammal before they can germinate. Others require burial—by, say, a forgetful jay or a squirrel which later leaves its cache uneaten—for eventual germination. Still others are equipped with sticky or sharp, spiny surfaces designed to catch onto a passing animal and go for a long ride before dropping or being rubbed off. Without thousands of animal species acting as seed dispersers, many plants would fail to reproduce successfully. For instance, the whitebark pine (Pinus albicaulis), a tree found in the Rockies and Sierra Nevada - Cascade Mountains, cannot reproduce successfully without a bird called Clark's Nutcracker (Nucifraga columbiana), which chisels pine seeds out of the tightly closed cones and disperses and buries them; without this service, the cones do not open far enough to let the seeds fall out on their own. Animal seed dispersers play a central role in the structure and regeneration of many pine forests (Lanner 1996). Disruption of these complex services may leave large areas of forest devoid of seedlings and younger age classes of trees, and thus unable to recover swiftly from human impacts such as land clearing.

Aesthetic Beauty and Intellectual and Spiritual Stimulation

Many human beings have a deep appreciation of natural ecosystems. That is apparent in the art, religions, and traditions of diverse cultures, as well as in activities such as gardening and pet-keeping, nature photography and film-making, bird feeding and watching, hiking and camping, ecotouring and mountaineering, river-rafting and boating, fishing and hunting, and in a wide range of other activities. For many, nature is an unparalleled source of wonderment and inspiration, peace and beauty, fulfillment and rejuvenation (e.g., Kellert and Wilson).

Threats to Ecosystem Services

Ecosystem services are being impaired and destroyed by a wide variety of human activities. Foremost among the immediate threats are the continuing destruction of natural habitats and the invasion of non-native species that often accompanies such disruption; in marine systems, overfishing is a major threat. The most irreversible of human impacts on ecosystems is the loss of native biodiversity. A conservative estimate of the rate of species loss is about one per hour, which unfortunately exceeds the rate of evolution of new species by a factor of 10,000 or more (Wilson 1989; Lawton and May 1995). But complete extinction of species is only the final act in the process. The rate of loss of local populations of species—the populations that generate ecosystem services in specific localities and regions—is orders of magnitude higher (Daily and Ehrlich 1995; Hughes et al., in prep.). Destroying other life forms also disrupts the web of interactions that could help us discover the potential usefulness of specific plants and animals (Thompson 1994). Once a pollinator or a predacious insect is on the brink of extinction, for instance, it would be difficult to discover its potential utility to farmers.

Other imminent threats include the alteration of the Earth's carbon, nitrogen, and other biogeochemical cycles through the burning of fossil fuels and heavy use of nitrogen fertilizer; degradation of farmland through unsustainable agricultural practices; squandering of freshwater resources; toxification of land and waterways; and overharvesting of fisheries, managed forests, and other theoretically renewable systems.

These threats to ecosystem services are driven ultimately by two broad underlying forces. One is rapid, unsustainable growth in the scale of the human enterprise: in population size, in per-capita consumption, and also in the environmental impacts that technologies and institutions generate as they produce and supply those consumables (Ehrlich et al. 1977). The other underlying driver is the frequent mismatch between short-term, individual economic incentives and long-term, societal well-being. Ecosystem services are generally greatly undervalued, for a number of reasons: many are not traded or valued in the marketplace; many serve the public good rather than provide direct benefits to individual landowners; private property owners often have no way to benefit financially from the ecosystem services supplied to society by their land; and, in fact, economic subsidies often encourage the conversion of such lands to other, market-valued activities. Thus, people whose activities disrupt ecosystem services often do not pay directly for the cost of those lost services. Moreover, society often does not compensate landowners and others who do safeguard ecosystem services for the economic benefits they lose by foregoing more lucrative but destructive land uses. There is a critical need for policy measures that address these driving forces and embed the value of ecosystem services into decision making frameworks.

Valuation of Ecosystem Services

Human society would cease to exist in the absence of ecosystem services. Thus, their immense value to humanity is unquestionable. Yet quantifying the value of ecosystem services in specific localities, and measuring their worth against that of competing land uses is no simple task. When tradeoffs must be made in the allocation of land and other resources to competing human activities, the resolution often requires a measure of what is known as the marginal value. In the case of ecosystem services, for example, the question that might be posed would be: By how much would the flow of ecosystem services be augmented (or diminished) with the preservation (or destruction) of the next hectare of forest or wetland? Estimation of marginal values is complex (e.g., Bawa and Gadgil 1997; Daily 1997b). Often a qualitative comparison of relative values is sufficient— that is, which is greater, the economic benefits of a particular development project or the benefits supplied by the ecosystem that would be destroyed, measured over a time period of interest to people concerned about the well-being of their grandchildren?

There are, and will remain, many cases in which ecosystem service values are highly uncertain. Yet the pace of destruction of natural ecosystems, and the irreversibility of most such destruction on a time scale of interest to humanity, warrants substantial caution. Valuing a natural ecosystem, like valuing a human life, is fraught with difficulties. Just as societies have recognized fundamental human rights, however, it may be prudent to establish fundamental ecosystem protections even though uncertainty over economic values remains. New institutions and agreements at the international and subnational level will be needed to encourage fair participation in such protections (see, e.g., Heal 1994).

The tremendous expense and difficulty of replicating lost ecosystem services is perhaps best illustrated by the results of the first Biosphere two "mission," in which eight people lived inside a 3.15-acre closed ecosystem for two years. The system featured agricultural land and replicas of several natural ecosystems such as forests and even a miniature ocean. In spite of an investment of more than $200 million in the design, construction, and operation of this model earth, it proved impossible to supply the material and physical needs of the eight Biospherians for the intended two years. Many unpleasant and unexpected problems arose, including a drop in atmospheric oxygen concentration to 14 percent (the level normally found at an elevation of 17,500 feet), high spikes in carbon dioxide concentrations, nitrous oxide concentrations high enough to impair the brain, an extremely high level of extinctions (including 19 of 25 vertebrate species and all pollinators brought into the enclosure, which would have ensured the eventual extinction of most of the plant species as well), overgrowth of aggressive vines and algal mats, and population explosions of crazy ants, cockroaches, and katydids. Even heroic personal efforts on the part of the Biospherians did not suffice to make the system viable and sustainable for either humans or many nonhuman species (Cohen and Tilman 1996).

Major Uncertainties

Society would clearly profit by further investigation into some of the following broad research questions so that we might avoid on Biosphere one, the earth, unpleasant surprises like those that plagued the Biosphere two project (Holdren 1991; Cohen and Tilman 1996; Daily 1997b):

- What is the relative impact of various human activities upon the supply of ecosystem services?
- What is the relationship between the condition of an ecosystem—that is, relatively pristine or heavily modified—and the quantity and quality of ecosystem services it supplies?
- To what extent do ecosystem services depend upon biodiversity at all levels, from genes to species to landscapes?

- To what extent have various ecosystem services already been impaired? And how are impairment and risk of future impairment distributed in various regions of the globe?
- How interdependent are different ecosystem services? How does exploiting or damaging one influence the functioning of others?
- To what extent, and over what time scale, are ecosystem services amenable to repair or restoration?
- How effectively, and at how large a scale, can existing or foreseeable human technologies substitute for ecosystem services? What would be the side effects of such substitutions?
- Given the current state of technology and the scale of the human enterprise, what proportion and spatial pattern of land must remain relatively undisturbed, locally, regionally, and globally, to sustain the delivery of essential ecosystem services?

Conclusions

The human economy depends upon the services performed "for free" by ecosystems. The ecosystem services supplied annually are worth many trillions of dollars. Economic development that destroys habitats and impairs services can create costs to humanity over the long term that may greatly exceed the short-term economic benefits of the development. These costs are generally hidden from traditional economic accounting, but are nonetheless real and are usually borne by society at large. Tragically, a short-term focus in land-use decisions often sets in motion potentially great costs to be borne by future generations. This suggests a need for policies that achieve a balance between sustaining ecosystem services and pursuing the worthy short-term goals of economic development.

References

Adams, R. McC. 1981. Heartland of Cities: Surveys of Ancient Settlement and Land Use on the Central Floodplain of the Euphrates, Chicago: University of Chicago Press.

Alexander, S., S. Schneider, and K. Lagerquist. 1997. Ecosystem services:Interaction of Climate and Life. Pages 71–92 in G. Daily, editor. Nature's Services: Societal Dependence on Natural Ecosystems. Island Press, Washington, D.C.

Allegre, C. and S. Schneider. 1994. The evolution of the earth. Scientific American 271: 44–51.

Bawa, K. and M. Gadgil. 1997. Ecosystem services, subsistence economies and conservation of biodiversity. Pages 295–310 in G. Daily, editor. Nature's Services: Societal Dependence on Natural Ecosystems. Island Press, Washington, D.C.

Bormann, F., G. Likens, D. Fisher, and R. Pierce. 1968. Nutrient loss accelerated by clear-cutting of a forest ecosystem. Science 159: 882–884.

Buchmann, S.L. and G.P. Nabhan. 1996. The Forgotten Pollinators. Island Press, Washington, D.C.

Chanway, C. 1993. Biodiversity at risk: soil microflora. Pages 229–238 in M. A. Fenger, E. H. Miller, J. A. Johnson, and E. J. R. Williams, editors. Our Living Legacy: Proceedings of a Symposium on Biological Diversity. Royal British Columbia Museum, Victoria, Canada.

Cohen, J.E. and D. Tilman. 1996. Biosphere 2 and biodiversity: The lessons so far. Science 274: 1150–1151.

Colborn, T, D. Dumanoski, and J. P. Myers. 1996. Our Stolen Future. Dutton, New York.

Daily, G.C. 1997a. Introduction: What are ecosystem services? Pages 1–10 in G. Daily, editor. Nature's Services: Societal Dependence on Natural Ecosystems. Island Press, Washington, D.C.

Daily, G.C. 1997b. Valuing and safeguarding Earth's life support systems. Pages 365–374 in G. Daily, editor. Natures Services: Societal Dependence on Natural Ecosystems. Island Press, Washington, D.C.

Daily, G.C., PA. Matson, and P.M. Vitousek. 1997. Ecosystem services supplied by soil. Pages 113–132 in G. Daily, editor. Nature's Services: Societal Dependence on Natural Ecosystems. Island Press, Washington, D.C.

Daily, G.C. and P.R. Ehrlich. 1995. Population diversity and the biodiversity crisis. Pp. 41–51 in C. Perrings, K.G. Maler, C. Folke, C.S. Holling and B.O. Jansson (eds.), Biodiversity Conservation: Problems and Policies, Dordrecht, Kluwer Academic Press.

DeBach, P. 1974. Biological Control by Natural Enemies. Cambridge University Press, London.

Diamond, J. 1991. The Rise and Fall of the Third Chimpanzee. Radius, London. Dobson, A. 1995. Biodiversity and human health. Trends in Ecology and Evolution 10: 390–391.

Dowling, H.F. 1977. Fighting Infection. Harvard Univ. Press, Cambridge, MA.

Ehrlich, P.R. and A.H. Ehrlich. 1981. Extinction. Ballantine, New York.

Ehrlich, P.R., A.H. Ehrlich, and J.P Holdren. 1977. Ecoscience: Population, Resources, Environment. Freeman and Co., San Francisco.

Ewel, K. 1997. Water quality improvement: evaluation of an ecosystem service. Pages 329–344 in G. Daily, editor. Nature's Services: Societal Dependence on Natural Ecosystems. Island Press, Washington, D.C.

Farnsworth, N.R., O. Akerele, A.S. Bingel, D.D. Soejarto, and Z.-G. Guo. 1985. Medicinal plants in therapy. Bulletin of the World Health Organization 63: 965–981.

Felder, A. J. and D. M. Nickum. 1992. The 1991 economic impact of sport fishing in the United States. American Sportfishing Association, Alexandria, Virginia.

FAO (United Nations Food and Agriculture Organization). 1990. Soilless Culture for Horticultural Crop Production. Rome: FAO. Grifo, F. and J. Rosenthal, editors. 1997. Biodiversity and Human Health. Island Press, Washington, D.C.

Hall, D.O., F. Rosillo-Calle, R.H. Williams, and J. Woods. 1993. Bio-mass for energy: supply prospects. Pages 593–651 in T. Johansson, H. Kelly, A. Reddy, and R. Williams, editors. Renewable Energy: Sources for Fuels and Electricity. Island Press, Washington, D.C.

Heal, G. 1994. Formation of international environmental agreements. Pages 301–332 in C. Carraro, editor, Trade, Innovation, Environment. Kluwer Academic Publishers, Dordrecht.

Hillel, D. 1991. Out of the Earth: Civilization and the Life of the Soil. The Free Press, New York.

Holdren, J. P. 1991. "Report of the planning meeting on ecological effects of human activities." National Research Council, 11–12 October, Irvine, California, mimeo.

Holdren, J.P. and P.R. Ehrlich. 1974. Human population and the global environment. American Scientist 62: 282–292.

Hughes, J.B., G.C. Daily, and P.R. Ehrlich. In prep. The importance, extent, and extinction rate of global population diversity.

Ives, J. and B. Messerli. 1989. The Himalayan Dilemma: Reconciling Development and Conservation, London: Routledge.

Jenny, H. 1980. The Soil Resource. Springer-Verlag, New York.

Kaufman, L. and P Dayton. 1997. Impacts of marine resource extraction on ecosystem services and sustainability. Pages 275–293 in G. Daily, editor. Nature's Services: Societal Dependence on Natural Ecosystems. Island Press, Washington, D.C.

Kellert, S.R. and E.O. Wilson, editors. 1993. The Biophilia Hypothesis. Island Press, Washington, D.C.

Lanner, R.M. 1996. Made for Eachother: A Symbiosis of Birds and Pines. Oxford University Press, New York.

Lawton, J. and R. May, editors. 1995. Extinction Rates. Oxford University Press, Oxford.

Lee, K. 1985. Earthworms: Their Ecology and Relationships with Soils and Land Use. Academic Press, New York.

Leung, A.Y. and S. Foster. 1996. Encyclopedia of Common Natural Ingredients Used in Food, Drugs, and Cosmetics. John Wiley & Sons, Inc., New York.

McCormick, K.D., M.A. Deyrup, E.S. Menges, S.R. Wallace, J. Meinwald, and T. Eisner. 1993. Relevance of chemistry to conservation of isolated populations: the case of volatile leaf components of Dicerandra mints. Proc. Nat. Acad. Sci. USA 90: 7701–7705.

Myers, N. 1983. A Wealth of Wild Species. Westview Press, Boulder, CO. Myers, N. 1997. The worlds forests and their ecosystem services. Pages 215–235 in G. Daily, editor. Nature's Services: Societal Dependence on Natural Ecosystems. Island Press, Washington, D.C.

Nabhan, G.P and S.L. Buchmann. 1997. Pollination services: Biodiversity's direct link to world food stability. Pages 133–150 in G. Daily, editor. Nature's Services: Societal Dependence on Natural Ecosystems. Island Press, Washington, D.C.

National Research Council (NRC). 1989. Alternative Agriculture. National Academy Press, Washington, D.C.

National Research Council (NRC). 1992. Managing Global Genetic Resources: The U.S. National Plant Germplasm System. National Academy Press, Washington, D.C.

Naylor, R. and P Ehrlich. The value of natural pest control services in agriculture. Pages 151–174 in G. Daily, editor. Nature's Services: Societal Dependence on Natural Ecosystems. Island Press, Washington, D.C.

Oldeman, L., V. van Engelen, and J. Pulles. 1990. "The extent of human-induced soil degradation, Annex 5" of L. R. Oldeman, R. T. A. Hakkeling, and W. G. Sombroek, World Map of the Status of Human-Induced Soil Degradation: An Explanatory Note, rev. 2d ed. Wageningen: International Soil Reference and Information Centre.

Overgaard-Nielsen, C. 1955. Studies on enchytraeidae 2: Field studies. Natura Jutlandica 4: 5–58.

Peterson, C.H. and J. Lubchenco. 1997. On the value of marine ecosystem services to society. Pages 177–194 in G. Daily, editor. Nature's Services: Societal Dependence on Natural Ecosystems. Island Press, Washington, D.C.

Pimentel, D., C. Harvey, P. Resosudarmo, K. Sinclair, D. Kurz, M. McNair, S. Crist, L. Shpritz, L. Fitton, R. Saffouri, and R. Blair. 1995. "Environmental and economic costs of soil erosion and conservation benefits." Science 267: 1117–1123.

Pimentel, D., L. McLaughlin, A. Zepp, B. Lakitan, T. Kraus, P. Kleinman, F. Vancini, W. Roach, E. Graap, W. Keeton, and G. Selig. 1989. Environmental and economic impacts of reducing U.S. agricultural pesticide use. Handbook of Pest Management in Agriculture 4: 223–278.

Postel, S. and S. Carpenter. 1997. Freshwater ecosystem services. Pages 195–214 in G. Daily, editor. Nature's Services: Societal Dependence on Natural Ecosystems. Island Press, Washington, D.C. Prescott-Allen, R. and C. Prescott-Allen. 1990. How many plants feed the world? Conservation Biology 4: 365–374.

Principe, P.P. 1989. The economic significance of plants and their constituents as drugs. Pages 1–17 in H. Wagner, H. Hikino, and N.R. Farnsworth, editors. Economic and Medicinal Plant Research, Vol. 3, Academic Press, London.

Reid, W.V., S.A. Laird, C.A. Meyer, R. Gamez, A. Sittenfeld, D. Janzen, M. Gollin, and C. Juma. 1996. Biodiversity prospecting. Pages 142–173 in M. Balick, E. Elisabetsky, and S. Laird, editors. Medicinal Resources of the Tropical Forest: Biodiversity and Its Importance to Human Health. Columbia Univ. Press, New York.

Rouatt, J. and H. Katznelson. 1961. A study of bacteria on the root surface and in the rhizosphere soil of crop plants. J. Applied Bacteriology 24: 164–171.

Sala, O.E. and J.M. Paruelo. 1997. Ecosystem services in grasslands. Pages 237–252 in G. Daily, editor. Nature's Services: Societal Dependence on Natural Ecosystems. Island Press, Washington, D.C.

Salati, E. 1987. The forest and the hydrological cycle. Pages 273–294 in R. Dickinson, editor. The Geophysiology of Amazonia. John Wiley and Sons, New York.

Schlesinger, W. 1991. Biogeochemistry: An Analysis of Global Change. Academic Press, San Diego.

Schneider, S. and P. Boston, editors. 1991. Scientists on Gaia. MIT Press, Boston.

Schneider, S. and R. Londer. 1984. The Coevolution of Climate and Life. Sierra Club Books, San Francisco.

Shiklomanov, I.A. 1993. World fresh water resources. Pp. 13–24 in P. Gleick, editor. Water in Crisis: A Guide to the World's Fresh Water Resources. Oxford University Press, New York.

Shukla, J., C. Nobre, and P Sellers. 1990. Amazon deforestation and climate change. Science. 247: 1322–1325.

Tilman, D. 1997. Biodiversity and ecosystem functioning. Pages 93–112 in G. Daily, editor. Nature's Services: Societal Dependence on Natural Ecosystems. Island Press, Washington, D.C.

Thompson, J.N. 1994. The Coevolutionary Process. Chicago Univ. Press, Chicago. United Nations Food and Agriculture Organization (UNFAO). 1993. Marine Fisheries and the Law of the Sea: A Decade of Change. Fisheries Circular No. 853, Rome.

United Nations Food and Agriculture Organization (UNFAO). 1994. FAO Yearbook of Fishery Statistics. Volume 17.

United States Department of Agriculture (USDA). 1993. World Agriculture: Trends and Indicators, 1970–91. Washington, DC: USDA.

Vitousek, P., P. Ehrlich, A. Ehrlich, and P. Matson. 1986. Human appropriation of the products of photosynthesis. BioScience 36: 368–373.

Vitousek, P., J. Aber, R. Howarth, G. Likens, P. Matson, D. Schindler, W. Schlesinger, and D. Tilman. 1997. Human alteration of the global nitrogen cycle: causes and consequences.

Wilson, E.O. 1987. The little things that run the world: The importance and conservation of invertebrates. Conservation Biology 1: 344–346. Wilson, E.O. 1989. Threats to biodiversity. Scientific American Sept: 108–116.

World Bank. 1991. Agricultural Biotechnology: The Next Green Revolution? World Bank Technical Paper no. 133. Washington, D.C.

World Resources Institute (WRI). 1994. World Resources: A Guide to the Global Environment. Oxford University Press, Oxford.

Critical Thinking

1. If, despite an investment of more than $200 million and the efforts of some of the most learned scientists in the world, we could not replicate the services provided by ecosystems in the Biosphere 2 project, how can we expect to maintain the life-support systems of Earth?
2. Identify 17 human activities that have harmful effects on ecosystems and their services. Rank order them according to the degree of their harmfulness. What do you think would be the findings if an analysis comparable to the previous reading were conducted; that is, instead of calculating the cost of ecosystem services, calculate the cost of human activities?
3. How important is it that humans maintain a certain portion of the planet where they do not dominate it?

How Have Ecosystems Changed?

MILLENNIUM ECOSYSTEM ASSESSMENT

Ecosystem Structure

The structure of the world's ecosystems changed more rapidly in the second half of the twentieth century than at any time in recorded human history, and virtually all of Earth's ecosystems have now been significantly transformed through human actions. The most significant change in the structure of ecosystems has been the transformation of approximately one quarter (24 percent) of Earth's terrestrial surface to cultivated systems. (See Box 1) More land was converted to cropland in the 30 years after 1950 than in the 150 years between 1700 and 1850.

Between 1960 and 2000, reservoir storage capacity quadrupled; as a result, the amount of water stored behind large dams is estimated to be three to six times the amount held by natural river channels (this excludes natural lakes). In countries for which sufficient multiyear data are available (encompassing more than half of the present-day mangrove area), approximately 35 percent of mangroves were lost in the last two decades. Roughly 20 percent of the world's coral reefs were lost and an additional 20 percent degraded in the last several decades of the twentieth century. Box 1 summarizes important characteristics and trends in different ecosystems.

Although the most rapid changes in ecosystems are now taking place in developing countries, industrial countries historically experienced comparable rates of change. Croplands expanded rapidly in Europe after 1700 and in North America and the former Soviet Union particularly after 1850. Roughly 70 percent of the original temperate forests and grasslands and Mediterranean forests had been lost by 1950, largely through conversion to agriculture. Historically, deforestation has been much more intensive in temperate regions than in the tropics, and Europe is the continent with the smallest fraction of its original forests remaining. However, changes prior to the industrial era seemed to occur at much slower rates than current transformations.

The ecosystems and biomes that have been most significantly altered globally by human activity include marine and freshwater ecosystems, temperate broadleaf forests, temperate grasslands, Mediterranean forests, and tropical dry forests. Within marine systems, the world's demand for food and animal feed over the last 50 years has resulted in fishing pressure so strong that the biomass of both targeted species and those caught incidentally (the "bycatch") has been reduced in much of the world to one tenth of the levels prior to the onset of industrial fishing. Globally, the degradation of fisheries is also reflected in the fact that the fish being harvested are increasingly coming from the less valuable lower trophic levels as populations of higher trophic level species are depleted.

Freshwater ecosystems have been modified through the creation of dams and through the withdrawal of water for human use. The construction of dams and other structures along rivers has moderately or strongly affected flows in 60 percent of the large river systems in the world. Water removal for human uses has reduced the flow of several major rivers, including the Nile, Yellow, and Colorado Rivers, to the extent that they do not always flow to the sea. As water flows have declined, so have sediment flows, which are the source of nutrients important for the maintenance of estuaries. Worldwide, although human activities have increased sediment flows in rivers by about 20 percent, reservoirs and water diversions prevent about 30 percent of sediments from reaching the oceans, resulting in a net reduction of sediment delivery to estuaries of roughly 10 percent.

Within terrestrial ecosystems, more than two thirds of the area of two of the world's 14 major terrestrial biomes (temperate grasslands and Mediterranean forests) and more than half of the area of four other biomes (tropical dry forests, temperate broadleaf forests, tropical grassland, and flooded grasslands) had been converted (primarily to agriculture) by 1990. Among the major biomes, only tundra and boreal forests show negligible levels of loss and conversion, although they have begun to be affected by climate change.

Globally, the rate of conversion of ecosystems has begun to slow largely due to reductions in the rate of expansion of cultivated land, and in some regions (particularly in temperate zones) ecosystems are returning to conditions and species compositions similar to their pre-conversion states. Yet rates of ecosystem conversion remain high or are increasing for specific ecosystems and regions. Under the aegis of the MA, the first systematic examination of the status and trends in terrestrial and coastal land cover was carried out using global and regional datasets. Opportunities for further expansion of cultivation are diminishing in many regions of the world as most of the land well-suited for intensive agriculture has been converted to cultivation. Increased agricultural productivity is also diminishing the need for agricultural expansion.

As a result of these two factors, a greater fraction of land in cultivated systems (areas with at least 30 percent of land cultivated) is actually being cultivated, the intensity of cultivation of land is increasing, fallow lengths are decreasing, and management practices are shifting from monocultures to polycultures.

Box 1
Characteristics of the World's Ecological Systems

We report assessment findings for 10 categories of the land and marine surface, which we refer to as "systems": forest, cultivated, dryland, coastal, marine, urban, polar, inland water, island, and mountain. Each category contains a number of ecosystems. However, ecosystems within each category share a suite of biological, climatic, and social factors that tend to be similar within categories and differ across categories. The MA reporting categories are not spatially exclusive; their areas often overlap. For example, transition zones between forest and cultivated lands are included in both the forest system and cultivated system reporting categories. These reporting categories were selected because they correspond to the regions of responsibility of different government ministries (such as agriculture, water, forestry, and so forth) and because they are the categories used within the Convention on Biological Diversity.

Marine, Coastal, and Island Systems

- Marine systems are the world's oceans. For mapping purposes, the map shows ocean areas where the depth is greater than 50 meters. Global fishery catches from marine systems peaked in the late 1980s and are now declining despite increasing fishing effort.
- Coastal systems refer to the interface between ocean and land, extending seawards to about the middle of the continental shelf and inland to include all areas strongly influenced by proximity to the ocean. The map shows the area between 50 meters below mean sea level and 50 meters above the high tide level or extending landward to a distance 100 kilometers from shore. Coastal systems include coral reefs, intertidal zones, estuaries, coastal aquaculture, and seagrass communities. Nearly half of the world's major cities (having more than 500,000 people) are located within 50 kilometers of the coast, and coastal population densities are 2.6 times larger than the density of inland areas. By all commonly used measures, the human well-being of coastal inhabitants is on average much higher than that of inland communities.
- Islands are lands (both continental and oceanic) isolated by surrounding water and with a high proportion of coast to hinterland. For mapping purposes, the MA uses the ESRI ArcWorld Country Boundary data-set, which contains nearly 12,000 islands. Islands smaller than 1.5 hectares are not mapped or included in the statistics. The largest island included is Greenland. The map includes islands within 2 kilometers of the mainland (e.g., Long Island in the United States), but the statistics provided for island systems in this report exclude these islands. Island states, together with their exclusive economic zones, cover 40 percent of the world's oceans. Island systems are especially sensitive to disturbances, and the majority of recorded extinctions have occurred on island systems, although this pattern is changing, and over the past 20 years as many extinctions have occurred on continents as on islands.

Urban, Dryland, and Polar Systems

- Urban systems are built environments with a high human density. For mapping purposes, the MA uses known human settlements with a population of 5,000 or more, with boundaries delineated by observing persistent night-time lights or by inferring areal extent in the cases where such observations are absent. The world's urban population increased from about 200 million in 1900 to 2.9 billion in 2000, and the number of cities with populations in excess of 1 million increased from 17 in 1900 to 388 in 2000.
- Dryland systems are lands where plant production is limited by water availability; the dominant human uses are large mammal herbivory, including livestock grazing, and cultivation. The map shows drylands as defined by the U.N. Convention to Combat Desertification, namely lands where annual precipitation is less than two thirds of potential evapo-transpiration—from dry subhumid areas (ratio ranges 0.50–0.65) through semiarid, arid, and hyperarid (ratio <0.05), but excluding polar areas. Drylands include cultivated lands, scrublands, shrublands, grasslands, savannas, semi-deserts, and true deserts. Dryland systems cover about 41 percent of Earth's land surface and are inhabited by more than 2 billion people (about one third of the total population). Croplands cover approximately 25 percent of drylands, and dryland rangelands support approximately 50 percent of the world's livestock. The current socioeconomic condition of people in dryland systems, of which about 90 percent are in developing countries, is worse than in other areas. Fresh water availability in drylands is projected to be further reduced from the current average of 1,300 cubic meters per person per year in 2000, which is already below the threshold of 2,000 cubic meters required for minimum human well-being and sustainable development. Approximately 10–20 percent of the world's drylands are degraded (*medium certainty*).
- Polar systems are high-latitude systems frozen for most of the year, including ice caps, areas underlain by permafrost, tundra, polar deserts, and polar coastal areas. Polar systems do not include high-altitude cold systems in low latitudes. Temperature in polar systems is on average warmer now than at any time in the last 400 years, resulting in widespread thaw of permafrost and reduction of sea ice. Most changes in feedback processes that occur in polar regions magnify trace gas–induced global warming trends and reduce the capacity of polar regions to act as a cooling system for Earth. Tundra constitutes the largest natural wetland in the world.

Forest Systems

- Forest systems are lands dominated by trees; they are often used for timber, fuel-wood, and non-wood forest products. The map shows areas with a canopy cover of at least 40 percent by woody plants taller than five meters. Forests include temporarily cut-over forests and plantations but exclude orchards and agroforests where the main products are food crops. The global area of forest systems has been reduced by one half over the past three centuries. Forests have effectively disappeared in 25 countries, and another 29 have lost more than 90 percent of their forest cover. Forest systems are associated with the regulation of 57 percent of total water runoff. About 4.6 billion people depend for all or some of their water on supplies from forest systems. From 1990 to 2000, the global area of temperate forest increased by almost 3 million hectares per year, while deforestation in the tropics occured at an average rate exceeding 12 million hectares per year over the past two decades.

Cultivated Systems

- Cultivated systems are lands dominated by domesticated species and used for and substantially changed by crop, agroforestry, or aquaculture production. The map shows areas in which at least 30 percent by area of the landscape comes under cultivation in any particular year. Cultivated systems, including croplands, shifting cultivation, confined livestock production, and freshwater aquaculture, cover approximately 24 percent of total land area. In the last two decades, the major areas of cropland expansion were located in Southeast Asia, parts of South Asia, the Great Lakes region of eastern Africa, the Amazon Basin, and the U.S. Great Plains. The major decreases of cropland occurred in the southeastern United States, eastern China, and parts of Brazil and Argentina. Most of the increase in food demand of the past 50 years has been met by intensification of crop, livestock, and aquaculture systems rather than expansion of production area. In developing countries, over the period 1961–99 expansion of harvested land contributed only 29 percent to growth in crop production, although in sub-Saharan Africa expansion accounted for two thirds of growth in production. Increased yields of crop production systems have reduced the pressure to convert natural ecosystems into cropland, but intensification has increased pressure on inland water ecosystems, generally reduced biodiversity within agricultural landscapes, and it requires higher energy inputs in the form of mechanization and the production of chemical fertilizers. Cultivated systems provide only 16 percent of global runoff, although their close proximity to humans means that about 5 billion people depend for all or some of their water on supplies from cultivated systems. Such proximity is associated with nutrient and industrial water pollution.

Inland Water and Mountain Systems

- Inland water systems are permanent water bodies inland from the coastal zone and areas whose properties and use are dominated by the permanent, seasonal, or intermittent occurrence of flooded conditions. Inland waters include rivers, lakes, floodplains, reservoirs, wetlands, and inland saline systems. (Note that the wetlands definition used by the Ramsar Convention includes the MA inland water and coastal system categories.) The biodiversity of inland waters appears to be in a worse condition than that of any other system, driven by declines in both the area of wetlands and the water quality in inland waters. It is *speculated* that 50 percent of inland water area (excluding large lakes) has been lost globally. Dams and other infrastructure fragment 60 percent of the large river systems in the world.
- Mountain systems are steep and high lands. The map is based on elevation and, at lower elevations, a combination of elevation, slope, and local topography. Some 20 percent (or 1.2 billion) of the world's people live in mountains or at their edges, and half of humankind depends, directly or indirectly, on mountain resources (largely water). Nearly all—90 percent—of the 1.2 billion people in mountains live in countries with developing or transition economies. In these countries, 7 percent of the total mountain area is currently classified as cropland, and people are often highly dependent on local agriculture or livestock production. About 4 billion people depend for all or some of their water on supplies from mountain systems. Some 90 million mountain people—almost all those living above 2,500 meters—live in poverty and are considered especially vulnerable to food insecurity.

Since 1950, cropland areas have stabilized in North America and decreased in Europe and China. Cropland areas in the Former Soviet Union have decreased since 1960. Within temperate and boreal zones, forest cover increased by approximately 2.9 million hectares per year in the 1990s, of which approximately 40 percent was forest plantations. In some cases, rates of conversion of ecosystems have apparently slowed because most of the ecosystem has now been converted, as is the case with temperate broadleaf forests and Mediterranean forests.

Ecosystem Processes

Ecosystem processes, including water, nitrogen, carbon, and phosphorus cycling, changed more rapidly in the second half of the twentieth century than at any time in recorded human history. Human modifications of ecosystems have changed not only the structure of the systems (such as what habitats or species are present in a particular location), but their processes and functioning as well. The capacity of ecosystems to provide services derives directly from the operation of natural biogeochemical cycles that in some cases have been significantly modified.

- *Water Cycle:* Water withdrawals from rivers and lakes for irrigation or for urban or industrial use doubled between 1960 and 2000. (Worldwide, 70 percent of water use is for agriculture). Large reservoir construction has doubled or tripled the residence time of river water—the average time, that is, that a drop of water takes to reach the sea. Globally, humans use slightly more than 10 percent of the available renewable freshwater supply through household, agricultural, and industrial activities, although in some regions such as the Middle East and North Africa, humans use 120 percent of renewable supplies (the excess is obtained through the use of groundwater supplies at rates greater than their rate of recharge).

- *Carbon Cycle:* Since 1750, the atmospheric concentration of carbon dioxide has increased by about 34 percent (from about 280 parts per million to 376 parts per million in 2003). Approximately 60 percent of that increase (60 parts per million) has taken place since 1959. The effect of changes in terrestrial ecosystems on the carbon cycle reversed during the last 50 years. Those ecosystems were on average a net source of CO_2 during the nineteenth and early twentieth centuries (primarily due to deforestation, but with contributions from degradation of agricultural, pasture, and forestlands) and became a net sink sometime around the middle of the last century (although carbon losses from land use change continue at high levels) (*high certainty*). Factors contributing to the growth of the role of ecosystems in carbon sequestration include afforestation, reforestation, and forest management in North America, Europe, China, and other regions; changed agriculture practices; and the fertilizing effects of nitrogen deposition and increasing atmospheric CO_2 (*high certainty*).
- *Nitrogen Cycle:* The total amount of reactive, or biologically available, nitrogen created by human activities increased ninefold between 1890 and 1990, with most of that increase taking place in the second half of the century in association with increased use of fertilizers. A recent study of global human contributions to reactive nitrogen flows projected that flows will increase from approximately 165 teragrams of reactive nitrogen in 1999 to 270 teragrams in 2050, an increase of 64 percent. More than half of all the synthetic nitrogen fertilizer (which was first produced in 1913) ever used on the planet has been used since 1985. Human activities have now roughly doubled the rate of creation of reactive nitrogen on the land surfaces of Earth. The flux of reactive nitrogen to the oceans increased by nearly 80 percent from 1860 to 1990, from roughly 27 teragrams of nitrogen per year to 48 teragrams in 1990. (This change is not uniform over Earth, however, and while some regions such as Labrador and Hudson's Bay in Canada have seen little if any change, the fluxes from more developed regions such as the northeastern United States, the watersheds of the North Sea in Europe, and the Yellow River basin in China have increased ten- to fifteenfold.)
- *Phosphorus Cycle:* The use of phosphorus fertilizers and the rate of phosphorus accumulation in agricultural soils increased nearly threefold between 1960 and 1990, although the rate has declined somewhat since that time. The current flux of phosphorus to the oceans is now triple that of back-ground rates (approximately 22 teragrams of phosphorus per year versus the natural flux of 8 teragrams).

Species

A change in an ecosystem necessarily affects the species in the system, and changes in species affect ecosystem processes.

The distribution of species on Earth is becoming more homogenous. By homogenous, we mean that the differences between the set of species at one location on the planet and the set at another location are, on average, diminishing. The natural process of evolution, and particularly the combination of natural barriers to migration and local adaptation of species, led to significant differences in the types of species in ecosystems in different regions. But these regional differences in the planet's biota are now being diminished.

Two factors are responsible for this trend. First, the extinction of species or the loss of populations results in the loss of the presence of species that had been unique to particular regions. Second, the rate of invasion or introduction of species into new ranges is already high and continues to accelerate apace with growing trade and faster transportation. For example, a high proportion of the roughly 100 non-native species in the Baltic Sea are native to the North American Great Lakes, and 75 percent of the recent arrivals of about 170 non-native species in the Great Lakes are native to the Baltic Sea. When species decline or go extinct as a result of human activities, they are replaced by a much smaller number of expanding species that thrive in human-altered environments. One effect is that in some regions where diversity has been low, the biotic diversity may actually increase—a result of invasions of non-native forms. (This is true in continental areas such as the Netherlands as well as on oceanic islands.)

Across a range of taxonomic groups, either the population size or range or both of the majority of species is currently declining. Studies of amphibians globally, African mammals, birds in agricultural lands, British butterflies, Caribbean corals, and fishery species show the majority of species to be declining in range or number. Exceptions include species that have been protected in reserves, that have had their particular threats (such as overexploitation) eliminated, or that tend to thrive in landscapes that have been modified by human activity.

Between 10 percent and 30 percent of mammal, bird, and amphibian species are currently threatened with extinction *(medium to high certainty),* based on IUCN–World Conservation Union criteria for threats of extinction. As of 2004, comprehensive assessments of every species within major taxonomic groups have been completed for only three groups of animals (mammals, birds, and amphibians) and two plant groups (conifers and cycads, a group of evergreen palm-like plants). Specialists on these groups have categorized species as "threatened with extinction" if they meet a set of quantitative criteria involving their population size, the size of area in which they are found, and trends in population size or area. (Under the widely used IUCN criteria for extinction, the vast majority of species categorized as "threatened with extinction" have approximately a 10 percent chance of going extinct within 100 years, although some long-lived species will persist much longer even though their small population size and lack of recruitment means that they have a very high likelihood of extinction.) Twelve percent of bird species, 23 percent of mammals, and 25 percent of conifers are currently threatened with extinction; 32 percent of amphibians are threatened with extinction, but information is more limited and this may be an underestimate. Higher levels of threat have been found in the cycads, where 52 percent are threatened. In general, freshwater habitats tend to have the highest proportion of threatened species.

Over the past few hundred years, humans have increased the species extinction rate by as much as 1,000 times background rates typical over the planet's history *(medium certainty)*. Extinction is a natural part of Earth's history. Most estimates of the total number of species today lie between 5 million and 30 million, although the overall total could be higher than 30 million if poorly known groups such as deep-sea organisms, fungi, and microorganisms including parasites have more species than currently estimated. Species present today only represent 2–4 percent of all species that have ever lived. The fossil record appears to be punctuated by five major mass extinctions, the most recent of which occurred 65 million years ago.

The average rate of extinction found for marine and mammal fossil species (excluding extinctions that occurred in the five major mass extinctions) is approximately 0.1–1 extinctions per million species per year. There are approximately 100 documented extinctions of birds, mammal, and amphibians over the past 100 years, a rate 50–500 times higher than background rates. Including possibly extinct species, the rate is more than 1,000 times higher than background rates. Although the data and techniques used to estimate current extinction rates have improved over the past two decades, significant uncertainty still exists in measuring current rates of extinction because the extent of extinctions of undescribed taxa is unknown, the status of many described species is poorly known, it is difficult to document the final disappearance of very rare species, and there are time lags between the impact of a threatening process and the resulting extinction.

Genes

Genetic diversity has declined globally, particularly among cultivated species. The extinction of species and loss of unique populations has resulted in the loss of unique genetic diversity contained by those species and populations. For wild species, there are few data on the actual changes in the magnitude and distribution of genetic diversity (C4.4), although studies have documented declining genetic diversity in wild species that have been heavily exploited. In cultivated systems, since 1960 there has been a fundamental shift in the pattern of intra-species diversity in farmers' fields and farming systems as the crop varieties planted by farmers have shifted from locally adapted and developed populations (landraces) to more widely adapted varieties produced through formal breeding systems (modern varieties). Roughly 80 percent of wheat area in developing countries and three quarters of the rice area in Asia is planted with modern varieties. (For other crops, such as maize, sorghum and millet, the proportion of area planted to modern varieties is far smaller.) The on-farm losses of genetic diversity of crops and livestock have been partially offset by the maintenance of genetic diversity in seed banks.

Critical Thinking

1. What is the most significant way in which ecosystems have been changed?
2. Can humanity rely on the oceans to provide enough food for its growing numbers?
3. If dryland systems are areas "where plant production is limited by water availability," then why do people primarily use these areas for livestock grazing and farming?

How Have Ecosystems Services and Their Uses Changed?

MILLENNIUM ECOSYSTEM ASSESSMENT

Ecosystem services are the benefits provided by ecosystems. These include provisioning services such as food, water, timber, fiber, and genetic resources; regulating services such as the regulation of climate, floods, disease, and water quality as well as waste treatment; cultural services such as recreation, aesthetic enjoyment, and spiritual fulfillment; and supporting services such as soil formation, pollination, and nutrient cycling.

Human use of all ecosystem services is growing rapidly. Approximately 60 percent (15 out of 24) of the ecosystem services evaluated in this assessment (including 70 percent of regulating and cultural services) are being degraded or used unsustainably. Of 24 provisioning, cultural, and regulating ecosystem services for which sufficient information was available, the use of 20 continues to increase. The use of one service, capture fisheries, is now declining as a result of a decline in the quantity of fish, which in turn is due to excessive capture of fish in past decades. Two other services (fuelwood and fiber) show mixed patterns. The use of some types of fiber is increasing and others decreasing; in the case of fuelwood, there is evidence of a recent peak in use.

Humans have enhanced production of three ecosystem services–crops, livestock, and aquaculture–through expansion of the area devoted to their production or through technological inputs. Recently, the service of carbon sequestration has been enhanced globally, due in part to the re-growth of forests in temperate regions, although previously deforestation had been a net source of carbon emissions. Half of provisioning services (6 of 11) and nearly 70 percent (9 of 13) of regulating and cultural services are being degraded or used unsustainably.

- *Provisioning Services:* The quantity of provisioning ecosystem services such as food, water, and timber used by humans increased rapidly, often more rapidly than population growth although generally slower than economic growth, during the second half of the twentieth century. And it continues to grow. In a number of cases, provisioning services are being used at unsustainable rates. The growing human use has been made possible by a combination of substantial increases in the absolute amount of some services produced by ecosystems and an increase in the fraction used by humans. World population doubled between 1960 and 2000, from 3 billion to 6 billion people, and the global economy increased more than sixfold. During this time, food production increased by roughly two-and-a-half times (a 160 percent increase in food production between 1961 and 2003), water use doubled, wood harvests for pulp and paper tripled, and timber production increased by nearly 60 percent. (Food production increased fourfold in developing countries over this period.)

The sustainability of the use of provisioning services differs in different locations. However, the use of several provisioning services is unsustainable even in the global aggregate. The current level of use of capture fisheries (marine and freshwater) is not sustainable, and many fisheries have already collapsed. Currently, one quarter of important commercial fish stocks are overexploited or significantly depleted (*high certainty*). From 5 percent to possibly 25 percent of global freshwater use exceeds long-term accessible supplies and is maintained only through engineered water transfers or the overdraft of groundwater supplies (*low to medium certainty*). Between 15 percent and 35 percent of irrigation withdrawals exceed supply rates and are therefore unsustainable (*low to medium certainty*). Current agricultural practices are also unsustainable in some regions due to their reliance on unsustainable sources of water, harmful impacts caused by excessive nutrient or pesticide use, salinization, nutrient depletion, and rates of soil loss that exceed rates of soil formation.

- *Regulating Services:* Humans have substantially altered regulating services such as disease and climate regulation by modifying the ecosystem providing the service and, in the case of waste processing services, by exceeding the capabilities of ecosystems to provide the service. Most changes to regulating services are inadvertent results of actions taken to enhance the supply of provisioning services. Humans have substantially modified the climate regulation service of ecosystems—first through land use changes that contributed to increases in the amount of carbon dioxide and other greenhouse gases

Ecosystem Services

Ecosystem services are the benefits people obtain from ecosystems. These include provisioning, regulating, and cultural services that directly affect people and the supporting services needed to maintain other services. Many of the services listed here are highly interlinked. (Primary production, photosynthesis, nutrient cycling, and water cycling, for example, all involve different aspects of the same biological processes.)

Provisioning Services

These are the products obtained from ecosystems, including:

Food. This includes the vast range of food products derived from plants, animals, and microbes.

Fiber. Materials included here are wood, jute, cotton, hemp, silk, and wool.

Fuel. Wood, dung, and other biological materials serve as sources of energy.

Genetic resources. This includes the genes and genetic information used for animal and plant breeding and biotechnology.

Biochemicals, natural medicines, and pharmaceuticals. Many medicines, biocides, food additives such as alginates, and biological materials are derived from ecosystems.

Ornamental resources. Animal and plant products, such as skins, shells, and flowers, are used as ornaments, and whole plants are used for landscaping and ornaments.

Fresh water. People obtain fresh water from ecosystems and thus the supply of fresh water can be considered a provisioning service. Fresh water in rivers is also a source of energy. Because water is required for other life to exist, however, it could also be considered a supporting service.

Regulating Services

These are the benefits obtained from the regulation of ecosystem processes, including:

Air quality regulation. Ecosystems both contribute chemicals to and extract chemicals from the atmosphere, influencing many aspects of air quality.

Climate regulation. Ecosystems influence climate both locally and globally. At a local scale, for example, changes in land cover can affect both temperature and precipitation. At the global scale, ecosystems play an important role in climate by either sequestering or emitting greenhouse gases.

Water regulation. The timing and magnitude of runoff, flooding, and aquifer recharge can be strongly influenced by changes in land cover, including, in particular, alterations that change the water storage potential of the system, such as the conversion of wetlands or the replacement of forests with croplands or croplands with urban areas.

Erosion regulation. Vegetative cover plays an important role in soil retention and the prevention of landslides.

Water purification and waste treatment. Ecosystems can be a source of impurities (for instance, in fresh water) but also can help filter out and decompose organic wastes introduced into inland waters and coastal and marine ecosystems and can assimilate and detoxify compounds through soil and subsoil processes.

Disease regulation. Changes in ecosystems can directly change the abundance of human pathogens, such as cholera, and can alter the abundance of disease vectors, such as mosquitoes.

Pest regulation. Ecosystem changes affect the prevalence of crop and livestock pests and diseases.

Pollination. Ecosystem changes affect the distribution, abundance, and effectiveness of pollinators.

Natural hazard regulation. The presence of coastal ecosystems such as mangroves and coral reefs can reduce the damage caused by hurricanes or large waves.

Cultural Services

These are the nonmaterial benefits people obtain from ecosystems through spiritual enrichment, cognitive development, reflection, recreation, and aesthetic experiences, including:

Cultural diversity. The diversity of ecosystems is one factor influencing the diversity of cultures.

Spiritual and religious values. Many religions attach spiritual and religious values to ecosystems or their components.

Knowledge systems (traditional and formal). Ecosystems influence the types of knowledge systems developed by different cultures.

Educational values. Ecosystems and their components and processes provide the basis for both formal and informal education in many societies.

Inspiration. Ecosystems provide a rich source of inspiration for art, folklore, national symbols, architecture, and advertising.

Aesthetic values. Many people find beauty or aesthetic value in various aspects of ecosystems, as reflected in the support for parks, scenic drives, and the selection of housing locations.

Social relations. Ecosystems influence the types of social relations that are established in particular cultures. Fishing societies, for example, differ in many respects in their social relations from nomadic herding or agricultural societies.

Sense of place. Many people value the "sense of place" that is associated with recognized features of their environment, including aspects of the ecosystem.

Cultural heritage values. Many societies place high value on the maintenance of either historically important landscapes ("cultural landscapes") or culturally significant species.

Recreation and ecotourism. People often choose where to spend their leisure time based in part on the characteristics of the natural or cultivated landscapes in a particular area.

Supporting Services

Supporting services are those that are necessary for the production of all other ecosystem services. They differ from provisioning, regulating, and cultural services in that their impacts on people are often indirect or occur over a very long time, whereas changes in the other categories have relatively direct and short-term impacts on people. (Some services, like erosion regulation, can be categorized as both a supporting and a regulating service, depending on the time scale and immediacy of their impact on people.) These services include:

Soil Formation. Because many provisioning services depend on soil fertility, the rate of soil formation influences human well-being in many ways.

Photosynthesis. Photosynthesis produces oxygen necessary for most living organisms.

Primary production. The assimilation or accumulation of energy and nutrients by organisms.

Nutrient cycling. Approximately 20 nutrients essential for life, including nitrogen and phosphorus, cycle through ecosystems and are maintained at different concentrations in different parts of ecosystems.

Water cycling. Water cycles through ecosystems and is essential for living organisms.

Table 1 Trends in the Human Use of Ecosystem Services and Enhancement or Degradation of the Service around the Year 2000 (See end of table for legend.)

Service	Sub-category	Human Use[a]	Enhanced or Degraded[b]
Provisioning Services			
Food	Crops	▲	▲
	Livestock	▲	▲
	Capture fisheries	▼	▼
	Aquaculture	▲	▲
	Wild plant and animal products	NA	▼
Fiber	Timber	▲	+/–
	Cotton, hemp, silk	+/–	+/–
	Wood fuel	+/–	▼
Genetic resources		▲	▼
Biochemicals, natural medicines, and pharmaceuticals		▲	▼
Ornamental resources		NA	NA
Fresh water		▲	▼
Regulating Services			
Air quality regulation		▲	▼
Climate regulation	Global	▲	▲
	Regional and local	▲	▼
Water regulation		▲	+/–
Erosion regulation		▲	▼
Water purification and waste treatment		▲	▼
Disease regulation		▲	+/–
Pest regulation		▲	▼
Pollination		▲	▼[c]
Natural hazard regulation		▲	▼
Cultural Services			
Cultural diversity		NA	NA
Spiritual and religious values		▲	▼
Knowledge systems		NA	NA
Educational values		NA	NA
Inspiration		NA	NA
Aesthetic values		▲	▼
Social relations		NA	NA
Sense of place		NA	NA
Cultural heritage values		NA	NA
Recreation and ecotourism		▲	+/–
Supporting Services			
Soil formation		†	†

(continued)

Table 1 Trends in the Human Use of Ecosystem Services and Enhancement or Degradation of the Service around the Year 2000 (See end of table for legend.) *(continued)*

Service	Sub-category	Human Use[a]	Enhanced or Degraded[b]
Supporting Services			
Photosynthesis		†	†
Primary production		†	†
Nutrient cycling		†	†
Water cycling		†	†

[a]For provisioning services, human use increases if the human consumption of the service increases (e.g., greater food consumption); for regulating and cultural services, human use increases if the number of people affected by the service increases. The time frame is in general the past 50 years, although if the trend has changed within that time frame, the indicator shows the most recent trend.

[b]For provisioning services, we define enhancement to mean increased production of the service through changes in area over which the service is provided (e.g., spread of agriculture) or increased production per unit area. We judge the production to be degraded if the current use exceeds sustainable levels. For regulating and supporting services, enhancement refers to a change in the service that leads to greater benefits for people (e.g., the service of disease regulation could be improved by eradication of a vector known to transmit a disease to people). Degradation of a regulating and supporting services means a reduction in the benefits obtained from the service, either through a change in the service (e.g., mangrove loss reducing the storm protection benefits of an ecosystem) or through human pressures on the service exceeding its limits (e.g., excessive pollution exceeding the capability of ecosystems to maintain water quality). For cultural services, degradation refers to a change in the ecosystem features that decreases the cultural (recreational, aesthetic, spiritual, etc.) benefits provided by the ecosystem. The time frame is in general the past 50 years, although if the trend has changed within that time frame the indicator shows the most recent trend.

[c]*Low to medium certainty.* All other trends are *medium to high certainty.*

Legend:

▲ = Increasing (for human use column) or enhanced (for enhanced or degraded column)

▼ = Decreasing (for human use column) or degraded (for enhanced or degraded column)

+/– = Mixed (trend increases and decreases over past 50 years or some components/regions increase while others decrease)

NA = Not assessed within the MA. In some cases, the service was not addressed at all in the MA (such as ornamental resources), while in other cases the service was included but the information and data available did not allow an assessment of the pattern of human use of the service or the status of the service.

† = The categories of "human use" and "enhanced or degraded" do not apply for supporting services since, by definition, these services are not directly used by people. (Their costs or benefits would be double-counted if the indirect effects were included.) Changes in supporting services influence the supply of provisioning, cultural, or regulating services that are then used by people and may be enhanced or degraded.

such as methane and nitrous oxide in the atmosphere and more recently by increasing the sequestration of carbon dioxide (although ecosystems remain a net source of methane and nitrous oxide). Modifications of ecosystems have altered patterns of disease by increasing or decreasing habitat for certain diseases or their vectors (such as dams and irrigation canals that provide habitat for schistosomiasis) or by bringing human populations into closer contact with various disease organisms.

Changes to ecosystems have contributed to a significant rise in the number of floods and major wildfires on all continents since the 1940s. Ecosystems serve an important role in detoxifying wastes introduced into the environment, but there are intrinsic limits to that waste processing capability. For example, aquatic ecosystems "cleanse" on average 80 percent of their global incident nitrogen loading, but this intrinsic self-purification capacity varies widely and is being reduced by the loss of wetlands.

- *Cultural Services*: Although the use of cultural services has continued to grow, the capability of ecosystems to provide cultural benefits has been significantly diminished in the past century (C17). Human cultures are strongly influenced by ecosystems, and ecosystem change can have a significant impact on cultural identity and social stability. Human cultures, knowledge systems, religions, heritage values, social interactions, and the linked amenity services (such as aesthetic enjoyment, recreation, artistic and spiritual fulfillment, and intellectual development) have always been influenced and shaped by the nature of the ecosystem and ecosystem conditions. Many of these benefits are being degraded, either through changes to ecosystems (a recent rapid decline in the numbers of sacred groves and other such protected areas, for example) or through societal changes (such as the loss of languages or of traditional knowledge) that reduce people's recognition or appreciation of those cultural benefits. Rapid loss of culturally valued ecosystems and landscapes can contribute to social disruptions and societal marginalization. And there has been a decline in the quantity and quality of aesthetically pleasing natural landscapes.

Global gains in the supply of food, water, timber, and other provisioning services were often achieved in the past century despite local resource depletion and local restrictions on resource use by shifting production and harvest to new underexploited regions, sometimes considerable distances away. These options are diminishing. This trend is most distinct in the case of marine fisheries. As individual stocks have been depleted, fishing pressure has shifted to less exploited stocks. Industrial fishing fleets have also shifted to fishing further offshore and in deeper water to meet global demand. A variety of drivers related to market demand, supply, and government policies have influenced patterns of timber harvest. For example, international trade in forest products increases when a nation's forests no longer can meet demand or when policies have been established to restrict or ban timber harvest.

Although human demand for ecosystem services continues to grow in the aggregate, the demand for particular services in specific regions is declining as substitutes are developed. For example, kerosene, electricity, and other energy sources are increasingly being substituted for fuelwood (still the primary source of energy for heating and cooking for some 2.6 billion people). The substitution of a variety of other materials for wood (such as vinyl, plastics, and metal) has contributed to relatively slow growth in global timber consumption in recent years. While the use of substitutes can reduce pressure on specific ecosystem services, this may not always have positive net environmental benefits. Substitution of fuelwood by fossil fuels, for example, reduces pressure on forests and lowers indoor air pollution, but it may increase net greenhouse gas emissions. Substitutes are also often costlier to provide than the original ecosystem services.

Both the supply and the resilience of ecosystem services are affected by changes in biodiversity. Biodiversity is the variability among living organisms and the ecological complexes of which they are part. When a species is lost from a particular location (even if it does not go extinct globally) or introduced to a new location, the various ecosystem services associated with that species are changed. More generally, when a habitat is converted, an array of ecosystem services associated with the species present in that location is changed, often with direct and immediate impacts on people. Changes in biodiversity also have numerous indirect impacts on ecosystem services over longer time periods, including influencing the capacity of ecosystems to adjust to changing environments (medium certainty), causing disproportionately large and sometimes irreversible changes in ecosystem processes, influencing the potential for infectious disease transmission, and, in agricultural systems, influencing the risk of crop failure in a variable environment and altering the potential impacts of pests and pathogens (medium to high certainty).

The modification of an ecosystem to alter one ecosystem service (to increase food or timber production, for instance) generally results in changes to other ecosystem services as well. Trade-offs among ecosystem services are commonplace. For example, actions to increase food production often involve one or more of the following: increased water use, degraded water quality, reduced biodiversity, reduced forest cover, loss of forest products, or release of greenhouse gases. Frequent cultivation, irrigated rice production, livestock production, and burning of cleared areas and crop residues now release 1,600 ± 800 million tons of carbon per year in CO_2. Cultivation, irrigated rice production, and livestock production release between 106 million and 201 million tons of carbon per year in methane. About 70 percent of anthropogenic nitrous oxide gas emissions are attributable to agriculture, mostly from land conversion and nitrogen fertilizer use. Similarly, the conversion of forest to agriculture can significantly change flood frequency and magnitude, although the amount and direction of this impact is highly dependent on the characteristics of the local ecosystem and the nature of the land cover change.

Many trade-offs associated with ecosystem services are expressed in areas remote from the site of degradation. For example, conversion of forests to agriculture can affect water quality and flood frequency downstream of where the ecosystem change occurred. And increased application of nitrogen fertilizers to croplands can have negative impacts on coastal water quality. These trade-offs are rarely taken fully into account in decisionmaking, partly due to the sectoral nature of planning and partly because some of the effects are also displaced in time (such as long-term climate impacts).

The net benefits gained through actions to increase the productivity or harvest of ecosystem services have been less than initially believed after taking into account negative trade-offs. The benefits of resource management actions have traditionally been evaluated only from the standpoint of the service targeted by the management intervention. However, management interventions to increase any particular service almost always result in costs to other services. Negative trade-offs are commonly found between individual provisioning services and between provisioning services and the combined regulating, cultural, and supporting services and biodiversity. Taking the costs of these negative trade-offs into account reduces the apparent benefits of the various management interventions. For example:

- Expansion of commercial shrimp farming has had serious impacts on ecosystems, including loss of vegetation, deterioration of water quality, decline of capture fisheries, and loss of biodiversity.
- Expansion of livestock production around the world has often led to overgrazing and dryland degradation, rangeland fragmentation, loss of wildlife habitat, dust formation, bush encroachment, deforestation, nutrient overload through disposal of manure, and greenhouse gas emissions.
- Poorly designed and executed agricultural policies led to an irreversible change in the Aral Sea ecosystem. By 1998, the Aral Sea had lost more than 60 percent of its area and approximately 80 percent of its volume, and ecosystem-related problems in the region now include excessive salt content of major rivers, contamination of agricultural products with agrochemicals, high levels of turbidity in major water sources, high levels of pesticides and phenols in surface waters, loss of soil fertility, extinctions of species, and destruction of commercial fisheries.
- Forested riparian wetlands adjacent to the Mississippi river in the United States had the capacity to store about 60 days of river discharge. With the removal of the wetlands through canalization, leveeing, and draining, the remaining wetlands have a storage capacity of less than 12 days discharge, an 80 percent reduction in flood storage capacity.

However, positive synergies can be achieved as well when actions to conserve or enhance a particular component of an ecosystem or its services benefit other services or stakeholders. Agroforestry can meet human needs for food and fuel, restore soils, and contribute to biodiversity conservation. Intercropping can increase yields, increase biocontrol, reduce soil erosion, and reduce weed invasion in fields. Urban parks and other urban green spaces provide spiritual, aesthetic, educational, and recreational benefits as well as such services such as water purification, wildlife habitat, waste management, and carbon sequestration. Protection of natural forests for biodiversity conservation can also reduce carbon emissions and protect water supplies. Protection of wetlands can contribute to flood control and also help to remove pollutants such as phosphorus and nitrogen from the water. For example, it is estimated that the nitrogen load from the heavily polluted Illinois River basin to the Mississippi River could be cut in half by converting 7 percent of the basin back to wetlands. Positive synergies often exist among regulating, cultural, and supporting services and with biodiversity conservation.

Critical Thinking

1. Which of the four categories of services do you think is the most important? The least important?
2. Which of the four categories of services do you think has been the most adversely affected? Why?
3. For each up arrow assign a value of +1, for each down arrow assign a value of −1, for each +/− assign a value of 0. Calculate a numerical score for each category. How does the ranking of the categories correspond to your answer for question 2?

The Competitive Exclusion Principle

An idea that took a century to be born has implications in ecology, economics, and genetics.

GARRETT HARDIN

On 21 March 1944 the British Ecological Society devoted a symposium to the ecology of closely allied species. There were about 60 members and guests present. In the words of an anonymous reporter[1], "a lively discussion . . . centred about Gause's contention (1934) that two species with similar ecology cannot live together in the same place. . . . A distinct cleavage of opinion revealed itself on the question of the validity of Gause's concept. Of the main speakers, Mr. Lack, Mr. Elton and Dr. Varley supported the postulate. . . . Capt. Diver made a vigorous attack on Gause's concept, on the grounds that the mathematical and experimental approaches had been dangerously over simplified. . . . Pointing out the difficulty of defining 'similar ecology' he gave examples of many congruent species of both plants and animals apparently living and feeding together."

Thus was born what has since been called "Gause's principle." I say "born" rather than "conceived" in order to draw an analogy with the process of mammalian reproduction, where the moment of birth, of exposure to the external world, of becoming a fully legal entity, takes place long after the moment of conception. With respect to the principle here discussed, the length of the gestation period is a matter of controversy: 10 years, 12 years, 18 years, 40 years, or about 100 years, depending on whom one takes to be the father of the child.

Statement of the Principle

For reasons given below, I here refer to the principle by a name already introduced[2]—namely, the "competitive exclusion principle," or more briefly, the "exclusion principle." It may be briefly stated thus: *Complete competitors cannot coexist.* Many published discussions of the principle revolve around the ambiguity of the words used in stating it. The statement given above has been very carefully constructed: every one of the four words is ambiguous. This formulation has been chosen not out of perversity but because of a belief that it is best to use that wording which is least likely to hide the fact that we still do not comprehend the exact limits of the principle. For the present, I think the "threat of clarity"[3] is a serious one that is best minimized by using a formulation that is *admittedly* unclear; thus can we keep in the forefront of our minds the unfinished work before us. The wording given has, I think, another point of superiority in that it seems brutal and dogmatic. By emphasizing the very aspects that might result in our denial of them were they less plain we can keep the principle explicitly present in our minds until we see if its implications are, or are not, as unpleasant as our subconscious might suppose. The meaning of these somewhat cryptic remarks should become clear further on in the discussion.

What does the exclusion principle mean? Roughly this: that (i) if two noninterbreeding populations "do the same thing"—that is, occupy precisely the same ecological niche in Elton's sense[4]—and (ii) if they are "sympatric"—that is, if they occupy the same geographic territory—and (iii) if population *A* multiplies even the least bit faster than population *B*, then ultimately *A* will completely displace *B*, which will become extinct. This is the "weak form" of the principle. Always in practice a stronger form is used, based on the removal of the hypothetical character of condition (iii). We do this because we adhere to what may be called the axiom of inequality, which states that no two things or processes, in a real world, are precisely equal. This basic idea is probably as old as philosophy itself but is usually ignored, for good reasons. With respect to the *things* of the world the axiom often leads to trivial conclusions. One postage stamp is as good as another. But with respect to competing *processes* (for example, the multiplication rates of competing species) the axiom is never trivial, as has been repeatedly shown[5–7]. No difference in rates of multiplication can be so slight as to negate the exclusion principle.

Demonstrations of the formal truth of the principle have been given in terms of the calculus (*5*, *7*) and set theory[8]. Those to whom the mathematics does not appeal may prefer the following intuitive verbal argument (*2*, pp. 84–85), which is based on an economic analogy that is very strange economics but quite normal biology.

"Let us imagine a very odd savings bank which has only two depositors. For some obscure reason the bank pays one of the depositors 2 percent compound interest, while paying the other 2.01 percent. Let us suppose further (and here the analogy is really strained) that whenever the sum of the combined funds of the two depositors reaches two million dollars, the bank arbitrarily appropriates one million dollars of it, taking from each depositor in proportion to his holdings at that time. Then both accounts are allowed to grow until their sum again equals two million dollars, at which time the appropriation process is repeated. If this procedure is continued indefinitely, what will happen to the wealth of these two depositors? A little intuition shows us (and mathematics verifies) that the man who receives the greater rate of interest will, in time, have all the money, and the other man none (we assume a penny cannot be subdivided). No matter how small the difference between the two interest rates (so long as there is a difference) such will be the outcome.

"Translated into evolutionary terms, this is what competition in nature amounts to. The fluctuating limit of one million to two million represents the finite available wealth (food, shelter, etc.) of any natural environment, and the difference in interest rates represents the difference between the competing species in their efficiency in producing offspring. No matter how small this difference may be, one species will eventually replace the other. In the scale of geological time, even a small competitive difference will result in a rapid extermination of the less successful species. Competitive differences that are so small as to be unmeasurable by direct means will, by virtue of the compound-interest effect, ultimately result in the extinction of one competing species by another."

The Question of Evidence

So much for the theory. Is it true? This sounds like a straightforward question, but it hides subtleties that have, unfortunately, escaped a good many of the ecologists who have done their bit to make the exclusion principle a matter of dispute. There are many who have supposed that the principle is one that can be proved or disproved by empirical facts, among them[9–10] Gause himself. Nothing could be farther from the truth. The "truth" of the principle is and can be established only by theory, not being subject to proof or disproof by facts, as ordinarily understood. Perhaps this statement shocks you. Let me explain.

Suppose you believe the principle is true and set out to prove it empirically. First you find two noninterbreeding species that seem to have the same ecological characteristics. You bring them together in the same geographic location and await developments. What happens? Either one species extinguishes the other, or they coexist. If the former, you say, "The principle is proved." But if the species continue to coexist indefinitely, do you conclude the principle is false? Not at all. You decide there must have been some subtle difference in the ecology of the species that escaped you at first, so you look at the species again to try to see how they differ ecologically, all the while retaining your belief in the exclusion principle. As Gilbert, Reynoldson, and Hobart[10] dryly remarked, "There is . . . a danger of a circular process here. . . ."

Indeed there is. Yet the procedure can be justified, both empirically and theoretically. First, empirically. On this point our argument is essentially an acknowledgement of ignorance. When we think of mixing two similar species that have previously lived apart, we realize that it is hardly possible to know enough about species to be able to say, in advance, which one will exclude the other in free competition. Or, as Darwin, at the close of chapter 4 of his *Origin of Species*[11] put it: "It is good thus to try in imagination to give any one species an advantage over another. Probably in no single instance should we know what to do. This ought to convince us of our ignorance on the mutual relations of all organic beings: a conviction as necessary, as it is difficult to acquire."

How profound our ignorance of competitive situations is has been made painfully clear by the extended experiments of Thomas Park and his collaborators[12]. For more than a decade Park has put two species of flour beetles (*Tribolium confusum* and *T. castaneum*) in closed universes under various conditions. In every experiment the competitive exclusion principle is obeyed—one of the species is completely eliminated, *but it is not always the same one.* With certain fixed values for the environmental parameters the experimenters have been unable to control conditions carefully enough to obtain an invariable result. Just how one is to interpret this is by no means clear, but in any case Park's extensive body of data makes patent our immense ignorance of the relations of organisms to each other and to the environment, even under the most carefully controlled conditions.

The theoretical defense for adhering come-hell-or-high-water to the competitive exclusion principle is best shown by apparently changing the subject. Consider Newton's first law: "Every body persists in a state of rest or of uniform motion in a straight line unless compelled by external force to change that state." How would one verify this law, by itself? An observer might (in principle) test Newton's first law by taking up a station out in space somewhere and then looking at all the bodies around him. Would any of the bodies be in a state of rest except (by definition) himself? Probably not. More important, would any of the bodies in motion be moving in a straight line? *Not one.* (We assume that the observer makes errorless measurements.) For the law says, ". . . in a straight line unless compelled by external force to change . . . ," and in a world in which another law says that "every body attracts every other body with a force that is inversely proportional to the square of the distance between them . . . ," the phrase in the first law that begins with the words *unless compelled* clearly indicates the hypothetical character of the law. So long as there are no sanctuaries from gravitation in space, every body is always "compelled." Our observer would claim that any body at rest or moving in a straight line verified the law; he would likewise claim that bodies moving in not-straight lines verified the law, too. In other words, any attempt to test Newton's first law *by itself* would lead to a circular argument of the sort encountered earlier in considering the exclusion principle.

The point is this: We do not test isolated laws, one by one. What we test is a whole conceptual model[13]. From the model we make predictions; these we test against empirical data. When we find that a prediction is not verifiable we then set about modifying the model. There is no procedural rule to tell us which element of the model is best abandoned or changed. (The scientific response to the results of the Michelson-Morley experiment was not in any sense *determined.*) Esthetics plays a part in such decisions.

The competitive exclusion principle is one element in a system of ecological thought. We cannot test it directly, by itself. What the whole ecological system is, we do not yet know. One immediate task is to discover the system, to find its elements, to work out their interactions, and to make the system as explicit as possible. (*Complete* explicitness can never be achieved.) The works of Lotka[14], Nicholson [15–16], and MacArthur[17] are encouraging starts toward the elaboration of such a theoretical system.

The Issue of Eponymy

That the competitive exclusion principle is often called "Gause's principle" is one of the more curious cases of eponymy in science (like calling human oviducts "Fallopian tubes," after a man who was not the first to see them and who misconstrued their significance). The practice was apparently originated by the English ecologists, among whom David Lack has been most influential. Lack made a careful study of *Geospiza* and other genera of finches in the Galápagos Islands, combining observational studies on location with museum work at the California Academy of Sciences. How his ideas of ecological principles matured during the process is evident from a passage in his little classic, *Darwin's Finches*[18].

"Snodgrass concluded that the beak differences between the species of *Geospiza* are not of adaptive significance in regard to food. The larger species tend to eat rather larger seeds, but this he considered to be an incidental result of the difference in the size of their beaks. This conclusion was accepted by Gilford (1919), Gulick (1932), Swarth (1934) and formerly by myself (Lack, 1945). Moreover, the discovery . . . that the beak differences serve as recognition marks, provided quite a different reason for their existence, and thus strengthened the view that any associated differences in diet are purely incidental and of no particular importance.

"My views have now completely changed, through appreciating the force of Gause's contention that two species with similar ecology cannot live in the same region (Gause, 1934). This is a simple consequence of natural selection. If two species of birds occur together in the same habitat in the same region, eat the same types of food and have the same other ecological requirements, then they should compete with each other, and since the chance of their being equally well adapted is negligible, one of them should eliminate the other completely. Nevertheless, three species of ground-finch live together in the same habitat on the same Galapagos islands, and this also applies to two species of insectivorous tree-finch. There must be some factor which prevents these species from effectively competing."

Implicit in this passage is a bit of warm and interesting autobiography. It is touching to see how intellectual gratitude led Lack to name the exclusion principle after Gause, calling it, in successive publications, "Gause's contention," "Gause's hypothesis," and "Gause's principle." But the eponymy is scarcely justified. As Gilbert, Reynoldson, and Hobart point out (*10*, p. 312): "Gause . . . draws no general conclusions from his experiments, and moreover, makes no statement which resembles any wording of the hypothesis which has arisen bearing his name." Moreover, in the very publication in which he discussed the principle, Gause acknowledged the priority of Lotka in 1932 (*5*) and Volterra in 1926 (*6*). Gause gave full credit to these men, viewing his own work merely as an empirical testing of their theory—a quite erroneous view, as we have seen. How curious it is that the principle should be named after a man who did not state it clearly, who misapprehended its relation to theory, and who acknowledged the priority of others!

Recently Udvardy[19], in an admirably compact note, has pointed out that Joseph Grinnell, in a number of publications, expressed the exclusion principle with considerable clarity. In the earliest passage that Udvardy found, Grinnell, in 1904[20], said: "Every animal tends to increase at a geometric ratio, and is checked only by limit of food supply. It is only by adaptations to different sorts of food, or modes of food getting, that more than one species can occupy the same locality. Two species of approximately the same food habits are not likely to remain long enough evenly balanced in numbers in the same region. One will crowd out the other."

Udvardy quotes from several subsequent publications of Grinnell, from all of which it is quite clear that this well-known naturalist had a much better grasp of the exclusion principle than did Gause. Is this fact, however, a sufficiently good reason for now speaking (as Udvardy recommends) of "Grinnell's axiom?" On the basis of present evidence there seems to be justice in the proposal, but we must remember that the principle has already been referred to, in various publications, as "Gause's principle," the "Volterra-Gause principle," and the "Lotka-Volterra principle." What assurance have we that some diligent scholar will not tomorrow unearth a predecessor of Grinnell? And if this happens, should we then replace Grinnell's name with another's? Or should we, in a fine show of fairness, use all the names? (According to this system, the principle would, at present, be called the Grinnell-Volterra-Lotka-Gause-Lack principle—and, even so, injustice would be done to A. J. Nicholson, who, in his wonderful gold mine of unexploited aphorisms (*15*), wrote: "For the steady state [in the coexistence of two or more species] to exist, each species must possess some advantage over all other species with respect to some one, or group, of the control factors to which it is subject." This is surely a corollary of the exclusion principle.)

In sum, I think we may say that arguments for pinning an eponym on this idea are unsound. But it does need a name of some sort; its lack of one has been one of the reasons (though not the only one) why this basic principle has trickled out of the scientific consciousness after each mention during the last half century. Like Allee *et al.*[21] we should wish "to avoid further implementation of the facetious definition of ecology as being that phase of biology primarily abandoned to terminology." But, on the other side, it has been pointed out[22]: "Not many recorded facts are lost; the bibliographic apparatus of science is fairly equal to the problem of recording melting points, indices of refraction, etc., in such a way that they can be recalled when needed. Ideas, more subtle and more diffusely expressed present a bibliographic problem to which there is no present solution." To solve the bibliographic problem some sort of handle is needed for the idea here discussed; the name "the competitive exclusion principle" is correctly descriptive and will not be made obsolete by future library research.

The Exclusion Principle and Darwin

In our search for early statements of the principle we must not pass by the writings of Charles Darwin, who had so keen an appreciation of the ecological relationships of organisms. I have been unable to find any unambiguous references to the exclusion principle in the "Essays" of 1842 and 1844[23], but in the *Origin* itself there are several passages that deserve recording. All the following passages are quoted from the sixth edition (*11*).

"As the species of the same genus usually have, though by no means invariably, much similarity in habits and constitution, and always in structure, the struggle will generally be more severe between them, if they come into competition with each other, than between the species of distinct genera. We see this in the recent extension over parts of the United States of one species of swallow having caused the decrease of another species. The recent increase of the missel-thrush in parts of Scotland has caused the decrease of the song-thrush. How frequently we hear of one species of rat taking the place of another species under the most different climates! In Russia the small Asiatic cockroach has everywhere driven before it its great congener. In Australia the imported hive-bee is rapidly exterminating the small, stingless native bee. One species of charlock has been known to supplant another species; and so in other cases. We can dimly see why the competition should be most severe between allied forms, which fill nearly the same place in the economy of nature; but probably in no one case could we precisely say why one species has been victorious over another in the great battle of life" (p. 71).

"Owing to the high geometrical rate of increase of all organic beings, each area is already fully stocked with inhabitants; and it follows from this, that as the favored forms increase in number, so, generally, will the less favored decrease and become rare. Rarity, as geology tells us, is the precursor to extinction. We can see that any form which is represented by few individuals will run a good chance of utter extinction, during great fluctuations in the nature or the seasons, or from a temporary increase in the number of its enemies. But we may go further than this; for, as new forms are produced, unless we admit that specific forms can go on indefinitely increasing in number, many old forms must become extinct" (p. 102).

"From these several considerations I think it inevitably follows, that as new species in the course of time are formed through natural selection, others will become rarer and rarer, and finally extinct. The forms which stand in closest competition with those undergoing modification and improvement, will naturally suffer most. And we have seen in the chapter on the Struggle for Existence that it is the most closely-allied forms—varieties of the same species, and species of the same genus or related genera—which, from having nearly the same structure, constitution and habits, generally come into the severest competition with each other consequently, each new variety or species, during the progress of its formation, will generally press hardest on its nearest kindred, and tend to exterminate them. We see the same process of extermination among our domesticated productions, through the selection of improved forms by man. Many curious instances could be given showing how quickly new breeds of cattle, sheep and other animals, and varieties of flowers, take the place of older and inferior kinds. In Yorkshire, it is historically known that the ancient black cattle were displaced by the long-horns, and that these 'were swept away by the short-horns' (I quote the words of an agricultural writer) 'as if by some murderous pestilence'" (p. 103).

"For it should be remembered that the competition will generally be most severe between those forms which are most nearly related to each other in habits, constitution and structure. Hence all the intermediate forms between the earlier and later states, that is between the less and more improved states of the same species, as well as the original parent species itself, will generally tend to become extinct" (p. 114).

Those passages are, we must admit, typically Darwinian; by turn clear, obscure, explicit, cryptic, suggestive, they have in them all the characteristics that litterateurs seek in James Joyce. The complexity of Darwin's work, however, is unintended; it is the result partly of his limitations as an analytical thinker, but in part also it is the consequence of the magnitude, importance, and intrinsic difficulty of the ideas he grappled with. Darwin was not one to impose premature clarity on his writings.

Origins in Economic Theory?

In chapter 3 of *Nature and Man's Fate* I have argued for the correctness of John Maynard Keynes' view that the biological principle of natural selection is just a vast generalization of Ricardian economics. The argument is based on the isomorphism of theoretical systems in the two fields of human thought. Now that we have at last brought the competitive exclusion principle out of the periphery of our vision into focus on the *fovea centralis* it is natural to wonder if this principle, too, originated in economic thought. I think it is possible. At any rate, there is a passage by the French mathematician J. Bertrand[24], published in 1883, which shows an appreciation of the exclusion principle as it applies to economic matters. The passage occurs in a review of a book of Cournot, published much earlier, in which Cournot discussed the outcome of a struggle between two merchants engaged in selling identical products to the public. Bertrand says: "Their interest would be to unite or at least to agree on a common price so as to extract from the body of customers the greatest possible receipts. But this solution is avoided by Cournot who supposes that one of the competitors will lower his price in order to attract the buyers to himself, and that the other, trying to regain them, will set his price still lower. The two rivals will cease to pursue this path only when each has nothing more to gain by lowering his price.

"To this argument we make a peremptory objection. Given the hypothesis, no solution is possible: there is no limit to the lowering of the price. Whatever common price might be initially adopted, if one of the competitors were to lower the price unilaterally he would thereby attract the totality of the business to himself. . . ."

This passage clearly antedates Grinnell, Lack, *et al.*, but it comes long after the *Origin of Species.* Are there statements of the principle in the economic literature before Darwin? It would be nice to know. I have run across cryptic references to the work of Simonde de Sismondi (1773–1842) which imply that he had a glimpse of the exclusion principle, but I have not tracked them down. Perhaps some colleague in the history of economics will someday do so. If it is true that Sismondi understood the principle, this fact would add a nice touch to the interweaving of the history of ideas, for this famous Swiss economist was related to Emma Darwin by marriage; he plays a prominent role in the letters published under her name[25].

Utility of the Exclusion Principle

"The most important lesson to be learned from evolutionary theory," says Michael Scriven in a brilliant essay recently published[26], "is a negative one: the theory shows us what scientific explanations need not do. In particular it shows us that one cannot regard explanations as unsatisfactory when they are not such as to enable the event in question to have been predicted." The theory of evolution is not one with which we can predict exactly the future course of species formation and extinction; rather, the theory "explains" the past. Strangely enough, we take mental satisfaction in this ex post facto explanation. Scriven has done well in showing why we are satisfied.

Much of the theory of ecology fits Scriven's description of evolutionary theory. Told that two formerly separated species are to be introduced into the same environment and asked to predict exactly what will happen, we are generally unable to do so. We can only make certain predictions of this sort: either *A* will extinguish *B*, or *B* will extinguish *A;* or the two species are (or must become) ecologically different—that is, they must come to occupy different ecological niches. The general rule may be stated in either of two different ways: *Complete competitors cannot coexist*—as was said earlier; or, *Ecological differentiation is the necessary condition for coexistence.*

It takes little imagination to see that the exclusion principle, to date stated explicitly only in ecological literature, has applications in many academic fields of study. I shall now point out some of these, showing how the principle has been used (mostly unconsciously) in the past, and predicting some of its applications in the future.

Economics

The principle unquestionably plays an indispensable role in almost all economic thinking, though it is seldom explicitly stated. Any competitor knows that unrestrained competition will ultimately result in but one victor. If he is confident that he is that one, he may plump for "rugged individualism." If, on the other hand, he has doubts, then he will seek to restrain or restrict competition. He can restrain it by forming a cartel with his competitors, or by maneuvering the passage of "fair trade" laws. (Laboring men achieve a similar end—though the problem is somewhat different—by the formation of unions and the passage of minimum wage laws.) Or he may restrict competition by "ecological differentiation," by putting out a slightly different product (aided by restrictive patent and copyright laws). All this may be regarded as individualistic action.

Society as a whole may take action. The end of unrestricted competition is a monopoly. It is well known that monopoly breeds power which acts to insure and extend the monopoly; the system has "positive feedback" and hence is always a threat to those aspects of society still "outside" the monopoly. For this reason, men may, in the interest of "society" (rather than of themselves as individual competitors), band together to insure continued competition; this they do by passing anti-monopoly laws which prevent competition from proceeding to its "naturally" inevitable conclusion. Or "society" may permit monopolies but seek to remove the power element by the "socialization" of the monopoly (expropriation or regulation).

In their actions both as individuals and as groups, men show that they have an implicit understanding of the exclusion principle. But the failure to bring this understanding to the level of consciousness has undoubtedly contributed to the accusations of bad faith ("exploiter of the masses," "profiteer," "nihilist," "communist") that have characterized many of the interchanges between competing groups of society during the last century. F. A. Lange[27], thinking only of laboring men, spoke in most fervent terms of the necessity of waging a "struggle against the struggle for existence"—that is, a struggle against the unimpeded working out of the exclusion principle. Groups with interests opposed to those of "labor" are equally passionate about the same cause, though the examples they have in mind are different.

At the present time, one of the great fields of economics in which the application of the exclusion principle is resisted is international competition (nonbellicose). For emotional reasons, most discussion of problems in this field is restricted by the assumption (largely implicit) that Cournot's solution of the *intra*national competition problem is correct and applicable to the *inter*national problem. On the less frequent occasions when it is recognized that Bertrand's, not Cournot's, reasoning is correct, it is assumed that the consequences of the exclusion principle can be indefinitely postponed by a rapid and endless multiplication of "ecological niches" (largely unprotected though they are by copyright and patent). If some of these assumptions prove to be unrealistic, the presently fashionable stance toward tariffs and other restrictions of international competition will have to be modified.

Genetics

The application of the exclusion principle to genetics is direct and undeniable. The system of discrete alleles at the same gene locus competing for existence within a single population of organisms is perfectly isomorphic with the system of different species of organisms competing for existence in the same habitat and ecological niche. The consequences of this have frequently been acknowledged, usually implicitly, at least since J. B. S. Haldane's work of 1924[28]. But in this field, also, the consequences have often been denied, explicitly or otherwise, and again for emotional reasons. The denial has most often been coupled with a "denial" (in the psychological sense) of the priority of the inequality axiom. As a result of recent findings in the fields of physiological genetics and population genetics, particularly as concerns blood groups, the applicability of both the inequality axiom and the exclusion principle is rapidly becoming accepted. William C. Boyd has recorded, in a dramatic way[29], his escape from the bondage of psychological denial. The emotional restrictions of rational discussion in this field are immense. How "the struggle against the struggle for existence" will be waged in the field of human genetics promises to make the next decade of study one of the most exciting of man's attempts to accept the implications of scientific knowledge.

Ecology

Once one has absorbed the competitive exclusion principle into one's thinking it is curious to note how one of the most popular problems of evolutionary speculation is turned upside down. Probably most people, when first taking in the picture of historical evolution, are astounded at the number of species of plants and animals that have become extinct. From Simpson's gallant "guesstimates"[30], it would appear that from 99 to 99.975 percent of all species evolved are now extinct, the larger percentage corresponding to 3999 million species. This seems like a lot. Yet it is even more remarkable that there should live at any one time (for example, the present) as many as a million species, more or less competing with each other. Competition is avoided between some of the species that coexist in time by separation in space. In addition, however, there are many ecologically more or less similar species that coexist. Their continued existence is a thing to wonder at and to study. As Darwin said (*11*, p. 363)—and this is one more bit of evidence that he appreciated the exclusion principle—"We need not marvel at extinction; if we must marvel, let it be at our own presumption in imagining for a moment that we understand the many complex contingencies on which the existence of each species depends."

I think it is not too much to say that in the history of ecology—which in the broadest sense includes the science of economics and the study of population genetics—we stand at the threshold of a renaissance of understanding, a renaissance made possible by the explicit acceptance of the competitive exclusion principle. This principle, like much of the essential theory of evolution, has (I think) long been psychologically denied, as the penetrating study of Morse Peckham[31] indicates. The reason for the denial is the usual one: admission of the principle to conciousness is painful. [Evidence for such an assertion is, in the nature of the case, difficult to find, but for a single clear-cut example see the letter by Krogman[32].] It is not sadism or masochism that makes us urge that the denial be brought to an end. Rather, it is a love of the reality principle, and recognition that only those truths that are admitted to the conscious mind are available for use in making sense of the world. To assert the truth of the competitive exclusion principle is not to say that nature is and always must be, everywhere, "red in tooth and claw." Rather, it is to point out that *every* instance of apparent coexistence must be accounted for. Out of the study of all such instances will come a fuller knowledge of the many prosthetic devices of coexistence, each with its own costs and its own benefits. On such a foundation we may set about the task of establishing a science of ecological engineering.

References

1. Anonymous, *J. Animal Ecol.* **13,** 176 (1944).
2. G. Hardin, *Nature and Man's Fate* (Rine-hart, New York, 1959).
3. _____, *Am. J. Psychiat.* **114,** 392 (1957).
4. C. Elton, *Animal Ecology* (Macmillan, New York, 1927).
5. A. J. Lotka, *J. Wash. Acad. Sci.* **22,** 469 (1932).
6. V. Volterra, *Mem. reale accad. nazl. Lincei, Classe sci. fis. mat. e nat. ser. 6, No.* 2 (1926).
7. _____, *Leçons sur la Théorie Mathématique de la Lutte pour la Vie* (Gauthier-Villars, Paris, 1931).
8. G. E. Hutchinson, *Cold Spring Harbor Symposia Quant. Biol.* **22,** 415 (1957).
9. G. F. Gause, *The Struggle for Existence* (Williams and Wilkins, Baltimore, 1934); H. H. Ross, *Evolution* **11,** 113 (1957).
10. O. Gilbert, T. B. Reynoldson, J. Hobart, *J. Animal Ecol.* **21,** 310–312 (1952).
11. C. Darwin, *On the Origin of Species by Means of Natural Selection* (Macmillan, New York, new ed. 6, 1927).
12. T. Park and M. Lloyd, *Am. Naturalist* **89,** 235 (1955).
13. R. M. Thrall, C. H. Coombs, R. L. Davis, *Decision Processes* (Wiley, New York, 1954), pp. 22–23.
14. A. J. Lotka, *Elements of Physical Biology* (Williams and Wilkins, Baltimore, 1925).
15. A. J. Nicholson, *J. Animal Ecol.* **2,** suppl., 132–178 (1933).
16. _____, *Australian J. Zool.* **2,** 9 (1954).
17. R. H. MacArthur, *Ecology* **39,** 599 (1958).
18. D. Lack, *Darwin's Finches* (University Press, Cambridge, 1947).
19. M. F. D. Udvardy, *Ecology* **40,** 725 (1959).
20. J. Grinnell, *Auk* **21,** 364 (1904).
21. W. C. Allee, A. E. Emerson, O. Park, T. Park, K. P. Schmidt, *Principles of Ecology* (Saunders, Philadelphia, 1949).
22. G. Hardin, *Sci. Monthly* **70,** 178 (1950).
23. F. Darwin, *The Foundations of the Origin of Species* (University Press, Cambridge, 1909).
24. J. Bertrand, *J. savants* (Sept. 1883), pp. 499–508.
25. H. Litchfield, *Emma Darwin, A Century of Family Letters, 1792–1896* (Murray, London, 1915).
26. M. Scriven, *Science* **130,** 477 (1959).
27. F. A. Lange, *History of Materialism* (Harcourt Brace, New York, ed. 3, 1925).
28. J. B. S. Haldane, *Trans. Cambridge Phil. Soc.* **23,** 19 (1924).
29. W. C. Boyd, *Am. J. Human Genet.* **11,** 397 (1959).
30. G. G. Simpson, *Evolution* **6,** 342 (1952).
31. M. Peckham, *Victorian Studies* **3,** 19 (1959).
32. W. M. Krogman, *Science* **111,** 43 (1950).

Critical Thinking

1. Identify at least two examples of how human activities are consistent with the competitive exclusion principle.
2. What, if any, is the connection between the competitive exclusion principle and the concept of sustainability?
3. Identify at least two examples of how humans might reduce the manner in which they competitively exclude other species for benefiting from ecosystem services.

UNIT 4

Why Do Humans Behave in Unsustainable Ways?

Unit Selections

Learning Outcomes

After reading this unit, you should be able to:

- Identify, by listing, the main points of White's presentation.
- Generate, by synthesizing the points of his argument, a paragraph supporting or opposing his position.
- Identify, by listing, the main points of Moncrief's presentation.
- Generate, by synthesizing the points of his argument, a paragraph supporting or opposing his position.
- Demonstrate, by relating to contemporary issues, understanding of the concept of a commons.
- Identify, by listing, as many communication options available today as can be named.
- Generate, by synthesizing the concept of dysfunction, a paragraph explaining how it is either aggravated or abated by the modern exposure to information.
- State in writing the major strengths and weaknesses of the various models associated with ecologically sustainable behaviors.
- Generate, by synthesizing relevant information, a paragraph examining why, if over 75 percent of Americans report believing that humans should coexist with nature, there are so many issues concerning human induced environmental degradation.
- Generate, by synthesizing the information in Articles 4 and 7, a paragraph describing the possible effect of video gaming on nature appreciation.

Student Website

www.mhhe.com/cls

Internet References

Post Carbon Institute—Energy Bulletin
www.energybulletin.net/node/46276

Sustainability Institute (SI)
www.sustainer.org

This unit focuses on the question, Why have humans behaved in ways that compromise the life support systems of Earth and why, despite repeated warnings about the potential ramifications of our actions, do we continue to do so? It should probably not come as a surprise that we behave as we do given that people continue to smoke cigarettes despite repeated warnings and unequivocal evidence of the harmful effects they have on human health. Maybe, it's as the late comedian George Carlin suggested, we are destined to be an "evolutionary cul-de-sac." The fact is that scholars have attempted to answer this question for decades. One of the first attempts to do so pointed the proverbial finger at religion. Because the Bible indicates that humans were given dominion over all on Earth, in Article 18, White argued that "Christianity is the most anthropocentric religion the world has seen" and that . . . Christianity made it possible to exploit nature. . . ."

© Photographer's Choice/Getty Images

Article 19 is what amounts to a rebuttal to White's proposition, where the focus is on the cultural basis for why people act as they do. If Christianity is not the root cause of the sustainability issue, then perhaps it is our cultures that are to blame. Just as it is vital for a physician to find the root cause of an illness before he or she can provide treatment, it is vital that humanity identify the root cause of our unsustainable attitudes, beliefs, and behaviors.

Article 20 is one of the most cited articles ever published in *Science*. It has informed the thinking of scholars virtually since it appeared in print. In one sense, it might represent the physical manifestation of the philosophies identified by White and Moncrief. The implications for sustainability are profound, for if Hardin is correct, more people will have to become what Daniel Quinn called "leavers." They will have to consider the impact of their action on others not after they have acted, but before.

Article 21, the oldest, and shortest, in this publication, may be more relevant today than when it was originally published. When Lazarsfeld and Merton wrote it, the communication/information options were essentially radio and newspapers. Today, the communication options are staggeringly more numerous. Is it possible that the dysfunction that was suggested as a result of such few options is even worse with so many more options? Is it possible that sustainability is still so far from being achieved because too many people are narcotizingly dysfunctional? If so, how do we address the issue?

Article 22 presents a brief description of the findings of a few research efforts that have attempted to identify, in a somewhat formulaic manner, the relationship between knowledge, awareness, and behavior. For decades, scholars have assumed that if people knew about the life support systems of Earth and about how human actions are degrading them, they would, in an almost knee-jerk fashion, modify their behaviors to behave in more ecologically benign ways. Clearly, this is not so. Why not? Apparently, sustainable behavior is instigated by factors that are far more complex than simply knowledge and awareness. If we could identify them, we could target them for activation through a variety of interventions. Perhaps even more important, if we could identify the factors that contribute to sustainable behaviors, we might be able to identify barriers to such behavior.

Article 23 examines the question, Do all denizens of Earth support the concept of sustainability? Are there data available to support an answer in either direction? What if there are not sufficient data to make an informed decision about the appropriate way to respond regarding sustainability? This reading provides the findings of some of the most current research that has explored these questions.

How are people to appreciate the objectives of sustainability, and more importantly, behave in ways that contribute to achieving it, if they are so disconnected from non-human nature that they have no comprehension of the unique characteristics of Earth that make life possible? Article 24 examines the trends in outdoor recreation, especially those that are nature-based.

The Historical Roots of Our Ecological Crisis

LYNN WHITE, JR.

A conversation with Aldous Huxley not infrequently put one at the receiving end of an unforgettable monologue. About a year before his lamented death he was discoursing on a favorite topic: Man's unnatural treatment of nature and its sad results. To illustrate his point he told how, during the previous summer, he had returned to a little valley in England where he had spent many happy months as a child. Once it had been composed of delightful grassy glades; now it was becoming overgrown with unsightly brush because the rabbits that formerly kept such growth under control had largely succumbed to a disease, myxomatosis, that was deliberately introduced by the local farmers to reduce the rabbits' destruction of crops. Being something of a Philistine, I could be silent no longer, even in the interests of great rhetoric. I interrupted to point out that the rabbit itself had been brought as a domestic animal to England in 1176, presumably to improve the protein diet of the peasantry.

All forms of life modify their contexts. The most spectacular and benign instance is doubtless the coral polyp. By serving its own ends, it has created a vast undersea world favorable to thousands of other kinds of animals and plants. Ever since man became a numerous species he has affected his environment notably. The hypothesis that his fire-drive method of hunting created the world's great grasslands and helped to exterminate the monster mammals of the Pleistocene from much of the globe is plausible, if not proved. For 6 millennia at least, the banks of the lower Nile have been a human artifact rather than the swampy African jungle which nature, apart from man, would have made it. The Aswan Dam, flooding 5000 square miles, is only the latest stage in a long process. In many regions terracing or irrigation, overgrazing, the cutting of forests by Romans to build ships to fight Carthaginians or by Crusaders to solve the logistics problems of their expeditions, have profoundly changed some ecologies. Observation that the French landscape falls into two basic types, the open fields of the north and the *bocage* of the south and west, inspired Marc Bloch to undertake his classic study of medieval agricultural methods. Quite unintentionally, changes in human ways often affect nonhuman nature. It has been noted, for example, that the advent of the automobile eliminated huge flocks of sparrows that once fed on the horse manure littering every street.

The history of ecologic change is still so rudimentary that we know little about what really happened, or what the results were. The extinction of the European aurochs as late as 1627 would seem to have been a simple case of overenthusiastic hunting. On more intricate matters it often is impossible to find solid information. For a thousand years or more the Frisians and Hollanders have been pushing back the North Sea, and the process is culminating in our own time in the reclamation of the Zuider Zee. What, if any, species of animals, birds, fish, shore life, or plants have died out in the process? In their epic combat with Neptune have the Netherlanders overlooked ecological values in such a way that the quality of human life in the Netherlands has suffered? I cannot discover that the questions have ever been asked, much less answered.

People, then, have often been a dynamic element in their own environment, but in the present state of historical scholarship we usually do not know exactly when, where, or with what effects man-induced changes came. As we enter the last third of the 20th century, however, concern for the problem of ecologic backlash is mounting feverishly. Natural science, conceived as the effort to understand the nature of things, had flourished in several eras and among several peoples. Similarly there had been an age-old accumulation of technological skills, sometimes growing rapidly, sometimes slowly. But it was not until about four generations ago that Western Europe and North America arranged a marriage between science and technology, a union of the theoretical and the empirical approaches to our natural environment. The emergence in widespread practice of the Baconian creed that scientific knowledge means technological power over nature can scarcely be dated before about 1850, save in the chemical industries, where it is anticipated in the 18th century. Its acceptance as a normal pattern of action may mark the greatest event in human history since the invention of agriculture, and perhaps in nonhuman terrestrial history as well.

Almost at once the new situation forced the crystallization of the novel concept of ecology; indeed, the word *ecology* first appeared in the English language in 1873. Today, less than a century later, the impact of our race upon the environment has so increased in force that it has changed in essence. When the first cannons were fired, in the early 14th century, they affected ecology by sending workers scrambling to the forests and mountains for more potash, sulfur, iron ore, and charcoal, with some resulting erosion and deforestation. Hydrogen bombs are of a different order: a war fought with them might alter the genetics of all life on this planet. By 1285 London had a smog problem arising from the burning of soft coal, but our present combustion of fossil fuels threatens to change the chemistry of the globe's atmosphere as a whole, with consequences which we are only beginning to guess. With the population explosion, the carcinoma of planless urbanism, the now geological deposits of sewage and garbage, surely no create other than man has ever managed to foul its nest in such short order.

There are many calls to action, but specific proposals, however worthy as individual items, seem too partial, palliative, negative: ban the bomb, tear down the billboards, give the Hindus contraceptives and tell them to eat their sacred cows. The simplest solution to any suspect change is, of course, to stop it, or, better yet, to revert to a romanticized past: make those ugly gasoline stations look like Anne Hathaway's cottage or (in the Far West) like ghost-town saloons. The "wilderness area" mentality invariably advocates deep-freezing an ecology, whether San Gimignano or the High Sierra, as it was before the first Kleenex was dropped. But neither atavism nor prettification will cope with the ecologic crisis of our time.

What shall we do? No one yet knows. Unless we think about fundamentals, our specific measures may produce new backlashes more serious than those they are designed to remedy.

As a beginning we should try to clarify our thinking by looking, in some historical depth, at the presuppositions that underlie modern technology and science. Science was traditionally aristocratic, speculative, intellectual in intent; technology was lower-class, empirical, action-oriented. The quite sudden fusion of these two, towards the middle of the 19th century, is surely related to the slightly prior and contemporary democratic revolutions which, by reducing social barriers, tended to assert a functional unity of brain and hand. Our ecologic crisis is the product of an emerging, entirely novel, democratic culture. The issue is whether a democratized world can survive its own implications. Presumably we cannot unless we rethink our axioms.

The Western Traditions of Technology and Science

One thing is so certain that it seems stupid to verbalize it: both modern technology and modern science are distinctively *Occidental*. Our technology has absorbed elements from all over the world, notably from China; yet everywhere today, whether in Japan or in Nigeria, successful technology is Western. Our science is the heir to all the sciences of the past, especially perhaps to the work of the great Islamic scientists of the Middle Ages, who so often outdid the ancient Greeks in skill and perspicacity: al Rāzi in medicine, for example; or ibn-al-Haytham in optics; or Omar Khayyám in mathematics. Indeed, not a few works of such geniuses seem to have vanished in the original Arabic and to survive only in medieval Latin translations that helped to lay foundations for later Western developments. Today, around the globe, all significant science is Western in style and method, whatever the pigmentation or language of the scientists.

A second pair of fact is less well recognized because they result from quite recent historical scholarship. The leadership of the West, both in technology and in science, is far older than the so-called Scientific Revolution of the 17th century or the so-called Industrial Revolution of the 18th century. These terms are in fact outmoded and obscure the true nature of what they try to describe—significant stages in two long and separate developments. By A.D. 1000 at the latest—and perhaps, feebly, as much as 200 years earlier—the West began to apply water power to industrial processes other than milling grain. This was followed in the late 12th century by the harnessing of wind power. From simple beginnings, but with remarkable consistency of style, the West rapidly expanded its skills in the development of power machinery, labor-saving devices, and automation. . . .

Since both our technological and our scientific movements got their start, acquired their character, and achieved world dominance in the Middle Ages, it would seem that we cannot understand their nature or their present impact upon ecology without examining fundamental medieval assumptions and developments.

Medieval View of Man and Nature

Until recently, agriculture has been the chief occupation even in "advanced" societies; hence, any change of methods of tillage has much importance. Early plows, drawn by two oxen, did not normally turn the sod but merely scratched it. Thus, cross-plowing was needed and fields tended to be squarish. In the fairly light soils and semiarid climates of the Near East and Mediterranean, this worked well. But such a plow was inappropriate to the wet climate and often sticky soils of northern Europe. By the latter part of the 7th century after Christ, however, following obscure beginnings, certain northern peasants were using an entirely new kind of plow, equipped with a vertical knife to cut the line of the furrow, a horizontal share to slice under the sod, and a moldboard to turn it over. The friction of this plow with the soil was so great that it normally required not two but eight oxen. It attacked the land with such violence that cross-plowing was not needed, and fields tended to be shaped in long strips.

In the days of the scratch-plow, fields were distributed generally in units capable of supporting a single family. Subsistence farming was the presupposition. But no peasant owned eight oxen: to use the new and more efficient plow, peasants pooled their oxen to form large plow-teams, originally receiving (it would appear) plowed strips in proportion to their contribution. Thus, distribution of land was based no longer on the needs of a family but, rather, on the capacity of a power machine to till the earth. Man's relation to the soil was profoundly changed. Formerly man had been part of nature; now he was the exploiter of nature. Nowhere else in the world did farmers develop any analogous agricultural implement. Is it coincidence that modern technology, with its ruthlessness toward nature, has so largely been produced by descendants of these peasants of northern Europe?

This same exploitive attitude appears slightly before A.D. 830 in Western illustrated calendars. In older calendars the months were shown as passive personifications. The new Frankish calendars, which set the style for the Middle Ages, are very different: they show men coercing the world around them—plowing, harvesting, chopping trees, butchering pigs. Man and nature are two things, and man is master.

These novelties seem to be in harmony with larger intellectual patterns. What people do about their ecology depends on what they think about themselves in relation to things around them. Human ecology is deeply conditioned by beliefs about our nature and destiny—that is, by religion. To Western eyes this is very evident in, say, India or Ceylon. It is equally true of ourselves and of our medieval ancestors.

The victory of Christianity over paganism was the greatest psychic revolution in the history of our culture. It has become fashionable today to say that, for better or worse, we live in "the post-Christian age." Certainly the forms of our thinking and language have largely ceased to be Christian, but to my eye the substance often remains amazingly akin to that of the past. Our daily habits of action, for example, are dominated by an implicit faith in perpetual progress which was unknown either to Greco-Roman antiquity or to the Orient. It is rooted in, and is indefensible apart from, Judeo-Christian teleology. The fact that Communists share it merely helps to show what can be demonstrated on many other grounds: that Marxism, like Islam, is a Judeo-Christian heresy. We continue today to live, as we have lived for about 1700 years, very largely in a context of Christian axioms.

What did Christianity tell people about their relations with the environment?

While many of the world's mythologies provide stories of creation, Greco-Roman mythology was singularly incoherent in this respect. Like Aristotle, the intellectuals of the ancient West denied that the visible world had had a beginning. Indeed, the idea of a beginning was impossible in the framework of their cyclical notion of time. In sharp contrast, Christianity inherited from Judaism not only a concept of time as nonrepetitive and linear but also a striking story of creation. By gradual stages a loving and all-powerful God had created light and darkness, the heavenly bodies, the earth and all its plants, animals, birds, and fishes. Finally, God had created Adam and, as an afterthought, Eve to keep man from being lonely. Man named all the animals, thus establishing his dominance over them. God planned all of this explicitly for man's benefit and rule: no item in the physical creation had any purpose save to serve man's purposes. And, although man's body is made of clay, he is not simply part of nature: he is made in God's image.

Especially in its Western form, Christianity is the most anthropocentric religion the world has seen. As early as the 2nd century both Tertullian and Saint Irenaeus of Lyons were insisting that when God shaped Adam he was foreshadowing the image of the incarnate Christ, the Second Adam. Man shares, in great measure, God's transcendence of nature. Christianity, in absolute contrast to ancient paganism and Asia's religions (except, perhaps, Zoroastrianism), not only established a dualism of man and nature but also insisted that it is God's will that man exploit nature for his proper ends. . . .

The Christian dogma of creation, which is found in the first clause of all the Creeds, has another meaning for our comprehension of today's ecologic crisis. By revelation, God had given man the Bible, the Book of Scripture. But since God had made nature, nature also must reveal the divine mentality. The religious study of nature for the better understanding of God was known as natural theology. In the early Church, and always in the Greek East, nature was conceived primarily as a symbolic system through which God speaks to men: the ant is a sermon to sluggards; rising flames are the symbol of the soul's aspiration. This view of nature was essentially artistic rather than scientific. While Byzantium preserved and copied great numbers of ancient Greek scientific texts, science as we conceive it could scarcely flourish in such an ambience.

However, in the Latin West by the early 13th century natural theology was following a very different bent. It was ceasing to be the decoding of the physical symbols of God's communication with man and was becoming the effort to understand God's mind by discovering how his creation operates. The rainbow was no longer simply a symbol of hope first sent to Noah after the Deluge: Robert Grosseteste, Friar Roger Bacon, and Theodoric of Freiberg produced startlingly sophisticated work on the optics of the rainbow, but they did it as a venture in religious understanding. From the 13th century onward, up to and including Leibnitz and Newton, every major scientist, in effect, explained his motivations in religious terms. Indeed, if Galileo had not been so expert an amateur theologian he would have got into far less trouble: the professionals resented his intrusion. And Newton seems to have regarded himself more as a theologian than as a scientist. It was not until the late 18th century that the hypothesis of God became unnecessary to many scientists.

It is often hard for the historian to judge, when men explain why they are doing what they want to do, whether they are offering real reasons or merely culturally acceptable reasons. The consistency with which scientists during the long formative centuries of Western science said that the task and the reward of the scientist was "to think God's thoughts after him" leads one to believe that this was their real motivation. If so, then modern Western science was cast in a matrix of Christian theology. The dynamism of religious devotion, shaped by the Judeo-Christian dogma of creation, gave it impetus.

An Alternative Christian View

We would seem to be headed toward conclusions unpalatable to many Christians. Since both *science* and *technology* are blessed words in our contemporary vocabulary, some may be happy at the notions, first, that, viewed historically, modern science is an extrapolation of natural theology and, second, that modern technology is at least partly to be explained as an Occidental voluntarist realization of the Christian dogma of man's transcendence of, and rightful mastery over, nature. But, as we now recognize, somewhat over a century ago science and technology—hitherto quite separate activities—joined to give mankind powers which, to judge by many of the ecologic effects, are out of control. If so, Christianity bears a huge burden of guilt.

I personally doubt that disastrous ecologic backlash can be avoided simply by applying to our problems more science and more technology. Our science and technology have grown out of Christian attitudes toward man's relation to nature which are almost universally held not only by Christians and neo-Christians but also by those who fondly regard themselves as post-Christians. Despite Copernicus, all the cosmos rotates around our little globe. Despite Darwin, we are *not*, in our hearts, part of the natural process. We are superior to nature, contemptuous of it, willing to use it for our slightest whim. The newly elected Governor of California, like myself a churchman but less troubled than I, spoke for the Christian tradition when he said (as is alleged), "when you've seen one redwood tree, you've seen them all." To a Christian a tree can be no more than a physical fact. The whole concept of the sacred grove is alien to Christianity and to the ethos of the West. For nearly 2 millennia Christian missionaries have been chopping down sacred groves, which are idolatrous because they assume spirit in nature.

What we do about ecology depends on our ideas of the man-nature relationship. More science and more technology are not going to get us out of the present ecologic crisis until we find a new religion, or rethink our old one. The beatniks, who are the basic revolutionaries of our time, show a sound instinct in their affinity for Zen Buddhism, which conceives of the man-nature relationship as very nearly the mirror image of the Christian view. Zen, however, is as deeply conditioned by Asian history as Christianity is by the experience of the West, and I am dubious of its viability among us.

Possibly we should ponder the greatest radical in Christian history since Christ: Saint Francis of Assisi. The prime miracle of Saint Francis is the fact that he did not end at the stake, as many of his left-wing followers did. He was so clearly heretical that a General of the Franciscan Order, Saint Bonaventura, a great and perceptive Christian, tried to suppress the early account of Franciscanism. The key to an understanding of Francis is his belief in the virtue of humility—not merely for the individual but for man as a species. Francis tried to depose man from his monarchy over creation and set up a democracy of all God's creatures. With him the ant is no longer simply a homily for the lazy, flames a sign of the thrust of the soul toward union with God; now they are Brother Ant and Sister Fire, praising the Creator in their own ways as Brother Man does in his.

Later commentators have said that Francis preached to the birds as a rebuke to men who would not listen. The records do not read so: he urged the little birds to praise God, and in spiritual ecstasy they flapped their wings and chirped rejoicing. Legends of saints, especially the Irish saints, had long told of their dealings with animals but always, I believe, to show their human dominance over creatures. With Francis it is different. The land around Gubbio in the Apennines was being ravaged by a fierce wolf. Saint Francis, says the legend, talked to the wolf and persuaded him of the error of his ways. The wolf repented, died in the odor of sanctity, and was buried in consecrated ground.

What Sir Steven Ruciman calls "the Franciscan doctrine of the animal soul" was quickly stamped out. Quite possible it was in part inspired, consciously or unconsciously, by the belief in reincarnation held by the Cathar heretics who at that time teemed in Italy and southern France, and who presumably had got it originally from India. It is significant that at just the same moment, about 1200, traces of metempsychosis are found also in western Judaism, in the Provençal *Cabbala*. But Francis held neither to transmigration of souls nor to pantheism. His view of nature and of man rested on a unique sort of pan-psychism of all things animate and inanimate, designed for the glorification of their transcendent Creator, who, in the ultimate gesture of cosmic humility, assumed flesh, lay helpless in a manger, and hung dying on a scaffold.

I am not suggesting that many contemporary Americans who are concerned about our ecologic crisis will be either able or willing to counsel with wolves or exhort birds. However, the present increasing disruption of the global environment is the product of a dynamic technology and science which were originating in the Western medieval world against which Saint Francis was rebelling in so original a way. Their growth cannot be understood historically apart from distinctive attitudes toward nature which are deeply grounded in Christian dogma. The fact that most people do not think of these attitudes as Christian is irrelevant. No new set of basic values has been accepted in our society to displace those of Christianity. Hence we shall continue to have a worsening ecologic crisis until we reject the Christian axiom that nature has no reason for existence save to serve man.

The greatest spiritual revolutionary in Western history, Saint Francis, proposed what he thought was an alternative Christian view of nature and man's relation to it: he tried to substitute the idea of the equality of all creatures, including man, for the idea of man's limitless rule of creation. He failed. Both our present science and our present technology are so tinctured with orthodox Christian arrogance toward nature that no solution for our ecologic crisis can be expected from them alone. Since the roots of our trouble are so largely religious, the remedy must also be essentially religious, whether we call it that or not. We must rethink and refeel our nature and destiny. The profoundly religious, but heretical, sense of the primitive Franciscans for the spiritual autonomy of all parts of nature may point a direction. I propose Francis as a patron saint for ecologists.

Critical Thinking

1. Do you believe that "nature has not existence save to serve [hu]man[s]?"
2. Do you believe that science and technology can resolve the issue of sustainability?
3. If humans have been modifying the Earth for thousands of years and we still have not gone extinct, isn't that proof that our actions are not as bad as some would have us believe?

The Cultural Basis for Our Environmental Crisis

Judeo-Christian tradition is only one of many cultural factors contributing to the environmental crisis.

LEWIS W. MONCRIEF

One hundred years ago at almost any location in the United States, potable water was no farther away than the closest brook or stream. Today there are hardly any streams in the United States, except in a few high mountainous reaches, that can safely satisfy human thirst without chemical treatment. An oft-mentioned satisfaction in the lives of urbanites in an earlier era was a leisurely stroll in late afternoon to get a breath of fresh air in a neighborhood park or along a quiet street. Today in many of our major metropolitan areas it is difficult to find a quiet, peaceful place to take a leisurely stroll and sometimes impossible to get a breath of fresh air. These contrasts point up the dramatic changes that have occurred in the quality of our environment.

It is not my intent in this article, however, to document the existence of an environmental crisis but rather to discuss the cultural basis for such a crisis. Particular attention will be given to the institutional structures as expressions of our culture.

Social Organization

In her book entitled *Social Institutions*[1], J. O. Hertzler classified all social institutions into nine functional categories: (i) economic and industrial, (ii) matrimonial and domestic, (iii) political, (iv) religious, (v) ethical, (vi) educational, (vii) communications, (viii) esthetic, and (ix) health. Institutions exist to carry on each of these functions in all cultures, regardless of their location or relative complexity. Thus, it is not surprising that one of the analytical criteria used by anthropologists in the study of various cultures is the comparison and contrast of the various social institutions as to form and relative importance[2].

A number of attempts have been made to explain attitudes and behavior that are commonly associated with one institutional function as the result of influence from a presumably independent institutional factor. The classic example of such an analysis is *The Protestant Ethic and the Spirit of Capitalism* by Max Weber[3]. In this significant work Weber attributes much of the economic and industrial growth in Western Europe and North America to capitalism, which, he argued, was an economic form that developed as a result of the religious teachings of Calvin, particularly spiritual determinism.

Social scientists have been particularly active in attempting to assess the influence of religious teaching and practice and of economic motivation on other institutional forms and behavior and on each other. In this connection, L. White[4] suggested that the exploitative attitude that has prompted much of the environmental crisis in Western Europe and North America is a result of the teachings of the Judeo-Christian tradition, which conceives of man as superior to all other creation and of everything else as created for his use and enjoyment. He goes on to contend that the only way to reduce the ecologic crisis which we are now facing is to "reject the Christian axiom that nature has no reason for existence save to serve man." As with other ideas that appear to be new and novel, Professor White's observations have begun to be widely circulated and accepted in scholarly circles, as witness the article by religious writer E. B. Fiske in the *New York Times* earlier this year[5]. In this article, note is taken of the fact that several prominent theologians and theological groups have accepted this basic premise that Judeo-Christian doctrine regarding man's relation to the rest of creation is at the root of the West's environmental crisis. I would suggest that the wide acceptance of such a simplistic explanation is at this point based more on fad than on fact.

Certainly, no fault can be found with White's statement that "Human ecology is deeply conditioned by beliefs about our nature and destiny—that is, by religion." However, to argue that it is the primary conditioner of human behavior toward the environment is much more than the data that he cites to support this proposition will bear. For example, White himself notes very early in his article that there is evidence for the idea that man has been dramatically altering his environment since antiquity. If this be true, and there is evidence that it is, then this mediates against the idea that the Judeo-Christian religion uniquely predisposes cultures within which it thrives to exploit their natural resources with indiscretion. White's own examples weaken his argument considerably. He points out that human intervention in the periodic flooding of the Nile River basin and the fire-drive method of hunting by prehistoric man have both probably wrought significant "unnatural" changes in man's environment. The absence of Judeo-Christian influence in these cases is obvious.

It seems tenable to affirm that the role played by religion in man-to-man and man-to-environment relationships is one of establishing a very broad system of allowable beliefs and behavior and of articulating and invoking a system of social and spiritual rewards for those who conform and of negative sanctions for individuals or groups who approach or cross the pale of the religiously unacceptable. In other words, it defines the ball park in which the game is played, and, by the very nature of the park, some types of games cannot be played. However, the kind of game that ultimately evolves is not itself defined by the ball park. For example, where animism is practiced, it is not likely that the believers will indiscriminately destroy objects of nature because such activity would incur the danger of spiritual and social sanctions. However, the fact that another culture does not associate spiritual beings with natural objects does not mean that such a culture will invariably ruthlessly exploit its resources. It simply means that there are fewer social and psychological constraints against such action.

In the remainder of this article, I present an alternative set of hypotheses based on cultural variables which, it seems to me, are more plausible and more defensible as an explanation of the environmental crisis that is now confronting us.

No culture has been able to completely screen out the egocentric tendencies of human beings. There also exists in all cultures a status hierarchy of positions and values, with certain groups partially or totally excluded from access to these normatively desirable goals. Historically, the differences in most cultures between the "rich" and the "poor" have been great. The many very poor have often produced the wealth for the few who controlled the means of production. There may have been no alternative where scarcity of supply and unsatiated demand were economic reality. Still, the desire for a "better life" is universal; that is, the desire for higher status positions and the achievement of culturally defined desirable goals is common to all societies.

The Experience in the Western World

In the West two significant revolutions that occurred in the 18th and 19th centuries completely redirected its political, social, and economic destiny[6]. These two types of revolutions were unique to the West until very recently. The French revolution marked the beginnings of widespread democratization. In specific terms, this revolution involved a redistribution of the means of production and a reallocation of the natural and human resources that are an integral part of the production process. In effect new channels of social mobility were created, which theoretically made more wealth accessible to more people. Even though the revolution was partially perpetrated in the guise of overthrowing the control of presumably Christian institutions and of destroying the influence of God over the minds of men, still it would be superficial to argue that Christianity did not influence this revolution. After all, biblical teaching is one of the strongest of all pronouncements concerning human dignity and individual worth.

At about the same time but over a more extended period, another kind of revolution was taking place, primarily in England. As White points out very well, this phenomenon, which began with a number of technological innovations, eventually consummated a marriage with natural science and began to take on the character that it has retained until today[7]. With this revolution the productive capacity of each worker was amplified by several times his potential prior to the revolution. It also became feasible to produce goods that were not previously producible on a commercial scale.

Later, with the integration of the democratic and the technological ideals, the increased wealth began to be distributed more equitably among the population. In addition, as the capital to land ratio increased in the production process and the demand grew for labor to work in the factories, large populations from the agrarian hinterlands began to concentrate in the emerging industrial cities. The stage was set for the development of the conditions that now exist in the Western world.

With growing affluence for an increasingly large segment of the population, there generally develops an increased demand for goods and services. The usual by-product of this affluence is waste from both the production and consumption processes. The disposal of that waste is further complicated by the high concentration of heavy waste producers in urban areas. Under these conditions the maxim that "Dilution is the solution to pollution" does not withstand the test of time, because the volume of such wastes is greater than the system can absorb and purify through natural means. With increasing population, increasing production, increasing urban concentrations, and increasing real median incomes for well over a hundred years, it is not surprising that our environment has taken a terrible beating in absorbing our filth and refuse.

The American Situation

The North American colonies of England and France were quick to pick up the technical and social innovations that were taking place in their motherlands. Thus, it is not surprising that the inclination to develop an industrial and manufacturing base is observable rather early in the colonies. A strong trend toward democratization also evidenced itself very early in the struggle for nationhood. In fact, Thistlewaite notes the significance of the concept of democracy as embodied in French thought to the framers of constitutional government in the colonies[8], pp. 33–34, 60.

From the time of the dissolution of the Roman Empire, resource ownership in the Western world was vested primarily with the monarchy or the Roman Catholic Church, which in turn bestowed control of the land resources on vassals who pledged fealty to the sovereign. Very slowly the concept of private ownership developed during the Middle Ages in Europe, until it finally developed into the fee simple concept.

In America, however, national policy from the outset was designed to convey ownership of the land and other natural resources into the hands of the citizenry. Thomas Jefferson was perhaps more influential in crystallizing this philosophy in the new nation than anyone else. It was his conviction that an agrarian society made up of small landowners would furnish the most stable foundation for building the nation (*8*, pp. 59–68). This concept has received support up to the present and, against growing economic pressures in recent years, through government programs that have encouraged the conventional family farm. This point is clearly relevant to the subject of this article because it explains how the natural resources of the nation came to be controlled not by a few aristocrats but by many citizens. It explains how decisions that ultimately degrade the environment are made not only by corporation boards and city engineers but by millions of owners of our natural resources. This is democracy exemplified!

Challenge of the Frontier

Perhaps the most significant interpretation of American history has been Fredrick Jackson Turner's much criticized thesis that the western frontier was the prime force in shaping our society[9]. In his own words,

> If one would understand why we are today one nation, rather than a collection of isolated states, he must study this economic and social consolidation of the country. . . . The effect of the Indian frontier as a consolidating agent in our history is important.

He further postulated that the nation experienced a series of frontier challenges that moved across the continent in waves. These included the explorers' and traders' frontier, the Indian frontier, the cattle frontier, and three distinct agrarian frontiers. His thesis can be extended to interpret the expansionist period of our history in Panama, in Cuba, and in the Philippines as a need for a continued frontier challenge.

Turner's insights furnish a starting point for suggesting a second variable in analyzing the cultural basis of the United States' environmental crisis. As the nation began to expand westward, the settlers faced many obstacles, including a primitive transportation system, hostile Indians, and the absence of physical and social security. To many frontiersmen, particularly small farmers, many of the natural resources that are now highly valued were originally perceived more as obstacles than as assets. Forests needed to be cleared to permit farming. Marshes needed to be drained. Rivers needed to be controlled. Wildlife often represented a competitive threat in addition to being a source of food. Sod was considered a nuisance—to be burned, plowed, or otherwise destroyed to permit "desirable" use of the land.

Undoubtedly, part of this attitude was the product of perceiving these resources as inexhaustible. After all, if a section of timber was put to the torch to clear it for farming, it made little difference because there was still plenty to be had very easily. It is no coincidence that the "First Conservation Movement" began to develop about 1890. At that point settlement of the frontier was almost complete. With the passing of the frontier era of American history, it began to dawn on people that our resources were indeed exhaustible. This realization ushered in a new philosophy of our national government toward natural resources management under the guidance of Theodore Roosevelt and Gifford Pinchot. Samuel Hays[10] has characterized this movement as the appearance of a new "Gospel of Efficiency" in the management and utilization of our natural resources.

The Present American Scene

America is the archetype of what happens when democracy, technology, urbanization, capitalistic mission, and antagonism (or apathy) toward natural environment are blended together. The present situation is characterized by three dominant features that mediate against quick solution to this impending crisis: (i) an absence of personal moral direction concerning our treatment of our natural resources, (ii) an inability on the part of our social institutions to make adjustments to this stress, and (iii) an abiding faith in technology.

The first characteristic is the absence of personal moral direction. There is moral disparity when a corporation executive can receive a prison sentence for embezzlement but be congratulated for increasing profits by ignoring pollution abatement laws. That the absolute cost to society of the second act may be infinitely greater than the first is often not even considered.

The moral principle that we are to treat others as we would want to be treated seems as appropriate a guide as it ever has been. The rarity of such teaching and the even more uncommon instance of its being practiced help to explain how one municipality can, without scruple, dump its effluent into a stream even though it may do irreparable damage to the resource and add tremendously to the cost incurred by downstream municipalities that use the same water. Such attitudes are not restricted to any one culture. There appears to be an almost universal tendency to maximize self-interests and a widespread willingness to shift production costs to society to promote individual ends.

Undoubtedly, much of this behavior is the result of ignorance. If our accounting systems were more efficient in computing the cost of such irresponsibility both to the present generation and to those who will inherit the environment we are creating, steps would undoubtedly be taken to enforce compliance with measures designed to conserve resources and protect the environment. And perhaps if the total costs were known, we might optimistically speculate that more voluntary compliance would result.

A second characteristic of our current situation involves institutional inadequacies. It has been said that "what belongs to everyone belongs to no one." This maxim seems particularly appropriate to the problem we are discussing. So much of our environment is so apparently abundant that it is considered a free commodity. Air and water are particularly good examples. Great liberties have been permitted in the use and abuse of these resources for at least two reasons. First, these resources have typically been considered of less economic value than other natural resources except when conditions of extreme scarcity impose limiting factors. Second, the right of use is more difficult to establish for resources that are not associated with a fixed location.

Government, as the institution representing the corporate interests of all its citizens, has responded to date with dozens of legislative acts and numerous court decisions which give it authority to regulate the use of natural resources. However, the decisiveness to act has thus far been generally lacking. This indecisiveness cannot be understood without noting that the simplistic models that depict the conflict as that of a few powerful special interests versus "The People" are altogether inadequate. A very large proportion of the total citizenry is implicated in environmental degradation; the responsibility ranges from that of the board and executives of a utility company who might wish to thermally pollute a river with impunity to that of the average citizen who votes against a bond issue to improve the efficiency of a municipal sanitation system in order to keep his taxes from being raised. The magnitude of irresponsibility among individuals and institutions might be characterized as falling along a continuum from highly irresponsible to indirectly responsible. With such a broad base of interests being threatened with every change in resource policy direction, it is not surprising, although regrettable, that government has been so indecisive.

A third characteristic of the present American scene is an abiding faith in technology. It is very evident that the idea that technology can overcome almost any problem is widespread in Western society. This optimism exists in the face of strong evidence that much of man's technology, when misused, has produced harmful results, particularly in the long run. The reasoning goes something like this: "After all, we have gone to the moon. All we need to do is allocate enough money and brainpower and we can solve any problem."

It is both interesting and alarming that many people view technology almost as something beyond human control. Rickover put it this way[11]:

It troubles me that we are so easily pressured by purveyors of technology into permitting so-called "progress" to alter our lives without attempting to control it—as if technology were an irrepressible force of nature to which we must meekly submit.

He goes on to add:

> It is important to maintain a humanistic attitude toward technology; to recognize clearly that since it is the product of human effort, technology can have no legitimate purpose but to serve man—man in general, not merely some men; future generations, not merely those who currently wish to gain advantage for themselves; man in the totality of his humanity, encompassing all his manifold interests and needs, not merely some one particular concern of his. When viewed humanistically, technology is seen not as an end in itself but a means to an end, the end being determined by man himself in accordance with the laws prevailing in his society.

In short, it is one thing to appreciate the value of technology; it is something else entirely to view it as our environmental savior—which will save us in spite of ourselves.

Conclusion

The forces of democracy, technology, urbanization, increasing individual wealth, and an aggressive attitude toward nature seem to be directly related to the environmental crisis now being confronted in the Western world. The Judeo-Christian tradition has probably influenced the character of each of these forces. However, to isolate religious tradition as a cultural component and to contend that it is the "historical root of our ecological crisis" is a bold affirmation for which there is little historical or scientific support.

To assert that the primary cultural condition that has created our environmental crisis is Judeo-Christian teaching avoids several hard questions. For example: Is there less tendency for those who control the resources in non-Christian cultures to live in extravagant affluence with attendant high levels of waste and inefficient consumption? If non-Judeo-Christian cultures had the same levels of economic productivity, urbanization, and high average household incomes, is there evidence to indicate that these cultures would not exploit or disregard nature as our culture does?

If our environmental crisis is a "religious problem," why are other parts of the world experiencing in various degrees the same environmental problems that we are so well acquainted with in the Western world? It is readily observable that the science and technology that developed on a large scale first in the West have been adopted elsewhere. Judeo-Christian tradition has not been adopted as a predecessor to science and technology on a comparable scale. Thus, all White can defensibly argue is that the West developed modern science and technology *first*. This says nothing about the origin or existence of a particular ethic toward our environment.

In essence, White has proposed this simple model:

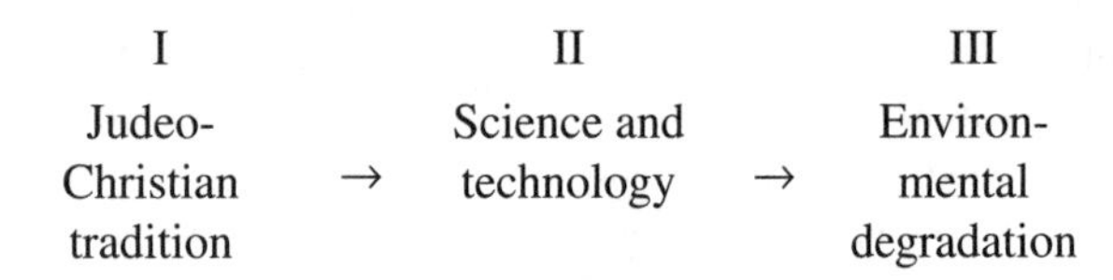

I have suggested here that, at best, Judeo-Christian teaching has had only an indirect effect on the treatment of our environment. The model could be characterized as follows:

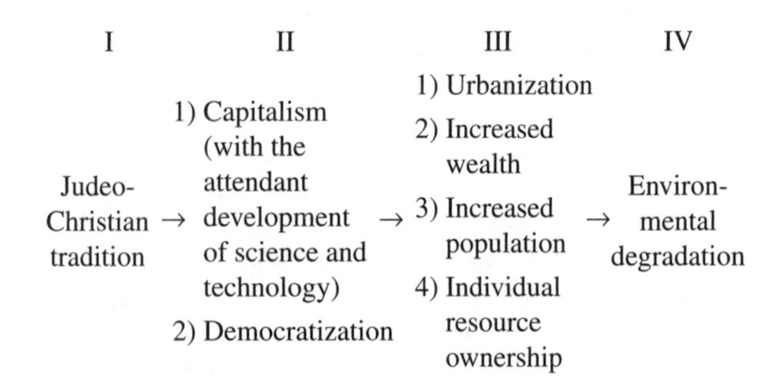

Even here, the link between Judeo-Christian tradition and the proposed dependent variables certainly have the least empirical support. One need only look at the veritable mountain of criticism of Weber's conclusions in *The Protestant Ethic and the Spirit of Capitalism* to sense the tenuous nature of this link. The second and third phases of this model are common to many parts of the world. Phase I is not.

Jean Mayer[12], the eminent food scientist, gave an appropriate conclusion about the cultural basis for our environmental crisis:

It might be bad in China with 700 million poor people but 700 million rich Chinese would wreck China in no time. . . . It's the rich who wreck the environment . . . occupy much more space, consume more of each natural resource, disturb ecology more, litter the landscape . . . and create more pollution.

References and Notes

1. J. O. Hertzler, *Social Institutions* (McGraw-Hill, New York, 1929), pp. 47–64.
2. L. A. White, *The Science of Culture* (Farrar, Straus & Young, New York, 1949), pp. 121–145.
3. M. Weber, *The Protestant Ethic and the Spirit of Capitalism,* translated by T. Parsons (Scribner's, New York, 1958).
4. L. White, Jr., *Science* **155,** 1203 (1967).
5. E. B. Fiske, "The link between faith and ecology," *New York Times* (4 January 1970), section 4, p. 5.
6. R. A. Nisbet, *The Sociological Tradition* (Basic Books, New York, 1966), pp. 21–44. Nisbet gives here a perceptive discourse on the social and political implications of the democratic and industrial revolutions to the Western world.
7. It should be noted that a slower and less dramatic process of democratization was evident in English history at a much earlier date than the French revolution. Thus, the concept of democracy was probably a much more pervasive influence in English than in French life. However, a rich body of philosophic literature regarding the rationale for democracy resulted from the French revolution. Its counterpart in English literature is much less conspicuous. It is an interesting aside to suggest that perhaps the industrial revolution would not have been possible except for the more broad-based ownership of the means of production that resulted from the long-standing process of democratization in England.
8. F. Thistlewaite, *The Great Experiment* (Cambridge Univ. Press, London, 1955).
9. F. J. Turner, *The Frontier in American History* (Henry Holt, New York, 1920 and 1947).
10. S. P. Hays, *Conservation and the Gospel of Efficiency* (Harvard Univ. Press, Cambridge, Mass., 1959).
11. H. G. Rickover, *Amer. Forests* **75,** 13 (August 1969).
12. J. Mayer and T. G. Harris, *Psychol. Today* **3,** 46 and 48 (January 1970).

Critical Thinking

1. What, if any, is the significance of the different adjective for the word crisis, namely environmental in place of ecological, in this article?
2. Which do you think comes first, culture or religion; that is, do you think culture determines religion or religion determines culture?
3. Explain the statement, "So much of our environment is so apparently abundant that it is considered a free commodity."

The Tragedy of the Commons

GARRETT HARDIN

An implicit and almost universal assumption of discussions published in professional and semipopular scientific journals is that the problem under discussion has a technical solution. A technical solution may be defined as one that requires a change only in the techniques of the natural sciences, demanding little or nothing in the way of change in human values or ideas of morality.

In our day (though not in earlier times) technical solutions are always welcome. Because of previous failures in prophecy, it takes courage to assert that a desired technical solution is not possible. . . . [T]he concern here is with the important concept of a class of human problems which can be called "no technical solution problems," and, more specifically, with the identification and discussion of one of these.

It is easy to show that the class is not a null class. Recall the game of tick-tack-toe. Consider the problem, "How can I win the game of tick-tack-toe?" It is well known that I cannot, if I assume (in keeping with the conventions of game theory) that my opponent understands the game perfectly. Put another way, there is no "technical solution" to the problem. I can win only by giving a radical meaning to the word "win." I can hit my opponent over the head; or I can drug him; or I can falsify the records. Every way in which I "win" involves, in some sense, an abandonment of the game, as we intuitively understand it. (I can also, of course, openly abandon the game—refuse to play it. This is what most adults do.)

The class of "No technical solution problems" has members. My thesis is that the "population problem," as conventionally conceived, is a member of this class. How it is conventionally conceived needs some comment. It is fair to say that most people who anguish over the population problem are trying to find a way to avoid the evils of overpopulation without relinquishing any of the privileges they now enjoy. They think that farming the seas or developing new strains of wheat will solve the problem—technologically. I try to show here that the solution they seek cannot be found. The population problem cannot be solved in a technical way, any more than can the problem of winning the game of tick-tack-toe. . . .

We can make little progress in working toward optimum population size until we explicitly exorcize the spirit of Adam Smith in the field of practical demography. In economic affairs, *The Wealth of Nations* (1776) popularized the "invisible hand," the idea that an individual who "intends only his own gain," is, as it were, "led by an invisible hand to promote. . . the public interest." Adam Smith did not assert that this was invariably true, and perhaps neither did any of his followers. But he contributed to a dominant tendency of thought that has ever since interfered with positive action based on rational analysis, namely, the tendency to assume that decisions reached individually will, in fact, be the best decisions for an entire society. If this assumption is correct it justifies the continuance of our present policy of laissez-faire in reproduction. If it is correct we can assume that men will control their individual fecundity so as to produce the optimum population. If the assumption is not correct, we need to reexamine our individual freedoms to see which ones are defensible.

Tragedy of Freedom in a Commons

The rebuttal to the invisible hand in population control is to be found in a scenario first sketched in a little-known pamphlet in 1833 by a mathematical amateur named William Forster Lloyd (1794–1852). We may well call it "the tragedy of the commons," using the word "tragedy" as the philosopher Whitehead used it: "The essence of dramatic tragedy is not unhappiness. It resides in the solemnity of the remorseless working of things." He then goes on to say, "This inevitableness of destiny can only be illustrated in terms of human life by incidents which in fact involve unhappiness. For it is only by them that the futility of escape can be made evident in the drama."

The tragedy of the commons develops in this way. Picture a pasture open to all. It is to be expected that each herdsman will try to keep as many cattle as possible on the commons. Such an arrangement may work reasonably satisfactorily for centuries because tribal wars, poaching, and disease keep the numbers of both man and beast well below the carrying capacity of the land. Finally, however, comes the day of reckoning, that is, the day when the long-desired goal of social stability becomes a reality. At this point, the inherent logic of the commons remorselessly generates tragedy.

As a rational being, each herdsman seeks to maximize his gain. Explicitly or implicitly, more or less consciously, he asks, "What is the utility *to me* of adding one more animal to my herd?" This utility has one negative and one positive component.

1. The positive component is a function of the increment of one animal. Since the herdsman receives all the proceeds from the sale of the additional animal, the positive utility is nearly +1.
2. The negative component is a function of the additional overgrazing created by one more animal. Since, however, the effects of overgrazing are shared by all the herdsmen, the negative utility for any particular decision-making herdsman is only a fraction of −1.

Adding together the component partial utilities, the rational herdsman concludes that the only sensible course for him to pursue is to add another animal to his herd. And another; and another. . . . But this is the conclusion reached by each and every rational herdsman sharing a commons. Therein is the tragedy. Each man is locked into a system that compels him to increase his herd without limit—in a world that is limited. Ruin is the destination toward which all men rush, each pursuing his own best interest in a society that believes in the freedom of the commons. Freedom in a commons brings ruin to all.

Some would say that this is a platitude. Would that it were! In a sense, it was learned thousands of years ago, but natural selection favors the forces of psychological denial. The individual benefits as an individual from his ability to deny the truth even though society as a whole, of which he is a part, suffers. Education can counteract the natural tendency to do the wrong thing, but the inexorable succession of generations requires that the basis for this knowledge be constantly refreshed.

A simple incident that occurred a few years ago in Leominster, Massachusetts, shows how perishable the knowledge is. During the Christmas shopping season the parking meters downtown were covered with plastic bags that bore tags reading: "Do not open until after Christmas. Free parking courtesy of the mayor and city council." In other words, facing the prospect of an increased demand for already scarce space, the city fathers reinstituted the system of the commons. (Cynically, we suspect that they gained more votes than they lost by this retrogressive act.)

In an approximate way, the logic of the commons has been understood for a long time, perhaps since the discovery of agriculture or the invention of private property in real estate. But it is understood mostly only in special cases which are not sufficiently generalized. Even at this late date, cattlemen leasing national land on the western ranges demonstrate no more than an ambivalent understanding, in constantly pressuring federal authorities to increase the head count to the point where overgrazing produces erosion and weed-dominance. Likewise, the oceans of the world continue to suffer from the survival of the philosophy of the commons. Maritime nations still respond automatically to the shibboleth of the "freedom of the seas." Professing to believe in the "inexhaustible resources of the oceans," they bring species after species of fish and whales closer to extinction.

The National Parks present another instance of the working out of the tragedy of the commons. At present, they are open to all, without limit. The parks themselves are limited in extent—there is only one Yosemite Valley—whereas population seems to grow without limit. The values that visitors seek in the parks are steadily eroded. Plainly, we must soon cease to treat the parks as commons or they will be of no value to anyone.

What shall we do? We have several options. We might sell them off as private property. We might keep them as public property, but allocate the right to enter them. The allocation might be on the basis of wealth, by the use of an auction system. It might be on the basis of merit, as defined by some agreed-upon standards. It might be by lottery. Or it might be on a first-come, first-served basis, administered to long queues. These, I think, are all the reasonable possibilities. They are all objectionable. But we must choose—or acquiesce in the destruction of the commons that we call our national parks.

Pollution

In a reverse way, the tragedy of the commons reappears in problems of pollution. Here it is not a question of taking something out of the commons, but of putting something in—sewage, or chemical, radioactive, and heat wastes into water; noxious and dangerous fumes into the air; and distracting and unpleasant advertising signs into the line of sight. The calculations of utility are much the same as before. The rational man finds that his share of the cost of the wastes he discharges into the commons is less than the cost of purifying his wastes before releasing them. Since this is true for everyone, we are locked into a system of "fouling our own nest," so long as we behave only as independent, rational, free-enterprisers.

The tragedy of the commons as a food basket is averted by private property, or something formally like it. But the air and waters surrounding us cannot readily be fenced, and so the tragedy of the commons as a cesspool must be prevented by different means, by coercive laws or taxing devices that make it cheaper for the polluter to treat his pollutants than to discharge them untreated. We have not progressed as far with the solution of this problem as we have with the first. Indeed, our particular concept of private property, which deters us from exhausting the positive resources of the earth, favors pollution. The owner of a factory on the bank of a stream—whose property extends to the middle of the stream—often has difficulty seeing why it is not his natural right to muddy the waters flowing past his door. The law, always behind the times, requires elaborate stitching and fitting to adapt it to this newly perceived aspect of the commons.

The pollution problem is a consequence of population. It did not much matter how a lonely American frontiersman disposed of his waste. "Flowing water purifies itself every 10 miles," my grandfather used to say, and the myth was near enough to the truth when he was a boy, for there were not too many people. But as population became denser, the natural chemical and biological recycling processes became overloaded, calling for a redefinition of property rights. . . .

Freedom to Breed Is Intolerable

The tragedy of the commons is involved in population problems in another way. In a world governed solely by the principle of "dog eat dog"—if indeed there ever was such a world—how many children a family had would not be a matter of public concern. Parents who bred too exuberantly would leave fewer descendants, not more, because they would be unable to care adequately for their children. David Lack and others have found that such a negative feedback demonstrably controls the fecundity of birds. But men are not birds, and have not acted like them for millenniums, at least.

If each human family were dependent only on its own resources; *if* the children of improvident parents starved to death; *if*, thus, overbreeding brought its own "punishment" to the germ line—*then* there would be no public interest in controlling the breeding of families. But our society is deeply committed to the welfare state, and hence is confronted with another aspect of the tragedy of the commons.

In a welfare state, how shall we deal with the family, the religion, the race, or the class (or indeed any distinguishable and cohesive group) that adopts overbreeding as a policy to secure its own aggrandizement? To couple the concept of freedom to breed with the belief that everyone born has an equal right to the commons is to lock the world into a tragic course of action.

Unfortunately this is just the course of action that is being pursued by the United Nations. In late 1967, some 30 nations agreed to the following:

> The Universal Declaration of Human Rights describes the family as the natural and fundamental unit of society. It follows that any choice and decision with regard to the size of the family must irrevocably rest with the family itself, and cannot be made by anyone else.

It is painful to have to deny categorically the validity of this right; denying it, one feels as uncomfortable as a resident of Salem, Massachusetts, who denied the reality of witches in the 17th century. At the present time, in liberal quarters, something like a taboo acts to inhibit criticism of the United Nations. There is a feeling that the United Nations is "our last and best hope," that we shouldn't find fault with it; we shouldn't play into the hands of the archconservatives. However, let us not forget what Robert Louis Stevenson said: "The truth that is suppressed by friends is the readiest weapon of the enemy." If we love the truth we must openly deny the validity of the Universal Declaration of Human Rights, even though it is promoted by the United Nations. We should also join with Kingsley Davis in attempting to get Planned Parenthood-World Population to see the error of its ways in embracing the same tragic ideal.

Conscience Is Self-Eliminating

It is a mistake to think that we can control the breeding of mankind in the long run by an appeal to conscience. Charles Galton Darwin made this point when he spoke on the centennial of the publication of his grandfather's great book. The argument is straightforward and Darwinian.

People vary. Confronted with appeals to limit breeding, some people will undoubtedly respond to the plea more than others. Those who have more children will produce a larger fraction of the next generation than those with more susceptible consciences. The difference will be accentuated, generation by generation.

In C. G. Darwin's words: "It may well be that it would take hundreds of generations for the progenitive instinct to develop in this way, but if it should do so, nature would have taken her revenge, and the variety *Homo contracipiens* would become extinct and would be replaced by the variety *Homo progenitivus.*"

The argument assumes that conscience or the desire for children (no matter which) is hereditary—but hereditary only in the most general formal sense. The result will be the same whether the attitude is transmitted through germ cells, or exosomatically, to use A. J. Lotka's term. (If one denies the latter possibility as well as the former, then what's the point of education?) The argument has here been stated in the context of the population problem, but it applies equally well to any instance in which society appeals to an individual exploiting a commons to restrain himself for the general good—by means of his conscience. To make such an appeal is to set up a selective system that works toward the elimination of conscience from the race. . . .

Mutual Coercion Mutually Agreed Upon

The social arrangements that produce responsibility are arrangements that create coercion, of some sort. Consider bank-robbing. The man who takes money from a bank acts as if the bank were a commons. How do we prevent such action? Certainly not by trying to control his behavior solely by a verbal appeal to his sense of responsibility. Rather than rely on propaganda we follow [Charles] Frankel's lead and insist that a bank is not a commons; we seek the definitive social arrangements that will keep it from becoming a commons. That we thereby infringe on the freedom of would-be robbers we neither deny nor regret.

The morality of bank-robbing is particularly easy to understand because we accept complete prohibition of this activity. We are willing to say "Thou shalt not rob banks," without providing for exceptions. But temperance also can be created by coercion. Taxing is a good coercive device. To keep downtown shoppers temperate in their use of parking space we introduce parking meters for short periods, and traffic fines for longer ones. We need not actually forbid a citizen to park as long as he wants to; we need merely make it increasingly expensive for him to do so. Not prohibition, but carefully biased options are what we offer him. A Madison Avenue man might call this persuasion; I prefer the greater candor of the word coercion.

Coercion is a dirty word to most liberals now, but it need not forever be so. As with the four-letter words, its dirtiness can be cleansed away by exposure to the light, by saying it over and over without apology or embarrassment. To many, the word coercion implies arbitrary decisions of distant and irresponsible bureaucrats; but this is not a necessary part of its meaning. The only kind of coercion I recommend is mutual coercion, mutually agreed upon by the majority of the people affected.

To say that we mutually agree to coercion is not to say that we are required to enjoy it, or even to pretend we enjoy it. Who enjoys taxes? We all grumble about them. But we accept compulsory taxes because we recognize that voluntary taxes would favor the conscienceless. We institute and (grumblingly) support taxes and other coercive devices to escape the horror of the commons.

An alternative to the commons need not be perfectly just to be preferable. With real estate and other material goods, the alternative we have chosen is the institution of private property coupled with legal inheritance. Is this system perfectly just? As a genetically trained biologist I deny that it is. It seems to me that, if there are to be differences in individual inheritance, legal possession should be perfectly correlated with biological inheritance—that those who are biologically more fit to be the custodians of property and power should legally inherit more. But genetic recombination continually makes a mockery of the doctrine of "like father, like son" implicit in our laws of legal inheritance. An idiot can inherit millions, and a trust fund can keep his estate intact. We must admit that our legal system of private property plus inheritance is unjust—but we put up with it because we are not convinced, at the moment, that anyone has invented a better system. The alternative of the commons is too horrifying to contemplate. Injustice is preferable to total ruin.

It is one of the peculiarities of the warfare between reform and the status quo that it is thoughtlessly governed by a double standard. Whenever a reform measure is proposed it is often defeated when its opponents triumphantly discover a flaw in it. As Kingsley Davis has pointed out, worshippers of the status quo sometimes imply that no reform is possible without unanimous agreement, an implication contrary to historical fact. As nearly as I can make out, automatic rejection of proposed reforms is based on one of two unconscious assumptions: (i) that the status quo is perfect; or (ii) that the choice we face is between reform and no action; if the proposed reform is imperfect, we presumably should take no action at all, while we wait for a perfect proposal.

But we can never do nothing. That which we have done for thousands of years is also action. It also produces evils. Once we are aware that the status quo is action, we can then compare its discoverable advantages and disadvantages with the predicted advantages and disadvantages of the proposed reform, discounting as best we can for our lack of experience. On the basis of such a comparison, we can make a rational decision which will not involve the unworkable assumption that only perfect systems are tolerable.

Recognition of Necessity

Perhaps the simplest summary of this analysis of man's population problems is this: the commons, if justifiable at all, is justifiable only under conditions of low-population density. As the human population has increased, the commons has had to be abandoned in one aspect after another.

First we abandoned the commons in food gathering, enclosing farm land and restricting pastures and hunting and fishing areas. These restrictions are still not complete throughout the world.

Somewhat later we saw that the commons as a place for waste disposal would also have to be abandoned. Restrictions on the disposal of domestic sewage are widely accepted in the Western world; we are still struggling to close the commons to pollution by automobiles, factories, insecticide sprayers, fertilizing operations, and atomic energy installations.

In a still more embryonic state is our recognition of the evils of the commons in matters of pleasure. There is almost no restriction on the propagation of sound waves in the public medium. The shopping public is assaulted with mindless music, without its consent. Our government is paying out billions of dollars to create supersonic transport which will disturb 50,000 people for every one person who is whisked from coast to coast 3 hours faster. Advertisers muddy the airwaves of radio and television and pollute the view of travelers. We are a long way from outlawing the commons in matters of pleasure. Is this because our Puritan inheritance makes us view pleasure as something of a sin, and pain (that is, the pollution of advertising) as the sign of virtue?

Every new enclosure of the commons involves the infringement of somebody's personal liberty. Infringements made in the distant past are accepted because no contemporary complains of a loss. It is the newly proposed infringements that we vigorously oppose; cries of "rights" and "freedom" fill the air. But what does "freedom" mean? When men mutually agreed to pass laws against robbing, mankind became more free, not less so. Individuals locked into the logic of the commons are free only to bring on universal ruin; once they see the necessity of mutual coercion, they become free to pursue other goals. I believe it was Hegel who said, "Freedom is the recognition of necessity."

The most important aspect of necessity that we must now recognize, is the necessity of abandoning the commons in breeding. No technical solution can rescue us from the misery of overpopulation. Freedom to breed will bring ruin to all. At the moment, to avoid hard decisions many of us are tempted to propagandize for conscience and responsible parenthood. The temptation must be resisted, because an appeal to independently acting consciences selects for the disappearance of all conscience in the long run, and an increase in anxiety in the short.

The only way we can preserve and nurture other and more precious freedoms is by relinquishing the freedom to breed, and that very soon. "Freedom is the recognition of necessity"—and it is the role of education to reveal to all the necessity of abandoning the freedom to breed. Only so, can we put an end to this aspect of the tragedy of the commons.

Critical Thinking

1. Discuss the similarities between the competitive exclusion principle and the tragedy of the commons.
2. What is meant by the statement, "natural selection favors the forces of psychological denial?"
3. Discuss the statement, "The great challenge facing us now is to invent the corrective feedbacks that are needed to keep custodians honest" in the context of the existence of the NEPA and assorted other laws and the USEPA.

The Narcotizing Dysfunction

PAUL F. LAZARSFELD AND ROBERT K. MERTON

The functions of status conferral and of reaffirmation of social norms are evidently well recognized by the operators of mass media. Like other social and psychological mechanisms, these functions tend themselves to diverse forms of application. Knowledge of these functions is power, and power may be used for special interests or for the general interest.

A third social consequence of the mass media has gone largely unnoticed. At least, it has received little explicit comment and, apparently, has not been systematically put to use for furthering planned objectives. This may be called the narcotizing dysfunction of the mass media. It is termed *dys*functional rather than functional on the assumption that it is not in the interest of modern complex society to have large masses of the population politically apathetic and inert. How does this unplanned mechanism operate?

Scattered studies have shown that an increasing proportion of the time of Americans is devoted to the products of the mass media. With distinct variations in different regions and among different social strata, the outpourings of the media presumably enable the twentieth-century American to "keep abreast of the world." Yet, it is suggested, this vast supply of communications may elicit only a superficial concern with the problems of society, and this superficiality often cloaks mass apathy.

Exposure to this flood of information may serve to narcotize rather than to energize the average reader or listener. As an increasing amount of time is devoted to reading and listening, a decreasing share is available for organized action. The individual reads accounts of issues and problems and may even discuss alternative lines of action. But this rather intellectualized, rather remote connection with organized social action is not activated. The interested and informed citizen can congratulate himself on his lofty state of interest and information, and neglect to see that he has abstained from decision and action. In short, he takes his secondary contact with the world of political reality, his reading and listening and thinking, as a vicarious performance. He comes to mistake *knowing* about problems of the day for *doing* something about them. His social conscience remains spotlessly clean. He *is* concerned. He *is* informed. And he has all sorts of ideas as to what should be done. But, after he has gotten through his dinner and after he has listened to his favored radio programs and after he has read his second newspaper of the day, it is really time for bed.

In this peculiar respect, mass communications may be included among the most respectable and efficient of social narcotics. They may be so fully effective as to keep the addict from recognizing his own malady.

That the mass media have lifted the level of information of large populations is evident. Yet, quite apart from intent, increasing dosages of mass communications may be inadvertently transforming the energies of men from active participation into passive knowledge.

Critical Thinking

1. How, if at all, has the potential increase in exposure to information affected this condition?
2. Analyze how much time you spend on the Internet compared to how much time to spend taking action on issues of importance to you. Are the two mutually exclusive? Can you accomplish the same amount of progress with both?
3. Analyze how much information related to sustainability you find in the contemporary mass media.

Mind the Gap: Why Do People Act Environmentally and What are the Barriers to Pro-environmental Behavior?

ANJA KOLLMUSS AND JULIAN AGYEMAN

Introduction

Environmental psychology, which developed in the US in the 1960s, looks at the range of complex interactions between humans and the environment. It is therefore a very broad field with many branches. The branch that looks at the psychological roots of environmental degradation and the connections between environmental attitudes and pro-environmental behaviors is part of environmental psychology but does not have a separate name in English. In German this field is called *Umweltpsychologie* [1].

Over the last 30 years many psychologists and sociologists have explored the roots of direct and indirect environmental action [2]. The answer to the questions: 'Why do people act environmentally and what are the barriers to pro-environmental behavior?' is extremely complex. By 'pro-environmental behavior' we simply mean behavior that consciously seeks to minimize the negative impact of one's actions on the natural and built world (e.g. minimize resource and energy consumption, use of non-toxic substances, reduce waste production).

Numerous theoretical frameworks have been developed to explain the gap between the possession of environmental knowledge and environmental awareness, and displaying pro-environmental behavior. Although many hundreds of studies have been done, no definitive answers have been found. Our article describes a few of the most influential and commonly used frameworks for analyzing pro-environmental behavior. These are: early US linear progression models; altruism, empathy and prosocial behavior models; and finally, sociological models. We then analyze the factors that have been found to have some influence, positive or negative, on pro-environmental behavior such as demographic factors, external factors (e.g. institutional, economic social and cultural factors) and internal factors (e.g. motivation, environmental knowledge, awareness, values, attitudes, emotion, locus of control, responsibilities and priorities). We present this article in order to give environmental educators a feel for some of the broader research findings which have informed current environmental education theory and practice. In doing so, we do not want to prescribe or constrain, but to open up a dialogue regarding the most effective ways environmental educators might help develop pro-environmental behavior at all levels in society.

In this article, we do not discuss recent (and very promising) advances in community social marketing for sustainability (see Agyeman and Angus, forth-coming). Social marketing techniques have been widely used in the field of public health, in anti-smoking campaigns, AIDS awareness campaigns, and to encourage the treatment of leprosy. The development of community-based social marketing specifically for sustainability arose out of concerns about the ineffectiveness of environmental campaigns that relied solely on providing information. The pragmatic approach of social marketing has been offered as an alternative to conventional campaigns, and, in contrast to traditional education methods, has been shown to be very effective at bringing about behavior change (McKenzie-Mohr & Smith 1999, p. 15). McKenzie-Mohr and Smith (1999) claim that the primary advantage of social marketing is that it starts with people's behavior and works backward to select a particular tactic suited for that behavior (McKenzie-Mohr & Smith 1999, p.7). The research on community-based social marketing indicates that the approach has been successful in transcending the gap between knowledge to action that has characterized many local environmental and sustainability projects to date.

Similarly, we do not discuss recent work by O'Riordan and Burgess (1999) and Owens (2000) on deliberative and inclusionary procedures (DIPS) which is showing that 'such [information-based] approaches have repeatedly been shown, by experience, and in research, to be flawed, and a growing body of opinion points instead towards the need for more deliberative and inclusionary procedures' (Owens, 2000, p. 1141). Bloomfield *et al.* argue that DIPS, which includes citizen's juries and round tables, should be seen as a significant, even essential ingredient in the development of more responsive forms of decision making capable of accounting for the diversity of values and opinions within societies (Bloomfield *et al.,* 1998, p. 2). The authors write that DIPS are not 'to be seen merely as a mechanism of achieving greater understanding, or even consensus, over environmental issues with in a fragmenting civil society . . . but to have "transformative" potential allowing those with no or weak voice to exert influence on decision making outcomes (Bloomfield *et al.,* 1998, p. 2).

In conclusion, we propose our own visual model based on the work of Fliegenschnee and Schelakovsky (1998) who were influenced by Fietkau and Kessel (1981).

Review of Selected Frameworks for Analyzing Pro-Environmental Behavior

Early US Linear Models

The oldest and simplest models of pro-environmental behavior were based on a linear progression of environmental knowledge leading to environmental awareness and concern (environmental attitudes), which in turn was thought to lead to pro-environmental behavior. These rationalist models assumed that educating people about environmental issues would automatically result in more pro-environmental behavior, and have been termed (information) 'deficit' models of public understanding and action by Burgess *et al.* (1998. p. 1447).

These models from the early 1970s were so on proven to be wrong. Research showed that in most cases, increases in knowledge and awareness did not lead to pro-environmental behavior. Yet today, most environmental Non-governmental Organisations (NGOs) still base their communication campaigns and strategies on the simplistic assumption that more knowledge will lead to more enlightened behavior. Owens (2000) points out that even governments use this assumption, for example the UK government's 'Save It' energy conservation campaign in the mid-1970s, and the 'Are You Doing Your Bit?'campaign which was launched in 1998 to develop public understanding of sustainable development. This reliance on information to drive change is surprising because common sense tells us that changing behavior is very difficult. Anyone who has ever tried to change a habit, even in a very minor way, will have discovered how difficult it is, even if the new behavior has distinct advantages over the old one.

As mentioned, quantitative research has shown that there is a discrepancy between attitude and behavior. Many researchers have tried to explain this gap. Rajecki (1982) defined four causes:

- *Direct versus indirect experience:* Direct experiences have a stronger influence on people's behavior than indirect experiences. In other words, indirect experiences, such as learning about an environmental problem in school as opposed to directly experiencing it (e.g. seeing the dead fish in the river) will lead to weaker correlation between attitude and behavior.
- *Normative influences:* Social norms [3], cultural traditions, and family customs influence and shape people's attitudes, e.g. if the dominant culture propagates a lifestyle that is unsustainable, pro-environmental behavior is less likely to occur and the gap between attitude and action will widen.
- *Temporal discrepancy:* Inconsistency in results occur when data collection for attitudes and data collection for the action lie far apart (e.g. after Chernobyl, an overwhelming majority of Swiss people were opposed to nuclear energy; yet a memorandum two years later that put a 10-year halt to building any new nuclear reactors in Switzerland was approved by only a very narrow margin). Temporal discrepancy refers to the fact that people's attitudes change over time.
- *Attitude-behavior measurement:* Often the measured attitudes are much broader in scope (e.g. Do you care about the environment?) than the measured actions (e.g. Do you recycle?). This leads to large discrepancies in results (Newhouse, 1991).

The last two items point out frequent flaws in research methodology and make it clear how difficult it is to design valid studies that measure and compare attitude and behavior. Ajzen and Fishbein addressed these issues of measurement discrepancies in their *Theory of Reasoned Action* and their *Theory of Planned Behavior* (Fishbein & Ajzen, 1975; Ajzen & Fishbein, 1980).

They pointed out that in order to find a high correlation between attitude and behavior the researcher has to measure the attitude toward that particular behavior. For example, comparing attitudes toward climate change and driving behavior usually shows no correlation. Even people who

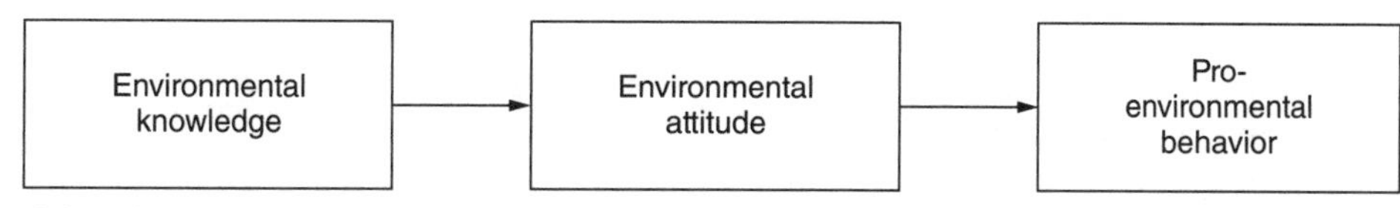

Figure 1 Early models of pro-environmental behavior

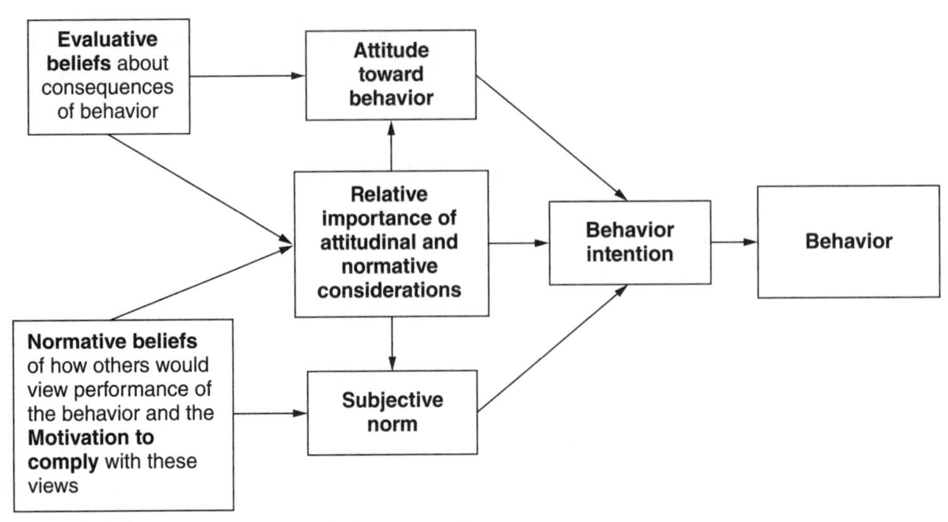

Figure 2 Theory of reasoned action (Ajzen & Fishbein, 1980)

are very concerned about climate change tend to drive. This is because the attitude toward climate change is not closely related to the behaviour (driving). More narrowly targeted attitude measurements lead to a higher correlation but much of the information is lost (Lehmann, 1999). In other words it is rather meaningless to discover that someone who has a negative attitude towards walking in the rain will choose to drive his car.

Fishbein and Ajzen maintain that people are essentially rational, in that they 'make systematic use of information available to them' and are not 'controlled by unconscious motives or overpowering desires', neither is their behavior 'capricious or thoughtless' (Ajzen & Fishbein, 1980, introduction; see also Fishbein & Ajzen, 1975, p. 15). Attitudes do not determine behavior directly, rather they influence behavioral intentions which in turn shape our actions. Intentions are not only influenced by attitudes but also by social ('normative') pressures. Thus 'the ultimate determinants of any behavior are the behavioral beliefs concerning its consequences and normative beliefs concerning the prescriptions of others' (Ajzen & Fishbein, 1980, p. 239).

Their model has been the most influential attitude-behavior model in social psychology—probably because they developed a mathematical equation that expressed their model which led researchers to conduct empirical studies. Although the model certainly has its limitations—for example the underlying assumption that people act rationally—it is useful because of if its clarity and simplicity (Regis, 1990).

In 1986, Hines, Hungerford and Tomera published their *Model of Responsible Environmental Behavior* which was based on Ajzen and Fishbein's theory of planned behavior (Hines *et al.*, 1986–87; Hungerford & Volk 1990; Sia *et al.* 1985–86). They did a meta-analysis of 128 pro-environmental behavior research studies and found the following variables associated with responsible pro-environmental behavior:

- *Knowledge of issues:* The person has to be familiar with the environmental problem and its causes.
- *Knowledge of action strategies:* The person has to know how he or she has to act to lower his or her impact on the environmental problem.
- *Locus of control:* This represents an individual's perception of whether he or she has the ability to bring about change through his or her own behavior. People with a strong internal locus of control believe that their actions can bring about change. People with an external locus of control, on the other hand, feel that their actions are insignificant, and feel that change can only be brought about by powerful others.
- *Attitudes:* People with strong pro-environmental attitudes were found to be more likely to engage in pro-environmental behavior, yet the relationship between attitudes and actions proved to be weak.
- *Verbal commitment:* The communicated willingness to take action also gave some indication about the person's willingness to engage in pro-environmental behavior.
- *Individual sense of responsibility:* People with a greater sense of personal responsibility are more likely to have engaged in environmentally responsible behavior.

Although the framework is more sophisticated than Ajzen and Fishbein's (1980), the identified factors do not sufficiently explain pro-environmental behavior. The relationship between knowledge and attitudes, attitudes and intentions, and intentions and actual responsible behavior, are weak at best. There seem to be many more factors that influence pro-environmental behavior. Hines *et al.* (1986–87) called these 'situational factors' which include economic constraints, social pressures, and opportunities to choose different actions.

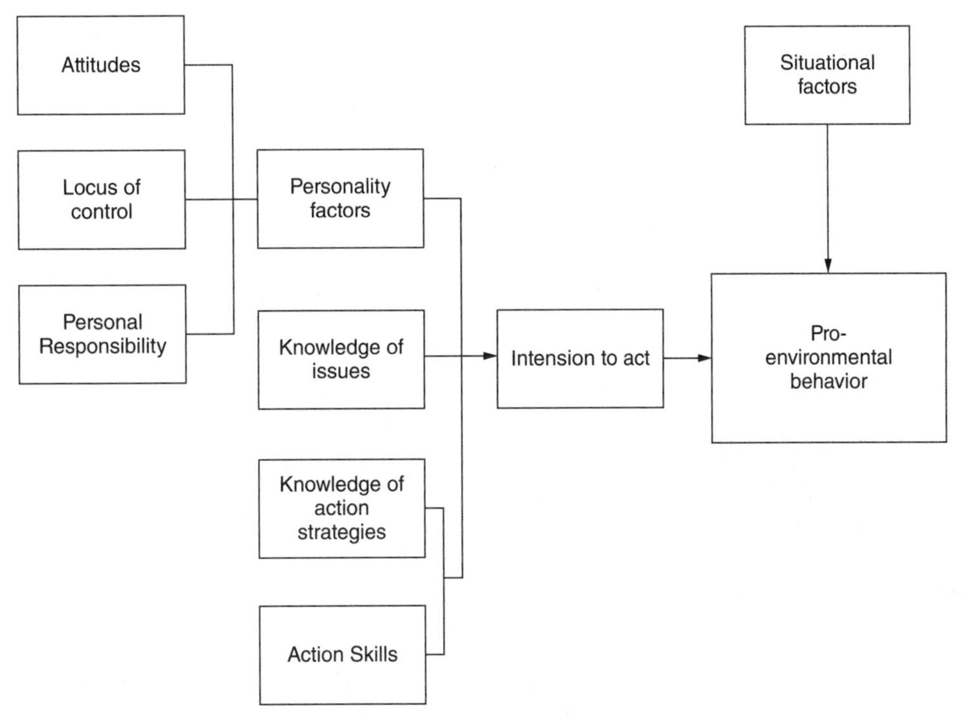

Figure 3 Models of predictors of environmental behavior (Hines *et al.*, 1986)

Altruism, Empathy, and Prosocial Behavior Models

Models of altruism, empathy, and prosocial behavior are another framework for analyzing pro-environmental behavior. Prosocial behavior is defined by Eisenberg and Miller (1987) as 'voluntary intentional behavior that results in benefits for another: the motive is unspecified and may be positive, negative, or both' (quoted in Lehmann, 1999, p. 33). Altruism is a subset of prosocial behavior. Borden and Francis (1978, as noted in Lehmann, 1999, p. 34) hypothesize that:

1. Persons with a strong selfish and competitive orientation are less likely to act ecologically;
2. People who have satisfied their personal needs are more likely to act ecologically because they have more resources (time, money, energy) to care about bigger, less personal social and pro-environmental issues.

The second assumption underlies many other studies and models (e.g. Maslow's hierarchy of human needs). For example, it is often claimed that people in poorer countries care less about the environment, yet the study by Diekmann and Franzen (1999) shows that the issue is more complicated. Using data from two different surveys they showed that when people from poorer countries are asked to *rank* the most pressing problems, environmental issues are indeed ranked lower. Yet if the people are asked to *rate* the severity of different problems, pro-environmental issues always rank high, no matter if the country is affluent or poor. Ranking therefore reflects more the reality of scarce economic resources and not the lack of environmental concern of less affluent people. In addition, 'ecological footprinting' (Wackernagel & Rees, 1997) and similar measures of resource consumption, such as 'environmental space' (McLaren *et al.,* 1998) show clearly that richer nations have a far greater negative environmental impact than poorer nations. This of course does not mean that poorer nations limit their ecological footprint out of environmental concern but it does show that more affluence does not lead to more ecological behavior (for an additional example see also endnote 4).

Several other researchers base their models and assumptions on theories of altruism, claiming that altruism is needed or at least supports pro-environmental behavior. Of note is the work of Allen and Ferrand (1999) who recently tested the 'actively caring' hypothesis of Geller. Similar to the altruism theory of Schwartz (1977), Geller hypothesized that in order to act pro-environmentally, individuals must focus beyond themselves and be concerned about the community at large. Geller suggested that this state of 'actively caring' can only occur if the need for self-esteem, belonging, personal control, self-efficacy, and optimism have been satisfied. In their study Allen and Ferrand (1999) found that self-esteem and belonging were not related to pro-environmental behavior but that there was a significant relationship between personal control and sympathy, their measure for 'actively caring'. They did not test for optimism or self-efficacy.

Stern *et al.*'s (1993) model is based on the altruism theory of Schwartz (1977). This theory assumes that altruistic behavior increases when a person becomes aware of other people's suffering and at the same time feels a responsibility of alleviating this suffering. Stern *et al.* expand this notion and include, next to this 'altruistic' orientation, which they call 'social orientation', an 'egoistic' and a 'biospheric orientation'. The social

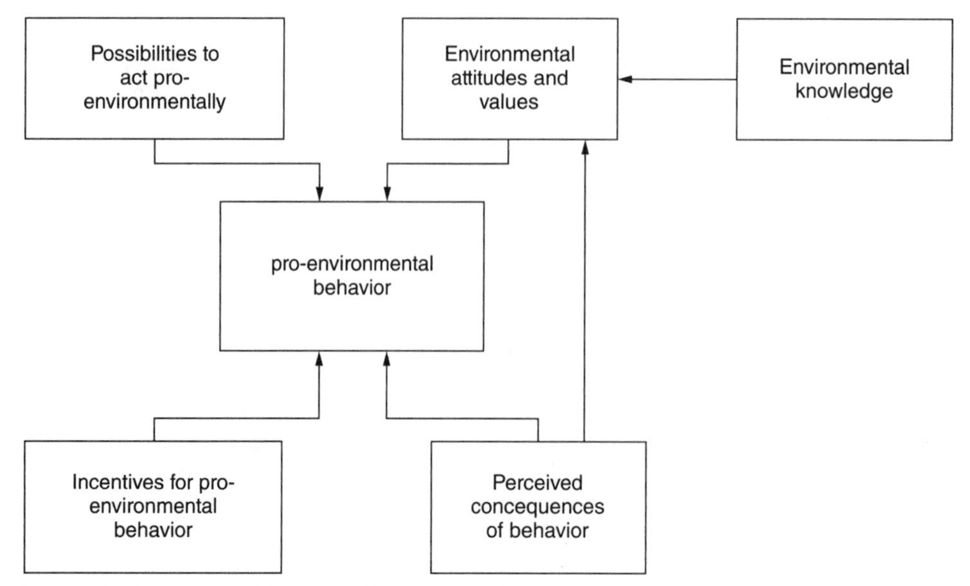

Figure 4 Model of ecological behavior (Fietkau & Kessel, 1981)

orientation is concerned with the removal of suffering of other people, the egoistic orientation is concerned with the removal of suffering and harm from oneself, and the biospheric orientation is concerned with the removal of destruction and suffering in the non-human world. Every person has all three orientations but in different strengths. Whereas a deep ecologist might have a very developed biospheric orientation, a physician might have a stronger social orientation. Stern *et al.* propose that environmental concern is caused by a combination of these three factors:

$$\text{Motivation} = V\,(\text{egoistic orientation}) + V\,(\text{social orientation}) + V(\text{biospheric orientation})$$

They found, not surprisingly, that the egoistic orientation is the strongest orientation, followed by social and then biospheric concern (Stern *et al.,* 1993, quoted in Lehmann, 1999). On the surface, their model therefore contradicts Borden and Francis's (1978) altruism hypothesis mentioned above since Stern *et al.* (1993) claim that the stronger the egoistic orientation the stronger the motivation for the behavior. Yet the egoistic orientation can only be a motivator for pro-environmental behavior as long as the action serves the person's needs and wants (e.g. taking the train instead of the car to have time to relax and read). A strong egoistic orientation is counter productive when the desired behavior negates a person's needs and desires (e.g. not flying to the tropics for a vacation). The models are therefore not contradictory; they just approach the issue from a different point.

Sociological Models for Analyzing Pro-environmental Behavior

Fietkau and Kessel (1981) use sociological as well as psychological factors to explain pro-environmental behavior or the lack of it. Their model comprises five variables that influence either directly or indirectly pro-environmental behavior. These variables are independent from each other and can be influenced and changed.

- *Attitude and values* (*Einstellung und Werte*).
- *Possibilities to act ecologically* [4] (*Verhaltensangebote*). These are external, infrastructural and economic factors that enable or hinder people to act ecologically.
- *Behavioral incentives* (*Handlungsanreize*). These are more internal factors that can reinforce and support ecological behavior (e.g. social desirability, quality of life, monetary savings).
- *Perceived feedback about ecological behavior* (*wahrgenommene Konsequenzen*). A person has to receive a positive reinforcement to continue a certain ecological behavior. This feedback can be intrinsic (e.g. satisfaction of 'doing the right thing'), or extrinsic (e.g. social: not littering or recycling are socially desirable actions; and economic: receiving money for collected bottles).
- *Knowledge* (*Wissen*). In Fietkau's model, knowledge does not directly influence behavior but acts as a modifier of attitudes and values.

Blake (1999) talks about the attitude–behavior gap as the *Value–Action Gap.* He points out that most pro-environmental behavior models are limited because they fail to take into account individual, social, and institutional constraints and assume that humans are rational and make systematic use of the information available to them. A new set of research, mostly by sociologists as opposed to psychologists, has tried to address these limitations. Blake uses a quote from Redclift and Benton to summarize this new approach:

> One of the most important insights which the social scientist can offer in the environmental debate is that the eminently rational appeals on the part of environmentalists

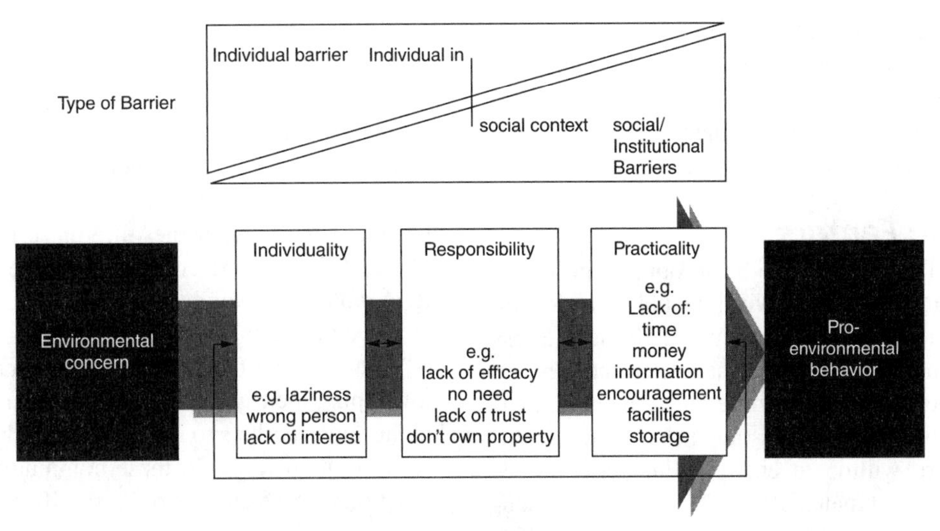

Figure 5 Barriers between environmental concern and action (Blake, 1999)

> for 'us' to change our attitudes or lifestyles, so as to advance a general 'human interest' are liable to be ineffective. This is not because . . . 'we' are irrational, but because the power to make a significant difference, one way or the other, to global or even local environmental change, is immensely unevenly distributed. This new body of research points out that people's values are 'negotiated, transitory, and sometimes contradictory'. (Redclift & Benton, 1994, pp. 7–8, quoted in Blake, 1999)

Blake identifies three barriers to action: individuality, responsibility, and practicality. Individual barriers are barriers lying within the person, having to do with attitude and temperament. He claims that these barriers are especially influential in people that do not have a strong environmental concern. Environmental concern is therefore out weighed by other conflicting attitudes. However, in our experience, even a strong environmental concern can be overcome by stronger desires and needs. For example, our need to fly from the US to visit our families in Europe each year overrides our feelings of responsibility about keeping our air travel to a minimum to minimize global warming. Blake's second set of barriers, responsibility, is very close to the psychologist's notion of 'locus of control'. People who don't act pro-environmentally feel that they cannot influence the situation or should not have to take the responsibility for it. He points out that in the particular community he is describing, a lack of trust in the institution often stops people from acting pro-environmentally—since they are suspicious of local and national government, they are less willing to follow the prescribed actions.

The third barrier, practicality, Blake defines as the social and institutional constraints that prevent people from acting pro-environmentally regardless of their attitudes or intentions. He lists such constraints as lack of time, lack of money, and lack of information. Although his model is very useful in that it combines external and internal factors and describes both in some detail, he does not account for social factors such as familial pressures and cultural norms nor does he explore in more depth the underlying psychological factors (e.g. what are the underlying factors of 'not having time'?).

Analysis: Commonalities, Contradictions and Omissions

We have discussed only a few of the many different models that have been developed to explain the attitude–action gap and investigate the barriers to pro-environmental behavior. All of the models we have discussed (and many of the ones we did not, such as economic models, psychological models that look at behavior in general, social marketing models and DIPS) have some validity in certain circumstances. This indicates that the question of what shapes pro-environmental behavior is such a complex one that it cannot be visualized in one single framework or diagram. Such a single diagram with all the factors that shape and influence behavior would be so complicated that it would lose its practicality and probably even its meaning. Yet, as we show, there are commonalties, contradictions, and omissions that can be found in the different models. In the following section we discuss in more detail the specific factors that have been established as having some influence (positive or negative) on the models of pro-environmental behavior which we have selected in this article.

The distinctions and the hierarchy between the different influential factors are to some extent arbitrary. For example, we distinguish between the following factors: demographic factors, external factors (e.g. institutional, economic, social, and cultural factors) and internal factors (e.g. motivation, environmental knowledge, awareness, values, attitudes, emotion, locus of control, responsibilities, and priorities). A valid argument could be made that environmental knowledge is a subcategory of environmental awareness (as does

Grob, 1991) and that emotional involvement is what shapes environmental awareness and attitude. This difficulty in defining and delimiting the different factors is due to the fact that most are broadly and vaguely defined, interrelated, and often do not have clear boundaries.

Demographic Factors

Two demographic factors that have been found to influence environmental attitude and pro-environmental behavior are gender and years of education. Women usually have a less extensive environmental knowledge than men but they are more emotionally engaged, show more concern about environmental destruction, believe less in technological solutions, and are more willing to change (Fliegenschnee & Schelakovsky, 1998; Lehmann, 1999). The longer the education, the more extensive is the knowledge about environmental issues. Yet more education does not necessarily mean increased pro-environmental behavior (see endnote 4).

External Factors

Institutional Factors

Many pro-environmental behaviors can only take place if the necessary infrastructure is provided (e.g. recycling, taking public transportation). The poorer such services are the less likely people are to use them. These institutional barriers (e.g. lack of public transportation) can be overcome primarily through people's actions as citizens (indirect environmental actions). Because of this, it is important to explore how environmental attitudes influence indirect environmental action. It might be true that environmental knowledge and environmental attitude have a more powerful influence on people's indirect actions than on people's direct pro-environmental behaviors. (See detailed discussion in the section on attitudes and values.)

Economic Factors

Economic factors have a strong influence on people's decisions and behavior. Some economic research indicates that people make purchasing decision using a 50 percent or higher interest rate. In otherwords, if the person decides between two possible items, one energy-efficient and the other not, he or she will only choose the energy efficient item if the payback time for the energy saved is very short. The economic factors that play into people's decision are very complex and only poorly understood. From our own experience, the economist's assumption that people act in an economically rational fashion is very often not true. Yet people can be influenced by economic incentives to behave pro-environmentally (e.g. the Massachusetts Bottle Bill is responsible for the very high recycling rate of bottles at over 80 percent compared to an overall recycling rate of less than 10 percent in Boston, Massachusetts). The opposite is also true. Until recently, very low prices for heating oil in the US prevented people from taking energy conservation measures.

Economic factors are clearly very important when designing new policies and strategies that are meant to influence and change people's behavior. Nevertheless, predicting people's behavior on purely economic grounds will not reveal the whole picture. Economic factors are intertwined with social, infrastructural, and psychological factors. How else could we explain the different effects of pay-per-bag policies [5]: In some communities, the bag fees did nothing to reduce the weight of disposed material and increased the recycling rates only slightly (Ackerman, 1997). In others, a similar bag fee led to a chain reaction: people started unwrapping their groceries in the supermarket which in turn led the supermarkets to redesign and reduce their packaging to a minimum level. In these communities, the per capita reduction of garbage was quite significant.

Social and Cultural Factors

Cultural norms play a very important role in shaping people's behavior. Boehmer-Christiansen and Skea (1991) explored the history of policy reactions to acid rain in Germany and the UK. They showed that the high cultural value of the forests in Germany, along with its geographic position and the Germans' strong need for security and stability, led to a drastically different approach to the problem. It would be very interesting to design a cross-cultural study that looks at pro-environmental behavior. We would hypothesize that cultures in small, highly populated countries such as Switzerland and the Netherlands tend to be more resource conscientious than societies in large, resource-rich countries such as the USA.

Internal Factors

Motivation

Motivation is the reason for a behavior or a strong internal stimulus around which behavior is organized (Wilkie, 1990, as quoted in Moisander, 1998). Motivation is shaped by intensity and direction (which determines which behavior is chosen from all the possible options). Motives for behavior can be overt or hidden—conscious or unconscious. Researchers distinguish between primary motives (the larger motives that let us engage in a whole set of behaviors, e.g. striving to live an environmental lifestyle and selective motives (the motives that influence one specific action), e.g. Should I bike to work today, even though it rains, or do I drive? (Moisander, 1998). Barriers, on the other hand, stifile certain behavior. Usually internal barriers to pro-environmental behavior are non-environmental motivations that are more intense and directed differently (e.g. I will drive to work because I'd rather be comfortable than environmentally sound). In this example, the primary motives (environmental values) are overridden by the selective motives (personal comfort).

As this example indicates, we hypothesize that primary motives, such as altruistic and social values, are often covered up by the more immediate, selective motives, which evolve around one's own needs (e.g. being comfortable, saving money and time). Similarly, Preuss distinguishes between an 'abstract willingness to act', based on values and knowledge and a 'concrete willingness to act', based on habits (Preuss, 1991).

Environmental Knowledge

Most researchers agree that only a small fraction of pro-environmental behavior can be directly linked to environmental knowledge and environmental awareness. There are a few studies that claim otherwise (e.g. Grob, 1991 and Kaiser *et al.,* 1999), yet these studies test only very specific behavior that does not seem to be generalizable. At least 80 percent of the motives for pro-environmental or non-environmental behavior seem to be situational factors and other internal factors (Fliegenschnee & Schelakovsky, 1998).

This argument is further strengthened by the study of Kempton *et al.* (1995). They surveyed different groups in the US, ranging from strong environmentalists to those they thought were strong anti-environmentalists. Kempton found the average knowledge about environmental issues to be low. Surprisingly, the lack of knowledge was equally strong among environmentalists and non-environmentalists. His study therefore implies that environmental knowledge *per se* is not a prerequisite for pro-environmental behavior.

It might be necessary to distinguish between different levels of knowledge. Clearly, people have to have a basic knowledge about environmental issues and the behaviors that cause them in order to act pro-environmentally in a conscious way. Whereas Kempton *et al.*'s study indicated that most people do not know enough about environmental issues to act in an environmentally responsible way, other studies have shown that very detailed technical knowledge does not seem to foster or increase pro-environmental behavior (Diekmann & Preisendoerfer, 1992; Fliegenschnee & Schelakovsky, 1998).

It is interesting to note that other incentives (e.g. economic advantages) and cultural values can motivate people to act pro-environmentally without doing it out of environmental concern. Ecological economists like to take advantage of this fact. By imposing taxes on environmentally harmful activities, people will automatically move away from these behaviors and look for less damaging alternatives. For example, in countries with high gasoline tax, people tend to drive significantly less than in countries with very low taxes (Von Weizaecker & Jesinghaus, 1992). Yet some people caution that such unconscious pro-environmental behavior can easily be reversed or changed to a more unsustainable pattern because it is not based on some fundamental values (Preuss, 1991). For instance, in China, people traveling in trains were used to disposing of their food and drinking utensils by throwing them out of the window. Formerly, this habit made perfect sense, since the drinking cups and the packaging were out of clay and other organic materials. More recently, these have been replaced by styrofoam and plastics. China now has a serious littering problem because people are still disposing of these new, non-degradable materials in the same way.

Values

Values are responsible for shaping much of our intrinsic motivation. The question of what shapes our values is a complex one. Fuhrer et al. (1995) proposed the following hypothesis: A person's values are most influenced by the 'microsystem', which is comprised of the immediate social net—family, neighbors, peer-groups, etc. Values are influenced to a lesser extent by the 'exosystem' such as the media and political organizations. Least strong, but nevertheless important, is the influence of the 'macrosystem', the cultural context in which the individual lives (Fuhrer et al., 1995, as quoted in Lehmann, 1999).

One way to explore the determining factors that shape environmental values is to study the life experiences that have shaped the beliefs and values of active environmentalists (see *Environmental Education Research* special issues on significant life experiences in Volumes 4(4) and 5(4)). A few researchers have approached the topic from this side and have studied environmentalist's life histories.

Chawla interviewed numerous professional environmentalists in the USA and in Norway about the experiences and people who shaped and influenced their decisions to become environmentalists. Furthermore, she reviewed previous studies that had been done on formative life experiences of environmentalists. In her study, she explored retrospectively what factors influenced people's environmental sensitivity. She defines environmental sensitivity as 'a predisposition to take an interest in learning about the environment, feeling concern for it, and acting to conserveit, on the basis of formative experiences' (Chawla, 1998). Not surprisingly, she finds that there is no single experience that sensitizes people's awareness but a combination of factors. Among the most frequently mentioned (decreasing in relevance) are:

- Childhood experiences in nature
- Experiences of pro-environmental destruction
- Pro-environmental values held by the family
- Pro-environmental organizations
- Role models (friends or teachers)
- Education.

During childhood, the most influential were experiences of natural areas and family; during adolescence and early adulthood, education and friends were mentioned most frequently; and during adulthood, it was pro-environmental organizations (Chawla, 1999).

It is important to note that Chawla did not explore the factors that foster direct pro-environmental behavior but indirect pro-environmental actions. Her interviewees were very

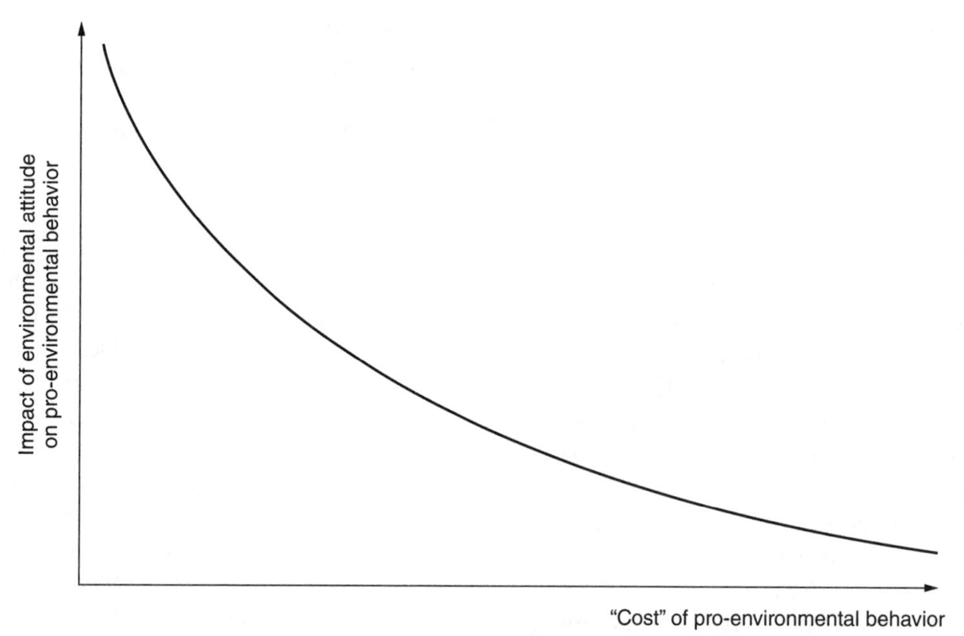

Figure 6 Low-cost high-cost model of pro-environmental behavior (Diekmann & Preisendoerfer)

active environmental professionals, yet their commitment to indirect environmental activism does not necessarily mean that these people exhibited increased direct pro-environmental behavior. Nevertheless her studies are valuable in that they show how important an emotional connection to the natural environment seems to be in fostering environmental awareness and environmental concern.

Attitudes

Attitudes are defined as the enduring positive or negative feeling about some person, object, or issue. Closely related to attitudes are beliefs, which refer to the information (the knowledge) a person has about a person, object, or issue (Newhouse, 1991).

Environmental attitudes have been found to have a varying, usually very small impact on pro-environmental behavior. This is unexpected because we tend to assume that people live according to their values. Diekmann and Preisendoerfer (1992) explain the discrepancy between environmental attitude and pro-environmental behavior by using a low-cost/high-cost model.

They propose that people choose the pro-environmental behaviors that demand the least cost. Cost in their model is not defined in a strictly economic sense but in a broader psychological sense that includes, among other factors, the time and effort needed to undertake a pro-environmental behavior. In their study they show that environmental attitude and low-cost pro-environmental behavior (e.g. recycling) do correlate significantly. People who care about the environment tend to engage in activities such as recycling but do not necessary engage in activities that are more costly and inconvenient such as driving or flying less. In other words, a positive environmental attitude can directly influence low-cost pro-environmental behavior [6]. These findings might be less disappointing than they might seem at first sight. Diekmann and Preisendoerfer (1992) point out that people with high levels of environmental awareness might not be willing to make bigger life style sacrifices, but they seem to be more willing to accept political changes that will enhance pro-environmental behavior such as higher fuel taxes or more stringent building codes (Diekmann & Franzen, 1996; Lehmann, 1999).

Attitudes can indirectly influence our pro-environmental behavior. A study of college students' willingness to engage in pro-environmental behavior found that those who believe technology and growth will solve environmental problems were less likely to make personal sacrifices. These findings indicate that people with a strong belief in growth and technological solutions might not see the need and will be less willing to engage in pro-environmental behavior with the implicit life style changes (Gigliotti, 1992, 1994). Other studies have confirmed these findings (Grob, 1991). Many barriers are responsible for the gap between environmental attitudes and pro-environmental behavior. Nevertheless, values and attitudes clearly play an important role in determining pro-environmental behavior.

Environmental Awareness

In this article, we define environmental awareness as 'knowing of the impact of human behavior on the environment'. Environmental awareness has both a cognitive, knowledge-based component and an affective, perception-based component (discussed in the next section on 'emotional involvement'). Environmental awareness is constrained by several cognitive and emotional limitations. Cognitive limitations of environmental awareness include:

1. *Non-immediacy of many ecological problems.* Most environmental degradation is not immediately tangible (Preuss, 1991). We cannot perceive nuclear radiation, the ozone hole, or the accumulation of greenhouse gases in the atmosphere. Even changes that would theoretically be noticeable, for example the loss of species, often go unnoticed by the lay person. We can only experience the effects of pollution and destruction (e.g. smelling the rotten odor of a water body that suffers from eutrophication caused by agricultural run-off). This implies a time lag: very often, we only perceive changes once the human impact has already caused severe damage. Also, more subtle changes and changes in remote are as escape our awareness.

Because most environmental degradation is not immediately tangible, the information about environmental damage has to be translated into understandable, perceivable information (language, pictures, graphs). Most of the time this information will further our intellectual understanding without making a link to our emotional involvement (Preuss, 1991). It is the rare exception that a vivid, provocative image can be found to explain a scientific concept that at the same time engages people emotionally (a good example of this is the 'ozone hole'). The reliance on secondary information about environmental destruction removes us emotionally from the issue and often leads to non-involvement (Preuss, 1991; Fliegenschnee & Schelakovsky, 1998). The need for emotional involvement also explains why campaigns to protect big mammals—aptly named 'charismatic mega-fauna'—enjoy much broader public support than more abstract issues such as climate change. They are much more immediate and 'real' than climate change, which is only really knowable through mathematical models.

2. *Slow and gradual ecological destruction.* Another cognitive barrier is the often very gradual, slow pace of environmental change (Preuss, 1991). Human beings are very good at perceiving drastic and sudden changes but are often unable to perceives low, incremental changes. We are, in many respects like the frogs in the famous experiment: when placed into hot water, they immediately jumped out but when put into cool water that was slowly heated, they did not react and boiled to death.
3. *Complex systems.* Most environmental problems are intricate and immensely complex. Yet we are often unable to comprehend such complex systems and tend to simplify them and think linearly (Preuss, 1991; Fliegenschnee & Schelakovsky, 1998). This prevents us from a deeper understanding of the consequences of natural destruction. It might also lead to under estimating the extent of the problem. Overall, our cognitive limitations to understanding environmental degradation seriously compromises our emotional engagement and our willingness to act.

Emotional Involvement

We define emotional involvement as the extent to which we have an affective relationship to the natural world. Chawla's (1998, 1999) work shows that such an emotional connection seems to be very important in shaping our beliefs, values, and attitudes towards the environment. Furthermore, we see emotional involvement as the ability to have an emotional reaction when confronted with environmental degradation. In otherwords, it is one's emotional investment in the problem. Research has shown that women tend to react more emotionally to environmental problems (Grob, 1991; Lehmann, 1999). Grob (1991) hypothesizes (and we agree with him) that the stronger a person's emotional reaction, the more likely that person will engage in pro-environmental behavior.

What makes us care? Why is it that some people care and others do not? The answers are extremely diverse, complex, and poorly understood. We all have areas that we are more passionate about than others. The question of why we are emotionally involved in one thing but not another is a very profound one. The following paragraphs cannot do justice to the enormous breadth and depth of the work that has been done in the fields of ethics, psychology, and sociology in an attempt to explore such questions.

1. *Emotional non-investment*
   a. *lack of knowledge and awareness.* As we argued in the previous section, because of the non-immediacy of ecological destruction, emotional involvement requires a certain degree of environmental knowledge and awareness. In many cases, emotional involvement is a learned ability to react emotionally to complex and sometimes very abstract environmental problems. Clearly, there are different degrees of abstraction: whereas most people understand and act emotionally to pictures of oil-covered seabirds, far fewer will feel saddened by the sight of a typical rhododendron-lawn-and-cedar-chip landscape surrounding the average New England home. Lack of knowledge about the causes and effects of ecological degradation can therefore lead to emotional non-involvement (Preuss, 1991; Fliegenschnee & Schelakovsky, 1998). Unfortunately, this does not mean that just providing this knowledge would be sufficient to create such emotional involvement.
   b. *Resistance against non-conforming information.* Festinger (1957) states in his theory of dissonance that we unconsciously seek consistency in our beliefs and mental frameworks and selectively

perceive information. Information that supports our existing values and mental frameworks is readily accepted whereas information that contradicts or undermines our beliefs is avoided or not perceived at all. Festinger's theory implies that we tend to avoid information about environmental problems because they contradict or threaten some of our basic assumption of quality of life, economic prosperity, and material needs.

2. *Emotional reactions.* Even if we are experiencing an emotional reaction to environmental degradation, we might still not act pro-environmentally. Faced with the effects and long-term implications of environmental degradation we can feel fear, sadness/pain, anger, and guilt. The emotional reaction is stronger when we experience the degradation directly (Newhouse, 1991; Chawla, 1999). We hypothesize that fear, sadness, pain, and anger are more likely to trigger pro-environmental behaviors than guilt. A decisive factor for action is locus of control (see below). Strong feelings together with a sense of helplessness will not lead to action.

The primary emotional reactions we experience when exposed to environmental degradation are distressing. They will lead to secondary psychological responses aimed at relieving us from these negative feelings. Very often those secondary responses prevent us from pro-environmental behavior. Psychologists distinguish between different defense mechanisms. These include denial, rational distancing, apathy, and delegation.

Denial is the refusal to accept reality. The person lives believing in a 'bright dream' (Mindell, 1988) and filters incoming information to fit his or her version of reality (e.g. climate sceptics have to ignore or reinterpret most of there search that comes out of the Intergovernmental Panel on climate change (IPCC), a panel of over 2500 reputable climate scientists). Denial will prevent a person from pro-environmental behavior because the person refuses to acknowledge the problem.

Rational distancing is another way of protecting oneself from painful emotions. The person who rationalizes is perfectly aware of the problems but has stopped to feel any emotions about it. This defense mechanism is especially common among scientists and environmentalists who are frequently exposed to 'bad news' [7]. We would hypothesize that people who have emotionally distanced themselves are less likely to engage in pro-environmental behavior, because their internal motivation to do so is much weaker.

Apathy and resignation are often the result of a person feeling pain, sadness, anger, and helplessness at the same time. If the person has a strong feeling that he or she cannot change the situation (see locus of control), he or she will very likely retreat into apathy, resignation, and sarcasm. A person might stop informing himself or herself about environmental issues and focus on different aspects of life. Such a person might still perform some pro-environmental actions out of a feeling of moral obligation but is very unlikely to become very proactive.

Delegation is a means to remove feelings of guilt. The person who delegates refuses to accept any personal responsibility and blames others for environmental destruction (e.g. the industries, the multi-nationals, the political establishment [8]). People who delegate are unlikely to take any pro-environmental behavior that asks for personal sacrifices.

Locus of Control

As defined earlier, locus of control represents an individual's perception of whether he or she has the ability to bring about change through his or her own behavior (Newhouse, 1991). People with a strong internal locus of control believe that their actions can bring about change. People with an external locus of control, on the other hand, feel that their actions are insignificant, and feel that change can only be brought about by powerful others (see paragraph on delegation). Such people are much less likely to act ecologically, since they feel that 'it does not make a difference anyway'.

Responsibility and Priorities

Our feelings of responsibility are shaped by our values and attitudes and are influenced by our locus of control. We prioritize our responsibilities. Most important to people is their own well-being and the well-being of their family (see Stern *et al.*'s (1993) model). When pro-environmental behaviors are in alignment with these personal priorities, the motivation to do them increases (e.g. buying organic food). If they contradict the priorities, the actions will less likely be taken (e.g. living in a smaller house, even though one could afford to live in a big one).

Conclusions

Many conflicting and competing factors shape our daily decisions and actions. Similarly, there are several factors that influence our decisions towards pro-environmental behavior that we have not elaborated on. We have omitted a discussion on our desires for comfort and convenience, two factors that certainly play an important role in shaping our pro-environmental behaviors. We have not discussed the influence of habits. If we want to establish a new behavior, we have to practice it (e.g. Fliegenschnee & Schelakovsky, 1998). We might be perfectly willing to change our behavior but still not do so, because we do not persist enough in practicing the new behavior until it has become a habit. Last but not least, we did not discuss the influence of personality traits and character on pro-environmental behavior.

Although we have already pointed out that developing a model that incorporates all the factors behind pro-environmental behavior might neither be feasible nor useful, we do find diagrams that serve as visual aides in clarifying and categorizing such factors helpful. We therefore conclude

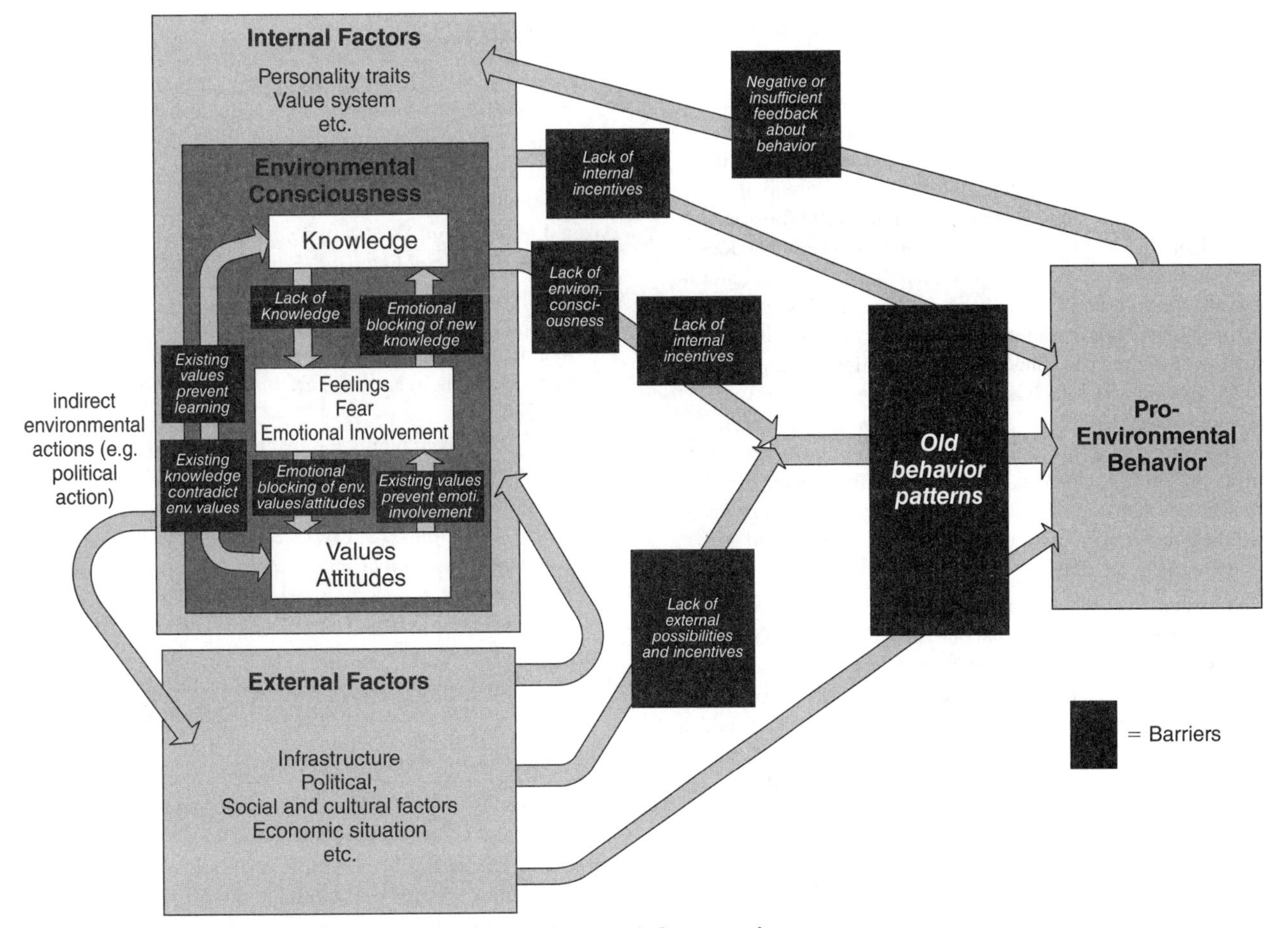

Figure 7 Model of pro-environmental behaviour (Kollmuss & Agyeman)

with our own graphic illustration of a possible model. As with the other models we have introduced, it has its advantages and short comings. We do not claim that this model is more sophisticated or inclusive than any of the other models. However, in designing it, we were influenced by many different authors, mostly Fliegenschnee and Schelakovsky (1998) who in turn based their diagram on the earlier discussed model of Fietkau and Kessel (1981).

As with Fietkau and Kessel (1981), we do not attribute a direct relationship to environmental knowledge and pro-environmental behavior. We see environmental knowledge, values, and attitudes, together with emotional involvement as making up a complex we call *'pro-environmental consciousness'*. This complex in turn is embedded in broader personal values and shaped by personality traits and other internal as well as external factors. We put social and cultural factors into the group of external factors even though it might be argued that social and cultural factors could be seen as a separate category which overlaps with internal and external factors. We also pondered if our model would differ at different stages in people's lives, and we agreed that it would not, but that the different factors inherent in it, and the synergies between them, would play greater or lesser roles during the development process. In addition, as we pointed out earlier, the longer the education, the more extensive is the knowledge about environmental issues. Yet more education does not necessarily mean increased pro-environmental behavior.

The arrows in Figure 7 indicate how the different factors influence each other and, ultimately, pro-environmental behavior. Most are self-explanatory. The two narrower arrows from internal and external factors directly to pro-environmental behavior indicate environmental actions that are taken for other than environmental reasons (e.g. consuming less because of a value system that promotes simplicity or because of external factors such as monetary constraints). The biggest positive influence on pro-environmental behavior, indicated by the larger arrow, is achieved when internal and external factors act synergistically.

The black boxes indicate possible barriers to positive influence on pro-environmental behavior. The model lists only a few of the most important barriers. In the diagram, the largest of them represents old behavior patterns. This is partly for graphical reasons—the barrier has to block all three arrows— but it is also because we want to draw attention to this aspect. We believe that old habits form a very strong barrier that is often overlooked in the literature on pro-environmental behavior.

Notes

1. Since we will analyze work in English and German publications, it is important to point out the subtle differences in meaning of the English *environment* and its German translation, *Umwelt. Environment* is defined as: 'The totality of circumstances surrounding an organism or a group of organisms' (*American Heritage Dictionary,* 1992, Boston, MA, Houghton-Mifflin). It is a very broad concept that does not have an *explicit* connection to the protection of the natural world. *Umwelt,* on the other hand, is almost exclusively used to describe natural environments and their destruction. It is a more narrow term that has a much stronger emotional component than *environment,* which is more abstract and scientific. *Umweltbewusstsein* (*environmental awareness*) has therefore a more emotional and ethical component to it in German than it has in English, where as the term *environmental awareness* emphasizes the cognitive awareness of environmental problems. *Umweltbewusssein* might more accurately be translated as 'environmental caring'.
2. Indirect environmental actions include donating money, political activities, educational outreach, environmental writing, etc. These activities, although extremely important, do not have a direct impact on the environment. Direct environmental actions include recycling, driving less, buying organic food, etc. These actions have a direct (admittedly sometimes very small) impact on the environment. We focus our study mostly on direct pro-environmental behavior.
3. Many of the tools and techniques that are used in community-based social marketing, such as norms, commitment, modeling, and social diffusion, all have at their core the interactions of individuals in a community. Norms develop as people interact and develop guidelines for their behavior (McKenzie-Mohr & Smith, 1999, p. 97).
4. We have made the assumption, that where an author uses 'ecologically', it is synonymous with 'environmentally'.
5. Pay per bag is a system in which garbage will only be collected if it is placed in pre-purchased bags. The theory is that if people have to purchase bags, they will cut down on their wastes, and recycle more.
6. Interestingly, in their study they found that driving correlates negatively with environmental attitude. This means that people drive more the more they care about the environment. This seemingly contradictory result can be explained when influences on environmental attitudes are explored. The more educated and affluent the people in the study were the more likely that they had a deeper environmental knowledge and a heightened sense of environmental awareness. At the same time, more affluent people tended to be more mobile, in other words, travel more.
7. Rational distancing is not always negative. It can be extremely important for people working in disaster areas. It allows the person not to be overwhelmed by the misery but react and plan cool-headedly.
8. We do note want to imply that everybody has the same influence or impact on environmental destruction. Some people have undoubtedly more influence, power, and ability to change things than others (see Blake, 1999).

References

Ackermann, F. (1997) *Why Do We Recyle? Markets, Values, and Public Policy* (Washington, DC, Island Press).

Agyeman, J. & Angus, B. (forthcoming) Community-based social marketing for sustainability: tools and approaches for changing personal transportation behaviour.

Ajzen, I. & Fishbein, M. (1980) *Understanding Attitudes and Predicting Social Behavior* (Englewood Cliffs, NJ, Prentice Hall).

Allen, J.B. & Ferrand, J. (1999) Environmental locus of control, sympathy, and pro-environmental behavior: a test of Geller's actively caring hypothesis, *Environment and Behavior,* 31(3), pp. 338–353.

Blake, J. (1999) Overcoming the 'value–action gap' in environmental policy: tensions between national policy and local experience, *Local Environment,* 4(3), pp. 257–278.

Bloomfield, D., Collins, K., Fry, C. & Munton, R. (1998) *Deliberative and Inclusionary Processes: their contribution to environmental governance* (London, Environment and Society Research Unit, Department of Geography, University College).

Boehmer-Christiansen, S. & Skea, J. (1991) *Acid Politics: environmental and energy policies in Britain and Germany* (New York, Belhaven Press).

Borden, D.& Francis, J.L. (1978) Who cares about ecology? Personality and sex difference in environmental concern, *Journal of Personality,* 46, pp. 190–203.

Burgess, J., Harrison, C. & Filius, P. (1998) Environmental communication and the cultural politics of environmental citizenship, *Environment and Planning A,* 30, pp. 1445–1460.

Chawla, L. (1998) Significant life experiences revisited: a review of research on sources of pro-environmental sensitivity, *The Journal of Environmental Education,* 29(3), pp. 11–21.

Chawla, L. (1999) Life paths into effective environmental action, *The Journal of Environmental Education,* 31(1), pp. 15–26.

Diekmann, A. & Franzen, A. (1996) Einsicht in ökologische Zusammenhänge und Umweltverhalten, in: R. Kaufmann-Hayoz & A. Di Giulio (Eds) *Umweltproblem Mensch: Humanwissenschaftliche Zusammenhänge zu umweltverantwortlichem Handeln* (Bern, Verlag Paul Haupt).

Diekmann, A. & Franzen, A. (1999) The wealth of nations, *Environment and Behavior,* 31(4), pp. 540–549.

Diekmann, A. & Preisendoerfer, P. (1992) Persoenliches Umweltverhalten: Die Diskrepanz zwischen Anspruch und Wirklichkeit *Koelner Zeitschrift fuer Soziologie und Sozialpsychologie,* 44, pp. 226–251.

Eisenberg, N. & Miller, P. (1987) The relation of empathy to prosocial and related behaviors, *Psychological Bulletin,* 101, pp. 91–119.

Festinger, L. (1957) *Theory of Cognitive Dissonance* (Stanford, CA, Stanford University Press).

Fietkau, H.-J. & Kessel, H. (1981) *Umweltlernen: Veraenderungsmoeglichkeite n des Umweltbewusstseins. Modell-Erfahrungen* (Koenigstein, Hain).

Fishbein, M. & Ajzen, I. (1975) *Belief, Attitude, Intention, and Behavior: an introduction to theory and research* (Reading, MA, Addison-Wesley).

Fliegenschnee, M. & Schelakovsky, M. (1998) *Umweltpsychologie und Umweltbildung: eine Einführung aus humanökologischer Sicht* (Wien, Facultas Universitäts Verlag).

Fuhrer, U., Kaiser, F.G., Seiler, J. & Maggi, M. (1995) From social representations to environmental concern: the influence of face to face versus mediated communication, in: U. Fuhrer (Ed.) *Oekologisches Handeln als sozialer Prozess* (Basel, Birkhaeuser).

Gigliotti, L.M. (1992) Environmental attitudes: 20 years of change?, *The Journal of Environmental Education,* 24(1), pp. 15–26.

Gigliotti, L.M. (1994) Environmental issues: Cornell students' willingness to take action, *The Journal of Environmental Education,* 25(1), pp. 34–42.

Grob, A. (1991) *Meinung, Verhalten, Umwelt* (Bern, Peter Lang Verlag).

Hines, J.M., Hungerford, H.R. & Tomera, A.N. (1986–87). Analysis and synthesis of research on responsible pro-environmental behavior: a meta-analysis, *The Journal of Environmental Education,* 18(2), pp. 1–8.

Hungerford, H.R. & Volk, T.L. (1990) Changing learner behavior through environmental education, *The Journal of Environmental Education,* 21(3), pp. 8–21.

Kaiser, F.G., Woelfing, S. & Fuhrer, U. (1999) Environmental attitude and ecological behavior, *Journal of Environmental Psychology,* 19, pp. 1–19.

Kempton, W., Boster, J.S. & Hartley, J.A. (1995) *Environmental Values in American Culture* (Cambridge, MA, MIT Press).

Lehmann, J. (1999) *Befunde empirischer Forschung zu Umweltbildung und Umweltbewusstsein* (Opladen, Leske und Budrich).

McKenzie-Mohr, D. & Smith, W. (1999) *Fostering Sustainable Behavior: an introduction to community-based social marketing* (Gabriola Island, Canada, New Society Publishers).

McLaren, D., Bullock, S. & Yousuf, N. (1998) *Tomorrow's World. Britain's Share in a Sustainable Future* (London, Earthscan).

Mindell, A. (1988) *City Shadow's: psychological interventions in psychiatry* (London, Routledge).

Moisander, J. (1998) *Motivation for Ecologically Oriented Consumer Behavior,* Workshop Proceedings, March. The European Science Foundation (ESF) TERM (Tackling Environmental Resource Management Phase II 1998–2000). www.lancs.ac.uk/users/scistud/esf/lind2.htm

Newhouse, N. (1991) Implications of attitude and behavior research for environmental conservation, *The Journal of Environmental Education,* 22(1), pp. 26–32.

O'Riordan, T. & Burgess, J. (Eds) (1999) *Deliberative and Inclusionary Processes: a report of two seminars* (Norwich, CSERGE, School of Environmental Sciences).

Owens, S. (2000) Engaging the public: information and deliberation in environmental policy, *Environment and Planning A,* 32, pp. 1141–1148.

Preuss, S. (1991) Umweltkatastrophe Mensch. Ueber unsere Grenzen und Moeglichkeiten, oekologisch bewusst zu handeln (Heidelberg, Roland Asanger Verlag).

Rajecki, D.W. (1982) *Attitudes: themes and advances* (Sunderland, MA, Sinauer).

Redclift, M. & Benton, T. (1994) Introduction, in: M. Redclift & T. Benton (Eds) *Social Theory and the Global Environment* (London, Routledge).

Regis, D. (1990) Self-concept and conformity in theories of health education, Doctoral disertation, School of Education, University of Exeter. http://helios.ex.ac.uk/~dregis/PhD/Contents.html

Sia, A.P., Hungerford, H.R. & Tomera, A.N. (1985–86) Selected predictors of responsible environmental behavior: an analysis, *The Journal of Environmental Education,* 17(2), pp. 31–40.

Schwartz, S.H. (1977) Normative influences on altruism, in: L. Berkowitz (Ed.) *Advances in Experimental Social Psychology,* Vol. 10 (New York, Academic Press).

Stern, P.S., Dietz, T. & Karlof, L. (1993) Values orientation, gender, and environmental concern, *Environment and Behavior,* 25(3), pp. 322–348.

Von Weizaecker, E.U. & Jesinghaus, J. (1992) *Ecological Tax Reform* (New Jersey, Zed Books).

Wackernagel, M. & Rees, W.(1997) *Unser oelogischer Fussabdruck: Wie der Mensch Einfluss auf die Umwelt nimmt [Our Ecological Footprint]* (Basel, Switzerland, Birkhaeuser Verlag).

Wilkie, W.L. (1990) *Consumer Behavior,* 2nd edn (New York, John Wiley & Sons).

Critical Thinking

1. Why is it that the question of what shapes sustainable behavior is so complex?
2. What do you think would be the findings of research designed to explore why people behave in unsustainable ways?
3. How important do you think it is to the quest for sustainability that we have models of sustainable behaviors such as the one proposed in this article?

Anja Kollmuss received her BA from Harvard Extension and her MA in Environmental Policy from Tufts University. She currently works as the outreach coordinator for the Tufts Climate Initiative, educating students, faculty, and staff about global warming and climate change mitigation strategies.

Julian Agyeman is Assistant Professor of Environmental Policy and Planning at Tufts University, Boston-Medford. His interests are in social marketing for sustainability, education for sustainability, community involvement in local environmental and sustainability policy, environmental justice and the development of sustainable communities. He is founder, and co-editor of the international journal *Local Environment* and his book, *Just Sustainabilities: Development in an Unequal World,* is due out later this year.

Do Global Attitudes and Behaviors Support Sustainable Development?

ANTHONY A. LEISEROWITZ, ROBERT W. KATES, AND THOMAS M. PARRIS

Many advocates of sustainable development recognize that a transition to global sustainability—meeting human needs and reducing hunger and poverty while maintaining the life-support systems of the planet—will require changes in human values, attitudes, and behaviors.[1] A previous article in *Environment* described some of the values used to define or support sustainable development as well as key goals, indicators, and practices.[2] Drawing on the few multinational and quasi-global-scale surveys that have been conducted,[3] this article synthesizes and reviews what is currently known about global attitudes and behavior that will either support or discourage a global sustainability transition.[4] (Table 1 provides details about these surveys.)

None of these surveys measured public attitudes toward "sustainable development" as a holistic concept. There is, however, a diverse range of empirical data related to many of the subcomponents of sustainable development: development and environment; the driving forces of population, affluence/poverty/consumerism, technology, and entitlement programs; and the gap between attitudes and behavior.

Development

Concerns for environment and development merged in the early concept of sustainable development, but the meaning of these terms has evolved over time. For example, global economic development is widely viewed as a central priority of sustainable development, but development has come to mean human and social development as well.

Economic Development

The desire for economic development is often assumed to be universal, transcending all cultural and national contexts. Although the surveys in Table 1 have no global-scale data on public attitudes toward economic development per se, this assumption appears to be supported by 91 percent of respondents from 35 developing countries, the United States, and Germany, who said that it is very important (75 percent) or somewhat important (16 percent) to live in a country where there is economic prosperity[5]. What level of affluence is desired, how that economic prosperity is to be achieved, and how economic wealth should ideally be distributed within and between nations, however, are much more contentious questions. Unfortunately, there does not appear to be any global-scale survey research that has tried to identify public attitudes or preferences for particular levels or end-states of economic development (for example, infinite growth versus steady-state economies) and only limited or tangential data on the ideal distribution of wealth (see the section on affluence below).

Data from the World Values Survey suggest that economic development leads to greater perceived happiness as countries make the transition from subsistence to advanced industrial

Table 1 Multinational Surveys

Name	Year(s)	Number of Countries
One-time Surveys		
Pew Global Attitudes Project	2002	43
Eurobarometer	2002	15
International Social Science Program	2000	25
Health of the Planet	1992	24
Repeated Surveys		
GlobeScan International Environmental Monitor	1997–2003	34
World Values Survey	1981–2002	79
Demographic and Health Surveys	1986–2002	17
Organisation for Economic Co-operation and Development	1990–2002	22

Note. Before November 2003, GlobeScan, Inc. was known as Environics International. Surveys before this time bear the older name.

Source: For more detail about these surveys and the countries sampled, see Appendix A in A. Leiserowitz, R. W. Kates, and T. M. Parris, *Sustainability Values, Attitudes and Behaviors: A Review of Multi-national and Global Trends,* CID Working Paper No. 113 (Cambridge, MA: Science, Environment and Development Group, Center for International Development, Harvard University, 2004), www.cid.harvard.edu/cidwp/113.htm.

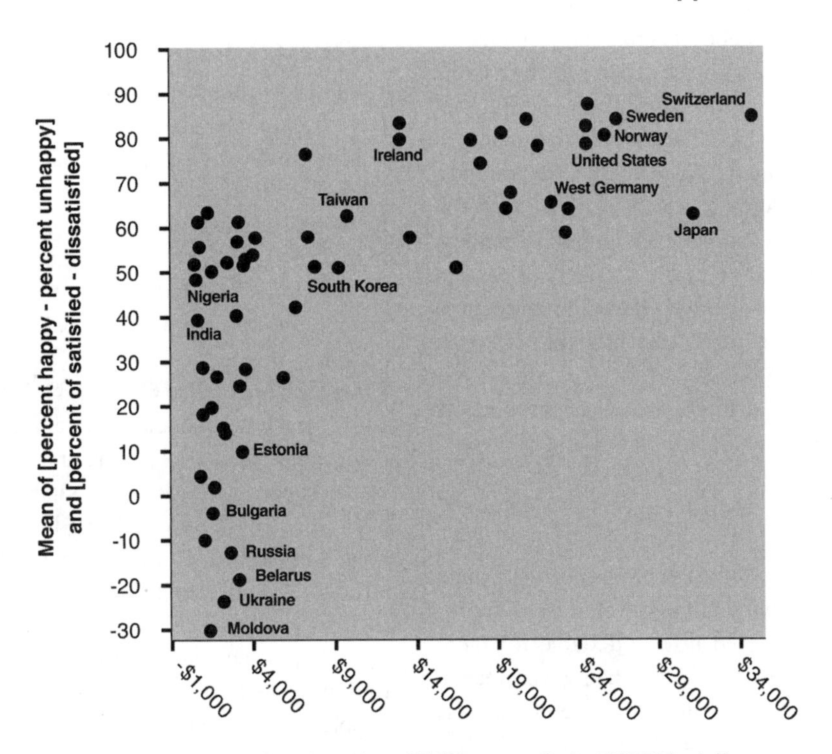

Figure 1 Subjective Well-being by Level of Economic Development.

Note. The subjective well-being index reflects the average of the percentage in each country who describe themselves as "very happy" or "happy" minus the percentage who describe themselves as "not very happy" or "unhappy"; and the percentage placing themselves in the 7–10 range, minus the percentage placing themselves in the 1–4 range, on a 10-point scale on which 1 indicates that one is strongly dissatisfied with one's life as a whole, and 10 indicates that one is highly satisfied with one's life as a whole.

Source: R. Inglehart, "Globalization and Postmodern Values," *Washington Quarterly* 23, no. 1 (1999): 215–228. Subjective well-being data from the 1990 and 1996 World Values Surveys. GNP per capita for 1993 data from *World Bank, World Development Report, 1995* (New York: Oxford University Press, 1995).

economies. But above a certain level of gross national product (GNP) per capita—approximately $14,000—the relationship between income level and subjective well-being disappears (see Figure 1). This implies that infinite economic growth does not lead to greater human happiness. Additionally, many of the unhappiest countries had, at the time of these surveys, recently experienced significant declines in living standards with the collapse of the Soviet Union. Yet GNP per capita remained higher in these ex-Soviet countries than in developing countries like India and Nigeria.[6] This suggests that relative trends in living standards influence happiness more than absolute levels of affluence, but the relationship between economic development and subjective well-being deserves more research attention.

Human Development

Very limited data is available on public attitudes toward issues of human development, although it can be assumed that there is near-universal support for increased child survival rates, adult life expectancies, and educational opportunities. However, despite the remarkable increases in these indicators of human well-being since World War II,[7] there appears to be a globally pervasive sense that human well-being has been deteriorating in recent years. In 2002, large majorities worldwide said that a variety of conditions had worsened over the previous five years, including the availability of well-paying jobs (58 percent); working conditions (59 percent); the spread of diseases (66 percent);the affordability of health care (60 percent); and the ability of old people to care for themselves in old age (59 percent). Likewise, thinking of their own countries, large majorities worldwide were concerned about the living conditions of the elderly (61 percent) and the sick and disabled (56 percent), while a plurality was concerned about the living conditions of the unemployed (42 percent).[8]

Development Assistance

One important way to promote development is to extend help to poorer countries and people, either through national governments or nongovernmental organizations and charities. There is strong popular support but less official support for development assistance to poor countries. In 1970, the United Nations General Assembly resolved that each economically advanced country would dedicate 0.7 percent of its gross national income (GNI) to official development assistance (ODA) by the middle of the 1970s—a target that has been reaffirmed in many subsequent international agreements.[9] As of 2004, only five countries had achieved this goal (Denmark, Norway, the Netherlands, Luxembourg, and Sweden). Portugal was close to the target at 0.63, yet all other countries ranged from a high of 0.42 percent (France) to lows of 0.16 and 0.15 percent (the United States and Italy respectively). Overall, the average ODA/GNI among the industrialized countries was only 0.25 percent—far below the UN target.[10]

By contrast, in 2002, more than 70 percent of respondents from 21 developed and developing countries said they would support paying 1 percent more in taxes to help the world's poor.[11] Likewise, surveys in the 13 countries of the Organisation for Economic Co-operation and Development's Development Assistance Committee (OECD-DAC) have found that public support for the principle of giving aid to developing countries (81 percent in 2003) has remained high and stable for more than 20 years.[12] Further, 45 percent said that their government's current (1999–2001) level of expenditure on foreign aid was too low, while only 10 percent said foreign aid was too high.[13] There is also little evidence that the public in OECD countries has developed "donor fatigue." Although surveys have found increasing public concerns about corruption, aid diversion, and inefficiency, these surveys also continue to show very high levels of public support for aid.

Public support for development aid is belied, however, by several factors. First, large majorities demonstrate little understanding of development aid, with most unable to identify their national aid agencies and greatly overestimating the percentage of their national budget devoted to development aid. For example, recent polls have found that Americans believed their government spent 24 percent (mean estimate) of the national budget on foreign assistance, while Europeans often estimated their governments spent 5 to 10 percent.[14] In reality, in 2004 the United States spent approximately 0.81 percent and the European Union member countries an average of approximately 0.75 percent of their national budgets on official development assistance, ranging from a low of 0.30 percent (Italy) to a high of 1.66 percent (Luxembourg).[15] Second, development aid is almost always ranked low on lists of national priorities, well below more salient concerns about (for example) unemployment, education, and health care. Third, "the overwhelming support for foreign aid is based upon the perception that it will be spent on remedying humanitarian crises," not used for other development-related issues like Third World debt, trade barriers, or increasing inequality between rich and poor countries—or for geopolitical reasons (for example, U.S. aid to Israel and Egypt).[16] Support for development assistance has thus been characterized as "a mile wide, but an inch deep" with large majorities supporting aid (in principle) and increasing budget allocations but few understanding what development aid encompasses or giving it a high priority.[17]

Environment

Compared to the very limited or nonexistent data on attitudes toward economic and human development and the overall concept of sustainable development, research on global environmental attitudes is somewhat more substantial. Several surveys have measured attitudes regarding the intrinsic value of nature, global environmental concerns, the trade-offs between environmental protection and economic growth, government policies, and individual and household behaviors.

Human-Nature Relationship

Most research has focused on anthropocentric concerns about environmental quality and natural resource use, with less attention to ecocentric concerns about the intrinsic value of nature. In 1967, the historian Lynn White Jr. published a now-famous and controversial article arguing that a Judeo-Christian ethic and attitude of domination, derived from Genesis, was an underlying historical and cultural cause of the modern environmental crisis.[18] Subsequent ecocentric, ecofeminist, and social ecology theorists have also argued that a domination ethic toward people, women, and nature runs deep in Western, patriarchal, and capitalist culture.[19] The 2000 World Values Survey, however, found that 76 percent of respondents across 27 countries said that human beings should "coexist with nature," while only 19 percent said they should "master nature" (see Figure 2). Overwhelming majorities of Europeans, Japanese, and North Americans said that human beings should coexist with nature, ranging from 85 percent in the United States to 96 percent in Japan. By contrast, only in Jordan, Vietnam, Tanzania, and the Philippines did more than 40 percent say that human beings should master nature.[20] In 2002, a national survey of the United States explored environmental values in more depth and found that Americans strongly agreed that nature has intrinsic value and that humans have moral duties and obligations to animals, plants, and non-living nature (such as rocks, water, and air). The survey found that Americans strongly disagreed that "humans have the right to alter nature to satisfy wants and desires" and that "humans are not part of nature" (see Figure 3).[21] This very limited data suggests that large majorities in the United States and worldwide now reject a domination ethic as the basis of the human-nature relationship, at least at an abstract level. This question, however, deserves much more cross-cultural empirical research.

Environmental Concern

In 2000, a survey of 11 developed and 23 developing countries found that 83 percent of all respondents were concerned a fair amount (41 percent) to a great deal (42 percent) about environmental problems. Interestingly, more respondents from developing countries (47 percent) were "a great deal concerned" about the environment than from developed countries (33 percent), ranging from more than 60 percent in Peru, the Philippines, Nigeria, and India to less than 30 percent in the Netherlands, Germany, Japan, and Spain.[22] This survey also

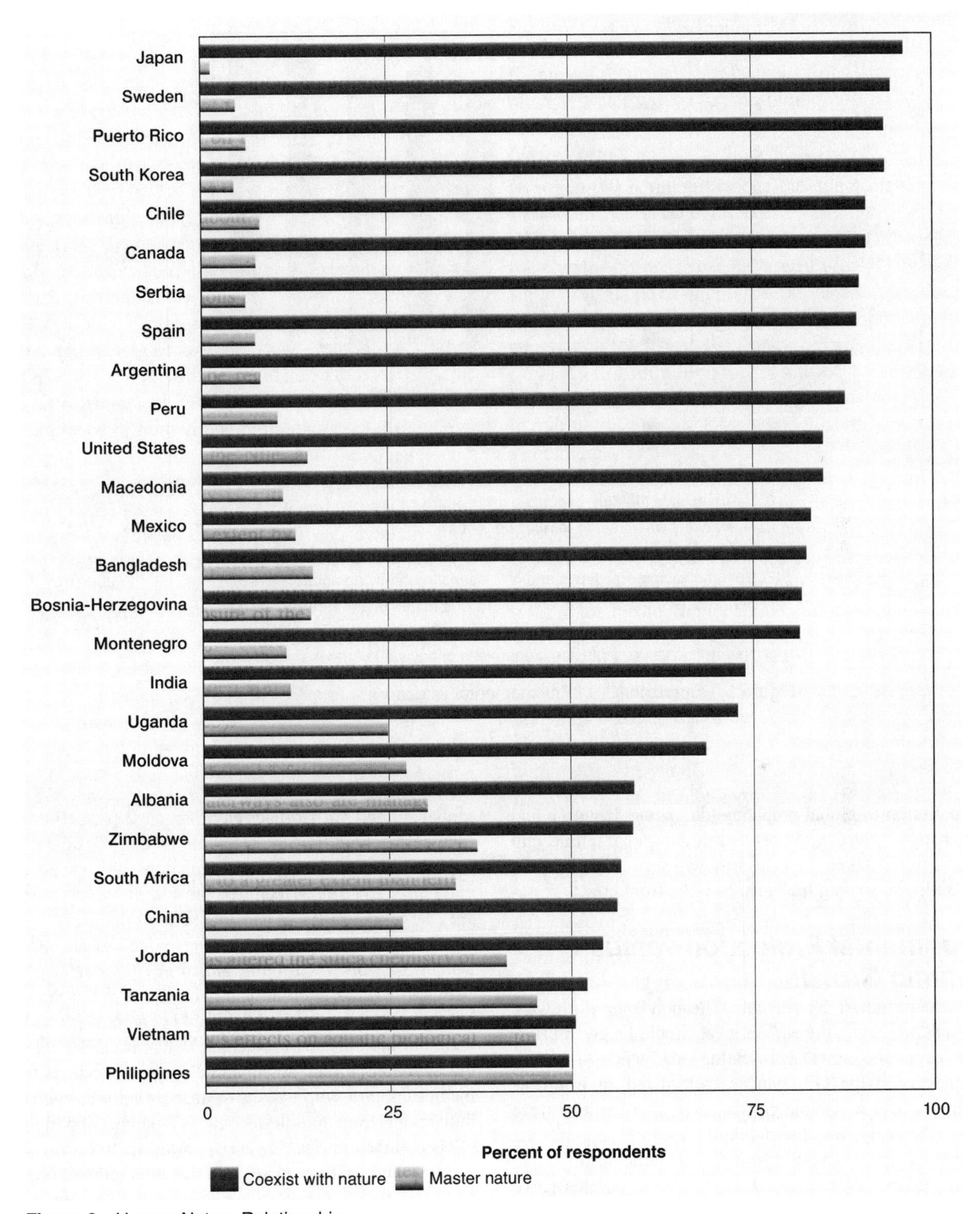

Figure 2 Human-Nature Relationship.

Note. The question asked, "Which statement comes closest to your own views: Human beings should master nature or humans should coexist with nature?"

Source: A. Leiserowitz, 2005. Data from world Values Survey, *The 1999–2002 Values Surveys Integrated Data File 1.0, CD-ROM in R. Inglehart, M. Basanez, J. Diez-Medrano, L. Halman, and R. Luijkx, eds., Human Beliefs and Values: A Cross-Cultural Sourcebook Based on the 1999–2002 Values Surveys, first edition* (Mexico City: Siglo XXI, 2004).

asked respondents to rate the seriousness of several environmental problems (see Figure 4). Large majorities worldwide selected the strongest response possible ("very serious") for seven of the eight problems measured. Overall, these results demonstrate very high levels of public concern about a wide range of environmental issues, from local problems like water

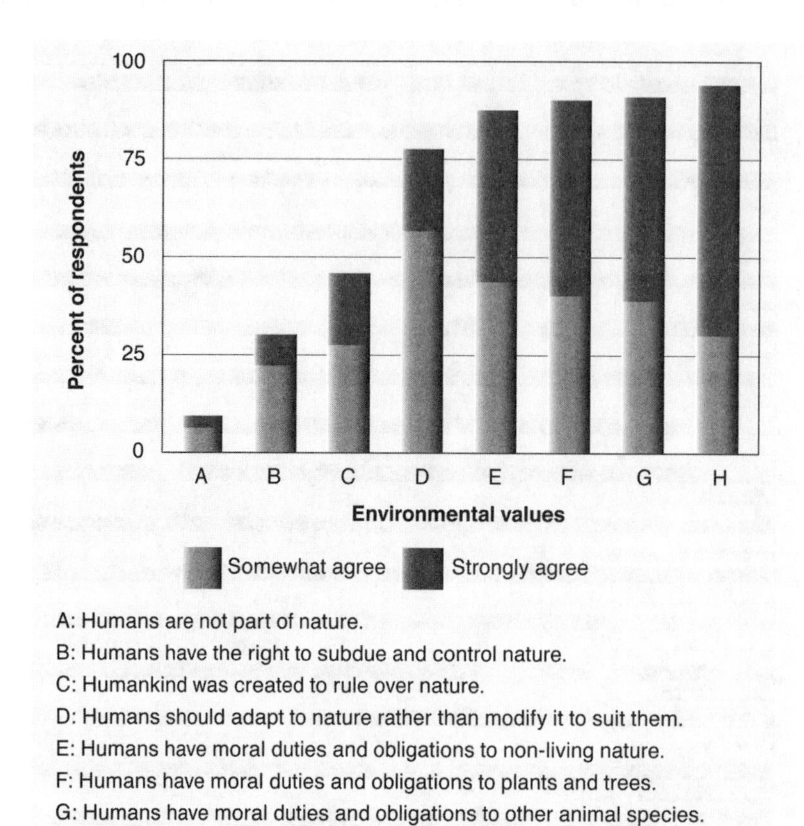

Figure 3 American (U.S.) Environmental Values.

Source: A. Leiserowitz, 2005.

and air pollution to global problems like ozone depletion and climate change.[23] Further, 52 percent of the global public said that if no action is taken, "species loss will seriously affect the planet's ability to sustain life" just 20 years from now.[24]

Environmental Protection versus Economic Growth

In two recent studies, 52 percent of respondents worldwide agreed that "protecting the environment should be given priority" over "economic growth and creating jobs," while 74 percent of respondents in the G7 countries prioritized environmental protection over economic growth, even if some jobs were lost.[25] Unfortunately, this now-standard survey question pits the environment against economic growth as an either/or dilemma. Rarely do surveys allow respondents to choose an alternative answer, that environmental protection can generate economic growth and create jobs (for example, in new energy system development, tourism, and manufacturing).

Attitudes toward Environmental Policies

In 1995, a large majority (62 percent) worldwide said they "would agree to an increase in taxes if the extra money were used to prevent environmental damage," while 33 percent said they would oppose them.[26] In 2000, there was widespread global support for stronger environmental protection laws and regulations, with 69 percent saying that, at the time of the survey, their national laws and regulations did not go at all far enough.[27] The 1992 Health of the Planet survey found that a very large majority (78 percent) favored the idea of their own national government "contributing money to an international agency to work on solving global environmental problems." Attitudes toward international agreements in this survey, however, were less favorable. In 1992, 47 percent worldwide agreed that "our nation's environmental problems can be solved without any international agreements," with respondents from low-income countries more likely to strongly agree (23 percent) than individuals from middle-income (17 percent) or high-income (12 percent) countries.[28] In 2001, however, 79 percent of respondents from the G8 countries said that international negotiations and progress on climate change was either "not good enough" (39 percent) or "not acceptable" (40 percent) and needed faster action. Surprisingly, this latter 40 percent supported giving the United Nations "the power to impose legally-binding actions on national governments to protect the Earth's climate."[29]

Environmental Behavior

Material consumption is one of the primary means by which environmental values and attitudes get translated into behavior. (For attitudes toward consumption per se, see the following section on affluence, poverty, and consumerism.)

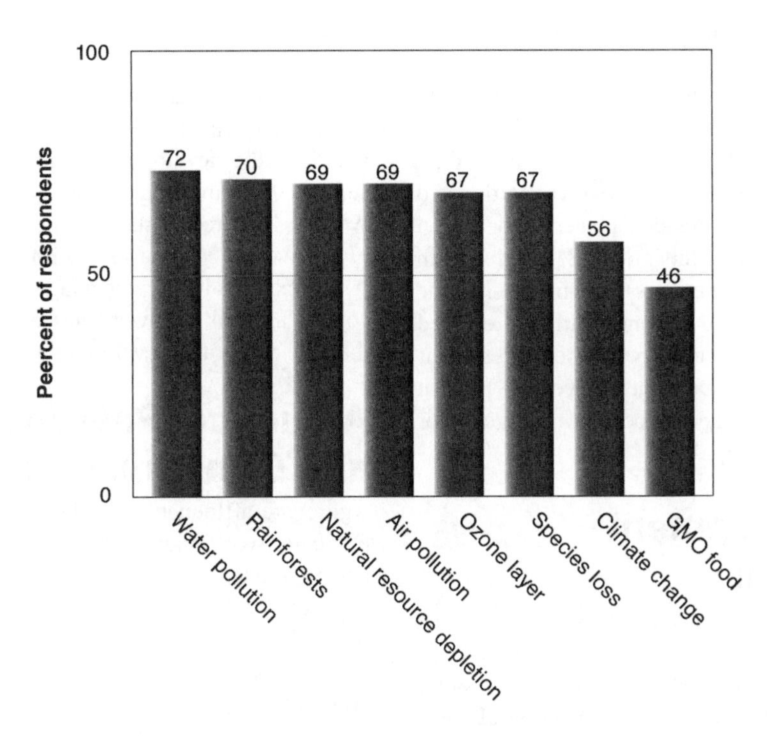

Figure 4 Percent of Global Public Calling Environmental Issues a "Very Serious Problem."

Source: A. Leiserowitz, 2005. Data from Environics International (Globe Scan), *Environics International Environmental Monitor Survey Dataset* (Kingston, Canada: Environics international, 2000), http://jeff-lab.queensu.ca/poadata/info/iem/iemlist.shtml (accessed 5 October 2004).

In 2002, Environics International (GlobeScan) found that 36 percent of respondents from 20 developed and developing countries stated that they had avoided a product or brand for environmental reasons, while 27 percent had refused packaging, and 25 percent had gathered environmental information.[30] Recycling was highly popular, with 6 in 10 people setting aside garbage for reuse, recycling, or safe disposal. These rates, however, reached 91 percent in North America versus only 36–38 percent in Latin America, Eastern Europe, and Central Asia,[31] which may be the result of structural barriers in these societies (for example, inadequate infrastructures, regulations, or markets). There is less survey data regarding international attitudes toward energy consumption, but among Europeans, large majorities said they had reduced or intended to reduce their use of heating, air conditioning, lighting, and domestic electrical appliances.[32]

In 1995, 46 percent of respondents worldwide reported having chosen products thought to be better for the environment, 50 percent of respondents said they had tried to reduce their own water consumption, and 48 percent reported that in the 12 months prior to the survey, they reused or recycled something rather than throwing it away. There was a clear distinction between richer and poorer societies: 67 percent of respondents from high-income countries reported that they had chosen "green" products, while only 30 percent had done so in low-income countries. Likewise, 75 percent of respondents from high-income countries said that they had reused or recycled something, while only 27 percent in low-income countries said this.[33] However, the latter results contradict the observations of researchers who have noted that many people in developing countries reuse things as part of everyday life (for example, converting oil barrels into water containers) and that millions eke out an existence by reusing and recycling items from landfills and garbage dumps.[34] This disparity could be the result of inadequate survey representation of the very poor, who are the most likely to reuse and recycle as part of survival, or, alternatively, different cultural interpretations of the concepts "reuse" and "recycle."

In 2002, 44 percent of respondents in high-income countries were very willing to pay 10 percent more for an environmentally friendly car, compared to 41 percent from low-income countries and 29 percent from middle-income countries.[35] These findings clearly mark the emergence of a global market for more energy-efficient and less-polluting automobiles. However, while many people appear willing to spend more to buy an environmentally friendly car, most do not appear willing to pay more for gasoline to reduce air pollution. The same 2002 survey found that among high-income countries, only 28 percent of respondents were very willing to pay 10 percent more for gasoline if the money was used to reduce air pollution, compared to 23 percent in medium-income countries and 36 percent in low-income countries.[36] People appear to generally oppose higher gasoline prices, although public attitudes are probably affected, at least in part, by the prices extant at the time of a given survey, the rationale given for the tax, and how the income from the tax will be spent.

Despite the generally pro-environment attitudes and behaviors outlined above, the worldwide public is much less likely to engage in political action for the environment. In 1995, only 13 percent of worldwide respondents reported having donated to an environmental organization, attended a meeting, or signed a petition for the environment in the prior 12 months, with more doing so in high-income countries than in low-income countries.[37] Finally, in 2000, only 10 percent worldwide reported having written a letter or made a telephone call to express their concern about an environmental issue in the past year, 18 percent had based a vote on green issues, and 11 percent belonged to or supported an environmental group.[38]

Drivers of Development and Environment

Many analyses of the human impact on life-support systems focus on three driving forces: population, affluence or income, and technology—the so-called I = PAT identity.[39] In other words, environmental impact is considered a function of these three drivers. In a similar example, carbon dioxide (CO_2) emissions from the energy sector are often considered a function of population, affluence (gross domestic product (GDP) per capita), energy intensity (units of energy per GDP), and technology (CO_2 emissions per unit of energy).[40] While useful, most analysts also recognize that these variables are not fundamental driving forces in and of themselves and are not independent from one another.[41] A similar approach has also been applied to human development (D = PAE), in which development is considered a function of population, affluence, and entitlements and equity.[42] What follows is a review of empirical trends in attitudes and behavior related to population, affluence, technology, and equity and entitlements.

Population

Global population continues to grow, but the rate of growth continues to decline almost everywhere. Recurrent Demographic and Health Surveys (DHS) have found that the ideal number of children desired is declining worldwide. Globally, attitudes toward family planning and contraception are very positive, with 67 percent worldwide and large majorities in 38 out of 40 countries agreeing that birth control and family planning have been a change for the better.[43] Worldwide, these positive attitudes toward family planning are reflected in the behavior of more than 62 percent of married women of reproductive age who are currently using contraception. Within the developing world, the United Nations reports that from 1990 to 2000, contraceptive use among married women in Asia increased from 52 percent to 66 percent, in Latin American and the Caribbean from 57 percent to 69 percent, but in Africa from only 15 percent to 25 percent.[44] Notwithstanding these positive attitudes toward contraception, in 1997, approximately 20 percent to 25 percent of births in the developing world were unwanted, indicating that access to or the use of contraceptives remains limited in some areas.[45]

DHS surveys have found that ideal family size remains significantly larger in western and middle Africa (5.2) than elsewhere in the developing world (2.9).[46] They also found that support for family planning is much lower in sub-Saharan Africa (44 percent) than in the rest of the developing world (74 percent).[47] Consistent with these attitudes, sub-Saharan Africa exhibits lower percentages of married women using birth control as well as lower rates of growth in contraceptive use than the rest of the developing world.[48]

Affluence, Poverty, and Consumerism

Aggregate affluence and related consumption have risen dramatically worldwide with GDP per capita (purchasing-power parity, constant 1995 international dollars) more than doubling between 1975 and 2002.[49] However, the rising tide has not lifted all boats. Worldwide in 2001, more than 1.1 billion people lived on less than $1 per day, and 2.7 billion people lived on less than $2 per day—with little overall change from 1990. However, the World Bank projects these numbers to decline dramatically by 2015—to 622 million living on less than $1 per day and 1.9 billion living on less than $2 per day. There are also large regional differences, with sub-Saharan Africa the most notable exception: There, the number of people living on less than $1 per day rose from an estimated 227 million in 1990 to 313 million in 2001 and is projected to increase to 340 million by 2015.[50]

Poverty

Poverty reduction is an essential objective of sustainable development.[51] In 1995, 65 percent of respondents worldwide said that more people were living in poverty than had been 10 years prior. Regarding the root causes of poverty, 63 percent blamed unfair treatment by society, while 26 percent blamed the laziness of the poor themselves. Majorities blamed poverty on the laziness and lack of willpower of the poor only in the United States (61 percent), Puerto Rico (72 percent), Japan (57 percent), China (59 percent), Taiwan (69 percent), and the Philippines (63 percent) (see Figure 5).[52] Worldwide, 68 percent said their own government was doing too little to help people in poverty within their own country, while only 4 percent said their government was doing too much. At the national level, only in the United States (33 percent) and the Philippines (21 percent) did significant proportions say their own government was doing too much to help people in poverty.[53]

Consumerism

Different surveys paint a complicated and contradictory picture of attitudes toward consumption. On the one hand, majorities around the world agree that, at the societal level, material and status-related consumption are threats to human cultures and the environment. Worldwide, 54 percent thought "less emphasis on money and material possessions" would be a good thing, while

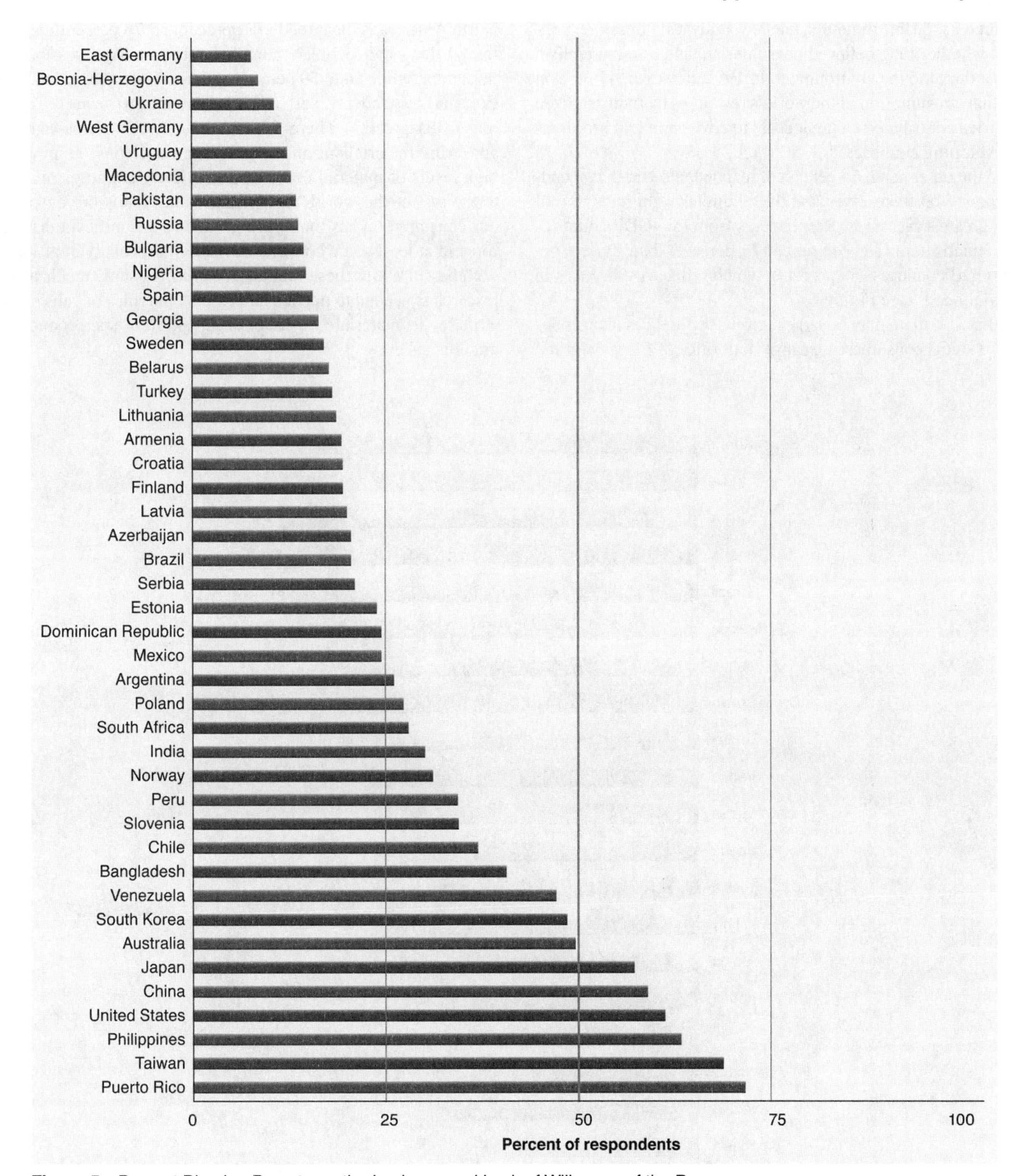

Figure 5 Percent Blaming Poverty on the Laziness and Lack of Willpower of the Poor.

Source: A. Leiserowitz, 2005. Data from R. Inglehart, et al., *World Values Surveys and European Values Surveys, 1981–1984, 1990–1993, and 1995–1997* [computer file], Inter-university Consortium for Political and Social Research (ICPSR) version (Ann Arbor, MI: Institute for Social Research [producer], 2000; Ann Arbor, MI: ICPSR [distributor], 2000).

only 21 percent thought this would be a bad thing.[54] Further, large majorities agreed that gaining more time for leisure activities or family life is their biggest goal in life.[55]

More broadly, in 2002 a global study sponsored by the Pew Research Center for the People & the Press found that 45 percent worldwide saw consumerism and commercialism as a threat to their own culture. Interestingly, more respondents from high-income and upper middle-income countries (approximately 51 percent) perceived consumerism as a threat than low-middle- and low-income countries (approximately

43 percent).[56] Unfortunately, the Pew study did not ask respondents whether they believed consumerism and commercialism were a threat to the environment. In 1992, however, 41 percent said that consumption of the world's resources by industrialized countries contributed "a great deal" to environmental problems in developing countries."[57]

On the other hand, 65 percent of respondents said that spending money on themselves and their families represents one of life's greatest pleasures. Respondents from low-GDP countries were much more likely to agree (74 percent) than those from high-GDP countries (58 percent), which reflects differences in material needs (see Figure 6).[58]

Likewise, there may be large regional differences in attitudes toward status consumerism. Large majorities of Europeans and North Americans disagreed (78 percent and 76 percent respectively) that other people's admiration for one's possessions is important, while 54 to 59 percent of Latin American, Asian, and Eurasian respondents, and only 19 percent of Africans (Nigeria only), disagreed.[59] There are strong cultural norms against appearing materialistic in many Western societies, despite the high levels of material consumption in these countries relative to the rest of the world. At the same time, status or conspicuous consumption has long been posited as a significant driving force in at least some consumer behavior, especially in affluent societies.[60] While these studies are a useful start, much more research is needed to unpack and explain the roles of values and attitudes in material consumption in different socioeconomic circumstances.

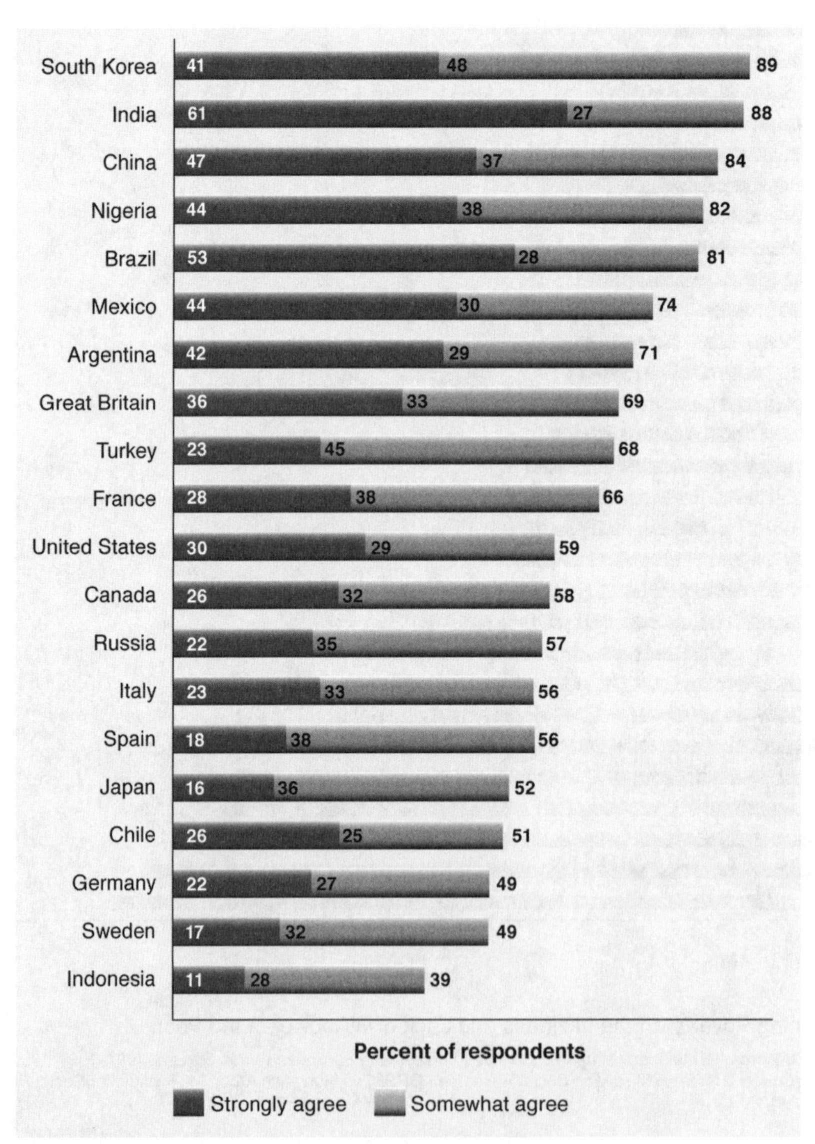

Figure 6 Purchasing for Self and Family Gives Greatest Pleasure ("Strongly" and "Somewhat" agree).

Note. The question was, "To spend money, to buy something new for myself or my family, is one of the greatest pleasures in my life."

Source: Environics International (GlobeScan), *Consumerism: A Special Report* (Toronto: Environics International, 2002). 6.

Science and Technology

Successful deployment of new and more efficient technologies is an important component of most sustainability strategies, even though it is often difficult to assess all the environmental, social, and public health consequences of these technologies in advance. Overall, the global public has very positive attitudes toward science and technology. The 1995 World Values Survey asked respondents, "In the long run, do you think the scientific advances we are making will help or harm mankind?" Worldwide, 56 percent of respondents thought science will help mankind, while 26 percent thought it will harm mankind. Further, 67 percent said an increased emphasis on technological development would be a good thing, while only 9 percent said it would be bad.[61] Likewise, in 2002, GlobeScan found large majorities worldwide believed that the benefits of modern technology outweigh the risks.[62] The support for technology, however, was significantly higher in countries with low GDPs (69 percent) than in high-GDP countries (56 percent), indicating more skepticism among people in technologically advanced societies. Further, this survey found dramatic differences in technological optimism between richer and poorer countries. Asked whether "new technologies will resolve most of our environmental challenges, requiring only minor changes in human thinking and individual behavior," 62 percent of respondents from low-GDP countries agreed, while 55 percent from high-GDP countries disagreed (see Figure 7).

But what about specific technologies with sustainability implications? Do these also enjoy strong public support? What follows is a summary of global-scale data on attitudes toward renewable energy, nuclear power, the agricultural use of chemical pesticides, and biotechnology.

Europeans strongly preferred several renewable energy technologies (solar, wind, and biomass) over all other energy sources, including solid fuels (such as coal and peat), oil, natural gas, nuclear fission, nuclear fusion, and hydroelectric power. Also, Europeans believed that by the year 2050, these energy sources will be best for the environment (67 percent), be the least expensive (40 percent), and will provide the greatest amount of useful energy (27 percent).[63] Further, 37 percent of Europeans and approximately 33 percent of respondents in 16 developed and developing countries were willing to pay 10 percent more for electricity derived from renewable energy sources.[64]

Nuclear power, however, remains highly stigmatized throughout much of the developed world.[65] Among respondents from 18 countries (mostly developed), 62 percent considered nuclear power stations "very dangerous" to "extremely dangerous" for the environment.[66] Whatever its merits or demerits as an alternative energy source, public attitudes about nuclear power continue to constrain its political feasibility.

Regarding the use of chemical pesticides on food crops, a majority of people in poorer countries believed that the benefits are greater than the risks (54 percent), while respondents in high-GDP countries were more suspicious, with only 32 percent believing the benefits outweigh the risks.[67] Since 1998, however, support for the use of agricultural chemicals has dropped worldwide. Further, chemical pesticides are now one of the top food-related concerns expressed by respondents around the world.[68]

Additionally, the use of biotechnology in agriculture remains controversial worldwide, and views on the issue are divided between rich and poor countries. Across the G7 countries, 70 percent of respondents were opposed to scientifically altered fruits and vegetables because of health and environmental concerns,[69] while 62 percent of Europeans and 45 percent of Americans opposed the use of biotechnology in agriculture.[70] While majorities in poorer countries (65 percent) believed the benefits of using biotechnology on food crops are greater than the risks, majorities in high-GDP countries (51 percent) believed the risks outweigh the benefits.[71]

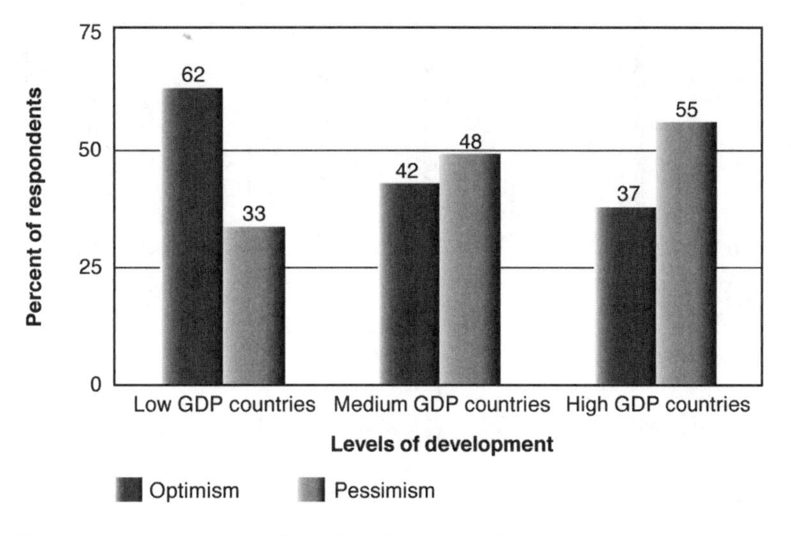

Figure 7 Technological Optimism Regarding Environmental Problems.

Source: A. Leiserowitz, 2005. Data from Environics International (GlobeScan), *International Environmental Monitor* (Toronto: Environics International, 2002), 135.

More broadly, public understanding of biotechnology is still limited, and slight variations in question wordings or framings can have significant impacts on support or opposition. For example, 56 percent worldwide thought that biotechnology will be good for society in the long term, yet 57 percent also agreed that "any attempt to modify the genes of plants or animals is ethically and morally wrong."[72] Particular applications of biotechnology also garnered widely different degrees of support. While 78 percent worldwide favored the use of biotechnology to develop new medicines, only 34 percent supported its use in the development of genetically modified food. Yet, when asked whether they supported the use of biotechnology to produce more nutritious crops, 61 percent agreed.[73]

Income Equity and Entitlements

Equity and entitlements strongly determine the degree to which rising population and affluence affect human development, particularly for the poor. For example, as global population and affluence have grown, income inequality between rich and poor countries has also increased over time, with the notable exceptions of East and Southeast Asia—where incomes are on the rise on a par with (or even faster than) the wealthier nations of the world.[74] Inequality within countries has also grown in many rich and poor countries. Similarly, access to entitlements—the bundle of income, natural resources, familial and social connections, and societal assistance that are key determinants of hunger and poverty[75]—has recently declined with the emergence of market-oriented economies in Eastern and Central Europe, Russia, and China; the rising costs of entitlement programs in the industrialized countries, including access to and quality of health care, education, housing, and employment; and structural adjustment programs in developing countries that were recommended by the International Monetary Fund. Critically, it appears there is no comparative data on global attitudes toward specific entitlements; however, there is much concern that living conditions for the elderly, unemployed, and the sick and injured are deteriorating, as cited above in the discussion on human development.

In 2002, large majorities said that the gap between rich and poor in their country had gotten wider over the previous 5 years. This was true across geographic regions and levels of economic development, with majorities ranging from 66 percent in Asia, 72 percent in North America, and 88 percent in Eastern Europe (excepting Ukraine) stating that the gap had gotten worse.[76] Nonetheless, 48 percent of respondents from 13 countries preferred a "competitive society, where wealth is distributed according to one's achievement," while 34 percent preferred an "egalitarian society, where the gap between rich and poor is small, regardless of achievement" (see Figure 8).[77]

More broadly, 47 percent of respondents from 72 countries preferred "larger income differences as incentives for individual effort," while 33 percent preferred that "incomes should be made more equal."[78] These results suggest that despite public perceptions of growing economic inequality, many accept it as an important incentive in a more individualistic and competitive economic system. These global results, however, are limited to just a few variables and gloss over many countries that strongly prefer more egalitarian distributions of wealth (such as India). Much more research is needed to understand how important the principles of income equality and equal economic opportunity are considered globally, either as global goals or as means to achieve other sustainability goals.

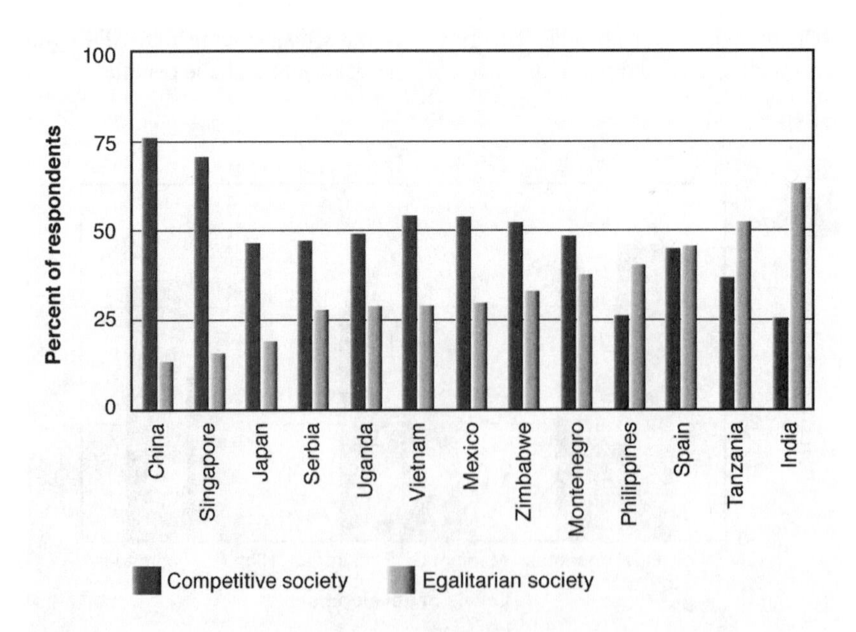

Figure 8 Multinational Preferences for a Competitive Versus Egalitarian Society.

Source: A. Leiserowitz, 2005. Data from World Values Survey. *The 1999–2002 Values Surveys Integrated Data File 1.0*, CD-ROM in R. Inglehart, M. Basanez, J. Diez-Medrano, L. Halman, and R. Luijkx, eds., *Human Beliefs and Values: A Cross-Cultural Sourcebook Based on the 1999–2002 Values Surveys*, first edition (Mexico City: Siglo XX[[]]I, 2004).

Does the Global Public Support Sustainable Development?

Surprisingly, the question of public support for sustainable development has never been asked directly, at least not globally. But two important themes emerge from the multinational data and analysis above. First, in general, the global public supports the main tenets of sustainable development. Second, however, there are many contradictions, including critical gaps between what people say and do—both as individuals and in aggregate. From these themes emerge a third finding: Diverse barriers stand between sustainability attitudes and action.

- *Large majorities worldwide appear to support environmental protection and economic and human development—the three pillars of sustainable development.* They express attitudes and have taken modest actions consonant with support for sustainable development, including support for environmental protection; economic growth; smaller populations; reduced poverty; improved technology; and care and concern for the poor, the marginal, the young, and the aged.
- *Amid the positive attitudes, however, are many contradictions.* Worldwide, all the components of the Human Development Index—life expectancy, adult literacy, and per capita income—have dramatically improved since World War II.[79] Despite the remarkable increases in human well-being, however, there appears to be a globally pervasive sense that human well-being has more recently been deteriorating. Meanwhile, levels of development assistance are consistently overestimated by lay publics, and the use of such aid is misunderstood, albeit strongly supported. Overall, there are very positive attitudes toward science and technology, but the most technologically sophisticated peoples are also the most pessimistic about the ability of technology to solve global problems. Likewise, attitudes toward biotechnology vary widely, depending on how the question is asked.
- Further, there are serious gaps between what people believe and what people do, both as individuals and as polities. Worldwide, the public strongly supports significantly larger levels of development assistance for poor countries, but national governments have yet to translate these attitudes into proportional action. Most people value the environment—for anthropocentric as well as ecocentric reasons—yet many ecological systems around the world continue to degrade, fragment, and lose resilience. Most favor smaller families, family planning, and contraception, but one-fifth to one-quarter of children born are not desired. Majorities are concerned with poverty and think more should be done to alleviate it, but important regions of the world think the poor themselves are to blame, and a majority worldwide accepts large gaps between rich and poor. Most people think that less emphasis on material possessions would be a good thing and that more time for leisure and family should be primary goals, but spending money often provides one of life's greatest pleasures. While many would pay more for fuel-efficient cars, fuel economy has either stagnated or even declined in many countries. Despite widespread public support for renewable energy, it still accounts for only a tiny proportion of global energy production.
- *There are diverse barriers standing between pro-sustainability attitudes and individual and collective behaviors.*[80] These include at least three types of barriers. First are the direction, strength, and priority of particular attitudes. Some sustainability attitudes may be widespread but not strongly or consistently enough relative to other, contradictory attitudes. A second type of barrier between attitudes and behavior relates to individual capabilities. Individuals often lack the time, money, access, literacy, knowledge, skills, power, or perceived efficacy to translate attitudes into action. Finally, a third type of barrier is structural and includes laws; regulations; perverse subsidies; infrastructure; available technology; social norms and expectations; and the broader social, economic, and political context (such as the price of oil, interest rates, special interest groups, and the election cycle).

Thus, each particular sustainability behavior may confront a unique set of barriers between attitudes and behaviors. Further, even the same behavior (such as contraceptive use) may confront different barriers across society, space, and scale—with different attitudes or individual and structural barriers operating in developed versus developing countries, in secular versus religious societies, or at different levels of decisionmaking (for example, individuals versus legislatures). Explaining unsustainable behavior is therefore "dauntingly complex, both in its variety and in the causal influences on it."[81] Yet bridging the gaps between what people believe and what people do will be an essential part of the transition to sustainability.

Promoting Sustainable Behavior

Our limited knowledge about global sustainability values, attitudes, and behaviors does suggest, however, that there are short and long-term strategies to promote sustainable behavior. We know that socially pervasive values and attitudes are often highly resistant to change. Thus, in the short term, leveraging the values and attitudes already dominant in particular cultures may be more practical than asking people to adopt new value orientations.[82] For example, economic values clearly influence and motivate many human behaviors, especially in the market and cash economies of the developed countries. Incorporating environmental and social "externalities" into prices or accounting for the monetary value of ecosystem services can thus encourage both individual and collective sustainable behavior.[83] Likewise, anthropocentric concerns about the impacts of environmental degradation and exploitative labor conditions on human health and social well-being remain strong motivators for action in both the developed and developing worlds.[84] Additionally, religious values are vital sources of meaning, motivation, and direction for much of the world, and many religions are actively re-evaluating and reinterpreting their traditions in support of sustainability.[85]

In the long term, however, more fundamental changes may be required, such as extending and accelerating the shift from materialist to post-materialist values, from anthropocentric to ecological worldviews, and a redefinition of "the good life."[86] These long term changes may be driven in part by impersonal forces, like changing economics (globalization) or technologies (for example, mass media and computer networks) or by broadly based social movements, like those that continue to challenge social attitudes about racism, environmental degradation, and human rights. Finally, sustainability science will play a critical role, at multiple scales and using multiple methodologies, as it works to identify and explain the key relationships between sustainability values, attitudes, and behaviors—and to apply this knowledge in support of sustainable development.

Notes

1. For example, see U.S. National Research Council, Policy Division, Board on Sustainable Development, *Our Common Journey: A Transition toward Sustainability* (Washington, DC: National Academy Press, 1999); and P. Raskin et al., *Great Transition: The Promise and Lure of the Times Ahead* (Boston: Stockholm Environment Institute, 2002).
2. R. W. Kates, T. M. Parris, and A. Leiserowitz, "What Is Sustainable Development? Goals, Indicators, Values, and Practice,"*Environment,* April 2005, 8–21.
3. For simplicity, the words "global" and "worldwide" are used throughout this article to refer to survey results. Please note, however, that there has never been a truly representative global survey with either representative samples from every country in the world or in which all human beings worldwide had an equal probability of being selected. Additionally, some developing country results are taken from predominantly urban samples and are thus not fully representative.
4. For more detail about these surveys and the countries sampled, see Appendix A in A. Leiserowitz, R. W. Kates, and T. M. Parris, *Sustainability Values, Attitudes and Behaviors: A Review of Multi-national and Global Trends* (No. CID Working Paper No. 113) (Cambridge, MA: Science, Environment and Development Group, Center for International Development, Harvard University, 2004), www.cid.harvard.edu/cidwp/113.htm
5. Pew Research Center for the People & the Press, *Views of a Changing World* (Washington, DC: The Pew Research Center for the People & the Press, 2003), T72.
6. See R. Inglehart, "Globalization and Postmodern Values," *Washington Quarterly* 23, no. 1 (1999): 215–28.
7. Leiserowitz, Kates, and Parris, note 4 above, page 8.
8. Pew Research Center for the People & the Press, *The Pew Global Attitudes Project Dataset* (Washington, DC: The Pew Research Center for the People & the Press, 2004).
9. Gross national income (GNI) is "[t]he total market value of goods and services produced during a given period by labor and capital supplied by residents of a country, regardless of where the labor and capital are located. [GNI] differs from GDP primarily by including the capital income that residents earn from investments abroad and excluding the capital income that nonresidents earn from domestic investment." Official development assistance (ODA) is defined as "[t]hose flows to developing countries and multilateral institutions provided by official agencies, including state and local governments, or by their executive agencies, each transaction of which meets the following tests: (a) it is administered with the promotion of the economic development and welfare of developing countries as its main objective; and (b) it is concessional in character and conveys a grant element of at least 25 percent." UN Millennium Project, *The O. 7 percent Target: An In-Depth Look,* www.unmillenniumproject.org/involved/action07.htm (accessed 24 August 2005). Official development assistance (ODA) does not include aid flows from private voluntary organizations (such as churches, universities, or foundations). For example, it is estimated that in 2000, the United States provided more than $4 billion in private grants for development assistance, versus nearly $10 billion in ODA. U.S. Agency for International Development (USAID), *Foreign Aid in the National Interest* (Washington. DC, 2002), 134.
10. Organisation for Economic Co-operation and Development (OECD), *Official Development Assistance Increases Further—But 2006 Targets Still a Challenge* (Paris: OECD, 2005), www.oecd.org/document/3/0,2340,en_2649_34447_34700611_ 1_1_1_1,00.html (accessed 30 July 2005).
11. Environics International (GlobeScan), *The World Economic Forum Poll: Global Public Opinion on Globalization* (Toronto: Environics International, 2002), www.globescan.com/brochures/WEF_Poll_Brief.pdf (accessed 5 October 2004), 3. Note that Environics International changed its name to Globe Scan Incorporated in November 2003.
12. OECD, *Public Opinion and the Fight Against Poverty* (Paris: OECD Development Centre, 2003), 17.
13. Ibid, page 19.
14. Program on International Policy Attitudes (PIPA), *Americans on Foreign Aid and World Hunger: A Study of U.S. Public Attitudes* (Washington, DC: PIPA, 2001), www.pipa.org/OnlineReports/BFW (accessed 17 November 2004); and OECD, note 12 above, page 22.
15. See OECD Development Co-operation Directorate, *OECD-DAC Secretariat Simulation of DAC Members' Net ODA Volumes in 2006 and 2010,* www.oecd.org/dataoecd/57/30/35320618.pdf; and Central Intelligence Agency, The World Factbook, www.cia.gov/cia/publications/factbook/.
16. OECD, note 12 above, page 20.
17. I. Smillie and H. Helmich, eds., *Stakeholders: Government-NGO Partnerships for International Development* (London: Earthscan, 1999).

18. L. White Jr., "The Historical Roots of Our Ecologic Crisis," *Science,* 10 March 1967, 1203–07.

19. See C. Merchant, *The Death of Nature: Women, Ecology, and the Scientific Revolution* (1st ed.) (San Francisco: Harper & Row 1980); C. Merchant, *Radical Ecology: The Search for a Livable Worm* (New York: Routledge, 1992); and G. Sessions, *Deep Ecology for the Twenty-First Century* (1st ed.) (New York: Shambhala Press 1995).

20. World Values Survey, *The 1999–2002 Values Surveys Integrated Data File 1.0,* CD-ROM in R. Inglehart, M. Basanez, J. Diez-Medrano, L. Halman, and R. Luijkx, eds., *Human Beliefs and Values: A Cross-Cultural Sourcebook Based on the 1999–2002 Values Surveys,* first edition (Mexico City: Siglo XXI, 2004).

21. These results come from a representative national survey of American Climate change risk perceptions, policy preferences, and behaviors and broader environmental and cultural values. From November 2002 to February 2003, 673 adults (18 and older) completed a mail-out, mail-back questionnaire, for a response rate of 55 percent. The results are weighted to bring them in line with actual population proportions. See A. Leiserowitz, "American Risk Perceptions: Is Climate Change Dangerous?" *Risk Analysis,* in press; and A. Leiserowitz, "Climate Change Risk Perception and Policy Preferences: The Role of Affect, Imagery, and Values," *Climatic Change,* in press.

22. These results support the argument that concerns about the envi-ronment are not "a luxury affordable only by those who have enough economic security to pursue quality-of-life goals." See R. E. Dunlap, G. H. Gallup Jr., and A. M. Gallup, "Of Global Concern: Results of the Health of the Planet Survey," *Environment,* November 1993, 7–15, 33–39 (quote at 37); R. E. Dunlap, A. G. Mertig, "Global Concern for the Environment: Is Affluence a Prerequisite?" *Journal of Social Issues* 511, no. 4 (1995): 121–37; S. R. Brechin and W. Kempton, "Global Environmentalism: A Challenge to the Postmaterialism Thesis?" *Social Science Quarterly* 75, no. 2 (1994): 245–69.

23. Environics International (GlobeScan), *Environics International Environmental Monitor Survey Data-set* (Kingston, Canada: Environics International, 2000), http://jeff-lab.queensu.ca/poadata/info/iem/iemlist.shtml (accessed 5 October 2004). These multinational levels of concern and perceived seriousness of environmental problems remained roughly equivalent from 1992 to 2000, averaged across the countries sampled by the 1992 Health of the Planet and the Environics surveys, although some countries saw significant increases in perceived seriousness of environmental problems (India, the Netherlands, the Philippines, and South Korea), while others saw significant decreases (Turkey and Uruguay). See R. E. Dunlap, G. H. Gallup Jr., and A. M. Gallup, *Health of the Planet: Results of a 1992 International Environmental Opinion Survey of Citizens in 24 Nations* (Princeton, N J: The George H. Gallup International Institute, 1993); and R. E. Dunlap, G. H. Gallup Jr., and A. M. Gallup, "Of Global Concern: Results of the Health of the Planet Survey," *Environment,* November 1993, 7–15, 33–39.

24. GlobeScan, *Results of First-Ever Global Poll on Humanity's Relationship with Nature* (Toronto: GlobeScan Incorporated, 2004), www.globescan.com/news_archives/IUCN_PR.html (accessed 30 July 2005).

25. World Values Survey, note 20 above; and Pew Research Center for the People & the Press, *What the World Thinks in 2002* (Washington, DC: The Pew Research Center for the People & the Press, 2002), T-9. The G7 includes Canada, France, Germany, Great Britain, Italy, Japan and the United States. It expanded to the G8 with the addition of Russia in 1998.

26. R. Inglehart, et al., *World Values Surveys and European Values Surveys, 1981–1984, 1990–1993, and 1995–1997* [computer file], Inter-university Consortium for Political and Social Research (ICPSR) version (Ann Arbor, MI: Institute for Social Research [producer], 2000; Ann Arbor, MI: ICPSR [distributor], 2000).

27. Environics International (GlobeScan), note 23 above.

28. Dunlap, Gallup Jr., and Gallup, *Health of the Planet: Results of a 1992 International Environmental Opinion Survey of Citizens in 24 Nations,* note 23 above.

29. Environics International (GlobeScan), *New Poll Shows G8 Citizens Want Legally-Binding Climate Accord* (Toronto: Environics International, 2001), www.globescan.com/news_archives/IEM_climatechange.pdf (accessed 30 July 2005).

30. Environics International (GlobeScan), *International Environmental Monitor* (Toronto: Environics International, 2002), 44.

31. Ibid., page 49.

32. The European Opinion Research Group, *Eurobarometer: Energy: Issues, Options and Technologies, Science and Society,* EUR 20624 (Brussels: European Commission, 2002), 96–99.

33. Inglehart, note 26 above.

34. C. M. Rogerson, "The Waste Sector and Informal Entrepreneurship in Developing World Cities," *Urban Forum* 12, no. 2 (2001): 247–59.

35. Environics International (GlobeScan), note 30 above, page 63. These results are based on the sub-sample of those who own or have regular use of a car.

36. Environics International (GlobeScan), note 30 above, page 65.

37. Inglehart, note 26 above.

38. Environics International (GlobeScan), note 23 above.

39. P. A. Ehrlich and J. P. Holdren, review of *The Closing Circle,* by Barry Commoner, *Environment,* April 1972, 24, 26–39.

40. Y. Kaya, "Impact of Carbon Dioxide Emission Control on GNP Growth: Interpretation of Proposed Scenarios," paper presented at the Intergovernmental Panel on Climate Change (IPCC) Energy and Industry Subgroup, Response Strategies Working Group, Paris, France, 1990; and R. York, E. Rosa, and T. Dietz, "STIRPAT, IPAT and ImPACT: Analytic Tools for Unpacking the Driving Forces of Environmental Impacts," *Ecological Economics* 46, no. 3 (2003): 351.

41. IPCC, *Emissions Scenarios* (Cambridge: Cambridge University Press, 2000); and E. F. Lambin, et al., "The Causes of Land-Use and Land-Cover Change: Moving Beyond the Myths," *Global Environmental Change: Human and Policy Dimensions* 11, no. 4 (2001):
42. T. M. Parris and R. W. Kates, "Characterizing a Sustainability Transition: Goals, Targets, Trends, and Driving Forces," *Proceedings of the National Academy of Sciences of the United States of* America 100, no. 14 (2003): 6.
43. Pew Research Center for the People & the Press, note 8 above, page T17.
44. United Nations, *Majority of World's Couples Are Using Contraception* (New York: United Nations Population Division, 2001).
45. J. Bongaarts, "Trends in Unwanted Childbearing in the Developing World," *Studies in Family Planning* 28, no. 4 (1997): 267–77.
46. Demographic and Health Surveys (DHS), *STATCompiler* (Calverton, MD: Measure DHS, 2004), www.measuredhs.com/ (accessed 5 October, 2004).
47. Ibid.
48. U.S. Bureau of the Census, *World Population Profile: 1998,* WP/98 (Washington, DC, 1999), 45.
49. World Bank, *World Development Indicators CD-ROM 2004* [computer file] (Washington, DC: International Bank for Reconstruction and Development (IBRD) [producer], 2004).
50. World Bank, *Global Economic Prospects 2005: Trade, Regionalism, and Development* [computer file] (Washington, DC: IBRD [producer] 2005).
51. For more information on poverty reduction strategies, see T. Banuri, review of *Investing in Development: A Practical Plan to Achieve the Millennium Goals,* by UN Millennium Project, *Environment,* November 2005 (this issue), 37.
52. Inglehart, note 26 above.
53. Inglehart, note 26 above.
54. Inglehart, note 26 above.
55. Environics International (GlobeScan), *Consumerism: A Special Report* (Toronto: Environics International, 2002), 6.
56. Pew Research Center for the People & the Press, note 25 above.
57. Dunlap, Gallup Jr., and Gallup, *Health of the Planet: Results of a 1992 International Environmental Opinion Survey of Citizens in 24 Nations,* note 23 above, page 57.
58. Environics International (GlobeScan), note 55 above, pages 3–4.
59. Environics International (GlobeScan), note 55 above, pages 3–4.
60. T. Veblen, *The Theory of the Leisure Class: An Economic Study of Institutions* (New York: Macmillan 1899).
61. Inglehart, note 26 above.
62. Environics International (GlobeScan), note 30 above, page 133.
63. The European Opinion Research Group, note 32 above, page 70.
64. Environics International (GlobeScan), note 23 above.
65. For example, see J. Flynn, P. Slovic, and H. Kunreuther, *Risk, Media and Stigma: Understanding Public Challenges to Modern Science and Technology* (London: Earthscan, 2001).
66. International Social Science Program, *Environment II,* (No. 3440) (Cologne: Zentralarchiv für Empirische Sozialforschung, Universitaet zu Koeln (Central Archive for Empirical Social Research, University of Cologne), 2000), 114.
67. Environics International (GlobeScan), note 30 above, page 139.
68. Environics International (GlobeScan), note 30 above, page 141.
69. Pew Research Center for the People & the Press, note 25 above, page T20.
70. Chicago Council on Foreign Relations (CCFR), *Worldviews 2002* (Chicago: CCFR, 2002), 26.
71. Environics International (GlobeScan), note 30 above, page 163.
72. Environics International (GlobeScan), note 30 above, page 156–57.
73. Environics International (GlobeScan), note 30 above, page 57.
74. W. J. Baumol, R. R. Nelson, and E. N. Wolff, *Convergence of Productivity: Cross-National Studies and Historical Evidence* (New York: Oxford University Press, 1994).
75. A. K. Sen, *Poverty and Famines: An Essay on Entitlement and Deprivation* (Oxford: Oxford University Press, 1981).
76. Pew Research Center for the People & the Press, note 5 above, page 37.
77. World Values Survey, note 20 above.
78. World Values Survey, note 20 above.
79. The human development index (HDI) measures a country's average achievements in three basic aspects of human development: longevity, knowledge, and a decent standard of living. Longevity is measured by life expectancy at birth; knowledge is measured with the adult literacy rate and the combined primary, secondary, and tertiary gross enrollment ratio; and standard of living is measured by gross domestic product per capita (purchase-power parity US$). The UN Development Programme (UNDP) has used the HDI for its annual reports since 1993. UNDP, *Questions About the Human Development Index (HDI),* www.undp.org/hdr2003/faq.html#21 (accessed 25 August 2005).
80. See, for example, J. Blake, "Overcoming the 'Value-Action Gap' in Environmental Policy: Tensions Between National Policy and Local Experience," *Local Environment* 4, no. 3 (1999): 257–78; A. Kollmuss and J. Agyeman, "Mind the Gap: Why Do People Act Environmentally and What Are the Barriers to Pro-EnvironmentalBehavior?" *Environmental Education Research* 8, no. 3 (2002): 239–60; and E C. Stem, "Toward a Coherent Theory of Environmentally Significant Behavior,"*Journal of Social Issues* 56, no. 3 (2000): 407–24.

81. Stern, ibid., page 421.
82. See, for example, P. W. Schultz and L. Zelezny, "Reframing Environmental Messages to Be Congruent with American Values," *Human Ecology Review* 10, no. 2 (2003): 126–36.
83. Millennium Ecosystem Assessment, *Ecosystems and Human Well-Being: Synthesis* (Washington, DC: Island Press, 2005).
84. Dunlap, Gallup Jr., and Gallup, *Health of the Planet: Results of a 1992 International Environmental Opinion Survey of Citizens in 24 Nations,* note 23 above, page 36.
85. See *The Harvard Forum on Religion and Ecology,* http://environment.harvard.edu/religion/main.html; R. S. Gottlieb, *This Sacred Earth: Religion, Nature, Environment* (New York: Routledge, 1996); and G. Gardner, *Worldwatch Paper # 164: Invoking the Spirit: Religion and Spirituality in the Quest for a Sustainable World* (Washington, DC: Worldwatch Institute, 2002).
86. R. Inglehart, *Modernization and Postmodernization: Cultural, Economic and Political Change in 43 Societies* (Princeton: Princeton University Press, 1997); T. O'Riordan, "Frameworks for Choice: Core Beliefs and the Environment," *Environment,* October 1995, 4–9, 25–29; and E Raskin and Global Scenario Group, *Great Transition: The Promise and Lure of the Times Ahead* (Boston: Stockholm Environment Institute, 2002).

Critical Thinking

1. Is it possible for the answer to either of the two factors identified in this question to be no?
2. Discuss the statement: "Despite the generally pro-environment attitudes and behaviors outlined above, the worldwide public is much less likely to engage in political action for the environment." Can you identify a possible explanation for this situation?
3. How much of an effect do you think a global-scale survey dataset that had sustainable development as its primary focus would have on the quest to achieve sustainability?

Anthony A. Leiserowitz is a research scientist at Decision Research and an adjunct professor of environmental studies at the University of Oregon, Eugene. He is also a principal investigator at the Center for Research on Environmental Decisions at Columbia University. Leiserowitz may be reached at (541) 485-2400 or by email at ecotone@uoregon.edu. **Robert W. Kates** is an independent scholar based in Trenton, Maine, and a professor emeritus at Brown University, where he served as director of the Feinstein World Hunger Program. He is also a former vice-chair of the Board of Sustainable Development of the U.S National Academy's National Research Council. In 1991, Kates was awarded the National Medal of Science for his work on hunger, environment, and natural hazards. He is an executive editor of *Environment* and may be contacted at rkates@acadia.net. **Thomas M. Parris** is a research scientist at and director of the New England office of ISCIENCES, LLC. He is a contributing editor of *Environment.* Parris may be reached at parris@isciences.com. The authors retain copyright.

The Latest On Trends In

Nature-Based Outdoor Recreation

H. KEN CORDELL

Considerable interest in better understanding current trends in nature-based outdoor recreation followed publication of Richard Louv's book, *Last Child in the Woods,* and a recent paper by Oliver R.W. Pergams and Patricia A. Zaradic titled "Evidence for a Fundamental and Pervasive Shift away from Nature-Based Recreation."[1] This latter paper attributed a decline in Americans' interest in nature-based recreation to the popularity of electronic entertainment. Is it really the case that nature-based recreation is declining in the United States? A recent national report, *Outdoor Recreation for 21st Century America,* indicated that Americans' participation in nature-based recreation activities had been rising up to the early part of this decade.[2] But is it still increasing?

Some of the speculation that nature-based outdoor recreation is declining is based on the well-documented growth of popularity of computers, home theaters, video games, and other electronic equipment by both adults and youth. These tools and entertainments compete for people's time and attention. Louv speculated that children are becoming less connected with nature. Others have written that recent changes in lifestyles have caused a significant shift away from nature-based recreation in the United States and abroad.[3] Because such conclusions could have important consequences—reducing federal, state, and other funding for natural resource conservation and for recreation management—a closer look at this speculation about declining interest in nature and recreating in the outdoors is warranted.

This paper gives an overview of outdoor recreation trends in the United States generally, and then looks at nature-based recreation specifically. Nature-based recreation is defined as outdoor activities in natural settings or otherwise involving in some direct way elements of nature—terrain, plants, wildlife, water bodies. Historical perspective is offered to help set the stage for looking at today's nature-based recreation. From this overview, insights for forest, other natural resource, and public land programs and policies are offered.

The Rise of Outdoor Recreation

Following the Great Depression and World War II, outdoor recreation emerged as a major component of many Americans' lifestyles. In 1960, the U.S. population reached about 131 million and was growing steadily. Along with increasing affluence, this population growth boosted demand for outdoor recreation. Family vacations and summer trips quickly became a significant part of the typical American's calendar. Recognizing and responding to this growing demand, the federal government initiated a study of the supply-and-demand situation for outdoor recreation in the 1950s and early 1960s. This study was conducted by the Outdoor Recreation Resources Review Commission (ORRRC), which reported its findings to the president and the Congress in 1962. That report set off a cascade of legislation, funding initiatives, and policy changes at both federal and state levels.[4] Congress passed a series of acts creating resources such as the National Wilderness Preservation System, the National Wild and Scenic Rivers System, the National Trails System, a system of National Recreation Areas, and the Land and Water Conservation Fund.

As reported by ORRRC, the most popular summertime outdoor recreation activities in 1960 were, in order, driving for pleasure, swimming, walking, playing outdoor games or sports, sightseeing, picnicking, fishing, bicycling, attending outdoor sports events, boating, nature walks, and hunting. Also popular were camping, horseback riding, water skiing, hiking, and attending concerts or other outdoor events. Though the technology of outdoor equipment and clothing has evolved dramatically over the years since then, all of the activities popular then are still popular with the American public. As we shall see, however, outdoor technology is not all that has changed since 1960.

Twenty years after the ORRRC report was published, the 1982–83 National Recreation Survey was conducted, as recommended by ORRRC. The survey found that the most popular activities at that time were, in order, swimming, walking, visiting zoos and parks, picnicking, driving for pleasure, sightseeing, attending outdoor sports events, fishing, and bicycling. The other activities enjoyed by Americans then were similar to those twenty years earlier. By the early 1980s, America's population had grown by 100 million, to about 231 million.

By 2000, the population had risen to around 284 million and was increasing at the rate of about 3 million per year. Viewing and photographing birds had become the fastest-growing activity in the country. As reported in *Outdoor Recreation for 21st Century America,* watching, photographing, and identifying wild birds had attracted more than 50 million new participants in less than twenty years.[5] Closely following the growth rate for birding were day hiking and backpacking, growing 193 and 182 percent, respectively, in less than twenty years. Snowmobiling increased 125 percent. Next fastest, increasing 50 to 100 percent since 1982, were attending outdoor concerts, plays, and other events; walking for pleasure; camping in developed sites; canoeing or kayaking; downhill skiing; and swimming in natural waters (i.e., streams, lakes, and oceans). Increasing 25 to 50 percent were ice skating, visiting nature centers and museums, picnicking, horseback riding, sightseeing, and driving for pleasure.

Because, in general, men participated in many of those activities at higher rates than women, they tended to have a heavy influence on overall activity trends. Through the 1990s and into this decade, however, women's participation in outdoor activities began to increase faster than in previous decades. In particular, more women engaged in horseback riding, pool swimming, fishing, and sailing. At this same time, participation in several activities by older Americans was moving up strongly as well.

In 2000, the most popular activities were walking for pleasure, outdoor family gatherings, and visiting a beach—the same activities that were at the top of the National Recreation Survey in 1994–95, when an update of the 1982–83 survey was done. When that update was conducted, the national survey was renamed the National Survey on Recreation and the Environment (NSRE). The most noticeable change by 2000 was growth in the proportion of the total population that participated in outdoor activities. In 1994–95, for example, 67 percent of the population participated in walking; by 2000, that rate had climbed to 83 percent, making it by far the most popular outdoor activity in America. The percentage of individuals who visited nature centers, nature museums, and similar nature study sites had risen substantially as well. Also noticeable was a continuing shift in the mix of activities people were turning to for outdoor experiences.

We see clear differences between what one would have witnessed at a typical outdoor area in the 1950s, 1960s, and 1970s and what one sees today, in the numbers of participants, types of clothing, and sophistication of equipment. In all likelihood, the changing composition of our population will stimulate further change. As well, one has to wonder how events like the terrorist attacks of September 11, 2001, have impacted outdoor recreation, particularly nature-based activities. Just after the attacks, respondents in an NSRE survey said they might modestly decrease their activities, but when asked about numbers of trips planned in the coming twelve months, people said they were planning about the same number of trips for recreation activities.

The NSRE

The data for this article were obtained from the National Survey on Recreation and the Environment, a federal survey of Americans' outdoor recreation activities.[6] The NSRE is conducted by the Forest Service research group in Athens, Georgia, with two partners, the University of Georgia and the University of Tennessee. The Athens research group has been collecting data and producing reports about the recreation activities, environmental attitudes, and natural resource values of Americans since the 1980s. The core of the survey covers outdoor activity participation and personal demographics. NSRE is a random-digit-dialed household telephone survey of a cross section of noninstitutionalized U.S. residents 16 years of age or older.

The most recent rounds of NSRE surveying were conducted between the summer of 2005 and spring of 2008 as part of a long-term data collection effort that began in the fall of 1999. The survey is statistically related to all previous National Recreation Surveys so that tracking of trends is possible. The current NSRE covers a much longer list of outdoor recreation activities than did ORRRC's first survey in 1960. Across all survey versions since 1999, more than 100,000 respondents were asked, "During the past 12 months, did you go [hiking, etc.] outdoors?" If the answer was yes, respondents were then asked, "On how many different days did you go [hiking, etc.] in the last 12 months?" Any amount of time spent on an activity on a given day, whether less than an hour or for several hours, was counted, whether that activity was the primary reason for recreating outdoors or not. The total number of people responding "yes" to any of the activities listed and the total days on which they participated, summed across activities and across all who participated, are the two primary statistics reported in this paper. The total days of participation across activities is a consistent index over time indicating the overall number of times people engage in outdoor activities.

Respondents were asked about more than eighty outdoor activities at varying times during the surveying to permit profiling the full scope of recreation activities. About twenty of these activities are competitive sports or fitness activities which are not addressed in this paper. The focus instead is on the other sixty outdoor activities, especially the fifty we define as nature-based. All data used in this report were weighted so that the demographic profiles of survey respondents matched those of the U.S. population over 16 years old at the time of surveying.

Nature-Based Recreation Trends Since 2000

Our overall picture of nature-based outdoor recreation includes number of people, number of days, and trends in participation. The time periods of primary interest are from 1999–2001 and 2005–2008—seven years across which we look for change in outdoor activities in general and change in nature-based activities especially. For the sake of simplification, each period will be referred to by a midpoint year, 2000 and 2007.

General Trends In Outdoor Recreation

Between 2000 and 2007, the total number of people who participated in one or more outdoor activities grew by 4.4 percent, from an estimated 208 million to 217 million. At the same time, the number of days of participation summed across all participants and activities increased from 67 billion to 84 billion, approximately 25 percent. The number of days of participation in walking for pleasure outdoors grew almost 14 percent, attending family gatherings outdoors grew almost 14 percent, visiting beaches grew more than 16 percent, and visiting farms and other agricultural settings grew by more than 100 percent. As will be discussed below, certain nature-based activities grew considerably as well.

For some of the sixty activities, on the other hand, days of participation (summed across all participants in each activity) decreased—examples include picnicking, driving for pleasure, visiting historic sites, and day hiking. Interestingly, for a few of these activities the number of people participating actually increased, while the opposite was observed for other activities. That is, we saw decreases in per capita days of participation for some activities, but increases in per capita days of participation for others. Over all sixty activities, per capita days of participation increased by almost 16 percent between 2000 and 2007. This is an important finding, especially in light of the study by Pergams and Zaradic.

General Trends In Nature-Based Outdoor Recreation

Within the list of sixty outdoor recreational activities are fifty natured-based activities associated with wildlife and birds, streams and lakes, snow and ice areas, hunting and fishing sites, different types of water bodies, trails, rugged terrain and caves, and other natural settings and resources on public or private forest, range, or other land and water. Some of these nature-based pursuits can be done near home (such as wildlife watching or swimming); others are enjoyed in more remote wildland areas (such as backpacking or mountain climbing). Here, too, between 2000 and 2007 we see discernible growth in the total number of participants and growth in the summed number of days they participated. Figure 2 summarizes these trends across the fifty nature-based activities. The total number of people who participated in these activities grew by 3.1 percent, from an estimated 197 million to 203 million. At the same time, the number of days of participation summed across all participants and activities grew about 32 percent, from an estimated 41 billion to nearly 55 billion. Over all fifty nature-based activities, per capita days of participation increased by more than 22 percent.

A few of the nature-based activities, such as mountain biking, coldwater fishing, whitewater rafting, and downhill skiing, experienced decreases in both the number of people who participated and total days of participation. Primitive camping (not in developed campgrounds), backpacking, and mountain climbing showed decreases in the number of people who participated, but increases in the number of days of participation. Visiting prehistoric sites, saltwater fishing, and snorkeling showed decreases in total days of participation, but increases in participants.

For a sizable number of nature-based activities, however, both the number of people participating and the summed days of participation increased. Prominent among these growth activities were viewing and photographing natural scenery, flowers, trees, wildlife, birds, and fish. This category of activities has contributed more than any other category to growth in nature-based recreation. Also growing in both number of participants and total days of participation were visiting nature centers,

Table 1 Fastest-growing U.S. nature-based outdoor activities, 2000–2007

Activity	Total Participants (Millions), 2007	Percentage Change In Participants, 2000–2007	Total Days of Participation (Billions), 2007	Percentage Change In Total Days, 2000–2007
Viewing or photographing flowers and trees	118.4	25.8	10.2	77.8
Viewing or photographing natural scenery	145.5	14.1	11.5	60.5
Driving off-road	44.2	18.6	1.3	56.1
Viewing or photographing other wildlife	114.8	21.3	5.3	46.9
Viewing or photographing birds	81.1	19.3	8.0	37.6
Kayaking	12.5	63.1	0.1	29.4
Visiting water (other than ocean beach)	55.5	1.6	1.1	28.1
Backpacking	22.1	-0.6	0.3	24.0
Snowboarding	11.3	7.3	0.1	23.9
Rock climbing	8.7	-5.5	0.1	23.8
Visiting nature centers, etc.	127.4	5.0	1.0	23.2
Big-game hunting	20.2	12.8	0.3	21.2
Mountain climbing	11.8	-12.5	0.1	20.5
Visiting ocean beach	96.0	10.5	1.4	16.3
Sightseeing	113.2	4.1	2.3	14.0
Visiting wilderness	70.6	3.0	1.1	12.8

sightseeing, visiting beaches, visiting wilderness, developed camping, boating, driving off-road motor vehicles, big-game hunting, kayaking, and snowboarding.

Fastest-Growing Nature-Based Activities

Overall, Americans' participation in nature-based outdoor recreation is on the rise, driven by several kinds of activities. Table 1 reports total participants and total days of participation in 2007 and rate of increase in both these measures between 2000 and 2007. The focus is on the seventeen fastest-growing activities—that is, those with days of participation growing by more than 10 percent.

Of these top seventeen activities, six involve viewing, photographing, identifying, visiting, or otherwise observing elements of nature—flowers, trees, natural scenery, birds, other wildlife, nature exhibits, and wilderness (wildlands generally). The growth in viewing and photographing plants and natural scenery has been most rapid, at about 78 and 60 percent, respectively. A motorized activity, driving motor vehicles off-road, occupies the number three slot; it grew by 56 percent between 2000 and 2007. Next are viewing, photographing, and identifying wildlife and birds. Thus, four of the top five activities are viewing, photographing, and otherwise observing nature.

Three water-oriented activities made the top seventeen: kayaking, visiting water areas or shores other than ocean beaches, and visiting ocean beaches. Four physically challenging activities are on the list: rock climbing, backpacking, snowboarding, and mountain climbing. Big-game hunting is also included, even though other data sources have reported that this activity is declining. (NSRE measures participation in an activity whether or not it is the only or primary motivation for an outdoor recreation occasion. Thus any participation at any level in big-game hunting is included.) Sightseeing and primitive camping round out the list.

Activities increasing at less than 10 percent in total participation days include gathering natural products (e.g., berries), motor-boating, developed camping, anadromous (migratory) fishing, warm-water fishing, swimming in natural waters, and visiting natural caves.

Water skiing, surfing, visiting prehistoric sites, small-game hunting, riding personal watercraft, rafting, rowing, cross-country skiing, and coldwater fishing have declined very modestly. Activities declining at somewhat greater rates, by 10 to 20 percent in total number of activity participation days, include snorkeling, saltwater fishing, migratory bird hunting, canoeing, sailing, and downhill skiing. Declining by 20 to 40 percent are day hiking, horseback riding on trails, snowmobiling, scuba diving, mountain biking, snowshoeing, and windsurfing.

Visiting Public Sites for Nature-Based Recreation

A recent examination of visitation to public lands found leveling for some, but moderate growth for other public lands.[7] For state parks, the number of visits per annum decreased modestly between 2000 and 2006. But that decrease began to turn around, and reported visitation in 2007 rose above the figure reported in 2001 (a 0.7 percentage increase). Similarly, although there were minor decreases in visits to national parks during the 2000s, for the most part visitation to national parks had been stable since 2001. For national wildlife refuges, visitation grew from about 33 million in 1998 to more than 40 million in 2007, nearly a 21 percent increase. Also, the U.S. Fish and Wildlife Service reported substantial increases in numbers of visitors watching wildlife in public parks and areas near their homes.[8] This survey, in fact, reported a 21 percent increase. According to NSRE data, growth in days visiting wilderness and other wildland areas increased more than 12 percent between 2000 and 2007.

Implications and Observations

The implications of these findings regarding outdoor recreation in contemporary America are many. First, things have changed since the 1950s and 1960s, when the federal government first began tracking recreation trends. The most popular activities then are not necessarily the most popular now. Snow skiing, day hiking, snowmobiling, horseback riding on trails, and some forms of fishing, for example, have begun to decline in popularity. On the other hand, viewing, photographing, and studying nature have shown rather spectacular growth. Some motorized recreation, especially off-road driving, has also been growing, as have some physically challenging activities. Generally, however, the greatest growth in participation is for activities that are physically not very challenging.

Our research suggests that Americans' interest in nature and nature-based recreation, though changing, is not declining; rather, it is strong and growing. The increase in the observation and study of nature is, in my view, a very healthy trend that apparently reflects rising and widespread interest in the future of natural resources, conservation, and public lands. Perhaps the interest in nature we see in our data represents more of an opportunity than we have realized. Two public policy implications come quickly to mind.

The first implication is that professional communities should seek to convert public interest in nature into active support of and engagement in conservation of forests, grasslands, and wetlands. A good example is the support for sustainable management of this nation's forests. No doubt today's youth live a very different lifestyle than did previous generations, but adults interested in nature can pass along and stimulate that interest in their children.

The second implication is that because outdoor activities can contribute to better physical conditioning, as well as better emotional health, perhaps there is an opportunity to use interest in nature as a means of stimulating greater physical activity. While still accommodating participants with disabilities, trails, overlooks, and wildlife observation sites could be designed to require some physical effort.

The recent spike in gasoline prices will very likely cause further change in the mix of outdoor activities that people choose, and perhaps reduce trips to more distant destinations. This may mean greater—not less—visitation to local parks, state parks, and federal lands near urban areas. In whatever ways the future unfolds, however, it is my opinion that this country's population in general will not lose interest in the forests, wildlife, and other natural resources of this country.

Notes

1. Richard Louv, *Last Child in the Woods: Saving Our Children from Nature-Deficit Disorder* (Chapel Hill, NC: Algonquin Books, 2005); and O.R.W Pergams and P.A. Zaradic, "Evidence for a Fundamental and Pervasive Shift away from Nature-Based Recreation," *Proceedings of the National Academy of Sciences* 105(7) (February 19, 2008): 2295–2300, www.pnas.org/cgi/reprint/0709893105v1 (accessed May 30, 2008).
2. H.K. Cordell, C.J. Betz, G.T. Green, S. Mou, VR. Leeworthy, P.C. Wiley, J.J. Barry, and D. Hellerstein, *Outdoor Recreation for 21st Century America* (State College, PA: Venture Publishing, Inc., 2004).
3. Pergrams and Zaradic, "Evidence for a Fundamental and Pervasive Shift away from Nature-Based Recreation," 2295—2300.
4. Citizens' Committee for the Outdoor Recreation Resources Review Commission, *Action for Outdoor Recreation in America: A Digest of Commission Findings and Recommendations* (Library of Congress No. 63-22303) (Washington, D.C., 1962).
5. Cordell et al., *Outdoor Recreation for 21st Century America.*
6. U.S. Forest Service, *National Survey on Recreation and the Environment* (Athens, GA: Southern Research Station, 2008). www.srs.fs.fed.us/trends.
7. H.K. Cordell, "Nature-Based Outdoor Recreation Trends and Wilderness," *International Journal of Wilderness* August 2008 (forthcoming).
8. U.S. Department of Interior Fish and Wildlife Service, *National Survey of Fishing, Hunting and Wildlife-Associated Recreation* (Washington, D.C., 2006).

Critical Thinking

1. Why do you think viewing and photographing birds had become the fastest-growing activity in the nation by 2000?
2. Has your participation in nature-based recreation changed during your life? If not, why not? If so, why?
3. The author states, ". . .the greatest growth in participation is for activities that are physically not very challenging." Yet, just two paragraphs later in the article, he states, ". . . perhaps there is an opportunity to use interest in nature as a means of stimulating greater physical activity." How would you capitalize on his suggested "opportunity" given the data interpretation he provides?

KEN CORDELL is a project leader and pioneering scientist in the U.S. Forest Service's Research and Development Branch. His offices are located at the University of Georgia in Athens. Appreciation is extended to research colleague Carter Betz for running the data for this paper.

UNIT 5

What Are the Impacts of Our Actions?

Unit Selections

Learning Outcomes

After reading this unit, you should be able to:

- Identify, by listing, the top 10 new consumers based on the number of new consumers in 2000.
- State, in writing, the major issue of increased automobile ownership throughout the world.
- State, in writing or orally, the most substantial human alteration of Earth.
- Identify, by constructing a diagram, the nitrogen cycle.
- State, in writing or orally, the predominant demand for phosphorus.
- Generate, by synthesizing the provided information, a paragraph explaining the statement, "We are effectively addicted to phosphate rock".
- Classify, by using a definition, the concept of biodiversity.
- Identify, by listing, the soil orders including the predominant one(s) in your home state.
- Identify, by listing, the number of rare, unique, and endangered soils in your home state.

Student Website

www.mhhe.com/cls

Internet References

Center for Science in the Public Interest (CSPI)
www.cspinet.org/EatingGreen/calculator.html

Conservation International (CI)
www.conservation.org/act/live_green/carboncalc/Pages/default.aspx

Global Footprint Network (GFN)
www.footprintnetwork.org

The Nature Conservancy (TNC)
www.nature.org/greenliving/carboncalculator

United States Department of Transportation (USDOT), Federal Transit Administration (FTA)
www.fta.dot.gov/planning/planning_environment_8523.html

Article 25 reinforces one of the main concerns regarding sustainability: Clearly citizens of some countries use more of Earth resources than those of other countries. Citizens of the United States use more resources than any other single nation on Earth. What will happen when other nations develop the infrastructure that allows them to consume like Americans? What are the implications of the Chinese, who represent roughly 15 percent of the human population, beginning to consume as much as Americans? What happens if citizens of the country of India, who also represent roughly 15 percent of the human population do likewise? It has been suggested that if all of humanity were to live the lifestyle of the average American, we would need at least three more Earths to provide all of the required resources. Obviously, that is not possible. Can penalties for overconsumption be far away?

That humans modify Earth is well established. What was not well established until relatively recently is the extent to which we have come to dominate virtually every ecosystem on the planet. Article 26 provides an overview of this issue. The impacts of our effect on the life support systems of the planet cannot be pushed off on future generations; they must be addressed now. In fact, they probably should have been addressed a few decades ago.

Article 27 concerns human alteration of the nitrogen cycle, one of the biogeochemical cycles most altered by humans. Because nitrogen is an essential nutrient for so many forms of life, any disturbance of its cycle is cause for alarm, especially since until relatively recently it was considered a limiting factor because it was not readily available in a biologically active form. Additionally, because nitrogen is a component of acidic deposition, it effects ecosystems in multiple ways.

Much attention has been paid to the coming phenomenon known as peak oil, or the top of the production curve for oil after which production will inexorably decline until depletion is reached. Far less attention has been paid to peak phosphorus, even though the implications are no less alarming. Article 28 addresses this issue. Phosphorus, like nitrogen, is a limiting factor in many ecosystems. This is especially true regarding agriculture. The challenge with reaching peak phosphorus is, like oil, it is considered a non-renewable resource because it is mined in the form of rock phosphate. The known reserves of rock phosphate could, at current rates of mining, be depleted within the next century.

Article 29 explores the connection between biodiversity and human well-being. Often, biodiversity is considered in the context of its importance to ecosystems; however, there is clear evidence that humans, because they benefit from ecosystem services, benefit from biodiversity. The reverse is also true: Humans will be adversely affected by a loss of biodiversity.

Article 30 addresses one of the least recognized, but quite possibly one of the most important, natural resources: soil. Throughout the world, soils are being degraded at an alarming rate. According to the Global Assessment of Soil Degradation, 100 percent of the world's soils exhibit some degree of degradation! At a time when the human population is increasing, food security is becoming an issue, and the oceans are badly overharvested, having soil that is not at the peak of its fertility is probably not beneficial to the quest for sustainability. Most people are surprised when they learn that there are so many different types of soil; many think it is just dirt.

New Consumers: The Influence of Affluence on the Environment

NORMAN MYERS AND JENNIFER KENT

Increasing consumption and especially its environmental impacts (1–5) are becoming all the more important now that the 850 million long-established consumers in rich countries have recently been joined by almost 1.1 billion new consumers in 17 developing and three transition countries. Most of these new consumers are far from possessing the spending capacity of the long-established consumers, but they have enough aggregate spending capacity, in terms of purchasing power parity (PPP), to match that of the U.S. Their numbers, consumption activities, and environmental impact are rising fast.

Of course, the new consumers should benefit from their affluence. This is a given, especially in light of the meager lifestyles that many of them have earlier experienced and the far greater consumption of long-affluent countries. But the environmental consequences of this new affluence are so significant that it will be in the self-interest of the 20 countries to restrict the damage with its economic penalties. In addition, and because of the global reach of certain environmental impacts, e.g., the CO_2 emissions from cars that accelerate climate change, the entire world community has an interest in the emergent phenomenon of the new consumers, on top of its even greater stake in long-established consumers with their far more pronounced environmental impacts. The world community also has an interest in those 1.3 billion people who endure abject poverty and whose basic needs demand far greater consumption forthwith. Their needs, however vital and urgent, lie outside the scope of this paper.

Who Are the New Consumers?

We define new consumers as people within typically four-member households with purchasing power of at least PPP $10,000 per year, i.e., at least PPP $2,500 per person, measured in PPP rather than conventional (international exchange) dollars (PPP dollars are between 1.3 and 5.3 times higher than conventional dollars in the 20 countries). From here on, we speak only of individual consumers, virtually all of whom possess purchasing power far above PPP $2,500. The PPP dollar levels themselves appear to mark a degree of affluence that enables wide-ranging purchases such as household appliances and televisions, air conditioners, personal computers, and other consumer electronics, among other perceived perquisites of an affluent lifestyle. More significantly for environmental purposes, many new consumers enjoy a strongly meat-based diet and buy cars.

The calculations reflect an analytic model by using data on country populations, economic growth, PPP equivalents, income distribution, and consumption patterns (6, 7). The model reveals percentiles of populations at various income levels for the year 2000. The data for this purpose, as for other calculations, have been drawn largely from >600 articles and books in the peer-reviewed literature and from reports and other documents of the World Bank/International Monetary Fund and the United Nations, among other leading agencies (see *Supporting References,* which is published as supporting information on the PNAS website, www.pnas.org). We believe these are the most reliable information sources available on a singularly wide-ranging issue and recognize that not all data may be as authoritative as one might wish. But the degree of credibility must be balanced against the need to address an emergent phenomenon that has exceptional significance for the world's environmental and economic future alike.

There are sizeable numbers of new consumers in 20 selected countries with records of strong economic growth and populations of at least 20 million people (Tables 1 and 2). They comprise 17 developing and three transition countries. Among them are four countries, South Korea, Mexico, Turkey, and Poland, that are members of the Organisation for Economic Cooperation and Development (OECD), the "rich nations club," even though their per-capita gross national products are far below that of 23 high-income members. They are generally listed in OECD, World Bank, and United Nations documents as "middle income" or "upper middle income" countries as opposed to "high income" countries, hence they are included here as developing or transition countries.

In 14 countries, new consumers make up 12–56 percent of the population, and in six countries they make up 61–96 percent. The new consumers' incomes are far greater than national averages because of income skewedness, a factor that applies

Abbreviation: PPP, purchasing power parity.

Table 1 New consumers, 2000

Country	Population, millions, 2000	New consumers, millions, 2000 (and percent of population)	Purchasing power, PPP $ billions (and percent of national total)*
China	1,262	303(24)	1,267(52)
India	1,016	132(13)	609(39)
South Korea	47	45(96)	502(99)
Philippines	76	33(43)	150(75)
Indonesia	210	63(30)	288(56)
Malaysia	23	12(53)	79(84)
Thailand	61	32(53)	179(79)
Pakistan	138	17(12)	62(31)
Iran	64	27(42)	136(71)
Saudi Arabia	21	13(61)	78(87)
South Africa	43	17(40)	202(83)
Brazil	170	74(44)	641(83)
Argentina†	37	31(84)	314(97)
Venezuela	24	13(56)	87(86)
Colombia	42	19(45)	136(83)
Mexico	98	68(69)	624(93)
Turkey	65	45(69)	265(85)
Poland	39	34(86)	206(95)
Ukraine	50	12(23)	44(45)
Russia	146	68(47)	436(79)
Totals	3,632	1,059(29)	6,305(67)‡

*Equates to household consumption.
†Argentina's figures do not reflect the recent economic recession.
‡Comparable to the U.S. (6).

in all 20 countries. In 16 countries, the top 20 percent of the population enjoy ≈ 50 percent or more of national income and in all 20 countries, 40 percent enjoy 62 percent or more (Table 2). In addition, the top quintiles generally show an increasing concentration of affluence and hence of consumption (6, 8).

New consumers have a far-reaching impact on economic activities nationwide, and hence on environmental repercussions. In India, for example, they accounted for less than one-eighth of the year 2000 population but two-fifths of the country's purchasing power (Table 1). With respect to the major sector of transportation, primarily cars, they accounted in the late 1990s for 85 percent of private spending. Their per-capita energy consumption has been causing CO_2 emissions 15 times greater than those of the rest of India's population (9).

Finally, the new consumers started to emerge in significant numbers only in the early 1980s, and their major increase in numbers occurred largely during the 1990s. True, there were some new consumers before the 1980s in countries such as Saudi Arabia, South Africa, Argentina, Brazil, and Mexico, but collectively they were generally few relative to the 2000 total.

Two Predominant Sectors

Meat

New consumers' diets are shifting toward meat, much of it raised in part on grain (Table 3). Raising 1 kg of beef can use 8 kg of grain, 4 kg of pork, and 2 kg of poultry (10). During the period 1990–2000 the amount of grain fed to livestock increased by 31 percent in China, 52 percent in Malaysia, and 63 percent in Indonesia. In nine of the 20 countries, two-fifths or more of grain consumed is now used for livestock (11), and the dietary change often leads to overloading of grainlands with resultant soil erosion and other forms of land degradation (12, 13). In addition, the 20 countries already account for nearly two-fifths of the world's grain imports; eight countries import one-fifth or more of their grain supply. In 2000, Malaysia imported 76 percent of its supply while feeding 41 percent to livestock, Saudi Arabia 78 percent and 65 percent, South Korea 75 percent and 44 percent, and Colombia 53 percent and 30 percent (Table 3). These imports serve to put pressure on international grain markets, to the detriment of poor countries that can hardly afford rising prices. Between 1997 and 2020, developing countries as a whole are forecast to increase their demand for meat, the great bulk to serve new consumers, by 92 percent, for grain by 50 percent, for food grain by 39 percent, and for feed grain by 85 percent (14).

Furthermore, demand for increased grain harvests aggravates water shortages. To produce 1 tonne (1 tonne = 10^3 kg) of grain can take 1,000 tonnes of water (15). Several sectors of China, a country with 29 percent of new consumers, experience water shortages, accentuated in part by the surging demand for grain. The North China Plain harbors two-fifths of the country's population and produces two-fifths of its grain but contains only one-fifth of its surface water. The region's aquifers have long been declining through overpumping by at least 1 m per year (16). In India, with 13 percent of new consumers, one-quarter of the grain harvest could be put at risk through groundwater depletion in its main breadbasket areas (17).

Cars

New consumers possess virtually all their countries' cars, and in most of the 20 countries, car totals have been expanding much more rapidly than national incomes. In 1990, these countries had 62 million cars, a total that by 2000 soared to 117 million or 21 percent of the global fleet (Table 4) (6). This 89 percent increase was led by China's 445 percent, South Korea's 319 percent, India's 259 percent, and Colombia's 217 percent. Five other countries registered increases of 100 percent or more, although the average annual increase for all 20 countries was only 6.4 percent. In particular, China and India, being the two countries with largest new consumer totals, registered average annual increases of 19 percent and 14 percent, respectively. Both these countries plan to push ahead vigorously with the "motorization" of their transport systems.

Allowing for growth in the proportionate numbers of people joining the affluent classes, the present decade could well see an average annual increase in car numbers at least matching the 6.4 percent rate of the 1990s. That would mean a total of at least 215 million cars in 2010, or one-quarter of the expected global

Table 2 Economic factors

Country	Conversion $/PPP$	Purchasing power, 2000, PPP $billions*	Income, top 40 % share	GDP growth, % 1990–1999	Household consumption growth, %, 1990–1999	Household consumption growth, %, 2000
China	4.67	2,434	69	10.7	8.8	6.3
India	5.20	1,554	65	6.1	4.9	4.2
South Korea	1.94	508	62	5.7	5.1	7.9
Philippines	4.06	199	73	3.2	3.7	3.1
Indonesia	4.96	511	66	4.7	6.2	3.6
Malaysia	2.46	94	74	6.3	5.4	12.2
Thailand	3.16	226	70	4.7	4.2	4.5
Pakistan	4.23	201	62	4.0	5.1	0.9
Iran	3.52	191	70†	3.4	2.9	6.9
Saudi Arabia	1.58	90	72†	1.6	n/a	n/a
South Africa	3.03	243	83	1.9	2.6	3.2
Brazil	2.04	760	82	2.9	4.3	9.9
Argentina	1.61	325	72	4.9	3.3	1.3
Venezuela	1.33	101	74	1.7	0.4	3.7
Colombia	3.00	164	79	3.2	3.0	6.5
Mexico	1.73	671	77	2.7	1.9	8.3
Turkey	2.27	312	69	4.1	3.7	6.2
Poland	2.15	217	63	4.7	5.2	4.9
Ukraine	5.29	98	63	−10.8	−8.0	5.2
Russia	4.83	553	74	−6.1	1.5	19.2
U.S.	0.00	6,269‡	69	3.4	3.2	5.6
World	1.43	26,914‡	—	2.5	2.6	3.6

n/a, not applicable.
*Equates to household consumption.
†Estimates.
‡Latest data (6).

fleet. Were the average annual increase to reach 10 percent due to disproportionately growing affluence on the part of new consumers both present and prospective, the 2010 total would approach 300 million, around one-third of the global fleet.

In 1997 the world's motor vehicles, of which 525 million, 74 percent, were passenger cars, emitted 73 percent of transport-related CO_2, for a 26 percent increase over 1990 and four times more than the CO_2 emissions increase overall. Cars are expected to make up the fastest-growing sector of energy use as far ahead as 2025 (18, 19).

Cars cause other forms of pollution such as urban smog and acid rain. They generate many other economic and social costs, notably road congestion, traffic accidents, and costly land use. In many cities of new consumer countries, road congestion is already acute and growing rapidly worse (20). In Bangkok, for instance, there are long periods every day when traffic moves at an average speed of just 3 km per hour. The problem has been costing an annual $1.6 billion of fuel wasted in idling car engines and at least $2.3 billion in lost worker productivity (21). Similar findings apply in Manila, Jakarta, and New Delhi, among other cities.

In addition to meat and cars, other new consumer purchases have harmful environmental impacts. A notable instance is electricity (generally derived from fossil fuels), which new consumers use at far above country-wide rates because of household appliances, air conditioners, and the like. The issue is not included in this article, because it is too difficult to pin down the amount used by new consumers, and because household electricity is not so environmentally significant as meat and cars.

Two Country Case Studies

In 2000, China had an estimated 303 million new consumers and India, 132 million, or two-fifths of the 20 countries' total. Let us look at these two countries in detail.

China

Not only does China possess the most new consumers today, but it offers the greatest scope for generating more new consumers in the future. During 1978–1998, its economic growth doubled per-capita income every 7 yr [compare South Korea at the height

Table 3 Meat

Country	Meat, kg per capita, 2000	Food grain, kg per capita increase, 1990–2000, %	Feed grain, kg per capita increase, 1990–2000	Feed grain as % of total grain, 2000	Grain imports as % of total grain, 2000
China	50	−9	20	23	3
India	5	2	0	1	<0.1
South Korea	46	−7	36	44	75
Philippines	27	−6	14	28	14
Indonesia	8	11	50	4	14
Malaysia	51	19	22	41	76
Thailand	24	11	11	34	9
Pakistan	12	5	33	4	4
Iran	22	3	10	32	44
Saudi Arabia	51	−7	84	65	78
South Africa	39	−1	−4	32	14
Brazil	77	−3	44	54	21
Argentina	98	−1	28	44	1
Venezuela	42	−9	−31	18	68
Colombia	34	15	47	30	53
Mexico	56	1	8	41	36
Turkey	20	−6	3	25	9
Poland	70	6	−23	58	9
Ukraine	31	−12	−40	50	5
Russia	40	−6	−44	48	9
U.S.	122	5	1	66	3
World	38	−2	−11	35	—

See ref. 11.

of its economic boom (11 yr) and Japan (34 yr)] (22). In 2000, China's gross national income [(GNI), a recent designation of the World Bank to replace gross national product (GNP)] totaled almost PPP $5 trillion, making it the world's second largest economy in PPP terms (seventh largest in conventional dollars) (6).

China's new consumers are enjoying strongly meat-based diets. With one-fifth of the world's population, the country accounts for 28 percent of the world's meat consumption (compare to the U.S., 15 percent) (6, 11), virtually all of it attributable to the new consumers. The 1990–2000 114 percent increase in meat consumption accounted for 25 million tonnes or four-fifths of the country's 7 percent growth in grain consumption; today 23 percent of grain is fed to livestock, up from 19 percent in 1990. Although per-capita meat consumption almost doubled, and per-capita feed grain consumption grew by 20 percent during 1990–2000, per-capita food grain consumption declined by 9 percent (11).

As we have seen above, surging demand for grain is aggravating water deficits in the extensive North China Plain. In common with several other new consumer countries, China's grain imports effectively amount in part to water imports. Whatever China's water shortages today, they could become much more acute given that during 1997–2020 the country is forecast to account for 40 percent of the increased global demand for meat and 27 percent for grain (14).

China's new consumers are also buying cars in significant numbers. The total has grown from 1.1 million in 1990 to at least 6 million in 2000 (6). If the country maintains its 1990–2000 average annual growth rate of 19 percent, it will have ≈ 34 million cars by 2010, an almost 6-fold increase. If the additional new consumers expected to come on stream (300 million, see below) were to markedly increase the 19 percent rate, conceivably to as high as 25 percent (19, 23, 24), the 2010 total could be as much as 56 million cars.

How far will the new consumers increase beyond their 303 million in 2000? Let us suppose that during the present decade, the country's annual economic growth averages the forecast 7.0 percent (less than the 10 percent of the past two decades), meaning that the economy will more than double to PPP $9.7 trillion (6, 7). Although many of the consumer benefits of the growing prosperity will accrue to the 303 million new consumers of 2000, their total will surely swell by 6 percent per year (household consumption increases at a rate less than that of economic growth) to ≈ 543 million in 2010. Because of income skewedness, however, which is likely to become yet more pronounced (25), the increasingly affluent new consumer class seems poised to forge still further ahead of the rest of the population, making for growth of their numbers more like 7 percent per year. Thus a more probable total in 2010 is ≈ 600 million, or 44 percent of the projected population (compare to 24 percent

Table 4 Cars, millions

Country	1990	2000 estimate	Percent change 1990–2000
China	1.1	6.0	445
India	1.7	6.1	259
South Korea	2.1	8.8	319
Philippines	0.4	0.8	100
Indonesia	1.3	2.9	123
Malaysia	1.8	4.1	128
Thailand	0.8	1.9	138
Pakistan	0.5	0.8	60
Iran	1.4	2.1	50
Saudi Arabia	1.6	1.9	19
South Africa	3.4	4.1	21
Brazil	11.8	18.5	57
Argentina	4.4	5.5	25
Venezuela	1.5	1.8	20
Colombia	0.6	1.9	217
Mexico	6.8	10.4	53
Turkey	1.9	4.5	137
Poland	5.3	9.9	87
Ukraine	3.3	5.5	67
Russia	10.1	19.5	93
Totals	62.0	117.0	89
Percent of world	13	21	62
U.S.	152	175*	15
World	478	560	

*Including sport utility vehicles (SUVs) (6).

in 2000). By then, their purchasing power could well climb to PPP $3.5 trillion, or over half that of the U.S. today.

All this means that China's continued economic advance will have a marked impact on environments both national and global, and a good part of these impacts will reflect the rise of the new consumers (26, 27). During the 1990s, environmental damage associated with economic growth (primarily air pollution and water deficits, but also deforestation and desertification) cost at least 8 percent and possibly 10–15 percent of gross domestic product, much of the damage being due to new consumers' activities (26, 28, 29). Given China's economic globalization and the spread of Western lifestyles, a potential doubling of the new consumer total within the present decade is a formidable prospect when linked with shortages of grain and water, plus loss of farmland for industry, urbanization, and transport networks (26, 30).

The prospect is further challenging in terms of CO_2 emissions (only a small proportion from cars but rising rapidly), which in 2000 placed China second to the U.S. with 49 percent as much, even though per capita they were only 2.2 tonnes per year by contrast with the U.S.'s 20.5 tonnes (31). Conversely, China has engaged in broad-scope policy reforms of its energy sector, resulting in a decline in CO_2 emissions estimated at somewhere between 6 percent and 14 percent during 1997–2000, although this decline applied far more to industry than to transportation (32–34).

India.

India's economic growth rate has averaged >6.0 percent per year during the 1990s, making its PPP $2.4 trillion economy the fourth largest in the world (12th in conventional dollars) (6–8). The new consumer total in 2000 is estimated at 132 million (Table 1).

India's per-capita meat consumption is still meager, only one-tenth as much as China's (11), although still significant given the large number of new consumers who eat the great bulk of the country's meat. In absolute terms, however, and given that India has the second-highest number of new consumers, the country is the fourth-largest meat eater among all new consumer countries.

In 2000, India's cars totaled ≈ 6.1 million (Table 4), little more than in Greater Chicago, yet there were enough cars to cause much pollution of several sorts. Motor vehicles of all kinds account for 70 percent of air pollution, which has increased eight times during the past 20 yr, compared with a 4-fold increase for industry. Of the world's 10 cities with the highest air pollution, three are in India. The health costs of air pollution in 36 Indian cities have amounted to at least $500 million per year and possibly four times as much (35). If the economy keeps expanding by at least 5.0 percent per year, car numbers can be expected to continue increasing by the annual 14 percent of 1990–2000 on the grounds that, as in the past, the expanding consumer classes will become relatively more affluent than the rest of the population. This prognosis postulates a total of 23 million cars by 2010.

Environmental damage of all kinds has been costing ≈ 5 percent of India' gross domestic product, due disproportionately to the activities of new consumers (36).

How many new consumers could there be by the end of this decade? An annual economic growth rate of 5.5 percent (reflecting the average for 1990–2000, 6.1 percent, and the expected growth rates for 2001–2003) means that India's economy will almost double to PPP $4.1 trillion. As in the case of China, where household consumption does not increase as fast as economic growth, the new consumer total of 2000 is likely to grow by 1 percent less than the economy's expansion, namely 4.5 percent per year (37, 38). This will bring their numbers to ≈ 205 million (18 percent of the projected population). In addition, income skewedness is likely to become more pronounced insofar as it is the top quintile that has benefited most from the country's economic advance of the past decade. So the new consumer total could readily soar by 5.5 percent per year to almost 225 million in 2010 or 19 percent of the population (compare to 13 percent in 2000). These new consumers would then have a purchasing power of >PPP $1.2 trillion, putting them on a par with Germany today.

By 2010, then, China and India alone could feature 825 million new consumers with a purchasing power approaching PPP $5.0 trillion (compare to the U.S. in 1999, $6.3 trillion).

Overview of 20 Countries

All new consumers in the 20 countries totaled 1.059 billion in 2000 (Table 1). China and India accounted for 41 percent of the total. The third-largest total was Brazil with 74 million (7 percent). Mexico and Russia had 68 million (6 percent) each and Indonesia had 63 million (6 percent). The smallest numbers were in Malaysia and Ukraine with 12 million (1 percent) each, then Saudi Arabia and Venezuela with 13 million (1 percent) each.

Equally revealing was the new consumers' share of each country's population. South Korea was top with 96 percent; second was Poland with 86 percent; and third, Argentina with 84 percent (the latter's figure will have dropped by today; see below). Lowest was Pakistan with 12 percent, next India with 13 percent, and then Ukraine with 23 percent.

All of the 20 countries' totals in 2000 reflect the latest statistics of the World Bank and International Monetary Fund (6, 7). There have been recent economic downturns in a number of countries, most notably in Argentina. Argentina's new consumer total in 2000 was only 31 million, so a subsequent decline of, say, one-third to 21 million will make <1.0 percent difference to the 20 countries' aggregate. No other country has registered such a severe and protracted economic decline (there have been transient dips in Saudi Arabia, Mexico, and Turkey), so no other new consumer totals will have dropped. As for the future, virtually all new consumer countries except Argentina are forecast to feature strong economic growth.

The overall purchasing power of the new consumers in 2000 amounted to PPP $6.3 trillion, matching the U.S.'s (where PPP dollars and conventional dollars are the same by definition). In South Korea, overall purchasing power constituted 99 percent of the country-wide total. Next was Argentina with 97 percent (although see qualifier above), followed by Poland with 95 percent and Mexico with 93 percent. The lowest was Pakistan with 31 percent, followed by India with 39 percent and Ukraine with 45 percent (Table 1).

Thus there is a sizeable "North" in the "South." The 2000 total of new consumers, approaching 1.1 billion, is to be compared with the collective populations of the 23 long-standing and much richer Organisation for Economic Cooperation and Development (OECD) countries, 850 million (6). True, the collective purchasing power of the new consumers, PPP $6.3 trillion, contrasts with the 23 OECD countries' PPP $15 trillion (6). All the same, the new consumers constitute a major consumer force in the global economy, just as they are becoming a front-rank factor in the global environment. China's environmental impact could eventually match that of the U.S.

An additional 14 countries, not considered here because their populations are <20 million or because their economies are not strong (or their data lacking), probably feature ≈ 140 million new consumers, meaning their omission does not markedly affect the overall situation.

Policy Responses

How can the new consumers be persuaded to reduce their environmental impacts and move toward sustainable consumption? Of course, the need to make consumption sustainable applies as well, only much more so, to the long-established consumers in the rich world, and the new consumers are unlikely to alter their consumption until the rich-world consumers take solid steps to modify their consumption.

Consider the scope for cleaner cars (39). A few recent models, notably the Toyota Prius and the Honda Insight, have hybrid engines that produce far less CO_2 among other pollutants. Compressed natural gas powers 10 percent of Argentina's car fleet, and India has introduced the same fuel gas for heavy vehicles in its major cities. The prospective hydrogen fuel-cell car would emit only water vapor. In addition, there are many alternatives to the conventional car culture. Cities in developing country can promote mass transit systems, bicycle networks, and restrictions on cars in congested areas and can make drivers pay the full cost of their activity (40, 41). These diverse routes into the future are already illustrated by Singapore, Bogota, and the Brazilian city of Curitiba, in all of which fewer cars bring benefits all around (42–44).

As for meat, prices are often held down through large subsidies for grain and water (45). Consumers are induced to move up the food chain through dietary fads, taught taste, and social status, all of which can be shifted toward healthier diets through fiscal incentives such as a "food conversion efficiency" tax. The least-efficient converters of grain, notably beef, could be highly taxed, whereas more efficient products, notably poultry, could be moderately taxed (46). Similarly, there are many opportunities to foster more efficient use of water for the growing of grain (47). Use of other natural resources can be improved through full-cost pricing (48, 49), shifts in tax systems (50, 51), substitutes for gross national product as an economic indicator (52, 53), elimination of "perverse" subsidies that foster both environmental and economic inefficiencies (45), and application of the many ecotechnologies available (54–57).

Above all is the need to establish sustainable consumption as a norm, which is not only about quantitative reductions in our use of materials and energy (58–60); it is also about ways in which we can achieve an acceptable quality of life for all in perpetuity and exemplify it throughout our lifestyles (61). How, for instance, can we attain a better balance between work, leisure, and consumption (63, 64)? How can we prevent yesterday's luxuries from becoming today's necessities and tomorrow's relics (65, 66)? How can we make fashion sustainable and sustainability fashionable?

For many helpful comments on an early version of this article, we thank G. C. Daily and D. M. Pimentel. T. L. Root and S. H. Schneider of Stanford University provided an economic model used to determine income distribution within national populations. We gratefully acknowledge financial support for our original research from the Winslow Foundation (Washington, DC).

References

1. Daily, G. C. & Ehrlich, P. R. (1996) *Ecol. Appl.* **6,** 991–1001.
2. Holdren, J. P. (2000) *Environment* **42,** 4–6.
3. Holdren, J. P. & Ehrlich, P. R. (1974) *Am. Sci.* **62,** 282–292.

4. Princen, T., Maniates, M. & Conca, K., eds. (2002) *Confronting Consumption* (MIT Press, Cambridge, MA).
5. Stern, P. Dietz, T., Ruttan, V., Socolow, R. & Sweeney, J., eds. (1997) *Environmentally Significant Consumption: Research Directions* (Natl. Acad. Press, Washington, DC).
6. The World Bank (2002) *World Development Indicators Online, May 2002* (World Bank, Washington, DC).
7. International Monetary Fund (2002) *World Economic Outlook April 2002* (International Monetary Fund, Washington, DC).
8. The World Bank (1994) *World Development Report 1994* (Oxford Univ. Press, New York).
9. Consumers International (1998) *A Discerning Middle Class? A Preliminary Enquiry of Sustainable Consumption Trends in Selected Countries in the Asia Pacific Region* (Consumers International, Penang, Malaysia).
10. Smil, V. (2000) *Feeding the World: A Challenge for the Twenty-First Century* (MIT Press, Cambridge, MA).
11. Food and Agriculture Organization (2002) *FAOSTAT Food Balance Sheets June 2002* (Food and Agriculture Organization, Rome).
12. World Resources Institute (1999) *Critical Consumption Trends and Implications: Degrading Earth's Ecosystems* (World Resources Institute, Washington, DC).
13. Tilman, D., Fargione, J., Wolff, B., D'Antonio, C., Dobson, A., Howarth, R., Schindler, D., Schlesinger, W. H., Simberloff, D. & Swackhamer, D. (2001) *Science* **292,** 281–284.
14. Rosegrant, M. W., Paisner. M. S., Meijer, S. & Witcover, J. (2001) *2020 Global Food Outlook: Trends, Alternatives, Choices* (International Food Policy Research Institute, Washington, DC).
15. Gleick, P. (2000) *The World's Water 2000–2001* (Island Press, Washington, DC).
16. Qingcheng, H. (2001) *The North China Plain and Its Aquifers* (Geological Environmental Monitoring Institute, Beijing, P.R. China).
17. Seckler, D., Molden, D. & Barker, R. (1998) *Water Scarcity in the Twenty-First Century* (International Water Management Institute, Colombo, Sri Lanka).
18. International Energy Agency (2000) *International Energy Outlook 2000* (International Energy Agency, Paris).
19. Schipper, L., Marie-Liliu, C. & Lewis-Davis. G. (1999) *Rapid Motorization in the Largest Countries in Asia: Implications for Oil, Carbon Dioxide and Transportation* (International Energy Agency, Paris).
20. Willoughby, C. (2000) *Managing Motorization* (The World Bank, Washington, DC).
21. DuPont, P. & Egan, K. (1997) *World Transport and Policy Practice* **3,** 25–37.
22. The World Bank (1999) *World Development Report, 1999* (Oxford Univ. Press, New York).
23. Schipper, L. (2001) *Designing "Effective" Solutions to the Urban Transport-Environment Dilemma* (Organisation for Economic Cooperation and Development, Paris).
24. Riley, K. (2002) *Popul. Environ.* **23,** 479–494.
25. Atinc, P. M. (1997) *Sharing Rising Incomes: Disparities in China* (The World Bank, Washington, DC).
26. The World Bank (2001) *China Air, Land and Water: Environmental Priorities to a New Milennium* (The World Bank, Washington, DC).
27. Palanivel, T. (2001) *Sustainable Development of China, India and Indonesia: Trends and Responses* (Institute of Advanced Studies, United Nations University, Tokyo).
28. Smil, V. & Yushi, M. (1998) *The Economic Costs of China's Environmental Degradation* (Am. Acad. Arts and Sciences, Cambridge, MA).
29. Takahiro, A. & Nakamura, Y. (2000) *Green GDP Estimates in China, Indonesia and Japan* (Institute for Advanced Studies, United Nations University, Tokyo).
30. Brown, L. R. (2001) *Eco-Economy: Building an Economy for the Earth* (Norton, New York).
31. Energy Information Administration, U.S. Department of Energy (2002) *International Energy Database, April 2002* (U.S. Department of Energy, Washington, DC).
32. Natural Resources Defence Council (2001) *Second Analysis Confirms Greenhouse Gas Reductions in China* (Natural Resources Defense Council, Washington, DC).
33. Streets, D. G., Jiang, K., Hu, X., Sinton, J. E., Zhang, X. Q., Xu, D., Jacobson, M. Z. & Hansen, J. E. (2001) *Science* **294,** 1835–1837.
34. Sinton, J. E., & Fridley, D. G. (2000) *Environ. Pol.* **28,** 671–687.
35. Agarwal, A. & Narain, S. (1997) *Economic Globalisation: Its Impact on Consumption, Equity and Sustainability* (Center for Science and the Environment, New Delhi, India).
36. Parikh, J. & Parikh, K. (2001) *Accounting for Environmental Degradation: A Case Study of India* (Institute for Advanced Studies, United Nations University, Tokyo).
37. Alagh, Y. K. (2001) *India's Sustainable Development Framework: 2020* (Institute for Advanced Studies, United Nations University, Tokyo).
38. Klein, L. R., & Palanivel, T. (2000) *Economic Reforms and Growth Prospects in India* (Institute for Advanced Studies, United Nations University, Tokyo).
39. Motavalli, J. (2002) *Forward Drive: The Race to Build Clean Cars for the Future* (Earthscan, London).
40. Satterthwaite, D., ed. (1999) *Sustainable Cities* (Earthscan, London).
41. Bose, R., Sperling, D., Tiwari, G. & Schipper, L. (2001) *Transportation in Developing Countries* (Pew Center on Global Climate Change, Arlington, VA).
42. Newman, P. (1999) *Automobile Dependence* (Island Press, Washington, DC).
43. Rabinovitch, J. (1996) *Land Use Pol.* **13,** 51–67.
44. Sheehan, M. (2001) *City Limits: Putting the Brakes on Sprawl* (Worldwatch Institute, Washington, DC).
45. Myers, N. & Kent, J. (2001) *Perverse Subsidies: How Tax Dollars Can Undercut the Environment and the Economy* (Island Press, Washington, DC).
46. Pimentel, D., Westra, L. & Noss, R. F. (2000) *Ecological Integrity: Integrating Environment, Conservation, and Health* (Island Press, Washington, DC).
47. Postel, S. (1999) *Pillar of Sand: Can the Irrigation Miracle Last?* (Norton, New York).
48. Pearce, D. & Barbier, E. B. (2001) *Blueprint for a Sustainable Economy* (Earthscan, London).
49. Roodman, D. (1998) *The Natural Wealth of Nations* (Norton, New York).
50. Ekins, P. & Speck, S. (2000) *J. Environ. Pol. Plann.* **2,** 93–114.
51. Wallart, N. (1999) *The Political Economy of Environmental Taxes* (Edward Elgar, Cheltenham, U.K.).
52. Cobb, C., Goodman, G. S. & Wackernagel, M. (1999) *Why Bigger Isn't Better: The Genuine Progress Indicator, 1999 Update* (Redefining Progress, San Francisco).
53. Dasgupta, P. & Maler, K.-G. (2000) *Environ. Dev. Econ.* **5,** 69–93.
54. DeSimone, L. D. & Popoff, F. (2000) *Eco-Efficiency: The Business Link to Sustainable Development* (MIT Press, Cambridge, MA).
55. Hawken, P., Lovins, A. B., & Lovins, L. H. (1999) *Natural Capitalism* (Little, Brown, Boston).

56. Russel, T. (2001) *Eco-Efficiency: A Management Guide* (Earthscan, London).
57. McDonough, W. & Braungart, M. (2002) *Cradle to Cradle: Remaking the Way We Make Things* (North Point Press, New York).
58. Arrow, K., Daily, G., Dasgupta, P., Ehrlich, P., Goulder, L., Heal, G., Levin, S., Maler, K.-G., Schneider, S., Starrett, D. & Walker, B. (2003) *J. Econ. Perspect.,* in press.
59. Pauli, G. (1998) *Up Sizing: The Road to Zero Emissions* (Greenleaf Publications, London).
60. Schmidt-Bleek, F. (2000) *Factor 10 Manifesto* (The Factor Ten Institute, Carnoules, France).
61. Frank, R. H. (1999) *Luxury Fever: Why Money Fails to Satisfy in an Era of Excess* (Free Press, New York).
62. Kates, R. W., Clark, W. C., Corell, R., Hall, J. M., Jaeger, C. C., Lowe, I., McCarthy, J. J., Schellnhuber, H. J., Bolin, B. & Dickson, N. M. (2001) *Science* **292,** 641–642.
63. Schor, J. B. (1998) *The Overspent American: Upscaling, Downshifting, and the New Consumer* (Basic Books, New York).
64. Kasser, T. (2002) *The High Price of Materialism* (MIT Press, Cambridge, MA).
65. Daly, H. E. (2000) *Ecological Economics and the Ecology of Economics* (Edward Elgar, Cheltenham, U.K.).
66. Robins, N. & Roberts, S. (1998) *Development (Cambridge, UK)* **41,** 28–36.

Critical Thinking

1. Is it ethical to propose: "it will be in the self-interest of the 20 countries to restrict the damage with its economic penalties?" Why or why not.
2. Does PPP $2,500 per person seem a reasonable definition of a new consumer to you? Why or why not.
3. Which is worse regarding the new consumers, meat consumption or driving cars?

Human Domination of Earth's Ecosystems

PETER M. VITOUSEK ET AL.

All organisms modify their environment, and humans are no exception. As the human population has grown and the power of technology has expanded, the scope and nature of this modification has changed drastically. Until recently, the term "human-dominated ecosystems" would have elicited images of agricultural fields, pastures, or urban landscapes; now it applies with greater or lesser force to all of Earth. Many ecosystems are dominated directly by humanity, and no ecosystem on Earth's surface is free of pervasive human influence.

This article provides an overview of human effects on Earth's ecosystems. It is not intended as a litany of environmental disasters, though some disastrous situations are described; nor is it intended either to downplay or to celebrate environmental successes, of which there have been many. Rather, we explore how large humanity looms as a presence on the globe—how, even on the grandest scale, most aspects of the structure and functioning of Earth's ecosystems cannot be understood without accounting for the strong, often dominant influence of humanity.

We view human alterations to the Earth system as operating through the interacting processes summarized in Figure 1. The growth of the human population, and growth in the resource base used by humanity, is maintained by a suite of human enterprises such as agriculture, industry, fishing, and international commerce. These enterprises transform the land surface (through cropping, forestry, and urbanization), alter the major biogeochemical cycles, and add or remove species and genetically distinct populations in most of Earth's ecosystems. Many of these changes are substantial and reasonably well quantified; all are ongoing. These relatively well-documented changes in turn entrain further alterations to the functioning of the Earth system, most notably by driving global climatic change (*1*) and causing irreversible losses of biological diversity (*2*).

Land Transformation

The use of land to yield goods and services represents the most substantial human alteration of the Earth system. Human use of land alters the structure and functioning of ecosystems, and it alters how ecosystems interact with the atmosphere, with aquatic systems, and with surrounding land. Moreover, land transformation interacts strongly with most other components of global environmental change.

The measurement of land transformation on a global scale is challenging; changes can be measured more or less straightforwardly at a given site, but it is difficult to aggregate these changes regionally and globally. In contrast to analyses of human alteration of the global carbon cycle, we cannot install instruments on a tropical mountain to collect evidence of land transformation. Remote sensing is a most useful technique, but only recently has there been a serious scientific effort to use high-resolution civilian satellite imagery to evaluate even the more visible forms of land transformation, such as deforestation, on continental to global scales (*3*).

Land transformation encompasses a wide variety of activities that vary substantially in their intensity and consequences. At one extreme, 10 to 15 percent of Earth's land surface is occupied by row-crop agriculture or by urban-industrial areas, and another 6 to 8 percent has been converted to pastureland (*4*); these systems are wholly changed by human activity. At the other extreme, every terrestrial ecosystem is affected by increased atmospheric carbon dioxide (CO_2), and most ecosystems have a history of hunting and other low-intensity resource extraction. Between these extremes lie grassland and semiarid ecosystems that are grazed (and sometimes degraded) by domestic animals, and forests and woodlands from which wood products have been harvested; together, these represent the majority of Earth's vegetated surface.

The variety of human effects on land makes any attempt to summarize land transformations globally a matter of semantics as well as substantial uncertainty. Estimates of the fraction of land transformed or degraded by humanity (or its corollary, the fraction of the land's biological production that is used or dominated) fall in the range of 39 to 50 percent (*5*). These numbers have large uncertainties, but the fact that they are large is not at all uncertain. Moreover, if anything these estimates understate the global impact of land transformation, in that land that has not been transformed

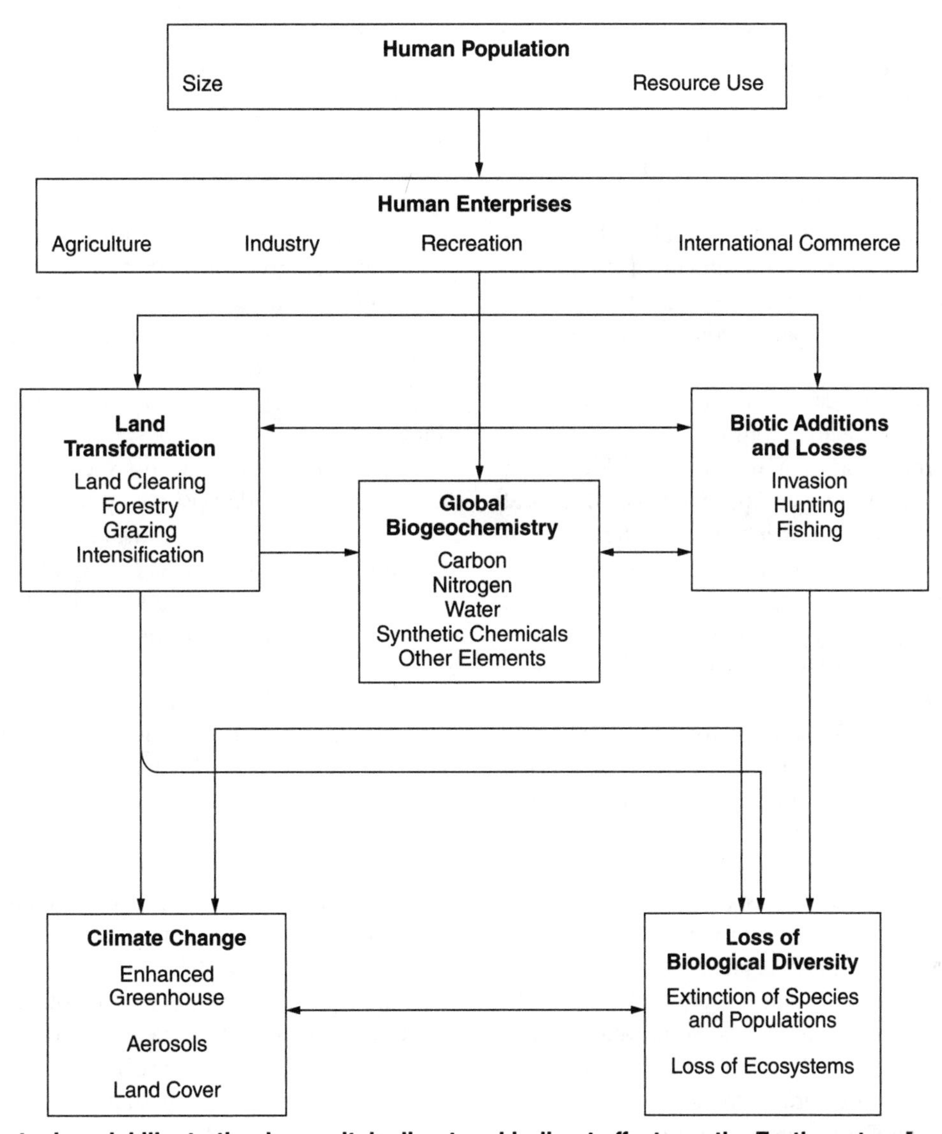

Figure 1 A conceptual model illustrating humanity's direct and indirect effects on the Earth system [modified from (*56*)]

often has been divided into fragments by human alteration of the surrounding areas. This fragmentation affects the species composition and functioning of otherwise little modified ecosystems (*6*).

Overall, land transformation represents the primary driving force in the loss of biological diversity worldwide. Moreover, the effects of land transformation extend far beyond the boundaries of transformed lands. Land transformation can affect climate directly at local and even regional scales. It contributes ~20 percent to current anthropogenic CO_2 emissions, and more substantially to the increasing concentrations of the greenhouse gases methane and nitrous oxide; fires associated with it alter the reactive chemistry of the troposphere, bringing elevated carbon monoxide concentrations and episodes of urban-like photochemical air pollution to remote tropical areas of Africa and South America; and it causes runoff of sediment and nutrients that drive substantial changes in stream, lake, estuarine, and coral reef ecosystems (*7–10*).

The central importance of land transformation is well recognized within the community of researchers concerned with global environmental change. Several research programs are focused on aspects of it (*9, 11*); recent and substantial progress toward understanding these aspects has been made (*3*), and much more progress can be anticipated. Understanding land transformation is a difficult challenge; it requires integrating the social, economic, and cultural causes of land transformation with evaluations of its biophysical nature and consequences. This interdisciplinary approach is essential to predicting the course, and to any hope of affecting the consequences, of human-caused land transformation.

Oceans

Human alterations of marine ecosystems are more difficult to quantify than those of terrestrial ecosystems, but several kinds of information suggest that they are substantial. The human population is concentrated near coasts—about

60 percent within 100 km—and the oceans' productive coastal margins have been affected strongly by humanity. Coastal wetlands that mediate interactions between land and sea have been altered over large areas; for example, approximately 50 percent of mangrove ecosystems globally have been transformed or destroyed by human activity (*12*). Moreover, a recent analysis suggested that although humans use about 8 percent of the primary production of the oceans, that fraction grows to more than 25 percent for upwelling areas and to 35 percent for temperate continental shelf systems (*13*).

Many of the fisheries that capture marine productivity are focused on top predators, whose removal can alter marine ecosystems out of proportion to their abundance. Moreover, many such fisheries have proved to be unsustainable, at least at our present level of knowledge and control. As of 1995, 22 percent of recognized marine fisheries were overexploited or already depleted, and 44 percent more were at their limit of exploitation (*14*). The consequences of fisheries are not restricted to their target organisms; commercial marine fisheries around the world discard 27 million tons of nontarget animals annually, a quantity nearly one-third as large as total landings (*15*). Moreover, the dredges and trawls used in some fisheries damage habitats substantially as they are dragged along the sea floor.

A recent increase in the frequency, extent, and duration of harmful algal blooms in coastal areas (*16*) suggests that human activity has affected the base as well as the top of marine food chains. Harmful algal blooms are sudden increases in the abundance of marine phytoplankton that produce harmful structures or chemicals. Some but not all of these phytoplankton are strongly pigmented (red or brown tides). Algal blooms usually are correlated with changes in temperature, nutrients, or salinity; nutrients in coastal waters, in particular, are much modified by human activity. Algal blooms can cause extensive fish kills through toxins and by causing anoxia; they also lead to paralytic shellfish poisoning and amnesic shellfish poisoning in humans. Although the existence of harmful algal blooms has long been recognized, they have spread widely in the past two decades (*16*).

Alterations of the Biogeochemical Cycles

Carbon

Life on Earth is based on carbon, and the CO_2 in the atmosphere is the primary resource for photosynthesis. Humanity adds CO_2 to the atmosphere by mining and burning fossil fuels, the residue of life from the distant past, and by converting forests and grasslands to agricultural and other low-biomass ecosystems. The net result of both activities is that organic carbon from rocks, organisms, and soils is released into the atmosphere as CO_2.

The modern increase in CO_2 represents the clearest and best documented signal of human alteration of the Earth system. Thanks to the foresight of Roger Revelle, Charles Keeling, and others who initiated careful and systematic measurements of atmospheric CO_2 in 1957 and sustained them through budget crises and changes in scientific fashions, we have observed the concentration of CO_2 as it has increased steadily from 315 ppm to 362 ppm. Analysis of air bubbles extracted from the Antarctic and Greenland ice caps extends the record back much further; the CO_2 concentration was more or less stable near 280 ppm for thousands of years until about 1800, and has increased exponentially since then (*17*).

There is no doubt that this increase has been driven by human activity, today primarily by fossil fuel combustion. The sources of CO_2 can be traced isotopically; before the period of extensive nuclear testing in the atmosphere, carbon depleted in ^{14}C was a specific tracer of CO_2 derived from fossil fuel combustion, whereas carbon depleted in ^{13}C characterized CO_2 from both fossil fuels and land transformation. Direct measurements in the atmosphere, and analyses of carbon isotopes in tree rings, show that both ^{13}C and ^{14}C in CO_2 were diluted in the atmosphere relative to ^{12}C as the CO_2 concentration in the atmosphere increased.

Fossil fuel combustion now adds 5.5 ± 0.5 billion metric tons of CO_2-C to the atmosphere annually, mostly in economically developed regions of the temperate zone (*18*). The annual accumulation of CO_2-C has averaged 3.2 ± 0.2 billion metric tons recently (*17*). The other major terms in the atmospheric carbon balance are net ocean-atmosphere flux, net release of carbon during land transformation, and net storage in terrestrial biomass and soil organic matter. All of these terms are smaller and less certain than fossil fuel combustion or annual atmospheric accumulation; they represent rich areas of current research, analysis, and sometimes contention.

The human-caused increase in atmospheric CO_2 already represents nearly a 30 percent change relative to the pre-industrial era, and CO_2 will continue to increase for the foreseeable future. Increased CO_2 represents the most important human enhancement to the greenhouse effect; the consensus of the climate research community is that it probably already affects climate detectably and will drive substantial climate change in the next century (*1*). The direct effects of increased CO_2 on plants and ecosystems may be even more important. The growth of most plants is enhanced by elevated CO_2, but to very different extents; the tissue chemistry of plants that respond to CO_2 is altered in ways that decrease food quality for animals and microbes; and the water use efficiency of plants and ecosystems generally is increased. The fact that increased CO_2 affects species differentially means that it is likely to drive substantial changes in the species composition and dynamics of all terrestrial ecosystems (*19*).

Water

Water is essential to all life. Its movement by gravity, and through evaporation and condensation, contributes to driving Earth's biogeochemical cycles and to controlling its climate. Very little of the water on Earth is directly usable by humans; most is either saline or frozen. Globally, humanity now uses more than half of the runoff water that is fresh and reasonably accessible, with about 70 percent of this use in agriculture (*20*). To meet increasing demands for the limited supply of fresh water, humanity has extensively altered river systems through diversions and impoundments. In the United States only 2 percent of the rivers run unimpeded, and by the end of this century the flow of about two-thirds of all of Earth's rivers will be regulated (*21*). At present, as much as 6 percent of Earth's river runoff is evaporated as a consequence of human manipulations (*22*). Major rivers, including the Colorado, the Nile, and the Ganges, are used so extensively that little water reaches the sea. Massive inland water bodies, including the Aral Sea and Lake Chad, have been greatly reduced in extent by water diversions for agriculture. Reduction in the volume of the Aral Sea resulted in the demise of native fishes and the loss of other biota; the loss of a major fishery; exposure of the salt-laden sea bottom, thereby providing a major source of windblown dust; the production of a drier and more continental local climate and a decrease in water quality in the general region; and an increase in human diseases (*23*).

Impounding and impeding the flow of rivers provides reservoirs of water that can be used for energy generation as well as for agriculture. Waterways also are managed for transport, for flood control, and for the dilution of chemical wastes. Together, these activities have altered Earth's freshwater ecosystems profoundly, to a greater extent than terrestrial ecosystems have been altered. The construction of dams affects biotic habitats indirectly as well; the damming of the Danube River, for example, has altered the silica chemistry of the entire Black Sea. The large number of operational dams (36,000) in the world, in conjunction with the many that are planned, ensure that humanity's effects on aquatic biological systems will continue (*24*). Where surface water is sparse or overexploited, humans use groundwater—and in many areas the groundwater that is drawn upon is nonrenewable, or fossil, water (*25*). For example, three-quarters of the water supply of Saudi Arabia currently comes from fossil water (*26*).

Alterations to the hydrological cycle can affect regional climate. Irrigation increases atmospheric humidity in semiarid areas, often increasing precipitation and thunderstorm frequency (*27*). In contrast, land transformation from forest to agriculture or pasture increases albedo and decreases surface roughness; simulations suggest that the net effect of this transformation is to increase temperature and decrease precipitation regionally (*7, 26*).

Conflicts arising from the global use of water will be exacerbated in the years ahead, with a growing human population and with the stresses that global changes will impose on water quality and availability. Of all of the environmental security issues facing nations, an adequate supply of clean water will be the most important.

Nitrogen

Nitrogen (N) is unique among the major elements required for life, in that its cycle includes a vast atmospheric reservoir (N_2) that must be fixed (combined with carbon, hydrogen, or oxygen) before it can be used by most organisms. The supply of this fixed N controls (at least in part) the productivity, carbon storage, and species composition of many ecosystems. Before the extensive human alteration of the N cycle, 90 to 130 million metric tons of N (Tg N) were fixed biologically on land each year; rates of biological fixation in marine systems are less certain, but perhaps as much was fixed there (28).

Human activity has altered the global cycle of N substantially by fixing N_2—deliberately for fertilizer and inadvertently during fossil fuel combustion. Industrial fixation of N fertilizer increased from <10 Tg/year in 1950 to 80 Tg/year in 1990; after a brief dip caused by economic dislocations in the former Soviet Union, it is expected to increase to >135 Tg/year by 2030 (*29*). Cultivation of soybeans, alfalfa, and other legume crops that fix N symbiotically enhances fixation by another ~ 40 Tg/year, and fossil fuel combustion puts >20 Tg/year of reactive N into the atmosphere globally—some by fixing N_2, more from the mobilization of N in the fuel. Overall, human activity adds at least as much fixed N to terrestrial ecosystems as do all natural sources combined, and it mobilizes >50 Tg/year more during land transformation (*28, 30*).

Alteration of the N cycle has multiple consequences. In the atmosphere, these include (i) an increasing concentration of the greenhouse gas nitrous oxide globally; (ii) substantial increases in fluxes of reactive N gases (two-thirds or more of both nitric oxide and ammonia emissions globally are human-caused); and (iii) a substantial contribution to acid rain and to the photochemical smog that afflicts urban and agricultural areas throughout the world (*31*). Reactive N that is emitted to the atmosphere is deposited downwind, where it can influence the dynamics of recipient ecosystems. In regions where fixed N was in short supply, added N generally increases productivity and C storage within ecosystems, and ultimately increases losses of N and cations from soils, in a set of processes termed "N saturation" (*32*). Where added N increases the productivity of ecosystems, usually it also decreases their biological diversity (*33*).

Human-fixed N also can move from agriculture, from sewage systems, and from N-saturated terrestrial systems to streams, rivers, groundwater, and ultimately the oceans. Fluxes of N through streams and rivers have increased markedly as human alteration of the N cycle has accelerated;

river nitrate is highly correlated with the human population of river basins and with the sum of human-caused N inputs to those basins (*8*). Increases in river N drive the eutrophication of most estuaries, causing blooms of nuisance and even toxic algae, and threatening the sustainability of marine fisheries (*16, 34*).

Other Cycles

The cycles of carbon, water, and nitrogen are not alone in being altered by human activity. Humanity is also the largest source of oxidized sulfur gases in the atmosphere; these affect regional air quality, biogeochemistry, and climate. Moreover, mining and mobilization of phosphorus and of many metals exceed their natural fluxes; some of the metals that are concentrated and mobilized are highly toxic (including lead, cadmium, and mercury) (*35*). Beyond any doubt, humanity is a major biogeochemical force on Earth.

Synthetic Organic Chemicals

Synthetic organic chemicals have brought humanity many beneficial services. However, many are toxic to humans and other species, and some are hazardous in concentrations as low as 1 part per billion. Many chemicals persist in the environment for decades; some are both toxic and persistent. Long-lived organochlorine compounds provide the clearest examples of environmental consequences of persistent compounds. Insecticides such as DDT and its relatives, and industrial compounds like polychlorinated biphenyls (PCBs), were used widely in North America in the 1950s and 1960s. They were transported globally, accumulated in organisms, and magnified in concentration through food chains; they devastated populations of some predators (notably falcons and eagles) and entered parts of the human food supply in concentrations higher than was prudent. Domestic use of these compounds was phased out in the 1970s in the United States and Canada, and their concentrations declined thereafter. However, PCBs in particular remain readily detectable in many organisms, sometimes approaching thresholds of public health concern (*36*). They will continue to circulate through organisms for many decades.

Synthetic chemicals need not be toxic to cause environmental problems. The fact that the persistent and volatile chlorofluorocarbons (CFCs) are wholly nontoxic contributed to their widespread use as refrigerants and even aerosol propellants. The subsequent discovery that CFCs drive the breakdown of stratospheric ozone, and especially the later discovery of the Antarctic ozone hole and their role in it, represent great surprises in global environmental science (*37*). Moreover, the response of the international political system to those discoveries is the best extant illustration that global environmental change can be dealt with effectively (*38*).

Particular compounds that pose serious health and environmental threats can be and often have been phased out (although PCB production is growing in Asia). Nonetheless, each year the chemical industry produces more than 100 million tons of organic chemicals representing some 70,000 different compounds, with about 1000 new ones being added annually (*39*). Only a small fraction of the many chemicals produced and released into the environment are tested adequately for health hazards or environmental impact (*40*).

Biotic Changes

Human modification of Earth's biological resources—its species and genetically distinct populations—is substantial and growing. Extinction is a natural process, but the current rate of loss of genetic variability, of populations, and of species is far above background rates; it is ongoing; and it represents a wholly irreversible global change. At the same time, human transport of species around Earth is homogenizing Earth's biota, introducing many species into new areas where they can disrupt both natural and human systems.

Losses

Rates of extinction are difficult to determine globally, in part because the majority of species on Earth have not yet been identified. Nevertheless, recent calculations suggest that rates of species extinction are now on the order of 100 to 1000 times those before humanity's dominance of Earth (*41*). For particular well-known groups, rates of loss are even greater; as many as one-quarter of Earth's bird species have been driven to extinction by human activities over the past two millennia, particularly on oceanic islands (*42*). At present, 11 percent of the remaining birds, 18 percent of the mammals, 5 percent of fish, and 8 percent of plant species on Earth are threatened with extinction (*43*). There has been a disproportionate loss of large mammal species because of hunting; these species played a dominant role in many ecosystems, and their loss has resulted in a fundamental change in the dynamics of those systems (*44*), one that could lead to further extinctions. The largest organisms in marine systems have been affected similarly, by fishing and whaling. Land transformation is the single most important cause of extinction, and current rates of land transformation eventually will drive many more species to extinction, although with a time lag that masks the true dimensions of the crisis (*45*). Moreover, the effects of other components of global environmental change—of altered carbon and nitrogen cycles, and of anthropogenic climate change—are just beginning.

As high as they are, these losses of species understate the magnitude of loss of genetic variation. The loss to land transformation of locally adapted populations within species, and of genetic material within populations, is a human-caused change that reduces the resilience of species and ecosystems while precluding human use of the library of natural products and genetic material that they represent (*46*).

Although conservation efforts focused on individual endangered species have yielded some successes, they are expensive—and the protection or restoration of whole ecosystems often represents the most effective way to sustain genetic, population, and species diversity. Moreover, ecosystems themselves may play important roles in both natural and human-dominated landscapes. For example, mangrove ecosystems protect coastal areas from erosion and provide nurseries for offshore fisheries, but they are threatened by land transformation in many areas.

Invasions

In addition to extinction, humanity has caused a rearrangement of Earth's biotic systems, through the mixing of floras and faunas that had long been isolated geographically. The magnitude of transport of species, termed "biological invasion," is enormous (*47*); invading species are present almost everywhere. On many islands, more than half of the plant species are nonindigenous, and in many continental areas the figure is 20 percent or more (*48*).

As with extinction, biological invasion occurs naturally—and as with extinction, human activity has accelerated its rate by orders of magnitude. Land transformation interacts strongly with biological invasion, in that human-altered ecosystems generally provide the primary foci for invasions, while in some cases land transformation itself is driven by biological invasions (*49*). International commerce is also a primary cause of the breakdown of biogeographic barriers; trade in live organisms is massive and global, and many other organisms are inadvertently taken along for the ride. In freshwater systems, the combination of upstream land transformation, altered hydrology, and numerous deliberate and accidental species introductions has led to particularly widespread invasion, in continental as well as island ecosystems (*50*).

In some regions, invasions are becoming more frequent. For example, in the San Francisco Bay of California, an average of one new species has been established every 36 weeks since 1850, every 24 weeks since 1970, and every 12 weeks for the last decade (*51*). Some introduced species quickly become invasive over large areas (for example, the Asian clam in the San Francisco Bay), whereas others become widespread only after a lag of decades, or even over a century (*52*).

Many biological invasions are effectively irreversible; once replicating biological material is released into the environment and becomes successful there, calling it back is difficult and expensive at best. Moreover, some species introductions have consequences. Some degrade human health and that of other species; after all, most infectious diseases are invaders over most of their range. Others have caused economic losses amounting to billions of dollars; the recent invasion of North America by the zebra mussel is a well-publicized example. Some disrupt ecosystem processes, altering the structure and functioning of whole ecosystems. Finally, some invasions drive losses in the biological diversity of native species and populations; after land transformation, they are the next most important cause of extinction (*53*).

Conclusions

The global consequences of human activity are not something to face in the future—they are with us now. All of these changes are ongoing, and in many cases accelerating; many of them were entrained long before their importance was recognized. Moreover, all of these seemingly disparate phenomena trace to a single cause—the growing scale of the human enterprise. The rates, scales, kinds, and combinations of changes occurring now are fundamentally different from those at any other time in history; we are changing Earth more rapidly than we are understanding it. We live on a human-dominated planet—and the momentum of human population growth, together with the imperative for further economic development in most of the world, ensures that our dominance will increase.

The papers in this special section summarize our knowledge of and provide specific policy recommendations concerning major human-dominated ecosystems. In addition, we suggest that the rate and extent of human alteration of Earth should affect how we think about Earth. It is clear that we control much of Earth, and that our activities affect the rest. In a very real sense, the world is in our hands—and how we handle it will determine its composition and dynamics, and our fate.

Recognition of the global consequences of the human enterprise suggests three complementary directions. First, we can work to reduce the rate at which we alter the Earth system. Humans and human-dominated systems may be able to adapt to slower change, and ecosystems and the species they support may cope more effectively with the changes we impose, if those changes are slow. Our footprint on the planet (*54*) might then be stabilized at a point where enough space and resources remain to sustain most of the other species on Earth, for their sake and our own. Reducing the rate of growth in human effects on Earth involves slowing human population growth and using resources as efficiently as is practical. Often it is the waste products and by-products of human activity that drive global environmental change.

Second, we can accelerate our efforts to understand Earth's ecosystems and how they interact with the numerous components of human-caused global change. Ecological research is inherently complex and demanding: It requires measurement and monitoring of populations and ecosystems; experimental studies to elucidate the regulation of ecological processes; the development, testing, and validation of regional and global models; and integration with a broad range of biological, earth, atmospheric, and marine sciences. The challenge of understanding a human-dominated planet further requires that the human dimensions of global change—the social, economic, cultural, and other drivers of human actions—be included within our analyses.

Finally, humanity's dominance of Earth means that we cannot escape responsibility for managing the planet. Our activities are causing rapid, novel, and substantial changes to Earth's ecosystems. Maintaining populations, species, and ecosystems in the face of those changes, and maintaining the flow of goods and services they provide humanity (*55*), will require active management for the foreseeable future. There is no clearer illustration of the extent of human dominance of Earth than the fact that maintaining the diversity of "wild" species and the functioning of "wild" ecosystems will require increasing human involvement.

References and Notes

1. Intergovernmental Panel on Climate Change, *Climate Change 1995* (Cambridge Univ. Press, Cambridge, 1996), pp. 9–49.
2. United Nations Environment Program, *Global Biodiversity Assessment,* V. H. Heywood, Ed. (Cambridge Univ. Press, Cambridge, 1995).
3. D. Skole and C. J. Tucker, *Science* **260,** 1905 (1993).
4. J. S. Olson, J. A. Watts, L. J. Allison, *Carbon in Live Vegetation of Major World Ecosystems* (Office of Energy Research, U.S. Department of Energy, Washington, DC, 1983).
5. P. M. Vitousek, P. R. Ehrlich, A. H. Ehrlich, P. A. Matson, *Bioscience* **36,** 368 (1986); R. W. Kates, B. L. Turner, W. C. Clark, in (*35*), pp. 1–17; G. C. Daily, *Science* **269,** 350 (1995).
6. D. A. Saunders, R. J. Hobbs, C. R. Margules, *Conserv. Biol.* **5,** 18 (1991).
7. J. Shukla, C. Nobre, P. Sellers, *Science* **247,** 1322 (1990).
8. R. W. Howarth *et al., Biogeochemistry* **35,** 75 (1996).
9. W. B. Meyer and B. L. Turner II, *Changes in Land Use and Land Cover: A Global Perspective* (Cambridge Univ. Press, Cambridge, 1994).
10. S. R. Carpenter, S. G. Fisher, N. B. Grimm, J. F. Kitchell, *Annu. Rev. Ecol. Syst.* **23,** 119 (1992); S. V. Smith and R. W. Buddemeier, *ibid.,* p. 89; J. M. Melillo, I. C. Prentice, G. D. Farquhar, E.-D. Schulze, O. E. Sala, in (*1*), pp. 449–481.
11. R. Leemans and G. Zuidema, *Trends Ecol. Evol.* **10,** 76 (1995).
12. World Resources Institute, *World Resources 1996–1997* (Oxford Univ. Press, New York, 1996).
13. D. Pauly and V. Christensen, *Nature* **374,** 257 (1995).
14. Food and Agricultural Organization (FAO), *FAO Fisheries Tech. Pap. 335* (1994).
15. D. L. Alverson, M. H. Freeberg, S. A. Murawski, J. G. Pope, *FAO Fisheries Tech. Pap. 339* (1994).
16. G. M. Hallegraeff, *Phycologia* **32,** 79 (1993).
17. D. S. Schimel *et al.,* in *Climate Change 1994: Radiative Forcing of Climate Change,* J. T. Houghton *et al.,* Eds. (Cambridge Univ. Press, Cambridge, 1995), pp. 39–71.
18. R. J. Andres, G. Marland, I. Y. Fung, E. Matthews, *Global Biogeochem. Cycles* **10,** 419 (1996).
19. G. W. Koch and H. A. Mooney, *Carbon Dioxide and Terrestrial Ecosystems* (Academic Press, San Diego, CA, 1996); C. Körner and F. A. Bazzaz, *Carbon Dioxide, Populations, and Communities* (Academic Press, San Diego, CA, 1996).
20. S. L. Postel, G. C. Daily, P. R. Ehrlich, *Science* **271,** 785 (1996).
21. J. N. Abramovitz, *Imperiled Waters, Impoverished Future: The Decline of Freshwater Ecosystems* (Worldwatch Institute, Washington, DC, 1996).
22. M. I. L'vovich and G. F. White, in (*35*), pp. 235–252; M. Dynesius and C. Nilsson, *Science* **266,** 753 (1994).
23. P. Micklin, *Science* **241,** 1170 (1988); V. Kotlyakov, *Environment* **33,** 4 (1991).
24. C. Humborg, V. Ittekkot, A. Cociasu, B. Bodungen, *Nature* **386,** 385 (1997).
25. P. H. Gleick, Ed., *Water in Crisis* (Oxford Univ. Press, New York, 1993).
26. V. Gornitz, C. Rosenzweig, D. Hillel, *Global Planet. Change* **14,** 147 (1997).
27. P. C. Milly and K. A. Dunne, *J. Clim.* **7,** 506 (1994).
28. J. N. Galloway, W. H. Schlesinger, H. Levy II, A. Michaels, J. L. Schnoor, *Global Biogeochem. Cycles* **9,** 235 (1995).
29. J. N. Galloway, H. Levy II, P. S. Kasibhatla, *Ambio* **23,** 120 (1994).
30. V. Smil, in (*35*), pp. 423–436.
31. P. M. Vitousek *et al., Ecol. Appl.,* in press.
32. J. D. Aber, J. M. Melillo, K. J. Nadelhoffer, J. Pastor, R. D. Boone, *ibid.* **1,** 303 (1991).
33. D. Tilman, *Ecol. Monogr.* **57,** 189 (1987).
34. S. W. Nixon *et al., Biogeochemistry* **35,** 141 (1996).
35. B. L. Turner II *et al.,* Eds., *The Earth As Transformed by Human Action* (Cambridge Univ. Press, Cambridge, 1990).
36. C. A. Stow, S. R. Carpenter, C. P. Madenjian, L. A. Eby, L. J. Jackson, *Bioscience* **45,** 752 (1995).
37. F. S. Rowland, *Am. Sci.* **77,** 36 (1989); S. Solomon, *Nature* **347,** 347 (1990).
38. M. K. Tolba *et al.,* Eds., *The World Environment 1972–1992* (Chapman & Hall, London, 1992).
39. S. Postel, *Defusing the Toxics Threat: Controlling Pesticides and Industrial Waste* (Worldwatch Institute, Washington, DC, 1987).
40. United Nations Environment Program (UNEP), *Saving Our Planet—Challenges and Hopes* (UNEP, Nairobi, 1992).
41. J. H. Lawton and R. M. May, Eds., *Extinction Rates* (Oxford Univ. Press, Oxford, 1995); S. L. Pimm, G. J. Russell, J. L. Gittleman, T. Brooks, *Science* **269,** 347 (1995).
42. S. L. Olson, in *Conservation for the Twenty-First Century,* D. Western and M. C. Pearl, Eds. (Oxford Univ. Press, Oxford, 1989), p. 50; D. W. Steadman, *Science* **267,** 1123 (1995).
43. R. Barbault and S. Sastrapradja, in (*2*), pp. 193–274.
44. R. Dirzo and A. Miranda, in *Plant-Animal Interactions,* P. W. Price, T. M. Lewinsohn, W. Fernandes, W. W. Benson, Eds. (Wiley Interscience, New York, 1991), p. 273.
45. D. Tilman, R. M. May, C. Lehman, M. A. Nowak, *Nature* **371,** 65 (1994).
46. H. A. Mooney, J. Lubchenco, R. Dirzo, O. E. Sala, in (*2*), pp. 279–325.
47. C. Elton, *The Ecology of Invasions by Animals and Plants* (Methuen, London, 1958); J. A. Drake *et al.,* Eds., *Biological Invasions. A Global Perspective* (Wiley, Chichester, UK, 1989).
48. M. Rejmanek and J. Randall, *Madrono* **41,** 161 (1994).
49. C. M. D'Antonio and P. M. Vitousek, *Annu. Rev. Ecol. Syst.* **23,** 63 (1992).
50. D. M. Lodge, *Trends Ecol. Evol.* **8,** 133 (1993).
51. A. N. Cohen and J. T. Carlton, *Biological Study: Nonindigenous Aquatic Species in a United States Esturary: A Case Study of the Biological Invasions of the San Francisco Bay and Delta* (U.S. Fish and Wildlife Service, Washington, DC, 1995).
52. I. Kowarik, in *Plant Invasions—General Aspects and Special Problems,* P. Pysek, K. Prach, M. Rejmánek, M. Wade, Eds. (SPB Academic, Amersterdam, 1995), p. 15.
53. P. M. Vitousek, C. M. D'Antonio, L. L. Loope, R. Westbrooks, *Am. Sci.* **84,** 468 (1996).
54. W. E. Rees and M. Wackernagel, in *Investing in Natural Capital: The Ecological Economics Approach to Sustainability,* A. M. Jansson, M. Hammer, C. Folke, R. Costanza, Eds. (Island, Washington, DC, 1994).

55. G. C. Daily, Ed., *Nature's Services* (Island, Washington, DC, 1997).
56. J. Lubchenco, *et al., Ecology* **72,** 371 (1991); P. M. Vitousek, *ibid.* **75,** 1861 (1994).
57. S. M. Garcia and R. Grainger, *FAO Fisheries Tech. Pap. 359* (1996).
58. We thank G. C. Daily, C. B. Field, S. Hobbie, D. Gordon, P. A. Matson, and R. L. Naylor for constructive comments on this paper, A. S. Denning and S. M. Garcia for assistance with illustrations, and C. Nakashima and B. Lilley for preparing text and figures for publication.

Critical Thinking

1. How does this article relate to the one about the competitive exclusion principle?
2. Should humans dominate Earth's ecosystem? If so, why. If not, why not?
3. What do think would happen if humans altered marine ecosystems as much as they have altered terrestrial ones?

Human Alteration of the Global Nitrogen Cycle: Causes and Consequences

PETER M. VITOUSEK ET AL.

Introduction

This report presents an overview of the current scientific understanding of human-driven changes to the global nitrogen cycle and their consequences. It also addresses policy and management options that could help moderate these changes in the nitrogen cycle and their impacts.

The Nitrogen Cycle

Nitrogen is an essential component of proteins, genetic material, chlorophyll, and other key organic molecules. All organisms require nitrogen in order to live. It ranks fourth behind oxygen, carbon, and hydrogen as the most common chemical element in living tissues. Until human activities began to alter the natural cycle, however, nitrogen was only scantily available to much of the biological world. As a result, nitrogen served as one of the major limiting factors that controlled the dynamics, biodiversity, and functioning of many ecosystems.

The Earth's atmosphere is 78 percent nitrogen gas, but most plants and animals cannot use nitrogen gas directly from the air as they do carbon dioxide and oxygen. Instead, plants—and all organisms from the grazing animals to the predators to the decomposers that ultimately secure their nourishment from the organic materials synthesized by plants—must wait for nitrogen to be "fixed," that is, pulled from the air and bonded to hydrogen or oxygen to form inorganic compounds, mainly ammonium (NH_4) and nitrate (NO_3), that they can use.

The amount of gaseous nitrogen being fixed at any given time by natural processes represents only a small addition to the pool of previously fixed nitrogen that cycles among the living and nonliving components of the Earth's ecosystems. Most of that nitrogen, too, is unavailable, locked up in soil organic matter—partially rotted plant and animal remains—that must be decomposed by soil microbes. These microbes release nitrogen as ammonium or nitrate, allowing it to be recycled through the food web. The two major natural sources of new nitrogen entering this cycle are nitrogen-fixing organisms and lightning.

Nitrogen-fixing organisms include a relatively small number of algae and bacteria. Many of them live free in the soil, but the most important ones are bacteria that form close symbiotic relationships with higher plants. Symbiotic nitrogen-fixing bacteria such as the Rhizobia, for instance, live and work in nodules on the roots of peas, beans, alfalfa and other legumes. These bacteria manufacture an enzyme that enables them to convert gaseous nitrogen directly into plant-usable forms.

Lightning may also indirectly transform atmospheric nitrogen into nitrates, which rain onto soil.

Quantifying the rate of natural nitrogen fixation prior to human alterations of the cycle is difficult but necessary for evaluating the impacts of human-driven changes to the global cycling of nitrogen. The standard unit of measurement for analyzing the global nitrogen cycle is the teragram (abbreviated Tg), which is equal to a million metric tons of nitrogen. Worldwide, lightning, for instance, fixes less than 10 Tg of nitrogen per year—maybe even less than 5 Tg. Microbes are the major natural suppliers of new biologically available nitrogen. Before the widespread planting of legume crops, terrestrial organisms probably fixed between 90 and 140 Tg of nitrogen per year. A reasonable upper bound for the rate of natural nitrogen fixation on land is thus about 140 Tg of N per year.

Human-Driven Nitrogen Fixation

During the past century, human activities clearly have accelerated the rate of nitrogen fixation on land, effectively doubling the annual transfer of nitrogen from the vast but unavailable atmospheric pool to the biologically available forms. The major sources of this enhanced supply include industrial processes that produce nitrogen fertilizers, the combustion of fossil fuels, and the cultivation of soybeans, peas, and other crops that host symbiotic nitrogen-fixing bacteria. Furthermore, human activity is also speeding up the release of nitrogen from long-term storage in soils and organic matter.

Nitrogen Fertilizer

Industrial fixation of nitrogen for use as fertilizer currently totals approximately 80 Tg per year and represents by far the largest human contribution of new nitrogen to the global cycle. That figure does not include manures and other organic nitrogen fertilizers, which represent a transfer of already-fixed nitrogen from one place to another rather than new fixation.

The process of manufacturing fertilizer by industrial nitrogen fixation was first developed in Germany during World War I, and fertilizer production has grown exponentially since the 1940s. In recent years, the increasing pace of production and use has been truly phenomenal. The amount of industrially fixed nitrogen applied to crops during the decade from 1980 to 1990 more than equaled all industrial fertilizer applied previously in human history.

Until the late 1970s, most industrially produced fertilizer was applied in developed countries. Use in these regions has now stabilized while fertilizer applications in developing countries have risen dramatically. The momentum of human population growth and increasing urbanization ensures that industrial fertilizer production will continue at high and likely accelerating rates for decades in order to meet the escalating demand for food.

Nitrogen-Fixing Crops

Nearly one third of the Earth's land surface is devoted to agricultural and pastoral uses, and humans have replaced large areas of diverse natural vegetation with monocultures of soybeans, peas, alfalfa, and other leguminous crops and forages. Because these plants support symbiotic nitrogen-fixers, they derive much of their nitrogen directly from the atmosphere and greatly increase the rate of nitrogen fixation previously occurring on those lands. Substantial levels of nitrogen fixation also occur during cultivation of some non-legumes, notably rice. All of this represents new, human-generated stocks of biologically available nitrogen. The quantity of nitrogen fixed by crops is more difficult to analyze than industrial nitrogen production. Estimates range from 32 to 53 Tg per year. As an average, 40 Tg will be used here.

Fossil Fuel Burning

The burning of fossil fuels such as coal and oil releases previously fixed nitrogen from long-term storage in geological formations back to the atmosphere in the form of nitrogen-based trace gases such as nitric oxide. High-temperature combustion also fixes a small amount of atmospheric nitrogen directly. Altogether, the operations of automobiles, factories, power plants, and other combustion processes emit more than 20 Tg per year of fixed nitrogen to the atmosphere. All of it is treated here as newly fixed nitrogen because it has been locked up for millions of years and would remain locked up indefinitely if not released by human action.

Mobilization of Stored Nitrogen

Besides enhancing fixation and releasing nitrogen from geological reservoirs, human activities also liberate nitrogen from long-term biological storage pools such as soil organic matter and tree trunks, contributing further to the proliferation of biologically available nitrogen. These activities include the burning of forests, wood fuels, and grasslands, which emits more than 40 Tg per year of nitrogen; the draining of wetlands, which sets the stage for oxidation of soil organic matter that could mobilize 10 Tg per year or more of nitrogen; and land clearing for crops, which could mobilize 20 Tg per year from soils.

There are substantial scientific uncertainties about both the quantity and the fate of nitrogen mobilized by such activities. Taken together, however, they could contribute significantly to changes in the global nitrogen cycle.

Human versus Natural Nitrogen Fixation

Overall, fertilizer production, legume crops, and fossil fuel burning deposit approximately 140 Tg of new nitrogen into land-based ecosystems each year, a figure that equals the upper estimates for nitrogen fixed naturally by organisms in these ecosystems. Other human activities liberate and make available half again that much nitrogen. From this evidence, it is fair to conclude that human activities have at least doubled the transfer of nitrogen from the atmosphere into the land-based biological nitrogen cycle.

This extra nitrogen is spread unevenly across the Earth's surface: Some areas such as northern Europe are being altered profoundly while others such as remote regions in the Southern Hemisphere receive little direct input of human-generated nitrogen. Yet no region remains unaffected. The increase in fixed nitrogen circulating around the globe and falling to the ground as wet or dry deposition is readily detectable, even in cores drilled from the glacial ice of Greenland.

Impacts on the Atmosphere

One major consequence of human-driven alterations in the nitrogen cycle has been regional and global change in the chemistry of the atmosphere—specifically, increased emissions of nitrogen-based trace gases such as nitrous oxide, nitric oxide, and ammonia (NH_3). Although such releases have received less attention than increased emissions of carbon dioxide and various sulfur compounds, the trace nitrogen gases cause environmental effects both while airborne and after they are deposited on the ground. For instance, nitrous oxide is long-lived in the atmosphere and contributes to the human-driven enhancement of the greenhouse effect that likely warms the Earth's climate. Nitric oxide is an important precursor of acid rain and photochemical smog.

Some of the human activities discussed above affect the atmosphere directly. For instance, essentially all of the more than 20 Tg per year of fixed nitrogen released in automobile exhausts and in other emissions from fossil fuel burning is emitted to the atmosphere as nitric oxide. Other activities indirectly enhance emissions to the atmosphere. Intensive fertilization of agricultural soils can increase the rates at which nitrogen in the form of ammonia is volatilized and lost to the air. It can also speed the microbial breakdown of ammonium and nitrates in the soil, enhancing the release of nitrous oxide. Even in wild or unmanaged lands downwind of agricultural or industrial areas, rain or windborne deposition of human-generated nitrogen can spur increased emissions of nitrogen gases from the soils.

Nitrous Oxide

Nitrous oxide is a very effective heat-trapping gas in the atmosphere, in part because it absorbs outgoing radiant heat from the Earth in infrared wavelengths that are not captured by the other major greenhouse gases, water vapor and carbon dioxide. By absorbing and reradiating this heat back toward the Earth, nitrous oxide contributes a few percent to overall greenhouse warming.

Although nitrous oxide is unreactive and long-lived in the lower atmosphere, when it rises into the stratosphere it can trigger reactions that deplete and thin the stratospheric ozone layer that shields the Earth from damaging ultraviolet radiation.

The concentration of nitrous oxide in the atmosphere is currently increasing at the rate of two- to three-tenths of a percent per year. While that rise is clearly documented, the sources of the increase remain unresolved. Both fossil fuel burning and the direct impacts of agricultural fertilization have been considered and rejected as the major source. Rather, there is a developing consensus that a wide array of human-driven sources contribute systematically to enrich the terrestrial nitrogen cycle. These "dispersed sources" include fertilizers, nitrogen-enriched groundwater, nitrogen-saturated forests, forest burning, land clearing, and even the manufacture of nylon, nitric acid, and other industrial products.

The net effect is increased global concentrations of a potent greenhouse gas that also contributes to the thinning of the stratospheric ozone layer.

Nitric Oxide and Ammonia

Unlike nitrous oxide, which is unreactive in the lower atmosphere, both nitric oxide and ammonia are highly reactive and therefore much shorter lived. Thus changes in their atmospheric concentrations can be detected only at local or regional scales.

Nitric oxide plays several critical roles in atmospheric chemistry, including catalyzing the formation of photochemical (or brown) smog. In the presence of sunlight, nitric oxide and oxygen react with hydrocarbons emitted by automobile exhausts to form ozone, the most dangerous component of smog. Ground-level ozone has serious detrimental effects on human health as well as the health and productivity of crops and forests.

Nitric oxide, along with other oxides of nitrogen and sulfur, can be transformed in the atmosphere into nitric acid and sulfuric acid, which are the major components of acid rain.

Although a number of sources contribute to nitric oxide emissions, combustion is the dominant one. Fossil fuel burning emits more than 20 Tg per year of nitric oxide. Human burning of forests and other plant material may add about 10 Tg, and global emissions of nitric oxide from soils, a substantial fraction of which are human-caused, total 5 to 20 Tg per year. Overall, 80 percent or more of nitric oxide emissions worldwide are generated by human activities, and in many regions the result is increased smog and acid rain.

In contrast to nitric oxide, ammonia acts as the primary acid-neutralizing agent in the atmosphere, having an opposite influence on the acidity of aerosols, cloudwater, and rainfall. Nearly 70 percent of global ammonia emissions are human-caused. Ammonia volatilized from fertilized fields contributes an estimated 10 Tg per year; ammonia released from domestic animal wastes about 32 Tg; and forest burning some 5 Tg.

Effects on the Carbon Cycle

Increased emissions of airborne nitrogen have led to enhanced deposition of nitrogen on land and in the oceans. Thanks to the fertilizer effects of nitrogen in stimulating plant growth, this deposition may be acting to influence the atmosphere indirectly by altering the global carbon cycle.

Over much of the Earth's surface, the lushness of plant growth and the accumulation of standing stocks of plant material historically have been limited by scanty nitrogen supplies, particularly in temperate and boreal regions. Human activity has substantially increased the deposition of nitrogen over much of this area, which raises important questions: How much extra plant growth has been caused by human-generated nitrogen additions? As a result, how much extra carbon has been stored in terrestrial ecosystems rather than contributing to the rising concentrations of carbon dioxide in the atmosphere?

Answers to these questions could help explain the imbalance in the carbon cycle that has come to be known as the 'missing sink.' The known emissions of carbon dioxide from human activities such as fossil fuel burning and deforestation exceed by more than 1,000 Tg the amount of carbon dioxide known to be accumulating in the atmosphere each year. Could increased growth rates in terrestrial vegetation be the 'sink' that accounts for the fate of much of that missing carbon?

Experiments in Europe and America indicate that a large portion of the extra nitrogen retained by forest, wetland, and tundra ecosystems stimulates carbon uptake and storage. On the other hand, this nitrogen can also stimulate microbial decomposition and thus releases of carbon from soil organic matter. On balance, however, the carbon uptake through new plant growth appears to exceed the carbon losses, especially in forests.

A number of groups have attempted to calculate the amount of carbon that could be stored in terrestrial vegetation thanks to plant growth spurred by added nitrogen. The resulting estimates range from 100 to 1,300 Tg per year. The number has tended to increase in more recent analyses as the magnitude of human-driven changes in the nitrogen cycle has become clearer. The most recent analysis of the global carbon cycle by the Intergovernmental Panel on Climate Change concluded that nitrogen deposition could represent a major component of the missing carbon sink.

More precise estimates will become possible when we have a more complete understanding of the fraction of human-generated nitrogen that actually is retained within various land-based ecosystems.

Nitrogen Saturation and Ecosystem Functioning

There are limits to how much plant growth can be increased by nitrogen fertilization. At some point, when the natural nitrogen deficiencies in an ecosystem are fully relieved, plant growth becomes limited by spar-sity of other resources such as phosphorus, calcium, or water. When the vegetation can no longer respond to further additions of nitrogen, the ecosystem reaches a state described as "nitrogen saturation." In theory, when an ecosystem is fully nitrogen-saturated and its soils, plants, and microbes cannot use or retain any more, all new nitrogen deposits will be dispersed to streams, groundwater, and the atmosphere.

Nitrogen saturation has a number of damaging consequences for the health and functioning of ecosystems. These impacts first became apparent in Europe almost two decades ago when scientists observed significant increases in nitrate concentrations in some lakes and streams and also extensive yellowing and loss of needles in spruce and other conifer forests subjected to heavy nitrogen deposition. These observations led to several field experiments in the U.S. and Europe that have revealed a complex cascade of effects set in motion by excess nitrogen in forest soils.

As ammonium builds up in the soil, it is increasingly converted to nitrate by bacterial action, a process that releases hydrogen ions and helps acidify the soil. The buildup of nitrate enhances emissions of nitrous oxides from the soil and also encourages leaching of highly water-soluble nitrate into streams or groundwater. As these negatively charged nitrates seep away, they carry with them positively charged alkaline minerals such as calcium, magnesium, and potassium. Thus human modifications to the nitrogen cycle decrease soil fertility by greatly accelerating the loss of calcium and other nutrients that are vital for plant growth. As calcium is depleted and the soil acidified, aluminum ions are mobilized, eventually reaching toxic concentrations that can damage tree roots or kill fish if the aluminum washes into streams. Trees growing in soils replete with nitrogen but starved of calcium, magnesium, and potassium can develop nutrient imbalances in their roots and leaves. This may reduce their photosynthetic rate and efficiency, stunt their growth, and even increase tree deaths.

Nitrogen saturation is much further advanced over extensive areas of northern Europe than in North America because human-generated nitrogen deposition is several times greater there than in even the most extremely affected areas of North America. In the nitrogen-saturated ecosystems of Europe, a substantial fraction of atmospheric nitrate deposits move from the land into streams without ever being taken up by organisms or playing a role in the biological cycle.

In contrast, in the northeastern U.S., increased leaching of nitrates from the soil and large shifts in the nutrient ratios in tree leaves generally have been observed only in certain types of forests. These include high-elevation sites that receive large nitrogen deposits and sites with shallow soils containing few alkaline minerals to buffer acidification. Elsewhere in the U.S., the early stages of nitrogen saturation have been seen in response to elevated nitrogen deposition in the forests surrounding the Los Angeles Basin and in the Front Range of the Colorado Rockies.

Some forests have a very high capacity to retain added nitrogen, particularly regrowing forests that have been subjected to intense or repeated harvesting, an activity that usually causes severe nitrogen losses. Overall, the ability of a forest to retain nitrogen depends on its potential for further growth and the extent of its current nitrogen stocks. Thus, the impacts of nitrogen deposition are tightly linked to other rapidly changing human-driven variables such as shifts in land use, climate, and atmospheric carbon dioxide and ozone levels.

Effects on Biodiversity and the Species Mix

Limited supplies of biologically available nitrogen are a fact of life in most natural ecosystems, and many native plant species are adapted to function best under this constraint. New supplies of nitrogen showered upon these ecosystems can cause a dramatic shift in the dominant species and also a marked reduction in overall species diversity as the few plant species adapted to take full advantage of high nitrogen out compete their neighbors. In England, for example, nitrogen fertilizers applied to experimental grasslands have led to increased dominance by a few nitrogen-responsive grasses and loss of many other plant species. At the highest fertilization rate, the number of plant species declined more than fivefold. In North America, similarly dramatic reductions in biodiversity have been created by fertilization of grasslands in Minnesota and California. In formerly species-rich heathlands across Western Europe, human-driven nitrogen deposition has been blamed for great losses of biodiversity in recent decades.

In the Netherlands, high human population density, intensive livestock operations, and industries have combined to generate the highest rates of nitrogen deposition in the world. One well-documented consequence has been the conversion of species-rich heathlands to species-poor grasslands and forest. Not only the species richness of the heath but also the biological diversity of the landscape has been reduced because the modified plant communities now resemble the composition of communities occupying more fertile soils. The unique species assemblage adapted to sandy, nitrogen-poor soils is being lost from the region.

Losses of biodiversity driven by nitrogen deposition can in turn affect other ecological processes. Recent experiments in Minnesota grasslands showed that in ecosystems made species-poor by fertilization, plant productivity was much less stable in the face of a major drought. Even in non-drought years, the normal vagaries of climate produced much more year-to-year variation in the productivity of species-poor grassland plots than in more diverse plots.

Effects on Aquatic Ecosystems

Historical Changes in Water Chemistry

Not surprisingly, nitrogen concentrations in surface waters have increased as human activities have accelerated the rate of fixed nitrogen being put into circulation. A recent study of the North Atlantic Ocean Basin by scientists from a dozen nations estimates that movements of total dissolved nitrogen into most of the temperate-zone rivers in the basin may have increased by two- to 20-fold since preindustrial times. For rivers in the North Sea region, the nitrogen increase may have been six- to 20-fold. The nitrogen increases in these rivers are highly correlated with human-generated inputs of nitrogen to their watersheds, and these inputs are dominated by fertilizers and atmospheric deposition.

For decades, nitrate concentrations in many rivers and drinking water supplies have been closely monitored in developed regions of the world, and analysis of these data confirms a historic rise in nitrogen levels in the surface waters. In 1,000 lakes in Norway, for example, nitrate levels doubled in less than a decade. In the Mississippi River, nitrates have more than doubled since 1965. In major rivers of the northeastern U.S., nitrate concentrations have risen three- to ten-fold since the early 1900s, and the evidence suggests a similar trend in many European rivers.

Again not surprisingly, nitrate concentrations in the world's large rivers rise along with the density of human population in the watersheds. Amounts of total dissolved nitrogen in rivers are also correlated with human population density, but total nitrogen does not increase as rapidly as the nitrate fraction. Evidence indicates that with increasing human disturbance, a higher proportion of the nitrogen in surface waters is composed of nitrate.

Increased concentrations of nitrate have also been observed in groundwater in many agricultural regions, although the magnitude of the trend is difficult to determine in all but a few well-characterized aquifers. Overall, the additions to groundwater probably represent only a small fraction of the increased nitrate transported in surface waters. However, groundwater has a long residence time in many aquifers, meaning that groundwater quality is likely to continue to decline as long as human activities are having substantial impacts on the nitrogen cycle.

High levels of nitrates in drinking water raise significant human health concerns, especially for infants. Microbes in an infant's stomach may convert high levels of nitrate to nitrite. When nitrite is absorbed into the bloodstream, it converts oxygen-carrying hemoglobin into an ineffective form called methemoglobin. Elevated methemoglobin levels—an anemic condition known as methemoglobinemia—can cause brain damage or death. The condition is rare in the U.S., but the potential exists whenever nitrate levels exceed U.S. Public Health Service standards (10 milligrams per liter).

Nitrogen and Acidification of Lakes

Nitric acid is playing an increasing role in the acidification of lakes and streams for two major reasons. One is that most efforts to control acid deposition—which includes acid rain, snow, fog, mist, and dry deposits—have focused on cutting emissions of sulfur dioxide to limit the formation of sulfuric acid in the atmosphere. In many areas, these efforts have succeeded in reducing inputs of sulfuric acid to soils and water while emissions of nitrogen oxides, the precursors of nitric acid, have gone largely unchecked. The second reason is that many watersheds in areas of moderate to high nitrogen deposition appear to be approaching nitrogen saturation, and the increasingly acidified soils have little capacity to buffer acid rain before it enters streams.

An additional factor in many areas is that nitric acid predominates among the pollutants that accumulate in the winter snowpack. Much of this nitric acid is flushed out with the first batch of spring meltwater, often washing a sudden, concentrated "acid pulse" into vulnerable lakes.

Adding inorganic nitrogen to freshwater ecosystems that are also rich in phosphorus can eutrophy as well as acidify the waters. Both eutrophication and acidification generally lead to decreased diversity of both plant and animal species. Fish populations, in particular, have been reduced or eliminated in many acidified lakes across Scandinavia, Canada, and northeastern United States.

Because the extent of nitrogen-saturated ecosystems continues to grow, along with human-caused nitrogen deposition, controls on sulfur dioxide emissions alone clearly will not be sufficient to decrease acid rain or prevent its detrimental effects on streams and lakes. European governments already have recognized the importance of nitrogen in acidifying soils and waters, and intergovernmental efforts are underway there to reduce emissions and deposition of nitrogen on a regional basis.

Eutrophication in Estuaries and Coastal Waters

One of the best documented and best understood consequences of human alterations of the nitrogen cycle is the eutrophication of estuaries and coastal seas. It is arguably the most serious human threat to the integrity of coastal ecosystems.

In sharp contrast to the majority of temperate-zone lakes, where phosphorus is the nutrient that most limits primary productivity by algae and other aquatic plants and controls eutrophication, these processes are controlled by nitrogen inputs in most temperate-zone estuaries and coastal waters. This is largely because the natural flow of nitrogen into these waters and the rate of nitrogen fixation by planktonic organisms are relatively low while microbes in the sea floor sediments actively release nitrogen back to the atmosphere.

When high nitrogen loading causes eutrophication in stratified waters—where a sharp temperature gradient prevents mixing of warm surface waters with colder bottom waters—the result can be anoxia (no oxygen) or hypoxia (low oxygen) in bottom waters. Both conditions appear to be becoming more prevalent in many estuaries and coastal seas. There is good evidence that since the 1950s or 1960s, anoxia has increased in the Baltic Sea, the Black Sea, and Chesapeake Bay. Periods of hypoxia have increased in Long Island Sound, the North Sea, and the Kattegat, resulting in significant losses of fish and shellfish.

Eutrophication is also linked to losses of diversity, both in the sea floor community—including seaweeds, seagrasses, and corals—and among planktonic organisms. In eutrophied waters, for example, "nuisance algae" may come to dominate the phytoplankton community. Increases in troublesome or toxic algal blooms have been observed in many estuaries and coastal seas worldwide in recent decades. During the 1980s, toxic blooms of dinoflagellates and brown-tide organisms caused extensive die-offs of fish and shellfish in many estuaries. Although the causes are not completely understood, there is compelling evidence that nutrient enrichment of coastal waters is at least partly to blame for such blooms.

Major Uncertainties

Although this report has focused on what is known about human-driven changes to the global nitrogen cycle, major uncertainties remain. Some of these have been noted in earlier sections. This section, however, focuses on important processes that remain so poorly understood that it is difficult to distinguish human-caused impacts or to predict their consequences.

Marine Nitrogen Fixation

Little is known about the unmodified nitrogen cycle in the open ocean. Credible estimates of nitrogen fixation by organisms in the sea vary more than ten-fold, ranging from less than 30 to more than 300 Tg per year. There is some evidence that human alteration of the nitrogen cycle could alter biological processes in the open ocean, but there is no adequate frame of reference against which to evaluate any potential human-driven change in marine nitrogen fixation.

Changes in Limiting Resources

One consequence of human-driven changes in the global nitrogen cycle is a shift in the resources that limit biological processes in many areas. Large amounts of nitrogen are now deposited on many ecosystems that were once nitrogen deficient. The dominant species in these systems may have evolved with nitrogen limitation, and the ways they grow and function and form symbiotic partnerships may reflect adaptations to this limit. With this limit removed, species must operate under novel constraints such as now-inadequate phosphorus or water supplies. How are the performance of organisms and the operation of larger ecological processes affected by changes in their chemical environment for which they have no evolutionary background and to which they are not adapted?

Capacity to Retain Nitrogen

Forests and wetlands vary substantially in their capacity to retain added nitrogen. Interacting factors that are known to affect this capacity include soil texture, degree of chemical weathering of soil, fire history, rate at which plant material accumulates, and past human land use. However, we still lack a fundamental understanding of how and why nitrogen-retention processes vary among ecosystems—much less how they have changed and will change with time.

Alteration of Denitrification

In large river basins, the majority of nitrogen that arrives is probably broken down by denitrifying bacteria and released to the atmosphere as nitrogen gas or nitrous oxide. Exactly where most of this activity takes place is poorly understood, although we know that stream-side areas and wetlands are important. Human activities such as increased nitrate deposition, dam building, and rice cultivation have probably enhanced denitrification, while draining of wetlands and alteration of riparian ecosystems has probably decreased it. But the net effect of human influence remains uncertain.

Natural Nitrogen Cycling

Information on the rate of nitrogen deposition and loss in various regions prior to extensive human alterations of the nitrogen cycle remains patchy. In part, this reflects the fact that all of the Earth already is affected to some degree by human activity. Nevertheless, studies in remote regions of the Southern Hemisphere illustrate that there is still valuable information to be gathered on areas that have been minimally altered by humans.

Future Prospects and Management Options

Fertilizer Use

The greatest human-driven increases in global nitrogen supplies are linked to activities intended to boost food production. Modern intensive agriculture requires large quantities of nitrogen fertilizer; humanity, in turn, requires intensive agriculture to support a growing population that is projected to double by the end of the next century. Consequently, the production and application of nitrogen fertilizer has grown exponentially, and the highest rates of application are now found in some developing countries with the highest rates of population growth. One study predicts that by the year 2020, global production of nitrogen fertilizer will increase from a current level of about 80 Tg to 134 Tg per year.

Curtailing this growth in nitrogen fertilizer production will be a difficult challenge. Nevertheless, there are ways to slow the growth of fertilizer use and also to reduce the mobility—and hence the regional and global impacts—of the nitrogen that is applied to fields.

One way to reduce the amount of fertilizer used is to increase its efficiency. Often at least half of the fertilizer applied to fields is lost to the air or water. This leakage represents an expensive waste to the farmer as well as a significant driver of environmental change. A number of management practices have been identified that can reduce the amounts of fertilizer used and cut losses of nitrogen to the air and water without sacrificing yields or profits (and in some cases, increasing them). For instance, one commercial sugar cane plantation in Hawaii was able to cut nitrogen fertilizer use by one third and reduce losses of nitrous oxide and nitric oxide ten-fold by dissolving the fertilizer in irrigation water, delivering it below the soil surface, and timing multiple applications to meet the needs of the growing crop. This knowledge-intensive system also proved more profitable than broadcasting fewer, larger applications of fertilizer onto the soil surface. The widespread implementation of such practices, particularly in developing regions, should be a high priority for agronomists as well as ecologists since improved practices provide an opportunity to reduce the costs of food production while slowing the rate of global change.

There are also ways to prevent the nitrogen that leaches from fertilized farmland from reaching streams, estuaries and coastal waters where it contributes to eutrophication. In many regions, agricultural lands have been expanded by channelizing streams, clearing riparian forests, and draining wetlands. Yet these areas serve as important natural nitrogen traps. Restoration of wetlands and riparian areas and even construction of artificial wetlands have been shown to be effective in preventing excess nitrogen from entering waters.

Fossil Fuel Burning

The second major source of human-fixed nitrogen is fossil fuel burning. It, too, will increase markedly as we enter the next century, particularly in the developing world. One study predicts that production of nitrogen oxides from fossil fuels will more than double in the next 25 years, from about 20 Tg per year to 46 Tg. Reducing these emissions will require improvements in the efficiency of fuel combustion as well as in the interception of airborne byproducts of combustion. As with improvements in fertilizer efficiency, it will be particularly important to transfer efficient combustion technologies to developing countries as their economies and industries grow.

Conclusions

Human activities during the past century have doubled the natural annual rate at which fixed nitrogen enters the land-based nitrogen cycle, and the pace is likely to accelerate. Serious environmental consequences are already apparent. In the atmosphere, concentrations of the greenhouse gas nitrous oxide and of the nitrogen-precursors of smog and acid rain are increasing. Soils in many regions are being acidified and stripped of nutrients essential for continued fertility. The waters of streams and lakes in these regions are also being acidified, and excess nitrogen is being transported by rivers into estuaries and coastal waters. It is quite likely that this unprecedented nitrogen loading has already contributed to long-term declines in coastal fisheries and accelerated losses of plant and animal diversity in both aquatic and land-based ecosystems. It is urgent that national and international policies address the nitrogen issue, slow the pace of this global change, and moderate its impacts.

For further information

Aber, J.D. 1992. Nitrogen cycling and nitrogen saturation in temperate forest ecosystems. Trends in Ecology and Evolution 7:220–223.

Berendse, F., R. Aerts, and R. Bobbink. 1993. Atmospheric nitrogen deposition and its impact on terrestrial ecosystems. Pp. 104–121 in C.C. Vos and P. Opdam (eds), Landscape Ecology of a Stressed Environment. Chapman and Hall, England.

Cole, J. J., B. L. Peierls, N. F. Caraco, and M. L. Pace. 1993. Nitrogen loadings of rivers as a human-driven process. Pages 141–157 in M. J. McDonnell and S. T. A. Picket (eds.), Humans as Components of Ecosystems: The Ecology of Subtle Human Effects and Populated Areas. Springer-Verlag, NY.

DOE (Department of Environment, UK). 1994. Impacts of Nitrogen Deposition in Terrestrial Ecosystems. Technical Policy Branch, Air Quality Div., London.

Galloway, J. N., W. H. Schlesinger, H. Levy II, A. Michaels, and J. L. Schnoor. 1995. Nitrogen fixation: atmospheric enhancement-environmental response. Global Bio-geochemical Cycles 9:235–252.

Howarth, R. W., G. Billen, D. Swaney, A. Townsend, N. Jaworski, K. Lajtha, J. A. Downing, R. Elmgren, N. Caraco, T. Jordan, F. Berendse, J. Freney, V. Kudeyarov, P Murdoch, and Zhu Zhao-liang. 1996. Regional nitrogen budgets and riverine N & P fluxes for the drainages to the North Atlantic Ocean: Natural and human influences. Biogeochemistry 35: 75–139.

Nixon, S. W., J. W. Ammerman, L. P. Atkinson, V. M. Berounsky, G. Billen, W. C. Boicourt, W. R. Boynton, T. M. Church, D. M. Ditoro, R. Elmgren, J. H. Garber, A. E. Giblin, R. A. Jahnke, N. P. J. Owens, M. E. Q. Pilson, and S. P. Seitzinger. The fate of nitrogen and phosphorus at the land-sea margin of the North Atlantic Ocean. Bio-geochemistry 35: 141–180.

NRC. 1994. Priorities for Coastal Ecosystem Science. National Research Council. Washington, D.C.

Prinn, R., D. Cunnold, R. Rasmussen, P. Simmonds, F. Alyca, A. Crawford, P. Fraser, and R. Rosen. 1990. Atmospheric emissions and trends of nitrous oxide deduced from 10 years of ALE-GAGE data. Journal of Geophysical Research 95:18, 369–18, 385.

Schindler, D. W. and S. E. Bayley. 1993. The biosphere as an increasing sink for atmospheric carbon:estimates from increasing nitrogen deposition. Global Biogeochemical Cycles 7:717–734.

Schlesinger, W. H. 1991. Biogeochemistry: An Analysis of Global Change. Academic Press, San Diego.

Smil, V. 1991. Population growth and nitrogen: an exploration of a critical existential link. Population and Development Review 17:569-601.

Tamm, C. O. 1991. Nitrogen in Terrestrial Ecosystems. Springer-Verlag, Berlin. 115 pp.

Tilman, D. 1987. Secondary succession and the pattern of plant dominance along experimental nitrogen gradients. Ecological Monographs 57(3):189-214.

Vitousek, P. M. and R. W. Howarth. 1991. Nitrogen limitation on land and in the sea: How can it occur? Bio-geochemistry 13:87–115.

Critical Thinking

1. Are there alternatives to industrially produced nitrogen fertilizers?
2. Do you believe that increasing the efficiency with which nitrogen fertilizers are used is the most appropriate way to address this sustainability issue?
3. Is it reasonable to expect human activities to have no effect on the nitrogen cycle?

The Story of Phosphorus: Global Food Security and Food for Thought

DANA CORDELL, JAN-OLOF DRANGERT, AND STUART WHITE

Introduction

Food production is fundamental to our existence, yet we are using up the world's supply of phosphorus, a critical ingredient in growing food. Today, phosphorus is mostly obtained from mined rock phosphate and is often combined in mineral fertilizers with sulphuric acid, nitrogen, and potassium. Existing rock phosphate reserves could be exhausted in the next 50–100 years (Steen, 1998; Smil, 2000b; Gunther, 2005). The fertilizer industry recognises that the quality of reserves is declining and the cost of extraction, processing and shipping is increasing (Runge-Metzger, 1995; Driver, 1998; Smil, 2000b; EcoSanRes, 2003). Box 1 outlines the key issues.

Common responses to resource scarcity problems include higher prices, more efficient resource use, the introduction of alternatives, and the recovery of the resource after use. The use of phosphorus is becoming more efficient, especially in Europe. Farmers in Europe and North America are increasingly avoiding over fertilization, and are ploughing straw and animal manure into agricultural soils, partly to recycle phosphorus (European Fertilizer Manufacturers Association, 2000). However, most of the discussion about efficient phosphorus use, and most of the measures to achieve this, have been motivated by concerns about toxic algal blooms caused by the leakage of phosphorus (and nitrogen) from agricultural land (Sharpley et al., 2005). While such measures are essential, they will not by themselves be sufficient to achieve phosphorus sustainability. A more integrated and effective approach to the management of the phosphorus cycle is needed—an approach which addresses future phosphorus scarcity and hence explores synergies that reduce leakage and recover and reuse phosphorus.

The following sections of this paper assess the historical, current and future availability of phosphorus in the context of global food security. Possible options for meeting the world's future phosphorus demand are outlined and institutional opportunities and obstacles are discussed.

Humanity's Addiction to Phosphate Rock

Historically, crop production relied on natural levels of soil phosphorus and the addition of locally available organic matter like manure and human excreta (Marald, 1998). To keep up with increased food demand due to rapid population growth in the 20th century, guano and later rock phosphate were applied extensively to food crops (Brink, 1977; Smil, 2000b).

The Chinese used human excreta ('night soil') as a fertilizer from the very early stages of their civilization, as did the Japanese from the 12th century onwards (Matsui, 1997). In Europe, soil degradation and recurring famines during the 17th and 18th centuries created the need to supplement animal and human

Box 1
Phosphorus (P): A Closer Look at an Emerging Crisis

- Plants require phosphorus to grow. Phosphorus is an element on the periodic table that cannot be substituted and is therefore vital for producing the food we eat (Steen, 1998).
- 90 percent of global demand for phosphorus is for food production, currently around 148 million tonnes of phosphate rock per year (Smil, 2000a,b; Gunther, 2005).
- The demand for phosphorus is predicted to increase by 50–100 percent by 2050 with increased global demand for food and changing diets (EFMA, 2000; Steen, 1998).
- Phosphorus is a non-renewable resource, like oil. Studies claim at current rates of extraction, global commercial phosphate reserves will be depleted in 50–100 years (Runge-Metzger, 1995; EcoSanRes, 2003; Steen, 1998). The remaining potential reserves are of lower quality or more costly to extract.
- Phosphate rock reserves are in the control of only a handful of countries (mainly Morocco, China and the US), and thus subject to international political influence. Morocco has a near monopoly on Western Sahara's reserves, China is drastically reducing exports to secure domestic supply, US has less than 30 years left of supplies, while Western Europe and India are totally dependent on imports (Jasinski, 2006; Rosmarin, 2004).

excreta with other sources of phosphorus (Mårald, 1998). In the early 19th century, for instance, England imported large quantities of bones from other European countries. In addition to the application of phosphorus from new sources, improved agricultural techniques enabled European agriculture to recover from the famines of the 18th century (Mårald, 1998). These improvements included crop rotation, improved handling of manure, and in particular, the introduction of new crops such as clover which could fix nitrogen from the atmosphere.

Liebig formulated his 'mineral theory' in 1840, which replaced the 'humus theory' that plants and animals were given life in a mysterious way from dead or decomposing plants and animals (Liebig, 1840; Mårald, 1998). Liebig provided a scientific explanation: nutrients such as nitrogen, phosphorus and potassium were elements circulating between dead and living material (Mårald, 1998). This discovery occurred during a period of rapid urbanization in Europe, when fertilizer factories were being established around growing cities. Food production was local and the factories manufactured phosphorus fertilizers from locally available organic waste products, such as human excreta, industrial organic waste by-products, animal dung, fish, ash, bones, and other slaughter-house by-products (Mårald, 1998; Neset et al., 2008).

However, around the mid-to-late 19th century, the use of local organic matter was replaced by phosphorus material from distant sources. The mining of guano (bird droppings deposited over previous millennia) and phosphate-rich rock had begun (Brink, 1977; Smil, 2000b). Guano was discovered on islands off the Peruvian coast and later on islands in the South Pacific. World trade in guano grew rapidly, but it relied on a limited resource which declined by the end of the 19th century (Stewart et al., 2005). Phosphate rock was seen as an unlimited source of concentrated phosphorus and the market for mineral fertilizers developed rapidly. At the same time, the introduction of flush toilets in towns meant that human waste was discharged into water bodies instead of being returned to the soil. There were protests among intellectuals that farmers were being robbed of human manure. Among them was Victor Hugo who wrote in *Les Miserables:*

> Science, after having long groped about, now knows that the most fecundating and the most efficacious of fertilizers is human manure. The Chinese, let us confess it to our shame, knew it before us. Not a Chinese peasant—it is Eckberg who says this—goes to town without bringing back with him, at the two extremities of his bamboo pole, two full buckets of what we designate as filth. Thanks to human dung, the earth in China is still as young as in the days of Abraham. Chinese wheat yields a hundredfold of the seed. There is no guano comparable in fertility with the detritus of a capital. A great city is the most mighty of dung-makers. Certain success would attend the experiment of employing the city to manure the plain. If our gold is manure, our manure, on the other hand, is gold (Hugo, 1862).

Trade in food grew steadily with urbanization and colonization, but insufficient amounts of nutrients were returned to the areas of food production to balance off-takes. By the late 19th century, processed mineral phosphorus fertilizer was routinely used in Europe and its use grew substantially in the 20th century (International Fertilizer Industry Association, 2006; Buckingham and Jasinski, 2004). Processed mineral fertilizers such as ordinary superphosphate (OSP) typically contained an order of magnitude greater concentration of phosphorus than did manures (Smil, 2000b). Application of such highly concentrated fertilizers helped rectify the phosphorus deficiency of soils. In the mid-20th century the Green Revolution improved agricultural output in many countries. As well as introducing new crop varieties, the Green Revolution involved the application of chemical fertilizers.[1] This new approach saved millions from starvation and the proportion of the world's population that was undernourished declined despite rapid population growth (IFPRI, 2002a). Today, food could not be produced at current global levels without the use of processed mineral fertilizers. We are effectively addicted to phosphate rock.

The Current Situation

Demand for Food, Demand for Fertilizers

Following more than half a century of generous application of inorganic high-grade phosphorus and nitrogen fertilizers, agricultural soils in Europe and North America are now said to have surpassed 'critical' phosphorus levels, and thus only require light applications to replace what is lost in harvest (FAO, 2006; European Fertilizer Manufacturers Association, 2000). Consequently, demand for phosphorus in these regions has stabilized or is decreasing.

However in developing and emerging economies the situation is different. Global demand for phosphorus is forecast to increase by around by 3–4 percent annually until 2010/11 (Maene, 2007; FAO, 2007a), with around two-thirds of this demand coming from Asia (FAO, 2007a), where both absolute and per capita demand for phosphate fertilizers is increasing. There will be an estimated 2–2.5 billion new mouths to feed by 2050 (IWMI, 2006), mainly in urban slums in the developing world. Meat and dairy products, which require higher phosphorus inputs than other foods, are becoming more popular in China and India. According to the International Water Management Institute (Fraiture, 2007) global food production will need to increase by about 70 percent by 2050 to meet global demand. Under these circumstances, acquiring enough phosphorus to grow food will be a significant challenge for humanity in the future.

In Sub-Saharan Africa, where at least 30 percent of the population is undernourished, fertilizer application rates are extremely low and 75 percent of agricultural soils are nutrient deficient,[2] leading to declining yields (IFDC, 2006; Smaling et al., 2006). The UN and the Alliance for a Green Revolution in Africa has called for a new Green Revolution in Sub-Saharan Africa, including increased access to fertilizers (Blair, 2008; AGRA, 2008) but there has been little discussion of the finiteness of phosphate fertilizer reserves.

In 2007–2008, the same pressures that caused the recent global food crisis led to phosphate rock and fertilizer demand exceeding supply and prices increased by 700 percent in a 14-month period (Minemakers Limited, 2008). Two significant contributors to the increased demand for phosphorus have been the increasing popularity of meat- and dairy-based diets, especially in growing economies like China and India, and the expansion of the biofuel industry. Increasing concern about oil scarcity and climate change led to the recent sharp increase in biofuel production. The biofuel industry competes with food production for grains and productive land and also for phosphorus fertilizers. The year 2007 was the first year a clear rise in phosphate rock demand could be attributed to ethanol production (USGS 2007, pers. comm., 5th September).

The International Fertilizer Industry Association expects the fertilizer market to remain tight for at least the next few years (IFA, 2008). It is therefore anticipated that the price of phosphate rock and related fertilizers will remain high in the near future, until new mining projects such as those planned in Saudi Arabia are commissioned (Heffer and Prud'homme, 2007). The sudden spike in the price of fertilizers in 2007–2008 took most of the world's farmers completely by surprise. In India, which is totally dependent on phosphate imports, there have been instances of farmer riots and deaths due to the severe national shortage of fertilizers (Bombay News, 2008). While this short-term crisis is not a direct consequence of the long-term scarcity issues outlined in this paper, the short-term situation can be seen as an indication of what is to come.

Global Food Security and Resource Scarcity

The UN's Food and Agricultural Organization (FAO) states that food security "exists when all people, at all times, have access to sufficient, safe and nutritious food to meet their dietary needs for an active and healthy life" (FAO, 2005b, p1). Securing future food security is now considered a global priority (UN, 2000; IFPRI, 2002b). At the turn of the Millennium, 191 nations formalised their commitment to the eight Millennium Development Goals (MDGs), one of which is to decrease poverty and hunger by 50 percent by 2015 (UN, 2000). Currently, there are over 800 million people without sufficient access to food (SOFI, 2005; UN, 2005). While over 40 percent of Africans today cannot secure adequate food on a day-to-day basis, many people in both the developed and developing world are suffering from obesity[3] (UN Millennium Project, 2005; SIWI-IWMI, 2004; Gardner and Halweil, 2000). Food security is a challenge that can only be met by addressing a number of relevant issues. The FAO's annual State of Food Insecurity (SOFI) reports, the International Food Policy Research Institute's (IFPRI) reports and the UN Millennium Development Project all stress that food insecurity is a consequence of numerous linked factors, including frequent illness, poor sanitation, limited access to safe water and lack of purchasing power (FAO, 2004a; Braun et al., 2004; UN Millennium Project, 2005).

Today it is acknowledged that addressing energy and water issues will be critical for meeting the future nutritional demands of a growing population (Smil, 2000a; Pfeiffer, 2006) but the need to address the issue of limited phosphorus availability has not been widely recognized. Approximately 70 percent of the world's demand for fresh water is for agriculture (SIWI-IWMI, 2004) and about 90 percent of worldwide demand for rock phosphate is for food production (Rosmarin, 2004; Smil, 2002). It is predicted that demand for both resources will outstrip supply in the coming decades. Experts suggest that a radical shift in the way we think about and manage water is required (Falkenmark and Rockstrom, 2002), to deal with the 'hydroclimatic realities' of water availability (SIWI-IWMI, 2004). In a similar way, food security faces the 'geochemical realities' of limited phosphate reserves.

Global food production is also highly dependent on cheap energy, particularly from fossil fuels like oil. Transporting food all over the world in addition to mining and manufacturing fertilizers is only possible while cheap oil exists. However a peak in global oil production is imminent (Royal Dutch Shell, 2008) and alternatives to fossil-fuel-dependent agricultural systems will be required in the future (Pfeiffer, 2006).

Global Phosphate Rock Reserves and Geopolitics

All modern agriculture is today dependent on regular inputs of phosphate fertilizer derived from mined rock to replenish the phosphorus removed from the soil by the growing and harvesting of crops. However, phosphate rock is a non-renewable resource and approximately 50–100 years remain of current known reserves (Steen, 1998; Smil, 2000b; Gunther, 2005). The world's remaining phosphate rock reserves are under the control of a handful of countries, including China, the US and Morocco. While China has the largest reported reserves, it has recently imposed a 135 percent export tariff on phosphate, effectively preventing any exports in order to secure domestic supply (Fertilizer Week, 2008). The US, historically the world's largest producer, consumer, importer and exporter of phosphate rock and phosphate fertilizers, has approximately 25 years left of domestic reserves (Stewart et al., 2005; Jasinski, 2008). US companies import significant quantities of phosphate rock from Morocco to feed their phosphate fertilizer factories (Jasinski, 2008). This is geopolitically sensitive as Morocco currently occupies Western Sahara and controls its phosphate rock reserves. The Western Sahara Resource Watch claims that *"extracting and trading with phosphates from Western Sahara are contrary to international law"* (WSRW, 2007) and such trade is highly condemned by the UN (Corell, 2002). Several Scandinavian firms have boycotted this trade in recent years (The Norwegian Support Committee for Western Sahara, 2007).

Together, Moroccan and Western Saharan reserves represent more than a third of the world's supply of high-quality phosphate rock (IFA, 2006). Ironically, the African continent is simultaneously the world's largest exporter of phosphate rock and the continent with the largest food shortage (FAO, 2006; Jasinski, 2006).

This highlights the importance of phosphorus *accessibility,* in addition to physical (and political) scarcity. Indeed, the average sub-Saharan farmer has less purchasing power to access fertilizer markets, yet phosphate fertilizers can cost an African farmer 2–6 times more than they cost a European farmer due to higher transport and storage costs (Runge-Metzger, 1995; Fresco, 2003).

Quantifying Today's Phosphorus Flows Through the Food System

A systems approach to understanding the phosphorus cycle, particularly in global food production and consumption, can help in locating and quantifying losses and inefficiencies and thus assist in identifying potential recovery points. A modification of the Substance Flows Analysis (SFA) tool from Industrial Ecology has been applied to track global phosphorus flows. SFA quantifies the material inputs and outputs from processes and stocks within a system to better understand pollution loads on a given environment, and determine places to intervene in a system to increase its efficiency, or reduce wastage and pollution (Brunner and Rechberge, 2004). The simplified SFA traces phosphorus through the global food production and consumption system, from the mine through to consumption, and identifies losses throughout the system. Unlike water (SIWI-IWMI, 2004; Lundqvist et al., 2007), carbon (GCP, 2008) and nitrogen (UNEP, 2007), there are no comprehensive studies analysing anthropogenic global flows of phosphorus.[4]

The inner white area termed the 'Anthroposphere' defines the human-activity system (in this case, food-related human activity), while the outer area termed 'Natural Environment' represents the 'natural' phosphorus biogeochemical system (in which the human activity system is embedded). The dotted arrows in the natural biogeochemical system occur at a rate of millions of years (for example, natural weathering and erosion of phosphate-bearing rock). The solid arrows within the human activity system indicate the approximate quantities of phosphorus (in millions of metric tonnes of phosphorus per year, MT P per year) in each key stage (the boxes) in the food production and consumption process. These stages are: mining, fertilizer production, the application of fertilizers to agricultural soils, the harvesting of crops, food and feed processing, consumption of food by animals and humans, excretion and leakage from the system to either the natural environment or recirculation back to the food system.

Mineral phosphorus in rock phosphate was formed 10–15 million years ago (White, 2000). Since the end of World War II, global extraction of phosphate rock has tripled to meet industrial agriculture's demand for NPK fertilizers (UNEP, 2005). Approximately 90 percent of society's use of phosphorus is for food production (including fertilizers, feed and food additives) (Smil, 2000b; European Fertilizer Manufacturers Association, 2000). Currently, phosphorus fertilizers sourced from mined phosphate rock accounts for around 15 MT P per year (Jasinski, 2006; Gumbo and Savenije, 2001; Rosmarin, 2004; Gumbo, 2005). Modern agricultural systems require annual applications of phosphorus-rich fertilizer. However, unlike the natural biochemical cycle, which recycles phosphorus back to the soil in situ' via dead plant matter, modern agriculture harvests crops prior to their decay phase, transporting them all over the world to food manufacturers and to consumers.

Because phosphate rock and phosphate fertilizers are both commodities on the international market, international data exists for mining, fertilizer production and application. However after fertilizer application, there is very little accurate data available for use in a global analysis. This is particularly true of sources of organic phosphorus, such as manure, crop residues and household organic waste, which are re-circulated or lost from the food system (FAO, 2006). These organic phosphorus sources are typically not commodities, but are applied informally and on an ad hoc basis, and so there is no formal tracking of their use and losses. Data that does exist is typically compiled at a farm or local level. The use of organic phosphorus sources is often not quantified in investigations of phosphorus flows in the food production and consumption process as researchers are presently more interested in losses to water bodies causing eutrophication. Calculations based on Smil (2000a, 2002) suggest the total phosphorus content in annual global agricultural harvests is approximately 12 MT P, of which 7 MT P is processed for feed and food and fibre, while 40 percent of the remaining 5 MT P of crop residues is returned to the land.[5]

Studies on post-harvest losses of food and embodied water from the global food production and consumption chain (Smil, 2000a; SIWI et al., 2008), can be used as a basis for estimating phosphorus losses. This suggests that approximately 55 percent of phosphorus in food is lost between 'farm and fork'. Smil (2000a) estimates that around 50 percent of the phosphorus consumed and hence excreted by livestock is returned to agriculture globally. However there are significant regional imbalances, such as an oversupply of manure in regions where a critical soil phosphorus level has already been surpassed (such as The Netherlands and parts of North America), and a lack of manure in regions where soils are most phosphorus-deficient (such as Sub-Saharan Africa or Australia) (Runge-Metzger, 1995; Smaling, 2005).

Close to 100 percent of phosphorus eaten in food is excreted (Jönsson et al., 2004). Working backwards using a mass balance, we can calculate that humans physically consume approximately 3 MT P globally.[6] Every year, the global population excretes around 3 million tonnes of phosphorus in urine and faeces. Given that more than half the world's population now lives in urban centres, and urbanization is set to increase (FAO, 2007b), cities are becoming phosphorus 'hotspots' and urine is the largest single source of phosphorus emerging from cities. While nutrient flows from food via human excreta typically found their way back to land in the past, today they more often end up in waterways via wastewater from urban centres or as sludge in landfills. Over-fertilization of agricultural soils has been a common practice in the northern hemisphere, and contributes to excess discharge into water bodies and environmental problems like eutrophication. Rosmarin (2004) estimates that close to 25 percent of the 1 billion tonnes of phosphorus mined since 1950 has ended up in water bodies, or is buried in landfills. It is estimated that on average, around 10 percent of human excreta is currently recirculated, either intentionally or unintentionally, back to agriculture or aquaculture. Examples of how this occurs include poor urban farmers in Pakistan diverting the city's untreated wastewater to irrigate and fertilize the crops (Ensink et al., 2004), and pit or composting toilets in rural China, Africa and other parts of the world (Esrey et al., 2001). Recirculating urban nutrients such as urine back to agriculture therefore presents an enormous opportunity for the future (see Section 5 for examples).

In addition to analysing the global use of phosphorus based on an average diet, it is also informative to analyse different scenarios of phosphorus demand by assessing likely phosphorus losses in the various phases of the food chain. By working backwards from human excreta to the field, we can calculate the required amount of phosphate rock for vegetarian and meat-based diets. A vegetarian excretes some 0.3 kg/(P year) (WHO, 2006), and if one-third of the phosphorus in a vegetarian's food is lost during food preparation, one can assume that the post-harvest material contained 0.45 kg P. Assuming that three quarters of the harvested crop ends up as organic waste, the average per capita annual harvest for a vegetarian would have contained 1.8 kg P originally. If one-third of the phosphorus taken up by plants is from mineral phosphate fertilizer and soil phosphorus provides the remaining two thirds, then 0.6 kg of mineral phosphate fertilizer is required annually for a vegetarian. It takes 4.2 kg of rock phosphate to produce 0.6 kg of phosphorus. If equivalent assumptions are made for meat production, it can be concluded that meat-eaters require some 11.8 kg of rock phosphate (for meat eaters, it is assumed that one-fifth of phosphorus uptake is from mineral phosphate fertilizer and four-fifths is from soil phosphorus).

This simple calculation using phosphorus losses in each phase highlights two things. Firstly, that a vegetarian diet demands significantly less phosphate fertilizer than a meat-based diet. And secondly, that returning biomass from plants to the soil is by far the most important measure to retain soil phosphorus in a meat-based diet. This also requires no transport back to the field. For the vegetarian diet, the use of human excreta is the most important recovery measure but this involves collection and transport back to the field.

Data from two recent material flow analyses (MFA) of phosphorus through urban centres in Sydney, Australia (Tangsubkul et al., 2005) and in Linköping, Sweden (Schmid-Neset et al., 2005) suggest that a change from the average western diet to a vegetarian diet could decrease phosphorus demand of fertilizers by at least 20–45 percent. Tangsubkul et al. (2005) further suggest a change in Sydney residents' current diet to one with no excess phosphorus consumption (i.e. recommended daily intake per person) could decrease the city's total phosphorus demand by 70 percent. On the other hand, a switch in the current Indian diet to meat would increase India's demand for phosphorus three-fold.

Globally, we are mining five times the amount of phosphorus that humans are actually consuming in food. This analysis tells us that to simultaneously address phosphate scarcity and water pollution due to phosphorus leakage, an integrated approach must be taken that considers:

- minimizing phosphorus losses from the farm (estimated at around 8 MT P),
- minimizing losses in the food commodity chain (losses estimated at 2 MT P),
- alternative renewable phosphorus sources, like manure (around 15 MT P), human excreta (3 MT P) and food residues (1.2 MT P),
- other important mechanisms to reduce overall demand (such as optimizing soil carbon to improve phosphate availability and influencing diets).

These options are covered further in Section 5.

The Environmental Costs of the Phosphate Rock Industry

As well as the problem of eutrophication due to the leakage of excess phosphorus into waterways, the production of fertilizers from rock phosphate involves significant carbon emissions, radioactive by-products and heavy metal pollutants.

Processing and transporting phosphate fertilizers from the mine to the farm gate, which up to now have relied on cheap fossil fuels, involve an ever-increasing energy cost. Phosphate rock is one of the most highly traded commodities on the international market. Each year around 30 million tonnes of phosphate rock and fertilizers are transported across the globe (IFA, 2006). With growing concern about oil scarcity and climate change, there is a need to reconsider the current production and use of phosphorus, particularly with respect to energy use and other environmental impacts.

Each tonne of phosphate processed from phosphate rock generates five tonnes of phosphogypsum, a toxic by-product of phosphate rock mining. Phosphogypsum cannot be used in most countries due to unacceptably high radiation levels (USGS, 1999). Global phosphogypsum stockpiles are growing by over 110 million tonnes each year and there is a risk of leakage to groundwater (Wissa, 2003). Phosphate rock naturally contains radionuclides of Uranium and Thorium, most of which end up in the phosphogypsum by-product and to a lesser extent in the processed phosphate fertilizers (Kratz and Schnug, 2006; Saueia et al., 2005). If crushed phosphate rock is applied directly to soils, radionuclides of the decay series are distributed to agricultural soils, risking overexposure to farmers and phosphate industry workers (Saueia et al., 2005). While radiation levels can vary above and below acceptable radiation limits, there are no standard procedures for measuring soil radioactivity due to applied phosphate rock (or phosphate fertilizers) (Saueia et al., 2005). Despite this, crushed rock phosphate is currently permitted as a fertilizer in organic agriculture in at least the European Union (EU, 2007), India (Department of Commerce, 2005) and Australia (Organic Federation of Australia, 2005). Similarly, associated heavy metals like cadmium can also be present in phosphate rock at levels which are either too toxic for soils or too costly and energy intensive to remove (Steen, 1998; Driver, 1998).

Peak Phosphorus—A Sequel to Peak Oil?

As first highlighted by Hubbert in 1949 (Hubbert, 1949), production of oil reserves will at some time reach a maximum rate or 'peak' based on the finite nature of non-renewable resources, after which point production will decline. In a similar way, the rate of global production of high-grade phosphate rock will eventually reach a maximum or peak. Hubbert and later others argue that the important period is not when 100 percent of the reserve is depleted, but rather when the high quality, highly accessible reserves have been depleted. At this point, production reaches its maximum. After this point, the quality of remaining reserves is lower and they are harder to access, making them uneconomical to mine and process. Therefore while demand continues to increase, supply decreases year upon year. A conservative analysis using industry data suggests that the peak in global phosphorus production could occur by 2033. This analysis of peak phosphorus is based on estimated P[7] content in remaining world phosphate rock reserves (approximately 2358 MT P[8]) and cumulative production between 1900 and 2007 (totaling 882 MT P) based on US Geological Survey data (Buckingham and Jasinski, 2006; Jasinski, 2007, 2008), data from the European Fertilizer Manufacturers Association (2000) and the International Fertilizer Industry Association (2006). The area under the Hubbert curve is set equal to the depleted plus current reserves, totaling approximately 3240 MT P.

The data for annual production is fitted using a Gaussian distribution (Laherrere, 2000), based on the depleted plus current reserves estimate of 3240 MT P, and a least squares optimization which results in a production at peak of 29 MT P/a and a peak year of 2033. However the actual timing may vary due to changes in production costs (such as the price of raw materials like oil), data reliability and changes in demand and supply.

The concept of the 'peak' production of non-renewable resources such as oil or phosphorus is the subject of limited dispute today, but the exact timeline for the peak in production is debated. According to Déry and Anderson (Déry and Anderson, 2007), global phosphorus reserves peaked around 1989.[9] However it is likely that this observed peak was not a true maximum production peak, and was instead a consequence of political factors such as the collapse of the Soviet Union (formerly a significant phosphate rock consumer) and decreased fertilizer demand from Western Europe and North America. Indeed, data from the International Fertilizer Association indicates that the 2004–2005 production exceeded the 1989–1990 production (IFA, 2006).

While the timing of the production peak may be uncertain, the fertilizer industry recognises that the quality of existing phosphate rock is declining, and cheap fertilizers will soon become a thing of the past. The average grade of phosphate rock has declined from 15 percent P in 1970s to less than 13 percent P in 1996 (Stewart et al., 2005; IFA, 2006; Smil, 2002).

While some scientists (such as Stewart et al., 2005) suggest market forces will stimulate new technologies to improve the efficiency of phosphate rock extraction and beneficiation in the future, there are no known alternatives to phosphate rock on the market today that could replace it on any significant scale. While small-scale trials of phosphorus recovery from excreta and other waste streams exist (CEEP, 2008), commercialisation and implementation on a global scale could take decades to develop. Significant adjustments in institutional arrangements will also be required to support these infrastructure changes.

While it is understood that phosphate rock, like oil and other key non-renewable resources, will follow a peak production curve, peak oil and peak phosphorus differ in at least two key ways. Firstly, while oil can be replaced with other forms of energy once it becomes too scarce, there is no substitute for phosphorus in food production. Phosphorus is an element and cannot be produced or synthesized in a laboratory. Secondly, oil is unavailable once it is used, while phosphorus can be recovered from the food production and consumption chain and reused within economic and technical limits. Shifting from importing phosphate rock to domestic production of renewable phosphorus fertilizers (such as human excreta and biomass) can increase countries' phosphorus security and reduce the reliance on increasingly inaccessible phosphate fertilizer markets.

Options for Sustainable Phosphorus Use and Management

There is no single 'quick fix' solution to current dependence on phosphate rock for phosphorus fertilizers. However there are a number of technologies and policy options that exist today at various stages of development—from research to demonstration and implementation—that together could meet future phosphate fertilizer needs for global food production. Implementing these measures will inevitably require an integrated approach that looks beyond the current focus on reducing agricultural phosphorus leakage into waterways. Such an approach, incorporating a combination of supply- and demand-side measures, is described below.

Conventional supply-side approaches look for solutions similar to those of the past 150 years, such as further exploration and more intensive exploitation of existing phosphate rock resources, including off-shore and/or lower grade deposits. Some advocates of conventional processed fertilizer production argue these potential reserves will become economically viable once all high-grade reserves have been depleted and prices have increased (FAO, 2004b; Stewart et al., 2005). However this approach fails to address several key issues, including the finiteness of phosphate rock reserves in the long term; poor farmers' limited access to globalised fertilizer markets, the energy intensity of the current production and use system, and the accumulation of phosphorus and associated toxic wastes in soils and waterways.

As discussed in Section 3, phosphorus can be recovered from the food production and consumption system and reused as a fertilizer either directly or after intermediate processing. These recovery measures include: ploughing crop residues back into the soil; composting food waste from households, food processing plants and food retailers; and using human and animal excreta. Such sources are renewable and are typically available locally. However, due to their lower phosphorus concentrations, they are also bulkier than fertilizers processed from phosphate rock. Leading-edge research and development on phosphorus recovery is increasingly focusing on recovery of struvite (ammonium magnesium phosphate crystals high in phosphorus) from both urban and livestock wastewater (Reindl, 2007; SCOPE, 2004). Struvite crystalisation and recovery is a promising technological process that has the potential to both remove phosphorus from wastewater byproducts more efficiently, and, provide an alternative source of phosphate fertilizer (Jaffer et al., 2002).

The International Fertilizer Industry Association (IFA) indicates it is committed to a sustainable fertilizer industry and while the industry does not explicitly advocate the reuse of human excreta as a potential alternative to mined phosphate rock, the European Fertilizer Manufacturers Association does state:

> Two major opportunities for increasing the life expectancy of the world's phosphorus resources lie in recycling by recovery from municipal and other waste products and in the efficient use in agriculture of both phosphatic mineral fertilizer and animal manure (European Fertilizer Manufacturers Association, 2000, p.9).

Already in some urban areas in Pakistan and elsewhere in Asia, more than 25 percent of urban vegetables are being fertilized with wastewater from cities (Ensink et al., 2004). The International Water Management Institute estimates that 200 million farmers worldwide use treated or untreated wastewater to irrigate crops (Raschid-Sally and Jayakody, 2008). Currently 67 percent of global yields of farmed fish are fertilized by wastewater (World Bank, 2005) because wastewater is a cheap and reliable source of water and nutrients for poor farmers. However it is essential that farmers and those working with wastewater take precautionary measures to avert associated health risks. The World Health Organization has recently developed comprehensive guidelines on the safe reuse of wastewater in agriculture (WHO, 2006). Another drawback is that wastewater-fed agriculture and aquaculture rely on water-borne sanitation systems, rather than on systems such as dry or ultra-low flow toilets.

Reuse is safer if sanitation service providers and urban planners avoid infrastructure that mixes human excreta with other wastewater streams, such as industrial wastewater. Industrial and non-residential wastewater may contain heavy metals and other toxic wastes. Moreover, if urine is not mixed with faecal matter in the toilet, the urine can be used safely through simple storage (WHO, 2006). Urine is essentially sterile and could provide more than half the phosphorus required to fertilize cereal crops (Drangert, 1998; WHO, 2006; Esrey et al., 2001). In Sweden for example, two municipalities have mandated that all new toilets must be urine-diverting (Kvarnström et al., 2006; Tanums Kommun, 2002). While there are numerous practical ways urine can be collected, stored, transported and reused, the typical arrangement in these Swedish cases involves either a dry or flush urine-diverting toilet to collect the urine. The urine is then piped and stored in a simple 1–3 kl storage tank under the house or piped to a communal urine storage tank. Local farmers then collect the urine approximately once a year for use as liquid fertilizer (see Kvarnström et al., 2006 for further details). Sanitized faecal matter can also be used as a soil conditioner (WHO, 2006).

There are numerous documented practical examples of ecological sanitation around the world in places such as Southern Africa, India, China, Vietnam, Mexico (Gumbo and Savenije, 2001; Drangert, 1998; Stockholm Environment Institute, 2004). According to the Stockholm Environment Institute (2005), the cost of such ecological sanitation systems globally could be offset by the commercial value of the phosphorus (and nitrogen) they yield.

Most of the projected 2 billion new mouths to feed in the coming decades are expected to reside in peri-urban areas of mega-cities in developing countries (FAO, 1999). Urban and peri-urban agriculture involves growing crops and raising livestock within urban areas and bordering urban settlements (FAO, 2007b). Fertilizing urban agriculture with phosphorus recovered from organic urban waste could be a significant step towards reaching the Millennium Development Goals on eradicating hunger and poverty, and providing access to safe sanitation. While this opportunity has been largely neglected at the global level to date (Cordell, 2007), a preliminary examination of the relationship between the land area required for food intake, the quantities of nutrients in human excreta, the capacity of the soil to absorb urine, crops' requirements for nutrients and population densities in peri-urban areas is outlined in Drangert (1998). Gumbo (2005) further studied the potential of reusing human excreta in urban Zimbabwe to 'short-cut' the urban phosphorus cycle. Gumbo found that the fertilizer value of the urine produced by urban dwellers in the case study catchment could sustain the agricultural activities in the surrounding area.

There is still significant scope to further explore the individual and combined potential for recovering organic urban waste products such as human excreta, food waste, garden waste, and manure. Bone meal, ash, and aquatic vegetation such as algae and seaweed are also potential sources of phosphorus.

Options aimed at reducing the demand for phosphorus in food production vary widely and can include: increasing agricultural efficiency to increase phosphorus uptake from the soil, reducing organic losses throughout the food chain and encouraging diets which contain fewer phosphorus-intensive foods.

Approaches to fertilizer efficiency range from high-tech solutions such as precision agriculture (FAO, 2000, 2008a; Johnston, 2000) through to organic farming techniques that seek to optimize soil conditions to increase soil phosphorus availability for plants (FAO, 2006, 2007c). Other approaches focus on the addition of microbial inoculants to increase soil phosphorus availability. The fertilizer industry, governmental institutes and research organizations have been actively supporting more efficient fertilizer application practices for over a decade (International Fertilizer Industry Association, 2006; European Fertilizer Manufacturers Association, 2000; Food21, 2005; FAO, 2006). Such initiatives have mainly been triggered by concerns about nutrient leakage to waterways causing eutrophication. However, much agricultural land is still subject to an over-application of phosphorus, resulting in unnecessary accumulation in soils in addition to runoff to water bodies (Steen, 1998; Gunther, 1997). Indeed, only 15–30 percent of applied phosphorus fertilizer is actually taken up by harvested crops[10] (FAO, 2006). At the same time, agricultural land in other regions are phosphorus-deficient due to naturally low soil phosphorus levels and fertilizer applications at rates which are far lower than would be required to replace the phosphorus lost through agriculture (Smaling et al., 2006).

Smil (2007) suggests that shifting to a 'smart vegetarian' diet, combined with reducing over-consumption, would be one of the most cost-effective measures to reduce agricultural resource inputs (including water, energy, land and fertilizers) and would also minimize greenhouse gas emissions and other forms of pollution. Food preferences are generally more strongly correlated with taste, advertisements and price than they are with nutritional value (SIWI-IWMI, 2004). Therefore, potential strategies to reduce the demand for phosphorus include encouraging the move to foods which require the input of less phosphorus, water and energy. This could be done through appropriate communication strategies or economic incentives in both the developed and developing worlds. In areas where there is a move away from vegetarian diets, communication strategies to combat this trend could be employed.

No analyses have yet been done that integrate such supply- and demand-side options in the same framework and assess the implications for global phosphate security. There is also a need to systematically assess potential options according to criteria[11] such as: economic cost; life cycle energy consumption; other environmental impacts; synergies between phosphorus and other resources (such as water, energy); logistics and technical feasibility, and cultural values and preferences.

Institutional and Attitudinal Barriers and Opportunities

Since a global phosphorus scarcity crisis is imminent, as we have demonstrated in the sections above, why is it not being discussed in relation to global food security or global environmental change? What are the current barriers to addressing a phosphorus 'crisis' and what are the underlying reasons for the lack of attention to nutrient recirculation options such as urine reuse?[12]

Despite increasing global demand for non-renewable phosphate rock, and phosphate rock's critical role in food production, global phosphate scarcity is missing from the dominant debates on global food security and global environmental change. For example, phosphorus scarcity has not received any explicit mention within official reports of the UN's Food and Agricultural Organization (FAO, 2005a, 2006, 2007a), the International Food Policy Research Institute (IFPRI, 2002b, 2005), the Millennium Ecosystem Assessment (Millennium Ecosystem Assessment, 2005), the Global Environmental Change and Food Systems programme (GECAFS, 2006), the International Assessment of Agricultural Knowledge, Science and Technology for Development (IAASTD, 2008) or the recent High-level Conference on World Food Security hosted by the FAO (FAO, 2008b). The implications of declining global phosphate availability and accessibility have been mentioned in a limited number of discussions by a few concerned scientists.[13]

We are entering a new and unprecedented era of global environmental change. As we are learning from climate change and global water scarcity, a long-term time frame is required to address phosphate scarcity. Decision-makers need to consider the next 50–100 years, rather than just the next 5–10 years. Young et al. (IDGEC, 2006) suggest that some global environmental problems occur due to the 'lack of fit between ecosystems and institutions' (IHDP, 2002). In the case of phosphorus, existing international institutional arrangements are inconsistent with the natural phosphorus cycle. This is most evident in the divide between the agricultural sector, where phosphorus is perceived as a fertilizer commodity, and the water and sanitation sector, where phosphorus is perceived as a pollutant in wastewater. This may hinder opportunities to find integrated solutions to the scarcity problem, since it is necessary for several sectors to be involved. In the case of phosphorus scarcity, part of the alternative resources and strategies are located in the sanitation sector (e.g. reuse of nutrients), whilst others are located in the household sector (e.g. the reduction of food waste, the reduction of meat and dairy consumption, etc.).

The recycling of urine is a socio-technical process that has no institutional or organizational home (Cordell, 2006; Livingston et al., 2005). Rather, a lack of institutional fit means it is seen as peripheral by all stakeholders and sectors (such as water service providers, town planners and farmers) and is not currently perceived as important enough for any single stakeholder group to make it a priority. Drangert suggests a 'urine-blindness' has prevented modern societies from tapping into this abundant source of plant nutrients in urine (Drangert, 1998). Both the professionals managing urban water and sanitation systems and residents using these systems avoid thinking about the character of individual fractions within wastewater and instead adhere to the routine of 'flush and discharge' (p157).

There are some significant similarities in the way in which the contemporaneous issues of climate change, water scarcity and phosphorus scarcity manifest themselves and can be addressed as potential solutions emerge. Climate change mitigation comprises a wide range of measures, and the same goes for water scarcity. World leaders have embraced the concept that limited water availability and accessibility is threatening food security, and discussions on solutions have followed. For example, it has been argued that reducing wastage in the entire food production and consumption chain will also reduce significantly the amount of water used to produce food (Lundqvist et al., 2008). The good news is that climate change, water and phosphorus scarcity can all be ameliorated with a concerted effort by the global community. In the extreme scenario where all wasted phosphorus would be recovered and recirculated back to agriculture, no additional phosphate rock inputs would be required. Scarcity of phosphate rock would then be of little concern. The crucial task however is to reduce the demand for phosphorus in addition to harnessing the measures needed to recirculate wasted phosphorus back to food production before it is dispersed into water bodies and non-agriculture soils. At present, there is a scarcity of management of phosphorus resources, rather than simply a physical scarcity of phosphate rock. With this in mind, institutional and other constraints can be better addressed.

The recent price spike in phosphate rock is likely to trigger further innovations in and adoption of phosphorus recovery and efficiency measures. However, the current market system alone is not adequate to manage phosphorus in a sustainable, equitable and timely manner in the longer term.

Opportunities also exist for integrating phosphorus management into existing discussions. For example, the issue of phosphorus scarcity could be given a higher profile in leading interdisciplinary international networks such as the Earth System Science Partnership (ESSP) which is addressing other important global biogeochemical cycles (GCP, 2008). The ESSP Global Environmental Change and Food Systems (GECAFS) program is an obvious place where this could occur.

The emergence of peak oil, the likelihood of a global emissions trading scheme for carbon, and the associated increases in energy costs will increase the cost of phosphate rock mining. This will provide an incentive for recirculating phosphorus found in organic sources, which will become more cost-effective relative to mining, processing and shipping rock phosphate. The energy required to produce mineral phosphate fertilizers is greater than that of organic phosphate fertilizers. The Earth Policy Institute reports that fertilizer production (including phosphorus) accounts for 29 percent of farm energy use in the US, excluding transporting chemicals to the field (Earth Policy Institute, 2005). A British study (Shepherd, 2003) indicated that organic agriculture uses less energy per crop output than industrial agriculture, mainly due to the significant amounts of energy required to produce mineral fertilizers. Johansson (2001) note that urine can be transported up to 100 km by truck and remain more energy-efficient than conventional systems of mineral fertilizer production, transportation and application.

Another incentive for increasing the reuse of phosphorus in this way is the avoidance of the environmental and financial costs associated with the discharge of phosphorus to waterways. The environmental cost of phosphate pollution of waterways is deemed unacceptable in many parts of the world and thus high levels of phosphorus must be removed from wastewater. Collecting urine, excreta and manure at the source will reduce phosphorus entering the wastewater treatment plant and thereby can achieve removal targets using less energy and at lower costs (Huang et al., 2007).

Sustainability initiatives in other sectors, such as materials manufacturing, can also be applied to the use of phosphorus. For example, concepts of 'design for the environment' and 'extended producer responsibility' involve capturing and reprocessing valuable substances directly after use, for reuse in production and manufacturing processes (OECD, 2001). Examples range from recovery and reuse of copper piping (Giurco and Petrie, 2007), to reusing vehicle parts under the European Union Directive for End-of-Life Vehicles (European Commission, 2000). In the case of nutrients, residents, local councils or entrepreneurs could be involved in recovering phosphorus from urban waste streams. Small and medium-scale examples already exist in sites around the world, including West Africa (Kvarnström et al., 2006), Inner Mongolia (EcoSanRes, 2008), and Stockholm (Kvarnström et al., 2006). There are clear synergies with sustainable sanitation strategies, which aim to decrease the mixing of water, faeces and urine in order to better contain, sanitise and reuse the water and nutrients. The World Health Organization is active in rethinking approaches to sanitation and has recently issued guidelines for the use of grey water, urine and faecal matter in agriculture (WHO, 2006). These guidelines map out ways that nutrients and water can be recovered, treated and reused. This is likely to play an important role in 'legitimizing' the use of human excreta among authorities and contribute to our understanding of the role of urban sanitation in the global nutrient and water cycles. For example, Sweden has recently proposed that 60 percent of phosphorus in sewage should be returned to land by 2015 (Swedish Environmental Objectives Council, 2007).[14]

Conclusions

This paper outlines how humanity became addicted to phosphate rock, and examines the current and future implications of this dependence on a non-renewable resource. Global demand for crops will continue to rise over the next half century, increasing the demand for phosphate fertilizers. However, modern agriculture is currently relying on a non-renewable resource and future phosphate rock is likely to yield lower quality phosphorus at a higher price. If significant physical and institutional changes are not made to the way we currently use and source phosphorus, agricultural yields will be severely compromised in the future. This will impact poor farmers and poor households first. However, there are opportunities to recover used phosphorus throughout the food production and consumption chain. Reducing losses in the food chain and increasing agricultural efficiency are also likely to contribute significantly to averting a future phosphate crisis.

Despite the depletion of global reserves and potential geopolitical tensions, future phosphate scarcity and reduced accessibility to farmers is not yet considered a significant problem by those who decide national or international policy. There are currently no international organizations or intentional governance structures to ensure the long-term, equitable use and management of phosphorus resources in the global food system. In order to avoid a future food-related crisis, phosphorus scarcity needs to be recognized and addressed in contemporary discussions on global environmental change and food security, alongside water, energy and nitrogen.

Notes

1. The Green Revolution in the early 1960s was enabled by the invention of the Haber-Bosch process decades earlier, which allowed the production of high volumes of artificial nitrogenous fertilizers (Brink, 1977).

2. Soil nutrient deficiency is due both to naturally low phosphate soils and to anthropogenic influences like soil mining and low fertilizer application rates which have resulted in net negative phosphorus budgets in many parts of Sub–Saharan Africa (Smaling et al., 2006).
3. For example, a recent FAO study found that in Egypt, there are currently more overweight than underweight children (see FAO, 2006, Fighting hunger–and obesity, Spotlight 2006, Agriculture 21, FAO, [Online], available: www.fao.org/ag/magazine/0602sp1.htm [accessed 4/6/06]).
4. A recently published paper by Liu et al. (2008) does provide an analysis of global anthropogenic phosphorus flows based on existing data.
5. This is fairly consistent with estimates by Liu et al. (2008), published after this analysis. Both analyses have drawn heavily from Smil, so this is not surprising. The actual amount lost from agricultural fields that is directly attributed to applied phosphate fertilizer is very difficult to calculate, as soil phosphate chemistry is complex and available phosphorus can move to unavailable forms and back again.
6. Human bodies require roughly 1.2 g/(person day) of phosphorus for healthy functions, which equates to approximately 3 MT P globally.
7. Units of phosphorus are presented as elemental P, rather than P_2O_5 (containing 44 percent P) or phosphate rock (containing 29–34 percent P_2O_5) as commonly used by industry.
8. Estimated from 18 000 MT phosphate rock (Jasinski, 2008).
9. If production is assumed to have been at maximum capacity in the period to about 1990, this would suggest that peak production would have occurred at about that time (Déry and Anderson, 2007), but that reserves are approximately half of the amount estimated by the USGS.
10. Because phosphorus is one of the most chemically reactive nutrients, it readily transforms to forms of phosphorus unavailable to plants.
11. Such an analysis is addressed further in Cordell et al. (in press).
12. A noteworthy exception to this lack of attention is the World Health Organisation issuing recommendations on safe use of excreta and greywater in agriculture (WHO, 2006).
13. However the situation is in a state of change and as recent as 2008 a published paper on "Long-term global availability of food: continued abundance or new scarcity?" (Koning et al., 2008) identifies phosphorus scarcity as a likely key factor limiting future food availability. Similarly, the closing ceremony of the 2008 World Water Week for the first time highlighted mineral phosphate scarcity, noting "in a time of rising peak oil, of rising costs of fertilisers, and of dwindling phosphorus-mineral sources" Falkenmark (2008).
14. This target was recommended to the Swedish government by a Swedish EPA Action Plan. See Sweden's National Environmental Objectives at www.miljomal.nu/englizh/englizh.php.

References

AGRA, 2008. About the Alliance for a Green Revolution in Africa, Alliance for a Green Revolution in Africa. Available: www.agra-alliance.org.

Blair, D., 2008. Green revolution needed to feed the poor: UN, Sydney Morning Herald, New York. Available: www.smh.com.au/articles/2008/05/19/1211182703408.html.

Bombay News, 2008. Farmer killed in stampede during fertiliser sale, Bombay News.Net, Wednesday 30th July, 2008 (IANS). Available: www.bombay-news.net/story/388149.

Braun, J., Swaminathan, M., Rosegrant, M., 2004. Agriculture, Food Security, Nutrition and the Millennium Development Goals, Essay in 2003–2004 IFPRI Annual Report. International Food Policy Research Institute.

Brink, J., 1977. World resources of phosphorus. Ciba Foundation Symposium 13–15, 23–48.

Brunner, P.H., Rechberge, H., 2004. Practical Handbook of Material Flow Analysis. Lewis Publishers, Boca Raton, FL, 2003, p. 332, ISBN 1-5667-0604–1.

Buckingham, D., Jasinski, S., 2004. Phosphate Rock Statistics 1900–2002. US Geological Survey.

Buckingham, D.A., Jasinski, S.M., 2006. Phosphate Rock Statistics, Historical Statistics for Mineral and Material Commodities in the United States, Data Series 140. US Geological Survey. Available: minerals.usgs.gov/ds/2005/140/.

CEEP, 2008. SCOPE Newsletter, Number 70. February.

Cordell, D., 2006. Urine Diversion and Reuse in Australia: A homeless paradigm or sustainable solution for the future?. February 2006, Masters Thesis, Masters of Water Resources & Livelihood Security, Department of Water & Environmental Studies, Linköping University Linköping. Available: www.ep.liu.se/undergraduate/abstract.xsql?dbid=8310.

Cordell, D., 2007. More Nutrition per Dropping: From Global Food Security to National 'Phosphorus Sovereignty', poster presented at workshop 'International Targets and National Implementation', World Water Week 2007, August 2007, Stockholm. www.worldwaterweek.org/stockholmwatersymposium/bestposteraward_07.asp.

Cordell, D., Neset, T.S.S., Drangert, J.-O., White, S., in press. Preferred future phosphorus scenarios: a framework for meeting long-term phosphorus needs for global food demand, International Conference on Nutrient Recovery from Wastewater Streams Vancouver, 2009. In: Don Mavinic, Ken Ashley, Fred Koch (Eds.). ISBN: 9781843392323. IWA Publishing, London, UK.

Corell, H., 2002. Letter dated 29 January 2002 from the Under-Secretary-General for Legal Affairs the Legal Counsel addressed to the President of the Security Council. In: United National Security Council, Under-Secretary-General for Legal Affairs The Legal Counsel.

Department of Commerce, 2005. National Programme for Organic Production, India Organic, 6th ed. Ministry of Commerce and Industry, New Delhi.

Déry, P., Anderson, B., 2007. Peak phosphorus. Energy Bulletin, 08/13/2007. Available: energybulletin.net/node/33164.

Drangert, J-O., 1998. Fighting the urine blindness to provide more sanitation options. Water SA 24, No 2.

Driver, J., 1998. Phosphates recovery for recycling from sewage and animal waste. Phosphorus and Potassium 216, 17–21.

Earth Policy Institute, 2005. Oil and Food: A Rising Security Challenge. Earth Policy Institute.

EcoSanRes, 2003. Closing the Loop on Phosphorus. Stockholm Environment Institute (SEI) funded by SIDA Stockholm.

EcoSanRes, 2008. Sweden-China Erdos Eco-Town Project Dongsheng, Inner Mongolia, Stockholm Environment Institute (SEI). www.ecosanres.org/asia.htm, Stockholm.

European Fertilizer Manufacturers Association, 2000. Phosphorus: Essential Element for Food Production. European Fertilizer Manufacturers Association (EFMA), Brussels.

Ensink, J.H.J., Mahmood, T., Hoek, W.v.d., Raschid-Sally, L., Amerasinghe, F.P., 2004. A nationwide assessment of wastewater use in Pakistan: an obscure activity or a vitally important one? Water Policy 6, 197–206.

Esrey, S., Andersson, I., Hillers, A., Sawyer, R., 2001. Closing the Loop: Ecological Sanitation for Food Security. UNDP & SIDA, Mexico.

EU, 2007. Council Regulation (EC) No. 834/2007 of 28 June 2007 on organic production and labelling of organic products and repealing Regulation (EEC) No. 2092/92. Official Journal of the European Union, L 189/2.

European Commission, 2000. Directive 2000/53/EC of the European Parliament and of the Council of 18th September 2000 on End-of-Life Vehicles—Commission Statements, document 300L0053. Official Journal L 269, 21 October 2000.

Falkenmark, M., 2008. Overarching summary of workshop contributons and personal reflections. World Water Week, August 2008, Stockholm.

Falkenmark, M., Rockström, J., 2002. Neither water nor food security without a major shift in thinking—a water-scarcity close-up. In: World Food Prize International Symposium Des Moines.

FAO, 1999. Urban and Peri-urban Agriculture. Food and Agriculture Organisation of the United Nations, Rome.

FAO, 2000. Fertilizer Requirements in 2015 and 2030. Food and Agriculture Organisation of the United Nations Rome.

FAO, 2004a. The State of Food Security in the World, Monitoring Progress Towards the World Food Summit and Millennium Development Goals. Food and Agricultural Organisation.

FAO, 2004b. The Use of Phosphate Rocks for Sustainable Agriculture Technical. In: F. Zapata (Ed.), Joint FAO/IAEA Division of Nuclear Techniques in Food and Agriculture, Vienna, Austria, R.N. Roy Land and Water Development Division, FAO, Rome, Italy. A joint publication of the FAO Land and Water Development Division and the International Atomic Energy Agency Food and Agriculture Organization Of The United Nations Rome.

FAO, 2005a. Assessment of the World Food Security Situation, Food and Agricultural Organisation of the United Nations. Committee on World Food Security, 23–26 May 2005, Rome. Available: www.fao.org/docrep/meeting/009/j4968e/j4968e00.htm.

FAO, 2005b. The Special Program for Food Security, Food and Agriculture Organisation of the United Nations. Available: www.fao.org/spfs.

FAO, 2006. Plant Nutrition for Food Security: A Guide for Integrated Nutrient Management, FAO Fertilizer And Plant Nutrition Bulletin 16. Food And Agriculture Organization Of The United Nations Rome.

FAO, 2007a. Current World Fertilizer Trends and Outlook to 2010/11. Food and Agriculture Organisation of the United Nations Rome.

FAO, 2007b. Food for the Cities Homepage. Rome, Food and Agriculture Organisation of the United Nations.

FAO, 2007c. Organic agriculture and food availability. International Conference on Organic Agriculture and Food Security, 3–5th May, 2007, Food and Agriculture Organisation of the United Nations Rome.

FAO, 2008a. Efficiency of soil and fertilizer phosphorus use: reconciling changing concepts of soils phosphorus behaviour with agronomic information. In: FAO Fertilizer and Plant Nutrition Bulletin 18, Food and Agriculture Organization of the United Nations Rome.

FAO, 2008b. High-level conference on world food security: the challenges of climate change and bioenergy. Soaring food prices: facts, perspectives, impacts and actions required, Rome, 3–5 June 2008.

Fertilizer Week, 2008. Industry ponders the impact of China's trade policy. Thursday Markets Report, 24th April 2008, British Sulphur Consultants, CRU.

Food21, 2005. Research Programme on sustainable food production, Swedish University of Agricultural Science, Uppsala, funded by MISTRA Foundation for Strategic Environmental Research.

Fraiture, C.D., 2007. Future Water Requirements for Food—Three Scenarios, International Water Management Institute (IWMI), SIWI Seminar: Water for Food, Bio-fuels or Ecosystems? World Water Week 2007, August 12th–18th 2007, Stockholm.

Fresco, L., 2003. Plant nutrients: what we know, guess and do not know, Assistant Director-General. Agriculture Department Food and Agriculture Organization of the United Nations (FAO) IFA/FAO AGRICULTURE CONFERENCE Rome.

Gardner, G., Halweil, B., 2000. Overfed and Underfed: The Global Epidemic of Malnutrition. In: Peterson, J. (Ed.), Worldwatch Institute, WORLDWATCH PAPER 150.

GCP, 2008. The Global Carbon Project. A programme of the Earth Systems Science Partnership, www.globalcarbonproject.org.

GECAFS, 2006. Conceptualising Food Systems for Global Environmental Change (GEC) Research, GECAFS Working Paper 2. P.J. Ericksen, ECI/OUCE, Oxford University, GECAFS International Project Office Wallingford, UK.

Giurco, D., Petrie, J., 2007. Strategies for reducing the carbon footprint of copper: new technologies. More Recycling Or Demand Management? Minerals Engineering. 20 (9), 842–853.

Gumbo, B., 2005. Short-Cutting the Phosphorus Cycle in Urban Ecosystems. PhD thesis, Delft University of Technology and UNESCO-IHE Institute for Water Education Delft.

Gumbo, B., Savenije, H.H.G., 2001. Inventory of phosphorus fluxes and storage in an urban-shed: options for local nutrient recycling. Internet Dialogue on Ecological Sanitation (15 November-20 December 2001). Delft.

Gunther, F., 1997. Hampered effluent accumulation process: phosphorus management and societal structure. Ecological Economics 21, 159–174.

Gunther, F., 2005. A solution to the heap problem: the doubly balanced agriculture: integration with population. Available: www.holon.se/folke/kurs/Distans/Ekofys/Recirk/Eng/balanced.shtml.

Heffer, P., Prud'homme, M., 2007. Medium-Term Outlook for Global Fertilizer Demand, Supply and Trade 2007–2011 Summary Report. International Fertilizer Industry Association, 75th IFA Annual Conference, 21–23 May 2007, Istanbul, Turkey.

Huang, D.-B., Bader, H.-P., Scheidegger, R., Schertenleib, R., Gujer, W., 2007. Confronting limitations: new solutions required for urban water management in Kunming City. Journal of Environmental Management 84 (1), p. 49–61.

Hubbert, M.K., 1949. Energy from fossil fuels. Science 109, 103.

Hugo, V., 1862. Les Miserables, ch.323, A. Lacroix, Verboeckhoven & Ce.

IAASTD, 2008. International Assessment of Agricultural Knowledge, Science and Technology for Development (IAASTD), agreed to at an Intergovernmental Plenary Session in Johannesburg, South Africa in April, 2008. www.agassessment.org.

IDGEC, 2006. Institutional Dimensions of Global Environmental Change programme, a core science programme of The

International Human Dimensions Programme on Global Environmental Change homepage.

IFA, 2006. Production and International Trade Statistics, International Fertilizer Industry Association Paris, available: www.fertilizer.org/ifa/statistics/pit_public/pit_public_statistics.asp (accessed 20 August 2007).

IFA, 2008. Feeding the Earth: Fertilizers and Global Food Security, Market Drivers and Fertilizer Economics. International Fertilizer Industry Association, Paris.

IFDC, 2005. Africa Fertilizer Situation, Fertilizer Situation Report Series. Market Development Division, International Centre for Soil Fertility and Agricultural Development Alabama.

IFDC, 2006. Global Leaders Launch Effort to Turn Around Africa's Failing Agriculture: New Study Reports Three-Quarters of African Farmlands Plagued by Severe Degradation, International Center for Soil Fertility and Agricultural Development, 30 March 2006, New York.

IFPRI, 2002a. GREEN REVOLUTION: Curse or Blessing? International Food Policy Research Institute, Washington, DC.

IFPRI, 2002b. Reaching Sustainable Food Security for All by 2020: Getting the Priorities and Responsibilities Right. International Food Policy Research Institute, Washington.

IFPRI, 2005. IFPRI's Strategy Toward Food and Nutritional Security: Food Policy Research, Capacity Strengthening and Policy Communications. International Food Policy Research Institute, Washington, DC.

IHDP, 2002. The Problem of Fit between Ecosystems and Institutions, IHDP Working Paper No. 2: The International Human Dimensions Programme on Global Environmental Change.

International Fertilizer Industry Association, 2006. Sustainable Development and the Fertilizer Industry IFA website.

IWMI, 2006. Comprehensive Assessment of water management in agriculture, Co-sponsers: FAO, CGIAR, CBD, Ramsar. www.iwmi.cgiar.org/assessment.

Jaffer, Y., et al., 2002. Potential phosphorus recovery by struvite formation. Water Research 36, 1834–1842.

Jasinski, S.M., 2006. Phosphate Rock, Statistics and Information. US Geological Survey.

Jasinski, S.M., 2007. Phosphate Rock, Mineral Commodity Summaries, U.S. Geological Survey minerals.usgs.gov/minerals/pubs/commodity/phosphate_ rock/.

Jasinski, S.M., 2008. Phosphate Rock, Mineral Commodity Summaries, U.S. Geological Survey minerals.usgs.gov/minerals/pubs/commodity/phosphate_ rock/.

Johansson, M., 2001. Urine Separation-Closing the Nutrient Cycle. Final Report On The R&D Project Source-Separated Human Urine—A Future Source Of Fertilizer For Agriculture In The Stockholm Region? Co-authors: Håkan Jönsson, Department of Agricultural Engineering, Swedish University of Agricultural Sciences, Caroline Hoglund, Swedish Institute for Infectious Disease Control, and Anna Richert Stintzing and Lena Rodhe, Swedish Institute of Agricultural and Environmental Engineering, prepared for Stockholm Water, Stockholmshem, HSB Stockholm.

Johnston, A.E., 2000. Soil and Plant Phosphate. International Fertilizer Industry Association (IFA), Paris.

Jönsson, H., Stintzing, A.R., Vinnerås, B., Salomon, E., 2004. Guidelines on the Use of Urine and Faeces in Crop Production. EcoSanRes, Stockholm Environment Institute, Stockholm.

Koning, N.B.J., Ittersum, M.K.v., Becx, G.A., Boekel, M.A.J.S.v., Brandenburg, W.A., Broek, J.A.v.d., Goudriaan, J., Hofwegen, G.v., Jongeneel, R.A., Schiere, J.B., Smies, M., 2008. Long-term global availability of food: continued abundance or new scarcity? NJAS Wageningen Journal of Life Sciences 55, 229–292.

Kratz, S., Schnug, E., 2006. Rock Phosphates and P Fertilizers as Sources of U Contamination in Agricultural Soils. Institute of Plant Nutrition and Soil Science, Federal Agricultural Research Center, Germany, pp. 57–67; in Uranium in the Environment: Mining Impact and Consequences, Springer Berlin Heidelberg.

Kvarnström, E., Emilsson, K., Stintzing, A.R., Johansson, M., Jönsson, H., Petersens, E.a., Schonning, C., Christensen, J., Hellström, D., Qvarnstrom, L., Ridderstolpe, P., Drangert, J.-O., 2006. Urine Diversion: One Step Towards Sustainable Sanitation. EcoSanRes programme, Stockholm Environment Institute Stockholm.

Laherrere, J.H., 2000. Learn strengths, weaknesses to understand Hubbert curve. Oil & Gas Journal 98 (16), 63.

Liebig, J., 1840. Die organische Chemie in ihrer Anwendung auf Agricultur und Physiologie (Organic Chemistry in its Applications to Agriculture and Physiology). Friedrich Vieweg und Sohn Publ. Co., Braunschweig, Germany.

Liu, Y., Villalba, G., Ayres, R.U., Schroder, H., 2008. Environmental impacts from a consumption perspective. Journal of Industrial Ecology 12 (2).

Livingston, D., Colebatch, H., Ashbolt, N., 2005. Sustainable Water Paradigm Shift: Does Changing Discourse Mean Change in Organisation? School of Civil and Environmental Engineering, UNSW, Sydney.

Lundqvist, J., Baron, J., Berndes, G., Berntell, A., Falkenmark, M., Karlberg, L., Rock-ström, J., 2007. Water pressure and increases in food & bio-energy demand -implications of economic growth and options for decoupling, Chapter 3. Scenarios on Economic Growth and Resource Demand—Background Report to the Swedish Environmental Advisory Council Memorandum 2007:1 (preliminary title) Stockholm.

Lundqvist, J., Fraiture, C.d., Molden, D., 2008. Saving Water: From Field to Fork—Curbing Losses and Wastage in the Food Chain, SIWI Policy Brief. Stockholm International Water Institute, Stockholm.

Maene, L.M., 2007. International Fertilizer Supply and Demand. In: Australian Fertilizer Industry Conference, International Fertilizer Industry Association, August.

Mårald, E., 1998. I mötet mellan jordbruk och kemi: agrikulturkemins framväxt p\{aa Lantbruksakademiens experimentalfält 1850–1907. Institutionen för idéehistoria, Univ Ume\{å.

Matsui, S., 1997. Nightsoil collection and treatment in Japan. In: Drangert, J.-O., Bew, J., Winblad, U. (Eds.). Ecological Alternatives in Sanitation. Publications on Water Resources: No 9. Sida, Stockholm.

Millennium Ecosystem Assessment, 2005. Chapter 12: Nutrient Cycling, Volume 1: Current State and Trends, Global Assessment Reports, Millennium Ecosystem Assessment. www.millenniumassessment.org/documents/document.281.aspx.pdf.

Minemakers Limited, 2008. ROCK PHOSPHATE PRICE ROCKETS TO US$200/TONNE. ASX and Press Release Perth.

Neset, T.S.S., Bader, H., Scheidegger, R., Lohm, U., 2008. The Flow of Phosphorus in Food Production and Consumption, Linköping, Sweden 1870–2000. Department of Water and Environmental

Studies, Linköping University and EAWAG Department S&E Dübendorf.

OECD, 2001. Extended Producer Responsibility: A Guidance Manual for Governments. OECD, Paris.

Organic Federation of Australia, 2005. National Standard for Organic and Bio-Dynamic Produce. Edition 3.1. As amended January 2005, Available: www.ofa.org.au/papers/2005 percent20Draft percent20NATIONAL percent20STANDARD.pdf.

Pazik, G., 1976. Our Petroleum Predicament, A Special Editorial Feature by George Pazik Editor & Publisher. Fishing Facts, November 1976.

Pfeiffer, D.A., 2006. Eating Fossil Fuels: Oil, Food and the Coming Crisis in Agriculture. New Society Publishers, Canada.

Prud'homme, M., 2006. Phosphate production in the new economics. In: International Fertilizer Industry Association (IFA), FIRP & IAEA Collabration Meeting, 3–4th October, Bartow, Florida.

Raschid-Sally, L., Jayakody, P., 2008. Drivers and Characteristics of Wastewater Agriculture in Developing Countries—Results from a Global Assessment. Comprehensive Assessment of water management in agriculture, International Water Management Institute.

Reindl, J., 2007. Phosphorus Removal from Wastewater and Manure Through Struvite Formation: An Annotated Bibliography. Dane County Dept ofHighway, Transportation and Public Works.

Rosmarin, A., 2004. The Precarious Geopolitics of Phosphorous, Down to Earth: Science and Environment Fortnightly, 2004, pp. 27–31.

Royal Dutch Shell, 2008. Shell Energy Scenarios 2050, Shell International BV. Available: www.shell.com/scenarios.

Runge-Metzger, A., 1995. Closing The Cycle: Obstacles To Efficient P Management For Improved Global Food Security. SCOPE 54–Phosphorus in the Global Environment–Transfers, Cycles and Management.

Saueia, C.H., Mazzilli, B.P., Fávaro, D.I.T., 2005. Natural radioactivity in phosphate rock, phosphogypsum and phosphate fertilizers in Brazil. Journal of Radio-analytical and Nuclear Chemistry 264, 445–448.

Schmid-Neset, T., Bader, H., Scheidegger, R., Lohm, U., 2005. The Flow of Phosphorus in Food Production and Consumption, Linköping, Sweden 1870–2000. Department of Water and Environmental Studies, Linköping University and EAWAG Department S&E Dübendorf.

SCOPE, 2004. Struvite: Its Role in Phosphorus Recovery and Recycling. International Conference 17th-18th June 2004. Cranfield University, Great Britain. Summary report in SCOPE 57.

SEPA, 1995. The content of wastewater from Swedish households'. Report 4425, Swedish Environmental Protection Agency (in Swedish) Stockholm.

Sharpley, A.N., Withers, P.J.A., Abdalla, C.W., Dodd, A.R., 2005. Strategies for the Sustainable Management of Phosphorus. Phosphorus: Agriculture and the Environment, Agronomy Monograph No. 46. Madison, American Society of Agronomy, Crop Science Society of America, Soil Science Society of America.

Shepherd, P., 2003. An Assessment of the Environmental Impact of Organic Farming. A review for DEFRA-funded Project OF0405, May 2003.

SIWI, IWMI, SEI, Chamlers, 2008. Saving Water: From Field to Fork, Curbing Losses and Wastage in the Food Chain, Paper 13. Draft report prepared for CSD, Stockholm International Water Institute Stockholm.

SIWI-IWMI, 2004. Water—More Nutrition Per Drop, Towards Sustainable Food Production and Consumption Patterns in a Rapidly Changing World. Stockholm International Water Institute, Stockholm.

Smaling, E., 2005. Harvest for the world, Inaugural address. International Institute for Geo-Information Science and Earth Observation, 2 November 2005, Enschede, The Netherlands.

Smaling, E., Toure, M., Ridder, N.d., Sanginga, N., Breman, H., 2006. Fertilizer Use and the Environment in Africa: Friends or Foes? Background Paper Prepared for the African Fertilizer Summit, June 9–13, 2006, Abuja, Nigeria.

Smil, V., 2000a. Feeding the World: A Challenge for the 21st Century. The MIT Press, Cambridge.

Smil, V., 2000b. Phosphorus in the environment: natural flows and human interferences. Annual Review of Energy and the Environment 25, 53–88.

Smil, V., 2002. Phosphorus: global transfers. In: Douglas, P.I. (Ed.), Encyclopedia of Global Environmental Change. John Wiley & Sons, Chichester.

Smil, V., 2007. Policy for Improved Efficiency in the Food Chain, SIWI Seminar: Water for Food, Bio-fuels or Ecosystems? World Water Week 2007, August 12th–18th 2007, Stockholm.

SOFI, 2005. The State of Food Insecurity in the World 2005. Eradicating world hunger-key to achieving the Millennium Development Goals, Economic and Social Department (ES), Food and Agricultural Organisation, FAO.

Steen, I., 1998. Phosphorus availability in the 21st Century: management of a non-renewable resource. Phosphorus and Potassium 217, 25–31.

Stewart, W., Hammond, L., Kauwenbergh, S.J.V., 2005. Phosphorus as a Natural Resource. Phosphorus: Agriculture and the Environment, Agronomy Monograph No. 46. Madison, American Society of Agronomy, Crop Science Society of America, Soil Science Society of America.

Stockholm Environment Institute, 2004. Ecological Sanitaiton: revised and enlarged edition, In: Uno Winblad, Mayling Simpson-Hébert (Eds.), SEI Stockholm.

Stockholm Environment Institute, 2005. Sustainable Pathways to Attain the Millennium Development Goals—Assessing the Role of Water, Energy and Sanitation (For the UN World Summit September 2005), SEI Stockholm.

Swedish Environmental Objectives Council, 2007. Sweden's environmental objectives, Swedish Government. www.miljomal.nu/english/english.php.

Tangsubkul, N., Moore, S., Waite, T.D., 2005. Phosphorus Balance and Water Recycling in Sydney. University of New South Wales, Sydney.

Tanums Kommun, 2002. Urine Separation, Tanum Municipality, Sweden. [Online], available: www.tanum.se/vanstermenykommun/miljo/toaletterochavlopp/urineseparation.4.8fc7a7104a93e5f2e8000595.html, accessed 30 November 2005.

The Norwegian Support Committee for Western Sahara, 2007. One more shipping company quits Western Sahara assignments. www.vest-sahara.no/index.php?parse_news=single&cat=49&art=949.

UN, 2000. Millennium Development Goals, Millennium Assembly.

UN, 2005. Millennium Development Goals Report, led by the Department of Economic and Social Affairs of the United Nations Secretariat New York.

UN Millennium Project, 2005. Fast Facts: The Face of Poverty. 2005 Millennium Project.

UNEP, 2005. Millennium Ecosystem Assessment Synthesis Report. Pre-publication Final Draft Approved by MA Board on March 23, 2005, The United Nations Environment Programme.

UNEP, 2007. Reactive Nitrogen in the Environment: Too Much or Too Little of a Good Thing, United Nations Environment Programme, Sustainable Consumption and Production (SCP) Branch, and The Woods Hole Research Center Paris.

USGS, 1999. Fertilizers—Sustaining Global Food Supplies, USGS Fact Sheet FS-155–99. US Geological Survey Reston, available: http://minerals.usgs.gov/minerals/pubs/commodity/phosphate_rock/.

Ward, J., 2008. Peak phosphorus: quoted reserves vs. production history, Energy Bulletin, Available: www.energybulletin.net/node/46386.

White, J., 2000. Introduction to Biogeochemical Cycles (Ch.4) Department of Geological Sciences, University of Colorado Boulder.

WHO, 2006. Guidelines for the safe use of wastewater, excreta and greywater. Volume 4: Excreta and Greywater Use in Agriculture, World Health Organisation.

Wissa, A.E.Z., 2003. Phosphogypsum Disposal and The Environment Ardaman & Associates Inc., Florida. Available: www.fipr.state.fl.us/pondwatercd/phosphogypsum_disposal.htm.

World Bank, 2005. Water Resources And Environment Technical Note F.3 Waste-water Reuse Series. In: Richard Davis, Rafik Hirji (Eds.), Washington, DC.

World Resources Institute, 2008. Agriculture and "Dead Zones". Available: www.wri.org/publication/content/7780.

WSRW, 2007. The Phosphate Exports, Western Sahara Resource Watch. www.wsrw.org/index.php?cat=117&art=521.

Critical Thinking

1. How familiar were you with this issue prior to reading this article? Why?
2. Answer the question, "Since a global phosphorus scarcity crisis is imminent . . . why is it not being discussed in relation to global food security to global environmental change?"
3. Evaluate the options identified in this article for sustainable phosphorus management. What do you see as obstacles? What do you see as opportunities?

Biodiversity Loss Threatens Human Well-Being

SANDRA DÍAZ ET AL.

The diversity of life on Earth is dramatically affected by human alterations of ecosystems [1]. Compelling evidence now shows that the reverse is also true: biodiversity in the broad sense affects the properties of ecosystems and, therefore, the benefits that humans obtain from them. In this article, we provide a synthesis of the most crucial messages emerging from the latest scientific literature and international assessments of the role of biodiversity in ecosystem services and human well-being.

Human societies have been built on biodiversity. Many activities indispensable for human subsistence lead to biodiversity loss, and this trend is likely to continue in the future. We clearly benefit from the diversity of organisms that we have learned to use for medicines, food, fibers, and other renewable resources. In addition, biodiversity has always been an integral part of the human experience, and there are many moral reasons to preserve it for its own sake. What has been less recognized is that biodiversity also influences human well-being, including the access to water and basic materials for a satisfactory life, and security in the face of environmental change, through its effects on the ecosystem processes that lie at the core of the Earth's most vital life support systems.

Three recent publications from the Millennium Ecosystem Assessment [2–4], an initiative involving more than 1,500 scientists from all over the world [5], provide an updated picture of the fundamental messages and key challenges regarding biodiversity at the global scale. Chief among them are: (a) human-induced changes in land cover at the global scale lead to clear losers and winners among species in biotic communities; (b) these changes have large impacts on ecosystem processes and, thus, human well-being; and (c) such consequences will be felt disproportionately by the poor, who are most vulnerable to the loss of ecosystem services.

What We Do Know: Functional Traits Matter Most

Biodiversity in the broad sense is the number, abundance, composition, spatial distribution, and interactions of genotypes, populations, species, functional types and traits, and landscape units in a given system. Biodiversity influences ecosystem services, that is, the benefits provided by ecosystems to humans, that contribute to making human life both possible and worth living [4]. As well as the direct provision of numerous organisms that are important for human material and cultural life, biodiversity has well-established or putative effects on a number of ecosystem services mediated by ecosystem processes. Examples of these services are pollination and seed dispersal of useful plants, regulation of climatic conditions suitable to humans and the animals and plants they consider important, the control of agricultural pests and diseases, and the regulation of human health. Also, by affecting ecosystem processes such as biomass production by plants, nutrient and water cycling, and soil formation and retention, biodiversity indirectly supports the production of food, fiber, potable water, shelter, and medicines. The links between biodiversity and ecosystem services have been gaining increasing attention in the scientific literature of the past few years [2–4, 6]. However, not until now has there been an effort to summarize those components of biodiversity that do, or should, matter the most for the provision of these services, and the underlying mechanisms explaining those links.

A few key messages can be drawn from existing theory and empirical studies. The first is that the number and strength of mechanistic connections between biodiversity and ecosystem processes and services clearly justify the protection of the biotic integrity of existing and restored ecosystems and its inclusion in the design of managed ecosystems. All components of biodiversity, from genetic diversity to the spatial arrangement of landscape units, may play a role in the long-term provision of at least some ecosystem services. However, some of these components are more important than others in influencing specific ecosystem services. The evidence available indicates that it is functional composition—that is, the identity, abundance, and range of species traits—that appears to cause the effects of biodiversity on many ecosystem services. At least among species within the same trophic level (e.g., plants), rarer species are likely to have small effects at any given point in time. Thus, in natural systems, if we are to preserve the services that ecosystems provide to humans, we should focus on preserving or restoring their biotic integrity in terms of species composition, relative abundance, functional organization, and species numbers (whether inherently species-poor or species-rich), rather than on simply maximizing the number of species present.

Another key message is that, precisely because ecosystem processes depend on the presence and abundance of organisms with particular functional traits, there is wide variation in how ecosystem services—that in turn depend on ecosystem processes—respond to changes in species number as particular species are lost from or get established in the system. So, to the question of how biodiversity matters to ecosystem services, we have to reply that it depends on what organisms there are. Daunting? Certainly, but not hopeless. We know from recent assessments [1, 2, 7, 8] that global biodiversity loss is not occurring at random. As a consequence of global change drivers, such as climate, biological invasions, and especially land use, not only is the total number of species on the planet decreasing, but there are also losers and winners. On average, the organisms that are losing out have longer lifespans, bigger bodies, poorer dispersal capacities, more specialized resource use, lower reproductive rates, and other traits that make them more susceptible to human activities such as nutrient loading, harvesting, and biomass removal by burning, livestock grazing, ploughing, clear-felling, etc. A small number of species with the opposite characteristics are becoming increasingly dominant around the world. Because there are well-established links between functional traits of locally abundant organisms and ecosystem processes, especially for plants [9–12], it may become possible to identify changes in ecosystem processes and in ecosystem services that depend on them under different biodiversity scenarios.

What We Do Not Know: Cascades, Surprises, and Megadiversity Hot-Spots

Some ecosystem services show a saturating relationship to species number—that is, the ecosystem-service response to additional species is large at low number of species and becomes asymptotic beyond a certain number of species. We seldom know what this threshold number is, but we suspect it differs among ecosystems, trophic levels, and services. The experimental evidence indicates that, in the case of primary production (e.g., for plant-based agricultural products), nutrient retention (which can reduce nutrient pollution and sustain production in the long term), and resistance to invasions (which incur damage and control costs in agricultural and other settings) by temperate, herbaceous communities, responses often do not show further significant increases beyond about ten plant species per square meter [3, 13]. But in order to achieve this number in a single square meter, a much higher number of species is needed at the landscape level [14]. What about slow-growing natural communities, or communities that consist of plant species with more contrasting biology? What about communities that typically include many more species—for example, the megadiverse forest hot-spots of the Amazon and Borneo, where species number can exceed 100 tree species per hectare [15]? To what extent are all those species essential for the maintenance of different ecosystem processes and services? Ecological theory [16] and traditional knowledge [17, 18] suggest that a large number of resident species per functional group, including those species that are rare, may act as "insurance" that buffers ecosystem processes and their derived services in the face of changes in the physical and biological environment (e.g., precipitation, temperature, pathogens), but these ideas have yet to be tested experimentally, and no manipulative experiment has been performed in any megadiversity hot-spot.

Most of the links between biodiversity and ecosystem services emerged from theory and manipulative experiments, involved biodiversity within a single trophic level (usually plants), and operated mostly at the level of local communities. However, the most dramatic examples of effects of small changes in biodiversity on ecosystem services have occurred at the landscape level and have involved alterations of food-web diversity through indirect interactions and trophic cascades. Most of these have been "natural experiments," that is, the unintended consequence of intentional or accidental removal or addition of certain predator, pathogen, herbivore, or plant species to ecosystems. These "ecological surprises" usually involve disproportionately large, unexpected, irreversible, and negative alterations of ecosystem processes, often with repercussions at the level of ecosystem services, with large environmental, economic, and cultural losses. Examples include the cascading effects of decreases in sea otter population that led to coastal erosion in the North Pacific [19], and a marked decrease in grassland productivity and nutritional quality in the Aleutian islands as a consequence of decreased nutrient flux from the sea by the introduction of Arctic foxes [20] (see [3] for a comprehensive list of examples). The vast literature on biological invasions and their ecological and socio-economic impacts [21] further illustrates this point. Ecological surprises are difficult to predict, since they usually involve novel interactions among species. They most often result from introductions of predators, herbivores, pathogens and diseases, although cases involving introduced plants are also known. They do not depend linearly on species number or on well-established links between the functional traits of the species in question and putative ecosystem processes or services [3, 22].

Uneven Impacts: Biodiversity and Vulnerable Peoples

People who rely most directly on ecosystem services, such as subsistence farmers, the rural poor, and traditional societies, face the most serious and immediate risks from biodiversity loss. First, they are the ones who rely the most on the "safety net" provided by the biodiversity of natural ecosystems in terms of food security and sustained access to medicinal products, fuel, construction materials, and protection from natural hazards such as storms and floods [4]. In many cases the provision of services to the most privileged sectors of society is subsidized but leaves the most vulnerable to pay most of the cost of biodiversity losses. These include, for example, subsistence farmers in the face of industrial agriculture [23] and subsistence fishermen in the face of intensive commercial fishing and aquaculture [24]. Second, because of their low economic and political power, the less privileged sectors cannot substitute purchased goods and services for the lost ecosystem benefits and they typically have little influence on national policy. When the quality of water deteriorates as a result of fertilizer and pesticide loading by industrial agriculture, the poor are unable to purchase safe water. When protein and vitamins from local sources, such as hunting and fruit, decrease as a result of habitat loss, the rich can still purchase them, whereas the poor cannot. When the capacity of natural ecosystems to buffer the effects of storms and floods is lost because of coastal development [25], it is usually the people who cannot flee—for example, subsistence fishermen—who suffer the most. In summary, the loss of biodiversity-dependent ecosystem services is likely to accentuate inequality and marginalization of the most vulnerable sectors of society, by decreasing their access to basic materials for a healthy life and by reducing their freedom of choice and action. Economic development that does not consider effects on these ecosystem services may decrease the quality of life of these vulnerable populations, even if other segments of society benefit. Biodiversity change is therefore inextricably linked to poverty, the largest threat to the future of humanity identified by the United Nations. This is a sobering conclusion for those who argue that biodiversity is simply an intellectual preoccupation of those whose basic needs and aspirations are fulfilled.

Future Directions

Most of the concrete actions to slow down biodiversity loss fall under the domain of policy making by governments and the civil society. However, the scientific community still needs to fill crucial knowledge gaps. First, we need to know more about the links between biodiversity and ecosystem services in species-rich ecosystems dominated by long-lived plants. Second, if we are to anticipate and avoid undesirable ecological surprises, better models and more empirical evidence are needed on the links between ecosystem services and interactions among different trophic levels. Third, we need to reinforce the systematic screening for functional traits of organisms likely to have ecosystem-level consequences. In this sense, our knowledge of how the presence and local abundance of organisms (especially plants) bearing certain attributes affect ecosystem processes has made considerable progress in the past few years. However, we know much less of how the range of responses to environmental change among species affecting the same ecosystem function contributes to the preservation of ecosystem processes and services in the face of environmental change and uncertainty [16, 26]. This is directly relevant to risk assessment of the sustained provision of ecosystem services. Fourth, experimental designs for studying links between biodiversity and ecosystem processes and services need to not only meet statistical criteria but also mimic biotic configurations that appear in real ecosystems as a result of common land-use practices (e.g., primary forest versus monospecific plantations versus enrichment planting, or grazing-timber agroforestry systems versus a diverse grazing megafauna versus a single grazer such as cattle). In pursuing this, traditional knowledge systems and common management practices provide a valuable source of inspiration to develop new designs and testable hypotheses [27, 28]. Finally, in order to assist policy decisions and negotiation among different local, national, and international stakeholders, considerable advance is needed in the evaluation and accounting of ecosystem services [29, 30]. The challenge here is to find ways to identify and monitor services that are as concrete as possible, but at the same time not alienate the view of less powerful social actors or bias the analysis against services that are difficult to quantify or grasp.

The Bottom Line

By affecting the magnitude, pace, and temporal continuity by which energy and materials are circulated through ecosystems, biodiversity in the broad sense influences the provision of ecosystem services. The most dramatic changes in ecosystem services are likely to come from altered functional compositions of communities and from the loss, within the same trophic level, of locally abundant species rather than from the loss of already rare species. Based on the available evidence, we cannot define a level of biodiversity loss that is safe, and we still do not have satisfactory models to account for ecological surprises. Direct effects of drivers of biodiversity loss (eutrophication, burning, soil erosion and flooding, etc.) on ecosystem processes and services are often more dramatic than those mediated by biodiversity change. Nevertheless, there is compelling evidence that the tapestry of life, rather than responding passively to global environmental change, actively mediates changes in the Earth's life-support systems. Its degradation is threatening the fulfillment of basic needs and aspiration of humanity as a whole, but especially, and most immediately, those of the most disadvantaged segments of society.

References

1. Baillie JEM, Hilton-Taylor C, Stuart SN (2004) IUCN Red List of Threatened Species: A Global Species Assessment. Gland (Switzerland): IUCN.
2. Mace G, Masundire H, Baillie J, Ricketts T, Brooks T, et al. (2005) Biodiversity. In: Hassan R, Scholes R, Ash N, editors. Ecosystems and human well-being: Current state and trends: Findings of the Condition and Trends Working Group. Washington (D. C.): Island Press. pp. 77–122.
3. Díaz S, Tilman D, Fargione J, Chapin FI, Dirzo R, et al. (2005) Biodiversity regulation of ecosystem services. In: Hassan R, Scholes R, Ash N, editors. Ecosystems and human well-being: Current state and trends: Findings of the Condition and Trends Working Group. Washington (D. C.): Island Press. pp. 297–329.
4. Millennium Ecosystem Assessment (2005) Ecosystems and human well-being: Biodiversity synthesis. Washington (D. C.): World Resources Institute. 86 p.
5. Stokstad E (2005) Ecology: Taking the pulse of earth's life-support systems. Science 308: 41–43.
6. Kremen C (2005) Managing ecosystem services: What do we need to know about their ecology? Ecol Lett 8: 468–479.
7. Kotiaho JS, Kaitala V, Komonen A, Paivinen J (2005) Predicting the risk of extinction from shared ecological characteristics. Proc Natl Acad Sci U S A 102: 1963–1967.

8. McKinney M, Lockwood J (1999) Biotic homogenization: A few winners replacing many losers in the next mass extinction. Trends Ecol Evol 14: 450–453.
9. Grime JP (2001) Plant strategies, vegetation processes, and ecosystem properties. New York: John Wiley & Sons. 417 p. Chichester (United Kingdom).
10. Eviner VT, Chapin FSFutuyma DJ (2003) Functional matrix: A conceptual framework for predicting multiple plant effects on ecosystem processes. Annual review of ecology evolution and systematics, volume 34. Palo Alto (California): Annual Reviews 455–485.
11. Díaz S, Hodgson JG, Thompson K, Cabido M, Cornelissen JHC, et al. (2004) The plant traits that drive ecosystems: Evidence from three continents. J Veg Sci 15: 295–304.
12. Garnier E, Cortez J, Billès G, Navas ML, Roumet C, et al. (2004) Plant functional markers capture ecosystem properties during secondary succession. Ecology 85: 2630–2637.
13. Hooper DU, Chapin FS, Ewel JJ, Hector A, Inchausti P, et al. (2005) Effects of biodiversity on ecosystem functioning: A consensus of current knowledge. Ecol Monogr 75: 3–35.
14. Tilman D (1999) Diversity and production in European grasslands. Science 286: 1099–1100.
15. Phillips OL, Hall P, Gentry AH, Sawyer SA, Vasquez R (1994) Dynamics and species richness of tropical rain-forests. Proc Natl Acad Sci U S A 91: 2805–2809.
16. Elmqvist T, Folke C, Nystrom M, Peterson G, Bengtsson J, et al. (2003) Response diversity, ecosystem change, and resilience. Frontiers in Ecology and the Environment 1: 488–494.
17. Trenbath B (1999) Multispecies cropping systems in India: Predictions of their productivity, stability, resilience and ecological sustainability. Agroforestry Systems 45: 81–107.
18. Altieri M (2004) Linking ecologists and traditional farmers in the search for sustainable agriculture. Frontiers in Ecology and the Environment 2: 35–42.
19. Estes JA, Tinker MT, Williams TM, Doak DF (1998) Killer whale predation on sea otters linking oceanic and nearshore ecosystems. Science 282: 473–476.
20. Maron JL, Estes JA, Croll DA, Danner EM, Elmendorf SC, et al. (2006) An introduced predator alters Aleutian Island plant communities by thwarting nutrient subsidies. Ecol Monogr 76: 3–24.
21. Mooney HA, Mack RN, McNeely J, Neville LE, Schei PJ, et al. (2005) Invasive alien species: A new synthesis. Washington (D. C.): Island Press. 368 p.
22. Walker B, Meyers JA (2004) Thresholds in ecological and social-ecological systems: A developing database. Ecology and Society 9. Available: www.ecologyandsociety.org/vol9/iss2/art3. Accessed 23 June 2006.
23. Lambin EF, Geist HJ, Lepers E (2003) Dynamics of land-use and land-cover change in tropical regions. Annual Review of Environment and Resources 28: 205–241.
24. Naylor RL, Goldburg RJ, Primavera JH, Kautsky N, Beveridge MCM, et al. (2000) Effect of aquaculture on world fish supplies. Nature 405: 1017–1024.
25. Danielsen F, Sorensen MK, Olwig MF, Selvam V, Parish F, et al. (2005) The Asian tsunami: A protective role for coastal vegetation. Science 310: 643.
26. Lavorel S, Garnier E (2002) Predicting changes in community composition and ecosystem functioning from plant traits: Revisiting the Holy Grail. Funct Ecol 16: 545–556.
27. Díaz S, Symstad AJ, Chapin FS, Wardle DA, Huenneke LF (2003) Functional diversity revealed by removal experiments. Trends Ecol Evol 18: 140–146.
28. Scherer-Lorenzen M, Potvin C, Koricheva J, Bornik Z, Hector A, et al. (2005) The design of experimental tree plantations for functional biodiversity research. In: Scherer-Lorenzen M, Körner C, Schulze ED, editors. The functional significance of forest diversity. Berlin: Springer-Verlag. pp. 377–389.
29. DeFries R, Pagiola S, Adamowicz W, Resit Akçakaya H, Arcenas A, et al. (2005) Ecosystems and human well-being Current state and trends: Findings of the Condition and Trends Working Group. In: Hassan R, Scholes R, Ash N, editors. Analytical approaches for assessing ecosystem conditions and human well-being. Washington (D. C.): Island Press. pp. 37–71.
30. Boyd J, Banzhaf S (2006) What are ecosystem services? The need for standardized environmental accounting units. Resources for the Future. Washington (D. C).

Critical Thinking

1. Do you believe that biodiversity loss is a threat to human well-being?
2. Evaluate the statement: "Ecological surprises are difficult to predict, since they usually involve novel interactions among species."
2. If it is not possible to "define of level of biodiversity loss that is safe," what options are available for humanity?

Sandra Díaz is principal researcher and associate professor of ecology and biogeography at Instituto Multidisciplinario de Biología Vegetal (CONICET-UNC) and FCEFyN, Universidad Nacional de Córdoba, Argentina. Joseph Fargione is research assistant faculty at the Department of Biology, University of New Mexico, Albuquerque, New Mexico, United States of America. F. Stuart Chapin III is professor of ecology at the Institute of Arctic Biology, University of Alaska at Fairbanks, Fairbanks, Alaska, United States of America. David Tilman is the McKnight Presidential Chair in Ecology at the Department of Ecology, Evolution and Behavior, University of Minnesota, St. Paul, Minnesota, United States of America.

Acknowledgments—We are grateful to W. Reid, H. A. Mooney, G. Orians, and S. Lavorel for encouragement, inspiration, and critical comments during the process that led to this article, and to the leading authors of Millennium Ecosystem Assessment's Current State and Trends, chapter 11.

Soil Diversity and Land Use in the United States

RONALD AMUNDSON, Y. GUO, AND P. GONG

"A town is saved, not more by the righteous men in it than by the woods and swamps that surround it."

—H. D. Thoreau (2001)

Introduction

Concern over the fate of terrestrial biotic diversity in the face of increasing human domination of the planet (Vitousek and others 1997a) has focused mainly on the aboveground flora and fauna. Yet soils, the foundation of terrestrial ecosystems (Yaalon 2000), are rarely explicitly considered in these discussions. Soils are biogeochemically dynamic bodies, formed by the combined effects of environmental and biological factors over (commonly) geological expanses of time (Amundson and Jenny 1997). The combination of oscillating glacial/interglacial climates and unique floras and faunas which have controlled soil formation during the Quaternary period is unique in Earth's history, which suggests that exact analogs of present soils have not existed in the past nor will they form again in the future. For these reasons and more, it seems prudent that undisturbed soils—and their values and services—be given careful consideration in the development of bio-and geodiversity planning (Amundson 1998, 2000; Ibáñez and others 1995).

Here we present the first quantitative analysis of the human impact upon soil diversity for the US. We then discuss the significance of these findings and the importance of maintaining natural soil diversity. We used a Geographical Information Systems (GIS)-based approach to the problem, combining digital data on soil distribution and land use in the US. The results reveal both the nation's original soil geography and the patterns of heavily impacted soils in the US, providing a basis for identifying target areas for biogeodiversity preservation in this country.

Methods

Definition of Soil Diversity

Soil is a continuum (Jenny 1941), having properties that vary enormously—and continuously—with depth and with horizontal distance. For both the purposes of scientific study and land management applications, it has been a practice to classify soils by breaking the continuum into discrete units of similar properties. Here, soil diversity is quantified within the framework of the USDA Soil Taxonomy (Soil Survey Staff 1999), an international system of soil classification. The system was designed to separate soils on the basis of properties important to potential land use. As such, the system differs from scientific taxonomies where genetic linkages between objects are emphasized. The system contains five hierarchical levels that proceed from the most generalized (the soil "order") to most specific (the soil "family"). In the US, a final and more detailed extra taxonomic level is referred to as the soil "series." Any soil mapped in the US is usually given a series name and a taxonomic designation in all higher levels of the taxonomy. In comparison to biological taxonomy, the levels of the soil classification system might be viewed as proceeding from the "kingdom" (order) through "species" (series). The major distinguishing attributes of the soil orders are given in Table 1. While soils (nonreplicating entities), and the soil classification system (practical, not scientific), differ from biological entities, this analogy is at least an organizing concept on which to begin this investigation.

Recently, Ibáñez and others (1995, 1998) have reviewed the concepts and definitions of soil diversity, exploring the possible application of biological diversity models to soil databases. Here, we use two simple numerical parameters to quantify US soil diversity: (1) "series density": number of series/area by state, and (2) "series abundance": total area of each soil series in a state. With respect to abundance criteria, we defined the

Table 1 Brief Description of the Characteristics of the Soil Orders Found in the STATSGO Database

Order	Characteristics
Alfisols	Clay-enriched B horizons with base saturation greater than 35 percent
Andisols	Formed from volcanic parent materials with unique chemical properties
Aridsols	Soils with observable weathering/chemical alteration in arid climates
Entisols	Soils lacking visible horizon development
Histosols	Composed primarily of organic materials
Inceptisols	Soils possessing some development not characteristic of other classes
Mollisols	Significant organic C accumulation and base saturation greater than 50 percent
Oxisols	Highly chemically altered soils of tropics
Spodosols	Coarse-textured soils of northern latitude forests bearing distinctive geochemical separation of Fe and Al compounds
Ultisols	Clay-enriched B horizons with base saturation greater than 35 percent
Vertisols	High concentrations of silicate clay exhibiting shrink–swell behavior

following categories of rare or uncommon soil series: (a) *rare soils*—less than 1,000 ha total area in US, (b) *unique soils* (for example, "endemic")—exist only in one state, and (c) *rare-unique soils*—occur only in one state, total area less than 10,000 ha. Finally, for those naturally rare soil series, we defined (d) *endangered soils* as those rare or rare-unique soil series that have lost more than 50 percent of their area to various land disturbances described below. The quantitative definition of these three classes is our first approximation for evaluating soil distribution, and as yet there is no accepted standard for defining soil rarity in the literature. Our analysis is focused on soil diversity by political boundary as opposed to ecosystem boundaries. We do this for several reasons. First, analyses of endangered plant and animal distributions are frequently made along political boundaries (Dobson and others 1997). Second, there are advantages to potential conservation planning, and public perception, when analyses are conducted by political boundary. Certainly, future analyses might also examine land use effects on soil diversity by ecosystem boundaries, such as the "Major Land Resource Area" which is embedded in the STATSGO soil database.

The criteria chosen for our definitions are first attempts to partition soils into categories deserving of attention, but, as we illustrate below, they appear to capture important features of land use effects on natural ecosystems. Our criteria for "endangered soils" could also be extended to all soils regardless of their original abundance.

GIS Data Acquisition and Use

The calculated area of each soil type in the State Soil Geographic Data Base (STATSGO)(1:250,000, compiled by the US National Resource Conservation Service; http:/www.ftw.usda.gov/stat_data.html) was used to calculate the area of soil types. The minimum map unit in STATSGO is 6.25 km^2, equivalent to square cells of 2.5 km size. There are 1–21 components (components are based on soil type, landscape characteristics, and other parameters) in each map unit, and the location of each component is not known. We summarized the component percentage and area (component percentage × polygon area) of soils at the different soil classification levels (order, suborder, great group, subgroup, family, series) in each map unit identifier (MUID). Finally, the area of soil in each MUID and in each state was tabulated to obtain the total area of the soil type in the US.

Soil disturbance in the US was determined using the National Land Cover Data (NLCD, 30 m resolution), interpreted from Landsat Thematic Mapper data acquired in the early 1990s as compiled by the USGS and EPA (http://mac.usgs.gov/mac/isb/pubs/factsheets/fs10800.html), to extract urban (low intensity, high intensity residential, commercial/industrial/transportation, and urban recreational grasses) and agricultural lands (orchards/vineyards/ non-natural woody, row crops, small grains, and fallow). The absence of a long-time series of these satellite data neglects or misclassifies lands that have revegetated from a previous agricultural use, which makes our assessment of land use conservative. Finally, we overlaid the disturbed land classes on each rare or rare-unique soil (defined above), on a state-by-state basis, to estimate the number and location of endangered soils.

Due to the nature of the STATSGO database, and soil mapping in general, there is an inherent uncertainty whether land use affects a given soil. First, because the exact location of soils in a MUID is not known, we have assumed all soils have an equal probability of being affected by the land-use types that apply to a polygon. This assumption can result in significant (but largely impossible to verify) errors in land-use status, particularly (we suspect) for soils of low occurrence. This error will obviously be reduced once future soil databases, which explicitly identify soil locations, become available on a state-or nationwide basis. They are not available now. A second source of error occurs because soil series with small areas are not included in the database. Thus, we likely underestimate the number of rare or endangered soils. Third, the number of soil series (or any other taxonomic class) present in an area increases with the detail at which the area is mapped. Parts of some states have been mapped at a reconnaissance level, and the soil diversity listed (in terms of series) is likely a substantial underestimate. Fourth, STATSGO polygons have a limit of 21 soils, and soils of small extent may not be fully represented in our analysis. Finally, all soil mapping is an inherently complex exercise, involving approximations due to scale limitations of the soil map (commonly 1:24,000) and some level of bias due to anticipated use of the survey. Soil mapping units used in the development of the STATSGO database may contain significant "inclusions" (that is different soils than the major soil type), and so the actual area of the named soil may be smaller than indicated. While all these inherent uncertainties are undesirable, there is currently no other means of quantitatively approaching the problem for the US. Therefore, all limitations considered, we emphasize that the uncertainty in our analysis is greatly outweighed by the insights that the results provide. Most of the uncertainties in our data err in the conservative direction, such that it is likely our assessment of endangered soils is actually an underestimate.

Results

Natural Soil Diversity

The spatial distribution of soil orders reflects the wide gradients of soil age, climate, and biota that systematically change across the nation. There are 11 soil orders, 52 suborders, 233 great groups, 1176 subgroups, 6226 families, and 13,129 series in the 50 states and Puerto Rico in the present STATSGO database (1997 edited version) (Table 1). The recently created 12th order, Gelisols, is present in the US but has not yet been incorporated into STATSGO (additionally, STATSGO data to the series level are not available for Alaska). At the order level, the most abundant (by area) are the Mollisols (soils that generally correlate with grassland vegetation) (207×10^6 ha) and the least abundant are the Oxisols (intensely weathered soils common to stable landforms in tropical environments) (0.2×10^6 ha). The relative abundance of soils in the US is not reflective of global patterns, given the nation's predominantly temperate setting.

Hawaii has soil representatives of all 11 orders, while California and Oregon have 10 orders. In terms of soil series, California has 1755 series, by far the largest number in the US, followed by Nevada (1354), Idaho (1083), Oregon (1075), and Utah (1006).

Land Use and Natural Soil Diversity

The USDA Soil Taxonomy is deliberately insensitive to land-use effects on soils (Soil Survey Staff 1999). Agricultural soils are intended to remain in their natural classification except under extreme cases of manipulation (deep ripping, chiseling, construction), in which case they may be grouped into Arents, a special suborder of Entisols ("recent" soils). In practice (K. Arroues personal communication), severely manipulated soils commonly remain in the same classification as their natural counterparts. Therefore, the mapped abundance of soils on soil maps is a reflection of predisturbance distribution and is not indicative of the present undisturbed areas.

Approximately 19 percent of the US is under intensive agriculture (Census of Agriculture 1997). Land use in the US is unevenly distributed, with agriculture particularly concentrated in the Midwest, Great Plains, Mississippi Valley, Snake River/ Palouse regions and California's Great Valley. A much smaller percentage of the US is urbanized (approximately 2–3 percent) (see for example Nizeyimana and others 2001), but urban growth poses a particular threat to soil resources in the loss of prime agricultural land (Sorenson and others 1997; Imhoff and others 1997; Nizeyimana and others 2001), an important issue, but a topic outside the focus of this article. In most areas, geologically young, level, and highly productive soils are preferentially used for both agriculture and urbanization, a situation that leads to drastic reductions in the area of certain soil types.

At the order and suborder levels of the US taxonomy, the results of development have resulted in certain soil types being more heavily affected by land use than others. The total undisturbed area of four soil orders has been reduced by more than 20 percent: Mollisols (28 percent), Histosols (24 percent), Vertisols (24 percent), and Alfisols (22 percent) (Table 2). At the suborder level, there is also an uneven effect of land use on soils. First, most Mollisol suborders are heavily utilized for agriculture, as is expected due to their inherent high fertility and suitable climate. Second, it is evident that virtually every "aquic" subclass of all orders (soils with at least seasonally high water tables and features indicative of saturation) are preferentially utilized for agriculture.

Rare, Unique, and Endangered Soils in the US

We found 4540 rare or rare-unique soil series, 35 percent of the total in the US. California has the largest number of unique soil series (1113), followed by Washington (712), Texas (630), Nevada (594), Oregon (573), and Idaho (547). The overall diversity and "soil endemism" within California is understandable in terms of the wide range and unique combinations of climate, flora, and geology within the state. It is likely these same combinations of factors have contributed to the biological diversity and high endemism of the region (Myers and others 2000; Cincotta and others 2000). California also leads the nation in terms of rare or rare-unique soil series (671), followed by Washington (462), Nevada (399), Idaho (361), and Oregon (301). In terms of the "density" of rare or unique soils (soils/area), the territory of Puerto Rico leads the states/territories (approximately 15 unique series/100,000 ha), followed by Hawaii (10), Washington (3), Idaho (2), and California (2).

There are 508 endangered soils in the US with a total area of 1,874,092 ha, about 0.3 percent of the US land area. With respect to endangered soils, California leads the nation with 104 endangered soil series, followed by Minnesota (65), Idaho (49), Indiana (36), and Illinois (29). In terms of endangered soil density, Indiana leads this category (0.4 series/100,000 ha), followed by Connecticut (0.3), Minnesota (0.3), California (0.3), and Idaho (0.2). In the US, there are presently 31 soils that may be considered "extinct" (90–100 percent land conversion) 27 converted to agriculture and 4 to urban uses.

Six states have more than 50 percent of their rare soil series in an endangered state, with Indiana leading the group at 82 percent, followed by Iowa (81 percent), Illinois (66 percent), Nebraska (61 percent), Minnesota (53 percent), and Connecticut (50 percent). In general, the corn and wheat belt states (plus Connecticut) comprise a group of states where more than 25 percent of their rare soils are endangered. California, in contrast, has only 15 percent of its rare soils in an endangered state. A large group of states have no endangered soils.

The data indicate that the "hotspots" of soil in danger of elimination reside in the agricultural heartland of the country, a land-use conflict that has been recently discussed in terms of its effect on biological diversity (Margules and Gaston 1994; and Huston 1994). We find that the prevalence of endangered soils increases with the annual value of agricultural products produced by the state, a relationship noted by Dobson and others (1997) for endangered plants. Considerable variability exists in this relationship, but the most notable exceptions are the states of California and Texas, both with high production values but relatively low percentage of rare soils endangered. In California, agriculture is concentrated primarily in the large structural basins of the Great Central, Salinas, and Imperial Valleys—geographically restricted areas of high-intensity agriculture and high-value products, characteristics which reduce the impact on remaining portions of the state. Connecticut is a second type of outlier (a relatively high percentage of endangered soils coupled with a low agricultural output), possibly due to the state's relatively small land area and the combined impact of agriculture and urbanization in its lowland corridor. Regardless of the exceptions, it is apparent that agricultural output and endangered soils are positively correlated. Although agriculture is the main mechanism reducing soil diversity, urbanization (despite its low total land area) is responsible for nearly 33 endangered soils nationally.

Discussion

The US has enormous natural soil diversity, yet our investigation indicates that these data are not representative of the extant, or undisturbed, soil resources. We argue that a change in land use results in profound changes in one or more soil properties and in a soil's biogeochemical functioning, to the degree that the soil is no longer representative of its undisturbed counterpart. Again, using a biological analogy, cultivated soils might be viewed as domesticated versions of their natural counterparts, with widely differing properties and functions. As we discuss below, these changes and others are important for an array of scientific and societal reasons.

Why Natural Soil Diversity Matters

Successful arguments for scientifically based conservation/preservation plans rely on both what we know and what we don't know about a habitat's values and services (Noss and others 1997). In terms of values that are cited to warrant the preservation of biodiversity, Ehrlich and Wilson (1991) identified aesthetic, ethical, economic, and ecosystem services as key areas of consideration and importance to human society. These reasons and more, including scientific and educational benefits, have a bearing on soil preservation efforts and their value to society. However, it is important to be clear that there is a dearth of information on why natural soils as a whole, or specific natural soil series individually, are quantitatively important to society. The primary reason for this is not due to an inherently low value of natural soils, but that these analyses have not been an imperative in a field (soil science and agronomy) whose driving forces have focused on the effective and efficient conversion of functions. As we discuss below, these changes and natural landscapes to farmland. Clearly, this focus will continue well into

Table 2 Percentage of Total Soil Orders and Suborders Affected by Development in the US

	%				%		
Order	**Urban**	**Agriculture**	**Total**	**Suborder**	**Urban**	**Agriculture**	**Total**
Alfisols	2.85	19.37	22.22	Aqualfs	4.13	28.62	32.75
				Boralfs	0.48	10.50	10.98
				Udalfs	4.02	24.84	28.86
				Ustalfs	1.07	16.88	17.95
				Xeralfs	3.27	8.00	11.27
Andisols	0.35	1.96	2.30	Andepts	0.36	0.49	0.85
				Aquands	0.52	11.18	11.70
				Cryands	0.21	0.75	0.95
				Torrands	1.30	0.65	1.94
				Udands	0.53	0.08	0.61
				Vitrands	0.16	1.15	1.31
				Xerands	0.32	1.06	1.38
Aridisols	0.61	5.46	6.07	Argids	0.61	3.94	4.55
				Orthids	0.63	5.12	5.75
Entisols	2.38	17.34	19.72	Aquents	3.89	22.24	26.13
				Arents	9.19	15.81	25.00
				Fluvents	1.94	21.24	23.18
				Orthents	1.59	13.27	14.86
				Psamments	3.49	12.79	16.28
Histosols	3.51	20.69	24.20	Fibrists	0.54	2.45	3.00
				Folists	0.28	0.35	0.63
				Hemists	2.36	8.13	10.49
				Saprists	3.60	22.44	26.05
Inceptisols	2.55	12.27	14.82	Aquepts	3.57	17.70	21.26
				Ochrepts	2.50	10.52	13.03
				Tropepts	2.28	0.21	2.49
				Umbrepts	1.32	1.02	2.34
Mollisols	1.81	25.69	27.50	Albolls	1.16	54.29	55.45
				Aquolls	2.99	44.95	47.94
				Borolls	0.38	23.61	23.99
				Rendolls	1.39	4.91	6.29
				Udolls	3.35	41.92	45.28
				Ustolls	1.14	23.52	24.65
				Xerolls	1.90	7.86	9.76
Spodosols	3.99	8.60	12.58	Aquods	5.98	12.43	18.42
				Cryods	0.29	0.18	0.46
				Humods	0.88	1.57	2.45
				Orthods	2.67	5.97	8.64
Ultisols	3.05	8.63	11.68	Aquults	3.63	15.93	19.56
				Humults	0.92	1.01	1.93
				Udults	3.15	9.06	12.21
				Ustults	3.00	2.23	5.23
				Xerults	1.30	1.80	3.10
Vertisols	2.35	21.31	23.67	Aquerts	1.77	49.12	50.89
				Torrcrts	0.38	3.11	3.49
				Uderts	5.65	18.16	23.81
				Usterts	1.67	25.62	27.29
							17.95
				Xererts	5.73	12.22	

this century as we enter societal reasons. what has been termed *"the final period of rapidly expanding, global human environmental impacts,"* particularly in the area of agricultural expansion to meet food demands (Tilman and others 2002). However, a fundamental institutional shift in these sciences is required to quantify and derive societal value from remaining natural soils and ecosystems and to provide the scientific basis to argue for their preservation.

Daily (1997) suggests that there are three ways in which science may contribute to a public understanding of biodiversity and ecosystem-related values: (a) establish standard metrics and systematic monitoring of the magnitude and rates of human change of ecosystems, (b) use the metrics to project the way in which change affects the functioning of ecosystems, and (c) translate the change into meaningful social terms of human health and economic well-being. In this article, which addresses the issue of soil diversity, we have thus far focused primarily on the first goal. Here, using a few examples, we attempt to provide a glimpse into the value of the "services" of natural soils as a population, for there are simply too few data to discuss specific soil types individually.

Daily (1997) lists the production of goods and services as a key ecosystem value. The economic value of soils has centered on agricultural lands. For example, the total value of US agricultural production (much derived from cultivated lands) in 1997 was $197 billion (Census of Agriculture 1997). The monetary value that can be derived from products from uncultivated soils (beyond grazing/timber) has not been fully examined. One potential source of income may lie in "bio-prospecting" for medicinal and industrial purposes. Antibiotics, such as streptomycin (Yaalon 2000), continue (for example, see Fielfer and others 2000) to be derived from soil microorganisms. However, there is a fundamental lack of knowledge on the geographical distribution of soil microbial diversity. Though it is commonly argued that global dispersal of microorganisms is *"rarely (if ever) restricted by geographical barriers"* (Findlay 2002), detailed genetic analyses of widely dispersed soils reveals a different pattern, one of *"strong endemicity suggesting that heterotrophic soil bacteria are not globally mixed"* (Cho and Tiedje 2000). The effect of land use on microbial biodiversity, using modern genetic tools, is also a poorly studied area. However, a recent study examining the conversion of a tropical rainforest to pasture in Hawaii showed that there was a 49 percent change in the microbial populations as a result of the land-use change (Nüsslein and Tiedje 1999). Given the uncertainty about below-ground soil microbial diversity (Adams and Wall 2000), these large reported changes warrant further research and ecological consideration but seem to justify concern over the effect of cultivation on below-ground genetic resources.

It is well recognized that undisturbed soils, and ecosystems, provide societal benefits from *"regenerative and stabilizing processes"* (Daily 1997; Daily and others 1997) in the form of water and elemental cycling on a global scale, regulating the climate, atmosphere, and hydrosphere conditions that allowed human society to expand so successfully during the Holocene. Tilman and others (2002) note that these services are *"difficult to quantify and have rarely been priced."* The magnitude of these services is sometimes most appreciated following their loss (Daily and others 2000). For example, in terms of C cycling, the cultivation of present agricultural lands worldwide has released approximately 55–70 Gt of C as CO2 (Paustian and others 1997; Amundson 2001), equivalent to approximately 12 years of present day fossil fuel burning. In terms of the N cycle, the mobilization of N from agricultural lands may be on the order of 4800 Tg N (10^{12}/g) (assuming N release during soil organic matter decomposition proportional to C). Currently, land clearing and drainage of wetlands may be releasing 30 Tg/y (Vitousek and others 1997b). While the "value" of pristine global soil conditions is possibly debatable, the costs (monetary and human health) associated with its disruption in terms of climate change, water quality, and disease is a focus of considerable international attention and analysis (IPCC 2001).

Two attributes of soil diversity that may be compelling to many are (a) the linkage between rare soils and rare plants and (b) the linkage between endangered soils and plants, relationships that intimately link soil and biodiversity preservation arguments and planning. As an example of the relationship of rare soils to plants, we discuss the annual grasslands of eastern Merced County, California, a region that is an integral part of the California Floristic Province, one of the top 25 biodiversity hot-spots on Earth (Cincotta and others 2000; Myers and others 2000). The region near the city of Merced is a complex mosaic of river/steam terraces or floodplains, that range in age from 10^2 to 10^6 years (Marchand and Allwardt 1981), with extensive areas of vernal pools. A sizable number of endemic species form in these pools (Volmar 2002). There is a systematic change in pool frequency and soil chemistry (Brenner and others 2001) with time that creates an edaphic gradient that is a major factor in influencing the plant species composition on a regional scale (Holland and Dains 1990). Holland and Dains (1990) found distinctive edaphic preferences of the regional flora related to soil age and parent material and concluded that *"attempts to mitigate the effects of proposed developments by transplanting whole vernal pool ecosystems to off site locations can not hope to succeed by use of blanket prescriptions to mimic soil profile conditions at the target site.*" Because of the plant preferences for specific soils near Merced, and the fact that the Merced area has the last remaining tracts of undisturbed soils peculiar to the oldest landforms, the initial science panel input into the Natural Community Conservation Plan (NCCP) and Habitat Conservation Plan (HCP) for the area emphasizes the need to establish a reserve design that explicitly includes a diversity of undisturbed soils (Noss and others unpublished). More generally, the relation between edaphic factors and biological diversity has been discussed in greater lengths elsewhere (Huston 1993; Margules and others 1994; Huston 1994), emphasizing both the linkages and the complexities.

Both plants and soils become endangered as a result of land use. The plants and soils clearly exhibit geographical differences, but also some similarities. Counties with endangered plants cover wide areas that contain few or no endangered soils. A major reason for this is that officially listed endangered plants are reportedly caused by grazing, logging, and other land uses (Flather and others 1994) which we have not deemed as affecting soil diversity. If our criteria for minimal disturbance that changes inherent natural soil processes had not been so conservative, the areas of endangered soils in the US would have certainly expanded and had greater overlap with the plants. However, even with our restricted approach, it is clear that both endangered plants (Flather and others 1994; Dobson and others 1997) and soils occur in heavily cultivated and urbanized areas. We note that Dobson and others (1997) reported that "agricultural activity is the key variable for plants (endangerment) ($r^2=0.61$, $P<0.01$)," a relationship with which the soil data are consistent.

While the discussion above gives a few scholarly and scientific reasons justifying the preservation of natural soils, it is equally valid to argue that a diversity of natural soils be maintained because we lack a scientific understanding of their full values and functions. This justification, called the "precautionary principle" in habitat conservation plans (Sharder–Frechette and McCoy 1993; Noss and others 1997) shifts the burden of proof for preservation from conservationists to developers in order to reduce the possibility that a "Type II" error occurs (acceptance of conclusion that no effect from land use occurs when one actually exists). The concern over Type II errors is argued to be particularly relevant to applied sciences (medicine, environmental engineering, and conservation biology) where such an error causes irreversible damage to the patient, ecosystem, or soil (Noss and others 1997). "*Nature is full of surprises*" (Noss and others 1997), and, in terms of natural soil types, numerous surprises and benefits will undoubtedly reveal themselves in the future. However, our present ignorance should not be an impediment to arguing that landscapes warrant preservation. The conservation of diverse soilscapes should proceed simultaneously with scientific research that fully explores their qualities, values, and functioning.

Conclusions

In less than two centuries, the landscape of the US has been transformed to a degree that would astound our 19th century predecessors. The change is not complete. Population growth and the on-going redistribution of the US population pose new and challenging issues to preservation efforts of all types. Soils, viewed during this expansion as an economic commodity, have in many cases become rare to the degree that we and others (Ibáñez and others 1995, 1998) suggest they become part of formal biodiversity planning.

This initial quantitative analysis of land-use effects on soil diversity raises many questions and opportunities for future research. Some key directions include:

1. *Use of improved soil databases:* Our analysis, made with the generalized STATSGO database, contains numerous uncertainties as discussed in the text. In the near future, the USDA–NRCS will likely release the SSURGO (Soil Survey Geographical) database, which will be a digital compilation of detailed 1:24,000 soil surveys. This dataset will greatly refine both the number and location of heavily impacted soils in the US. However, the general method of our initial query using STATSGO will apply to the use of this improved database.
2. *Examination of soil distribution by ecological rather than political boundaries:* As a result of our own interest in soil distribution by state, and for our study to parallel national analyses of endangered flora and faunas (Dobson and others 1997), we have used political boundaries to quantify soil diversity. The STATGO/SSURGO databases offer the opportunity to examine soil distribution along general ecological boundaries—Major Land Resources Areas (MRLAs). Future work, which we are already initiating (Guo and others 2003), should further explore soil diversity by ecological region.
3. *Monitor changes in soil diversity in response to changing land-use:* Here we provided a snapshot of a land-use/soil overlay based on 1990s land-use patterns in the US. As Daily (1997) notes, it will

be important to periodically monitor changes of agriculture and urbanization in the future in order to quantify the rate of soil change.

4. *Extend the method to global scales:* While the US is certain to experience land-use changes in the future, it might be argued that the present areas of intensive agricultural expansion are the tropics and subtropics. Unfortunately, geographically referenced soil data are unavailable for the globe, highlighting the unique value that the STATSGO database provides for the US.
5. *Establish the societal value of undisturbed soils:* Following the lead of ecologists (Daily and others 1997, 2000), other scientists and economists must begin to assign value to undisturbed soils capes and make the value known to the general public. This must also include the value of present farmland in rapidly urbanizing areas, since the loss of this land increases the pressure to agriculturally develop native landscapes elsewhere.
6. *Focus conservation efforts in soil diversity "hotspots":* Our analysis clearly shows that certain areas—the Midwestern states for example—have been severely impacted by human activity. Efforts to locate and maintain tracts of undisturbed soils and ecosystems there deserve immediate and special attention.

Soils are integral components of terrestrial ecosystems, providing global-scale services in elemental cycling, water purification, genetic diversity, and more (Daily and others 1997). The importance of soil as an agricultural resource has been successfully cast to capture the public imagination and support (Sorenson and others 1997). The key now is to extend these arguments and to direct attention to the Earth's remaining natural soils. We conclude by recognizing that even these "pristine" sites commonly support an array of invasive species and experience atmospheric N inputs, fire regimes, and now (and into the future) climatic conditions different than in preindustrial society (for example, Vitousek and others 1997a). At this stage in human history, simply minimizing the human footprint on these regions of the landscape constitutes a modest environmental obligation to, and inheritance for, future generations.

References

Adams GA, Wall DH. 2000. Biodiversity above and below the surface of soils and sediments: linkages and implications for global change. Bioscience 50:1043–8.

Amundson R. 1998. Do soils need our protection? Geotimes March: 16–20.

Amundson R. 2000. Are Soils Endangered? In: Schneiderman J, editor. The Earth Around Us, Maintaining a Livable Planet. New York: WH Freeman. pp 144–53.

Amundson R. 2001. The carbon budget in soils. Annu Rev Earth Planet Sci 29:535–62.

Amundson R, Jenny H. 1997. On a state factor model of ecosystems. Bioscience 47:536–43.

Brenner DL, Amundson R, Baisden WT, Kendall C, Harden J. 2001. Soil N and ^{15}N variation with time in a California annual grassland ecosystem. Geochim Cosmochim Acta 65:4171–86.

Census of Agriculture. 1997: www.census.gov/econ/www/ag0100.html.

Cho J-C, Tiedje JM. 2000. Biogeography and degree of endemicity of fluorescent *Pseudomonas* strains in soil. Appl Envir Microbiol 66:5448–56.

Cincotta RP, Wisnewski J, Engelman R. 2000. Human population in the biodiversity hotspots. Nature 404:990–2.

Daily GC. 1997. Developing a scientific basis for managing Earth's life support systems. Conserv Ecol 3:14. [online] URL: www.consecol.org/vol3/iss2/art14.

Daily GC, Matson PA, Vitousek PM. 1997. Ecosystem services supplied by soil. In: Daily GC, editor. Natures Services. Washington, DC: Island Press. pp 113–32.

Daily GC, Söderqvist T, Aniyar S, Arrow K, Dasgupta P, Ehrlich PR, Folke C, Jansson A, Jansson B-O, Kautsky N, Levin S, Lubchenco J, Mäler K-G, Simpson D, Starrett D, Tilman D, Walker B. 2000. The value of nature and the nature of value. Science 289:395–96.

Dobson AP, Rodriguez JP, Roberts WM, Wilcove DS. 1997. Geographic distribution of endangered species in the United States. Science 275:550–3.

Ehrlich PR, Wilson EO. 1991. Biodiversity studies: science and policy. Science 2:758–62.

Finlay BJ. 2002. Global dispersal of free-living microbial eukaryote species. Science 296:1061–63.

Flather CH, Joyce LA, Bloomgarden CA. 1994. Species endangerment in the United States. RM-241. General Technical Report, Rocky Mountain Forest and Range Experimental Station. Ft. Collins, CO: US Department of Agriculture.

Guo Y, Amundson R, Gong P, Ahrens R (2003) Taxonomic structure, distribution, and abundance of the soils in the United States (in review).

Holland RF, Dains VI. 1990. The edaphic factor in vernal pool vegetation. In: Ikeda DH, Schlising RA, editors. Vernal Pool Plants: Their Habitat and Biology. Studies from the Herbarium, Number 8. Chico, CA: CSU. pp 31–48.

Huston M. 1993. Biological diversity, soils, and economics. Science 262:1676–80.

Huston M. 1994. Biological diversity and agriculture. Science 265:458–59.

Ibáñez JJ, De-Alba S, Bermúdez FF, García–Álvarez A. 1995. Pedodiversity: concepts and measures. Catena 24:215–32.

Ibáñez JJ, De-Alba S, Lobo A, Zucarello V. 1998. Pedodiversity and global soil patterns at coarse scales. Geoderma 83:171–92.

Imhoff ML, Lawrence WT, Elvidge CD, Paul T, Levine E, Privalsky MV, Brown V. 1997. Using nighttime DMSP/OLS images of city lights for estimating the impacts of urban land use on soil resources in the United States. Remote Sens Environ 59:105–17.

IPCC. 2000. Land Use, Land-use Change, and Forestry. Summary for Policy Makers. Intergovernmental Panel on Climate Change, World Meteorological Organization/United Nations Environment Programme.

Jenny H. 1941. Factors of Soil Formation. New York: McGraw-Hill.

Marchand DE, Allwardt A. 1981. Later Cenozoic stratigraphic units in northeastern San Joaquin Valley, California. US Geological Survey Bulletin 170. Washington, DC: US G P U.

Margules CR, Gaston KJ. 1994. Biological diversity and agriculture. Science 265:457.

Myers N, Mittermier RA, Mittermeier CG, da Fonseca GAB, Kent J. 2000. Biodiversity hotspots for conservation priorities. Nature 403:853–8.

Nizeyimana EL, Peterson GW, Imhoff ML, Sinclair HR Jr., Walt-man SW, Reed– Margetan DS, Levine ER, Russo JM. 2001. Assessing the impact of land conversion to urban use on soils with different productivity levels in the USA. Soil Sci Soc Am J 65:391–402.

Noss R, O'Connell MA, Murphy DD. 1997. The Science of Conservation Planning. Habitat Conservation Under the Endangered Species Act. Washington, DC: Island Press.

Nüsslein K, Tiedje JM. 1999. Soil bacterial community shift correlated with change from forest to pasture vegetation in a tropical soil. Appl Environ Microbiol 65:3622–6.

Paustian K, Andrèn O, Janzen HH, Lal R, Smith P, Tian G, Tiessen H, Van Noordwijk M, Woomer PL. 1997. Agricultural soils as a sinck to mitigate CO2 emissions. Soil Use Manage 13:230–244.

Pfeifer BA, Admiraal SJ, Gramajo H, Cane DE, Khosla C. 2000. Biosynthesis of complex polyketides in a metabolically engineered strain of E. coli. Science 291:1790–2.

Shrader–Frechette KS, McCoy ED. 1993. Method in Ecology: Strategies for Conservation. Cambridge, UK: Cambridge University Press.

Soil Survey Staff. 1999. Soil Taxonomy. A Basic System of Soil Classification for Making and Interpreting Soil Surveys. Agric. Handbk. 436. US Department of Agriculture., Nat Res Cons Serv. Washington, DC: US GPO.

Sorenson AA, Greene RP, Russ K. 1997. Farming on the edge. DeKalb, IL: American Farmland Trust, Center for Agricultural Development, Northern Illinois University.

Thoreau HD. 2001. Collected Essay and Poems. Library of America.

Tilman D, Cassman KG, Matson PA, Naylor R, Polasky. 2002. Agricultural sustainability and intensive production practices. Nature 418:671–7.

Vitousek PM, Mooney HA, Lubchenco J, Mellilo JM. 1997a. Human domination of earth's ecosystems. Science 277:494–9.

Vitousek PM, Aber JD, Howarth RW, Likens GE, Matson PA, Schindler DW, Schlesinger WH, Tilman DG. 1997b. Human alterations of the global nitrogen cycle: sources and consequences. Ecol Appl 7:737–50.

Vollmar JE. 2002. Wildlife and Rare Plant Ecology of Eastern Merced County's Vernal Pool Grasslands. Berkeley, CA: Vollmar Consulting.

Yaalon D. 2000. Down to earth. Why soil—and soil science—matters. Nature 407:301.

Critical Thinking

1. Why do you think that soils have not been "explicitly considered" in analyses of Earth systems?
2. Evaluate the statement. ". . . there is a dearth of information on why natural soils as a whole, or specific natural soil series individually, are quantitatively important to society."
3. Do you agree that soils should "become part of formal biodiversity planning?"

Acknowledgments—We thank Kit Paris and Eric Vinson of the NRCS in California for assistance and advice on the STATSGO database. Stephen Howard at the USGS EROS Data Center provided assistance in describing the NLCD data structure. Mu Lan provided technical assistance, and Cristina Castanha, Stephanie Ewing, and Kyungsoo Yoo provided comments on an earlier draft of the paper. The research was supported by the Kearney Foundation of Soil Science.

UNIT 6

How Do We Correct Our Actions and Embrace Sustainablility?

Unit Selections

Learning Outcomes

After reading this unit, you should be able to:

- State, orally or in writing, the fundamental aspects of the prisoner's dilemma.
- Identify, by listing, ways to reward ecologically sustainable behaviors and ways to punish ecologically unsustainable behaviors.
- Identify, by listing, three belief changes that will support the transition to sustainability.
- Classify, by using essential characteristics, alternatives to the GDP.
- Identify, by listing, the strengths and weaknesses of alternatives to the GDP and of the GDP.
- State, orally or in writing, the basic concept of the Jevons paradox.
- State, orally or in writing, the implications of increasing carbon concentrations in Earth's atmosphere.
- Generate, by synthesizing relevant information, a paragraph arguing for and a paragraph arguing against holding consumers responsible for carbon emissions generated during production of a product.
- Demonstrate, by using available data and appropriate mathematics, the calculation of expiration dates for domestic and foreign crude, domestic natural gas, and domestic coal.
- Generate, by synthesizing relevant information, a paragraph distinguishing the difference between using less energy and using the same amount of energy from renewable sources.
- State, orally or in writing, the dilemma presented by the possibility of three billion people [moving] from low-impact to high-impact lifestyles.

Student Website

www.mhhe.com/cls

Internet References

GoingGreen
www.goinggreen.com

The Green Guide
http://environment.nationalgeographic.com/environment/green-guide

Original Green
www.originalgreen.org

Sustainable Communities Online
www.sustainable.org

YouSustain
www.yousustain.com/footprint/actions

As paradoxical as it may sound, apparently people must learn how to live sustainably on Earth. One might think that it should come naturally, but clearly it does not. So, how do we move forward and adjust our lives so that we don't degrade the life support systems of the planet to the extent that our own survival is jeopardized? Do we "appeal to the heart or to the head"? Article 31 explores this complex question.

Article 32 includes the suggestion that the transition to sustainability "would be a modification of society comparable in scale to only two other changes: the agricultural revolution . . . and the Industrial Revolution." If this is correct, then clearly societal changes will have to be instigated from multiple directions. This reading explores policies that can create behavior change.

If unsustainable actions are not accounted for when calculating the economic activity of a nation and if environmental degradation actually improves the perception of economic activity because of the costs associated with pollution clean-up, then how can the concept of sustainability ever gain a firm foothold? The Gross Domestic Product is the "monetary, market value of all final goods and services produced by a country over the period of one year." Whether it is the appropriate model for such considerations has been seriously debated. Article 33 presents and argument for the abolishment of it and provides suggestions for its replacement.

Famed economist Herman Daly suggested, "to do more efficiently that which should not be done in the first place is no cause for rejoicing." Curiously, the idea of improved efficiency is consistently offered as a solution to the challenge of sustainability. Even the rallying call of reduce, reuse, recycle advocates the least preferred option, recycling, over the most desirable option, reducing. The mindset seems to be that if we simply reduce the energy demands of our lifestyle without really changing it, then we are moving in the direction of sustainability. Is this perception valid? Article 34 examines this dilemma using the Jevons paradox to examine this question of efficiency versus frugality.

Insofar as sustainability is concerned, does it matter who is most responsible for degradation of Earth's life support systems? For example, if one nation produces a product that contributes to such degradation, but that product is used by another nation, who should be held accountable, the nation of origin or the nation of use? This is a challenging question. Article 35 examines it in the context of CO_2 emissions. This issue is of critical importance as nations wrestle with the question of how best to curb anthropogenic sources of carbon in the atmosphere.

One of the biggest issues facing humanity in the context of sustainability is how the growing population will be fed. Agriculture consumes enormous amounts of land and water; uses enormous amounts of fossil-fuel-derived synthetic fertilizers, pesticides, fungicides, herbicides, etc.; and enormous amounts of energy in the form of fossil fuels. A food item may travel over 1,000 miles before it reaches a person's plate. Is any of this sustainable? Article 36 introduces the idea of vertical farming as a possible model for agriculture of the future, a model that might represent a second agricultural revolution.

One of the challenges associated with sustainability is the perception that it will require substantial and unacceptable sacrifice; that we will need to return to the time when we lived in caves and ate grubs. That is not the case, but if we are to live more harmoniously on the planet and if we are to continue to enjoy the creature comforts that we currently enjoy and that we might enjoy in the future, we will need to figure out how to do so using less energy. What is needed are directions for how to reach that destination. What is needed is a blueprint for how to construct such a society. Article 37 is a contribution by the Union of Concerned Scientists that provides just that.

Article 38, the final reading of this publication provides the exclamation point for the one preceding it. It places sustainability in the context of the past, the present, and the future. It places it in the context of the contribution that the United States should, perhaps must, be making to the transition to it. It places it in the context of global affairs and current events. It provides the perfect concluding remark for this publication by offering, ". . . our parents [perhaps your grandparents] rose to . . . a challenge in World War II—when an entire generation mobilized to preserve our way of life. That is why they [are] called the Greatest Generation. Our [your] kids will only call us the Greatest Generation if we rise to our challenge and become the Greenest [most sustainable] Generation."

Can Selfishness Save the Environment?

Conventional wisdom has it that the way to avert global ecological disaster is to persuade people to change their selfish habits for the common good. A more sensible approach would be to tap a boundless and renewable resource: the human propensity for thinking mainly of short term self-interest

MATT RIDLEY AND BOBBI S. LOW

John Hildebrand, who has lived in the Artesian Valley, near Fowler, Kansas, since he was two years old, remembers why the valley has the name it does. "There were hundreds of natural springs in this valley. If you drilled a well for your house, the natural water pressure was enough to go through your hot-water system and out the shower head." There were marshes in Fowler in the 1920s, where cattle sank to their bellies in mud. And the early settlers went boating down Crooked Creek, in the shade of the cottonwoods, as far as Meade, twelve miles away.

Today the creek is dry, the bogs and the springs have gone, and the inhabitants of Fowler must dig deeper and deeper wells to bring up water. The reason is plain enough: seen from the air, the surrounding land is pockmarked with giant discs of green—quarter-section pivot-irrigation systems water rich crops of corn, steadily depleting the underlying aquifer. Everybody in Fowler knows what is happening, but it is in nobody's interest to cut down his own consumption of water. That would just leave more for somebody else.

Five thousand miles to the east, near the Spanish city of Valencia, the waters of the River Turia are shared by some 15,000 farmers in an arrangement that dates back at least 550 years and probably longer. Each farmer, when his turn comes, takes as much water as he needs from the distributory canal and wastes none. He is discouraged from cheating—watering out of turn—merely by the watchful eyes of his neighbors above and below him on the canal. If they have a grievance, they can take it to the Tribunal de las Aguas, which meets on Thursday mornings outside the Apostles' door of the Cathedral of Valencia. Records dating back to the 1400s suggest that cheating is rare. The *huerta* of Valencia is a profitable region, growing at least two crops a year.

Two irrigation systems: one sustainable, equitable, and long-lived, the other a doomed free-for-all. Two case histories cited by political scientists who struggle to understand the persistent human failure to solve "common-pool resource problems." Two models for how the planet Earth might be managed in an age of global warming. The atmosphere is just like the aquifer beneath Fowler or the waters of the Turia: limited and shared. The only way we can be sure not to abuse it is by self restraint. And yet nobody knows how best to persuade the human race to exercise self-restraint.

At the center of all environmentalism lies a problem: whether to appeal to the heart or to the head—whether to urge people to make sacrifices in behalf of the planet or to accept that they will not, and instead rig the economic choices so that they find it rational to be environmentalist. It is a problem that most activists in the environmental movement barely pause to recognize. Good environmental practice is compatible with growth, they insist, so it is rational as well as moral. Yet if this were so, good environmental practice would pay for itself, and there would be no need to pass laws to deter polluters or regulate emissions. A country or a firm that cut corners on pollution control would have no cost advantage over its rivals.

Those who do recognize this problem often conclude that their appeals should not be made to self-interest but rather should be couched in terms of sacrifice, selflessness, or, increasingly, moral shame.

We believe they are wrong. Our evidence comes from a surprising convergence of ideas in two disciplines that are normally on very different tracks: economics and biology. It is a convergence of which most economists and biologists are still ignorant, but a few have begun to notice. "I can talk to evolutionary biologists," says Paul Romer, an economist at the University of California at Berkeley and the Canadian Institute for Advanced Research, in Toronto, "because, like me, they think individuals are important. Sociologists still talk more of the action of classes rather than individuals." Gary Becker, who won the Nobel Prize in economics last year, has been reading biological treatises for years; Paul Samuelson, who won it more than twenty years ago, has published several papers recently applying economic principles to biological problems. And biologists such as John Maynard Smith and William Hamilton have been raiding economics for an equally long time. Not that all economists and biologists agree—that would be impossible. But there are emerging orthodoxies in both disciplines that are strikingly parallel.

The last time that biology and economics were engaged was in the Social Darwinism of Herbert Spencer and Francis Galton. The precedent is not encouraging. The economists used the biologists' idea of survival of the fittest to justify everything from inequalities of wealth to racism and eugenics. So most academics are likely to be rightly wary of what comes from the new entente. But they need not fear. This obsession is not with struggle but with cooperation.

For the Good of the World?

Biologists and economists agree that cooperation cannot be taken for granted. People and animals will cooperate only if they as individuals are given reasons to do so. For economists that means economic incentives; for biologists it means the pursuit of short-term goals that were once the means to reproduction. Both think that people are generally not willing to pay for the long-term good of society or the planet. To save the environment, therefore, we will have to find a way to reward individuals for good behavior and punish them for bad. Exhorting them to self sacrifice for the sake of "humanity" or "the earth" will not be enough.

This is utterly at odds with conventional wisdom. "Building an environmentally sustainable future depends on restructuring the global economy, major shifts in human reproductive behavior, and dramatic changes in values and lifestyles," wrote Lester Brown, of the Worldwatch Institute, in his *State of the World* for 1992, typifying the way environmentalists see economics. If people are shortsighted, an alien value system, not human nature, is to blame.

Consider the environmental summit at Rio de Janeiro last year. Behind its debates and agreements lay two entirely unexamined assumptions: that governments could deliver their peoples, and that the problem was getting people to see the global forest beyond their local trees. In other words, politicians and lobbyists assume that a combination of international treaties and better information can save the world. Many biologists and economists meanwhile assert that even a fully informed public, whose governments have agreed on all sorts of treaties, will still head blindly for the cliff of oblivion.

Three decades ago there was little dissonance between academic thinking and the environmentalists' faith in the collective good. Biologists frequently explained animal behavior in terms of the "good of the species," and some economists were happy to believe in the Great Society, prepared to pay according to its means for the sake of the general welfare of the less fortunate. But both disciplines have undergone radical reformations since then. Evolutionary biology has been transformed by the "selfish gene" notion, popularized by Richard Dawkins, of Oxford University, which essentially asserts that animals, including man, act altruistically only when it brings some benefit to copies of their own genes. This happens under two circumstances: when the altruist and the beneficiary are close relatives, such as bees in a hive, and when the altruist is in a position to have the favor returned at a later date. This new view holds that there simply are no cases of cooperation in the animal kingdom except these. It took root with an eye-opening book called *Adaptation and Natural Selection* (1966), by George Williams, a professor of biological sciences at the State University of New York at Stony Brook. Williams's message was that evolution pits individuals against each other far more than it pits species or groups against each other.

By coincidence (Williams says he was unaware of economic theory at the time), the year before had seen the publication of a book that was to have a similar impact on economics. Mancur Olson's *Logic of Collective Action* set out to challenge the notion that individuals would try to further their collective interest rather than their short-term individual interests. Since then economics has hewed ever more closely to the idea that societies are sums of their individuals, each acting in rational self interest, and policies that assume otherwise are doomed. This is why it is so hard to make a communist ideal work, or even to get the American electorate to vote for any of the sacrifices necessary to achieve deficit reduction.

And yet the environmental lobby posits a view of the human species in which individual self-interest is not the mainspring of human conduct. It proposes policies that assume that when properly informed of the long term collective consequences of their actions, people will accept the need for rules that impose restraint. One of the two philosophies must be wrong. Which?

We are going to argue that the environmental movement has set itself an unnecessary obstacle by largely ignoring the fact that human beings are motivated by self-interest rather than collective interests. But that does not mean that the collective interest is unobtainable: examples from biology and economics show that there are all sorts of ways to make the individual interest concordant with the collective—so long as we recognize the need to.

The environmentalists are otherwise in danger of making the same mistakes that Marxists made, but our point is not political. For some reason it is thought conservative to believe that human nature is inherently incapable of ignoring individual incentives for the greater good, and liberal to believe the opposite. But in practice liberals often believe just as strongly as conservatives in individual incentives that are not monetary. The threat of prison, or even corporate shame, can be incentives to polluters. The real divide comes between those who believe it is necessary to impose such incentives, and those who hope to persuade merely by force of argument.

Wherever environmentalism has succeeded, it has done so by changing individual incentives, not by exhortation, moral reprimand, or appeals to our better natures. If somebody wants to dump a toxic chemical or smuggle an endangered species, it is the thought of prison or a fine that deters him. If a state wants to avoid enforcing the federal Clean Air Act of 1990, it is the thought of eventually being "bumped up" to a more stringent nonattainment category of the act that haunts state officials. Given that this is the case, environmental policy should be a matter of seeking the most enforceable, least bureaucratic, cheapest, most effective incentives. Why should these always be sanctions? Why not some prizes, too? Nations, states, local jurisdictions, and even firms could contribute to financial rewards for the "greenest" of their fellow bodies.

Biologists and Economists agree that cooperation cannot be taken for granted. Both think that people are generally not willing to pay for the long-term good of society or the planet.

Playing Games with Life

The new convergence of biology and economics has been helped by a common methodology—game theory. John Maynard Smith, a professor of biology at the University of Sussex, in Britain, was the first effectively to apply the economist's habit of playing a "game" with competing strategies to evolutionary enigmas, the only difference being that the economic games reward winners with money while evolutionary games reward winners with the chance to survive and breed. One game in particular has proved especially informative in both disciplines: the prisoner's dilemma.

A dramatized version of the game runs as follows: Two guilty accomplices are held in separate cells and interrogated by the police. Each is faced with a dilemma. If they both confess (or "defect"), they will both go to jail for three years. If they both stay silent (or "cooperate"), they will both go to jail for a year on a lesser charge that the police can prove. But if one confesses and the other does not, the defector will walk free on a plea bargain, while the cooperator, who stayed silent, will get a five-year sentence.

Assuming that they have not discussed the dilemma before they were arrested, can each trust his accomplice to stay silent? If not, he should defect and reduce his sentence from five to three years. But even if he can rely on his partner to cooperate, he is still better off if he defects, because that reduces his sentence from three years to none at all. So each will reason that the right thing to do is to defect, which results in three years for each of them. In the language of game theorists, individually rational strategies result in a collectively irrational outcome.

Biologists were interested in the prisoner's dilemma as a model for the evolution of cooperation. Under what conditions, they wanted to know, would it pay an animal to evolve a strategy based on cooperation rather than defection? They discovered that the bleak message of the prisoner's dilemma need not obtain if the game is only one in a long series—played by students, researchers, or computers, for points rather than years in jail. Under these circumstances the best strategy is to cooperate on the first trial and then do whatever the other guy did last time. This strategy became known as tit-for-tat. The threat of retaliation makes defection much less likely to pay. Robert Axelrod, a political scientist, and William Hamilton, a biologist, both at the University of Michigan, discovered by public tournament that there seems to be no strategy that beats tit-for tat. Tit-for-two-tats—that is, cooperate even if the other defects once, but not if he defects twice—comes close to beating it, but of hundreds of strategies that have been tried, none works better. Field biologists have been finding tit-for-tat at work throughout the animal kingdom ever since. A female vampire bat, for example, will regurgitate blood for another, unrelated, female bat that has failed to find a meal during the night—but not if the donee has refused to be similarly generous in the past.

Such cases have contributed to a growing conviction among biologists that reciprocity is the basis of social life in animals like primates and dolphins, too. Male dolphins call in their debts when collecting allies to help them abduct females from other groups. Baboons and chimpanzees remember past favors when coming to one another's aid in fights. And human beings? Kim Hill and Hillard Kaplan, of the University of New Mexico, have discovered that among the Ache people of Paraguay, successful hunters share spare meat with those who have helped them in the past or might help them in the future.

The implication of these studies is that where cooperation among individuals does evolve, surmounting the prisoner's dilemma, it does so through tit-for-tat. A cautious exchange of favors enables trust to be built upon a scaffolding of individual reward. The conclusion of biology, in other words, is a hopeful one. Cooperation can emerge naturally. The collective interest can be served by the pursuit of selfish interests.

The Tragedy of the Commons

Economists are interested in the prisoner's dilemma as a paradoxical case in which individually rational behavior leads to collectively irrational results—both accomplices spend three years in jail when they could have spent only one. This makes it a model of a "commons" problem, the archetype of which is the history of medieval English common land. In 1968 the ecologist Garrett Hardin wrote an article in Science magazine that explained "the tragedy of the commons"—why common land tended to suffer from overgrazing, and why every sea fishery suffers from overfishing. It is because the benefits that each extra cow (or netful of fish) brings are reaped by its owner, but the costs of the extra strain it puts on the grass (or on fish stocks) are shared among all the users of what is held in common. In economic jargon, the costs are externalized. Individually rational behavior deteriorates into collective ruin.

The ozone hole and the greenhouse effect are classic tragedies of the commons in the making: each time you burn a gallon of gas to drive into town, you reap the benefit of it, but the environmental cost is shared with all five billion other members of the human race. You are a "free rider." Being rational, you drive, and the atmosphere's capacity to absorb carbon dioxide is "overgrazed," and the globe warms. Even if individuals will benefit in the long run from the prevention of global warming, in the short run such prevention will cost them dear. As Michael McGinnis and Elinor Ostrom, of Indiana University at Bloomington, put it in a recent paper, global warming is a "classic dilemma of collective action: a large group of potential beneficiaries facing diffuse and uncertain gains is much harder to organize for collective action than clearly defined groups who are being asked to suffer easily understandable costs."

Hardin recognized two ways to avoid overexploiting commons. One is to privatize them, so that the owner has both costs and benefits. Now he has every incentive not to overgraze. The other is to regulate them by having an outside agency with the force of law behind it—a government, in short—restrict the number of cattle.

At the time Hardin published his article, the latter solution was very popular. Governments throughout the world reacted to the mere existence of a commons problem by grabbing powers of regulation. Most egregiously, in the Indian subcontinent communally exploited forests and grasslands were nationalized and put under the charge of centralized bureaucracies far away. This might have worked if governments were competent and incorruptible, and had bottomless resources to police their charges. But it made problems worse, because the forest was no longer the possession of the local village even collectively. So the grazing, poaching, and logging intensified—the cost had been externalized not just to the rest of the village but to the entire country.

The whole structure of pollution regulation in the United States represents a centralized solution to a commons problem. Bureaucrats decide, in response to pressure from lobbyists, exactly what levels of pollution to allow, usually give no credit for any reductions below the threshold, and even specify the technologies to be used (the so-called "best available technology" policy). This creates perverse incentives for polluters, because it makes pollution free up to the threshold, and so there is no encouragement to reduce pollution further. Howard Klee, the director of regulatory affairs at Amoco Corporation, gives a dramatic account of how topsy-turvy this world of "command and control" can become. "If your company does voluntary control of pollution rather than waiting for regulation, it is punished by putting itself at a comparative disadvantage. The guy who does nothing until forced to by law is rewarded." Amoco and the Environmental Protection Agency did a thorough study of one refinery in Yorktown, Virginia, to discover what pollutants came out from it and how dangerous each was. Their conclusion was startling. Some of the things that Amoco and other refiners were required to do by EPA regulations were less effective than alternatives; meanwhile, pollution from many sources that government does not regulate could have been decreased. The study group concluded that for one fourth of the amount that it currently spends on pollution control, Amoco could achieve the same effect in protection of health and the environment—just by spending money where it made a difference, rather than where government dictated.

A more general way, favored by free-market economists, of putting the same point is that regulatory regimes set the value of cleanliness at zero: if a company wishes to produce any pollutant, at present it can do so free, as long as it produces less than the legal limit. If, instead, it had to buy a quota from the government, it would have an incentive to drive emissions as low as possible to keep costs down, and the government would have a source of revenue to spend on environmental protection. The 1990 Clean Air Act set up a market in tradable pollution permits for sulfur-dioxide emissions, which is a form of privatization.

The Pitfalls of Privatization

Because privatizing a common resource can internalize the costs of damaging it, economists increasingly call for privatization as the solution to commons problems. After all, the original commons—common grazing land in England—were gradually "enclosed" by thorn hedges and divided among private owners. Though the reasons are complex, among them undoubtedly was the accountability of the private landowner. As Sir Anthony Fitzherbert put it in *The Boke of Husbandrie* (1534): "And thoughe a man be but a farmer, and shall have his farm XX [20] yeres, it is lesse coste to hym, and more profyte to quyckeset [fence with thorns], dyche and hedge, than to have his cattell goo before the herdeman [on common land]." The hawthorn hedge did for England what barbed wire did for the prairies—it privatized a common.

It would be possible to define private property rights in clean air. Paul Romer, of Berkeley, points out that the atmosphere is not like the light from a lighthouse, freely shared by all users. One person cannot use a given chunk of air for seeing through—or comfortably breathing—after another person has filled it with pollution any more than two people in succession can kill the same whale. What stands in the way of privatizing whales or the atmosphere is that enforcement of a market would require as large a bureaucracy as if the whole thing had been centralized in the first place.

The privatization route has other drawbacks. The enclosure movement itself sparked at least three serious rebellions against the established order by self-employed yeomen dispossessed when commons were divided. It would be much the same today. Were whale-killing rights to be auctioned to the highest bidder, protectors (who would want to buy rights in order to let them go unused) would likely be unable to match the buying power of the whalers. If U.S. citizens were to be sold shares in their national parks, those who would rather operate strip mines or charge access might be prepared to pay a premium for the shares, whereas those who would keep the parks pristine and allow visitors free access might not.

Moreover, there is no guarantee that rationality would call for a private owner of an environmental public good to preserve it or use it sustainably. Twenty years ago Colin Clark, a mathematician at the University of British Columbia, wrote an article in *Science* pointing out that under certain circumstances it might make economic sense to exterminate whales. What he meant was that because interest rates could allow money to grow faster than whales reproduce, even somebody who had a certain monopoly over the world's whales and could therefore forget about free riders should not, for reasons of economic self-interest, take a sustainable yield of the animals. It would be more profitable to kill them all, bank the proceeds, sell the equipment, and live off the interest.

So until recently the economists had emerged from their study of the prisoner's dilemma more pessimistic than the biologists. Cooperation, they concluded, could not be imposed by a central bureaucracy, nor would it emerge from the allocation of private property rights. The destructive free-for-all of Fowler, Kansas, not the cooperative harmony of Valencia's *huerta,* was the inevitable fate of common-pool resources.

The Middle Way

In the past few years, however, there has been a glint of hope amid the gloom. And it bears an uncanny similarity to tit-for-tat, in that it rewards cooperators with cooperation and punishes defectors with defection—a strategy animals often use. Elinor Ostrom and her colleagues at Indiana University have made a special study of commons problems that were solved, including the Valencia irrigation system, and she finds that the connective thread is neither privatization nor centralization. She believes that local people can and do get together to solve their difficulties, as long as the community is small, stable, and communicating, and has a strong concern for the future. Among the examples she cites is a Turkish inshore fishery at Alanya. In the 1970s the local fishermen fell into the usual trap of heavy fishing, conflict, and potential depletion. But they then developed an ingenious and complicated set of rules, allocating by lot each known fishing location to a licensed fisher in a pattern that rotates through the season. Enforcement is done by the fishermen themselves, though the government recognizes the system in law.

Valencia is much the same. Individuals know each other and can quickly identify cheaters. Just as in tit-for-tat, because the game is played again and again, any cheater risks ostracism and sanction in the next round. So a small, stable community that interacts repeatedly can find a way to pursue the collective interest—by altering the individual calculation.

"There's a presumption out there that users always overexploit a common resource," Ostrom says, "and therefore governments always have to step in and set things right. But the many cases of well-governed and -managed irrigation systems, fisheries, and forests show this to be an inadequate starting point. A faraway government could never have found the resources to design systems like Alanya." Ostrom is critical of the unthinking application of oversimplified game-theory models because, she says, economists and biologists alike frequently begin to believe that people who have depended on a given economic or biological resource for centuries are incapable of communicating, devising rules, and monitoring one another. She admits that cooperation is more likely in small groups that have common interests and the autonomy to create and enforce their own rules.

Where cooperation among individuals does evolve, it does so through tit-for-tat. A cautious exchange of favors enables trust to be built upon a scaffolding of individual reward.

Some biologists go further, and argue that even quite big groups can cooperate. Egbert Leigh, of the Smithsonian Tropical Research Institute, points out that commons problems go deep into the genetics of animals and plants. To run a human body, 75,000 different genes must "agree" to cooperate and suppress free-riders (free-riding genes, known as outlaw genes, are increasingly recognized as a major force in evolution). Mostly they do, but why? Leigh found the answer in Adam Smith, who argued, in Leigh's words, that "if individuals had sufficient common interest in their groups good, they would combine to suppress the activities of members acting contrary to the group's welfare." Leigh calls this idea a "parliament of genes," though it is crucial to it that all members of such a parliament would suffer if cooperation broke down—as the members of real national parliaments do not when they impose local solutions.

What Changed Du Pont's Mind?

For all these reasons, cooperation ought not to be a problem in Fowler, Kansas—a community in which everybody knows everybody else and shares the immediate consequences of a tragedy of the commons. Professor Kenneth Oye, the director of the Center for International Studies at the Massachusetts Institute of Technology, first heard about Fowler's sinking water table when his wife attended a family reunion there.

Oye's interest was further piqued when he subsequently heard rumors that the state had put a freeze on the drilling of new wells in the Fowler area: such a move might be the beginning of a solution to the water depletion, but it was also a classic barrier to the entry of new competitors in an industry. Oye had been reflecting on the case of Du Pont and chlorofluorocarbons, wondering why a corporation would willingly abandon a profitable business by agreeing to phase out the chemicals that seem to damage the ozone layer. Du Pont's decision stands out as an unusually altruistic gesture amid the selfish strivings of big business. Without it the Montreal protocol on ozone-destroying chemicals, a prototype for international agreements on the environment, might not have come about. Why had Du Pont made that decision? Conventional wisdom, and Du Pont's own assertions, credit improved scientific understanding and environmental pressure groups. Lobbyists had raised public consciousness about the ozone layer so high that Du Pont's executives quickly realized that the loss of public good will could cost them more than the products were worth. This seems to challenge the logic of tit-for-tat. It suggests that appeals to the wider good can be effective where appeals to self interest cannot.

Oye speculates that this explanation was incomplete, and that the company's executives may have been swayed in favor of a ban on CFCs by the realization that the CFC technology was mature and vulnerable. Du Pont was in danger of losing market share to its rivals. A ban beginning ten years hence would at least make it worth no potential rival's while to join in; Du Pont could keep its market share for longer and meanwhile stand a chance of gaining a dominant market share of the chemicals to replace CFCs. Again self-interest was part of the motive for environmental change. If consciousness-raising really changes corporate minds, why did the utility industry fight the Clean Air Act of 1990 every step of the way? The case of Du Pont is not, after all, an exception to the rule that self-interest is paramount.

The Intangible Carrots

Besides, environmentalists cannot really believe that mere consciousness-raising is enough or they would not lobby so hard in favor of enforceable laws. About the only cases in which they can claim to have achieved very much through moral suasion are the campaigns against furs and ivory. There can be little doubt that the world's leopards breathe easier because of the success of campaigns in recent decades against the wearing of furs. There was no need to bribe rich socialites to wear fake furs—they were easily shamed into it. But then shame can often be as effective an incentive as money.

Certainly the environmental movement believes in the power of shame, but it also believes in appealing to people's better natures. Yet the evidence is thin that normative pressures work for necessities. Furs are luxuries; and recycling works better with financial incentives or legal sanctions attached. Even a small refund can dramatically increase the amount of material that is recycled in household waste. In one Michigan study recycling rates were less than 10 percent for nonrefundable glass, metal, and plastic, and more than 90 percent for refundable objects. Charities have long known that people are more likely to make donations if they are rewarded with even just a tag or a lapel pin. Tit for tat.

The issue of normative pressure versus material incentive comes into sharp focus in the ivory debate. Western environmentalists and East African governments argue that the only hope for saving the elephant is to extinguish the demand for ivory by stifling supply and raising environmental consciousness. Many economists and southern African governments argue otherwise: that local people need incentives if they are to tolerate and protect elephants, incentives that must come from a regulated market for ivory enabling sustained production. Which is right depends on two things: whether it is possible to extinguish the demand for ivory in time to save the elephants, and whether the profits from legal ivory trading can buy sufficient enforcement to prevent poaching at home.

Even if it proved possible to make ivory so shameful a purchase that demand died, this would be no precedent for dealing with global warming. By giving up ivory, people are losing nothing. By giving up carbon dioxide, people are losing part of their standard of living.

Yet again and again in recent years environmentalists have persisted in introducing an element of mysticism and morality into the greenhouse debate, from Bill McKibben's nostalgia about a nature untouched by man in *The End of Nature* to James Lovelock's invention of the Gaia hypothesis. Others have often claimed that a mystical and moral approach works in Asia, so why not here? The reverence for nature that characterizes the Buddhist, Jain, and Hindu religions stands in marked contrast to the more exploitative attitudes of Islam and Christianity. Crossing the border from India to Pakistan, one is made immediately aware of the difference: the peacocks and monkeys that swarm, tame and confident, over every Indian temple and shrine are suddenly scarce and scared in the Muslim country.

In surveying people's attitudes around the Kosi Tappu wildlife reserve in southeastern Nepal, Joel Heinen, of the University of Michigan, discovered that Brahmin Hindus and Buddhists respect the aims of conservation programs much more than Muslims and low-caste Hindus. Nonetheless, religious reverence did not stand in the way of the overexploitation of nature. Heinen told us, "Sixty-five percent of the households in my survey expressed negative attitudes about the reserve, because the reserve took away many rights of local citizens." Nepal's and India's forests, grasslands, and rivers have suffered tragedies of the commons as severe as any country's. The eastern religious harmony with nature is largely lip service.

The Golden Age That Never Was

In recent years those who believe that the narrow view of selfish rationalism expressed by economists and biologists is a characteristically Western concept have tended to stress not Buddhist peoples but pre-industrial peoples living close to nature. Indeed, so common is the view that all environmental problems stem from man's recent and hubristic attempt to establish dominion over nature, rather than living in harmony with it, that this has attained the status of a cliche, uttered by politicians as diverse as Pope John Paul II and Albert Gore. It is a compulsory part of the preface in most environmental books.

If the cliché is true, then the biologists and economists are largely wrong. Individuals can change their attitudes and counteract selfish ambitions. If the cliché is false, then it is the intangible incentive of shame, not the appeal to collective interest, that changes people's minds.

Evidence bearing on this matter comes from archaeologists and anthropologists. They are gradually reaching the conclusion that pre-industrial people were just as often capable of environmental mismanagement as modern people, and that the legend of an age of environmental harmony—before we "lost touch with nature"—is a myth. Examples are now legion. The giant birds of Madagascar and New Zealand were almost certainly wiped out by man. In 2,000 years the Polynesians converted Easter Island, in the eastern Pacific, from a lush forest that provided wood for fishing canoes into a treeless, infertile grassland where famine, warfare, and cannibalism were rife. Some archaeologists believe that the Mayan empire reduced the Yucatán peninsula to meager scrub, and so fatally wounded itself. The Anasazi Indians apparently deforested a vast area.

History abounds with evidence that limitations of technology or demand, rather than a culture of self-restraint, are what has kept tribal people from overgrazing their commons. The Indians of Canada had the technology to exterminate the beaver long before white men arrived; at that point they changed their behavior not because they lost some ancient reverence for their prey but because for the first time they had an insatiable market for beaver pelts. The Hudson's Bay Company would trade a brass kettle or twenty steel fishhooks for every pelt.

Cause for Hope

We conclude that the cynicism of the economist and the biologist about man's selfish, shortsighted nature seems justified. The optimism of the environmental movement about changing that nature does not. Unless we can find a way to tip individual incentives in favor of saving the atmosphere, we will fail. Even in a pre-industrial state or with the backing of a compassionate, vegetarian religion, humanity proves incapable of overriding individual greed for the good of large, diverse groups. So must we assume that we are powerless to avert the tragedy of the aerial commons, the greenhouse effect?

Fortunately not. Tit-for-tat can come to the rescue. If the principles it represents are embodied in the treaties and legislation that are being written to avert global warming, then there need be no problem in producing an effective, enforceable, and acceptable series of laws.

Care will have to be taken that free-rider countries don't become a problem. As Robert Keohane, of Harvard University's Center for International Affairs, has stressed, the commons problem is mirrored at the international level. Countries may agree to treaties and then try to free-ride their way past them. Just as in the case of local commons, there seem to be two solutions: to privatize the issue and leave it to competition between sovereign states (that is, war), or to centralize it and enforce obedience (that is, world government). But Keohane's work on international environmental regimes to control such things as acid rain, oil pollution, and overfishing came to much the same conclusion as Ostrom's; a middle way exists. Trade sanctions, blackmail, bribes, and even shame can be used between sovereign governments to create incentives for cooperation as long as violations can be easily detected. The implicit threat of trade sanctions for CFC manufacture is "a classic piece of tit-for-tat," Paul Romer observes.

Local governments within the nation can play tit-for-tat as well. The U.S. government is practiced at this art: it often threatens to deprive states of highway construction funds, for example, to encourage them to pass laws. States can play the same game with counties, or cities, or firms, and so on down to the level of the individual, taking care at each stage to rig the incentives so that obedience is cheaper than disobedience. Any action that raises the cost of being a free-rider, or raises the reward of being a cooperator, will work. Let the United States drag its feet over the Rio conventions if it wants, but let it feel the sting of some sanction for doing so. Let people drive gas-guzzlers if they wish, but tax them until it hurts. Let companies lobby against anti-pollution laws, but pass laws that make obeying them worthwhile. Make it rational for individuals to act green.

If this sounds unrealistic, remember what many environmental lobbyists are calling for. "A fundamental restructuring of many elements of society," Lester Brown proposes; "a wholly new economic order." "Modern society will find no solution to the ecological problem unless it takes a serious look at its lifestyle," the Pope has said. These are hardly realistic aims.

We are merely asking governments to be more cynical about human nature. Instead of being shocked that people take such a narrow view of their interests, use the fact. Instead of trying to change human nature, go with the grain of it. For in refusing to put group good ahead of individual advantage, people are being both rational and consistent with their evolutionary past.

Critical Thinking

1. Which do you think is preferable: to appeal to the heart or to the head?
2. Do you agree with the statement: "people are generally not willing to pay for the long-term good of society or the planet?"
3. Evaluate the statement: "Modern society will find no solution to the ecological problem unless it takes a serious look at its lifestyle."

Toward A Sustainable World

What policies can lead to the changes in behavior—of individuals, industries and governments—that will allow development and growth to take place within the limits set by ecological imperatives?

WILLIAM D. RUCKELSHAUS

The difficulty of converting scientific findings into political action is a function of the uncertainty of the science and the pain generated by the action. Given the current uncertainties surrounding just one aspect of the global environmental crisis—the predicted rise in greenhouse gases—and the enormous technological and social effort that will be required to control that rise, it is fair to say that responding successfully to the multi-faceted crisis will be a difficult political enterprise. It means trying to get a substantial proportion of the world's people to change their behavior in order to (possibly) avert threats that will otherwise (probably) affect a world most of them will not be alive to see.

The models that predict climatic change, for example, are subject to varying interpretations as to the timing, distribution and severity of the changes in store. Also, whereas models may convince scientists, who understand their assumptions and limitations, as a rule projections make poor politics. It is hard for people—hard even for the groups of people who constitute governments—to change in response to dangers that may not arise for a long time or that just might not happen at all.

How, then, can we make change happen? The previous articles in this single-topic issue have documented the reality of the global ecological crisis and have pointed to some specific ameliorative measures. This article is about how to shape the policies, launch the programs and harness the resources that will lead to the adoption of such measures—and that will actually convince ordinary people throughout the world to start doing things differently.

Insurance is the way people ordinarily deal with potentially serious contingencies, and it is appropriate here as well. People consider it prudent to pay insurance premiums so that if catastrophe strikes, they or their survivors will be better off than if there had been no insurance. The analogy is clear. Current resources foregone or spent to prevent the buildup of greenhouse gases are a kind of premium. Moreover, as long as we are going to pay premiums, we might as well pay them in ways that will yield dividends in the form of greater efficiency, improved human health or more widely distributed prosperity. If we turn out to be wrong on greenhouse warming or ozone depletion, we still retain the dividend benefits. In any case, no one complains to the insurance company when disaster does not strike.

That is the argument for some immediate, modest actions. We can hope that if shortages or problems arise, there will turn out to be a technological fix or set of fixes, or that technology and the normal workings of the market will combine to solve the problem by product substitution. Already, for example, new refrigerants that do not have the atmospheric effects of the chlorofluorocarbons are being introduced; perhaps a cheap and non-polluting source of energy will be discovered.

It is comforting to imagine that we might arrive at a more secure tomorrow with little strain, to suppose with Dickens's Mr. Micawber that something will turn up. Imagining is harmless, but counting on such a rescue is not. We need to face up to the fact that something enormous may be happening to our world. Our species may be pushing up against some immovable limits on the combustion of fossil fuels and damage to ecosystems. We must at least consider the possibility that, besides those modest adjustments for the sake of prudence, we may have to prepare for far more dramatic changes, changes that will begin to shape a sustainable world economy and society.

Sustainability is the nascent doctrine that economic growth and development must take place, and be maintained over time, within the limits set by ecology in the broadest sense—by the interrelations of human beings and their works, the biosphere and the physical and chemical laws that govern it. The doctrine of sustainability holds too that the spread of a reasonable level of prosperity and security to the less developed nations is essential to protecting ecological balance and hence essential to the continued prosperity of the wealthy nations. It follows that environmental protection and economic development are complementary rather than antagonistic processes.

Can we move nations and people in the direction of sustainability? Such a move would be a modification of society comparable in scale to only two other changes: the agricultural revolution of the late Neolithic and the Industrial Revolution of the past two centuries. Those revolutions were gradual, spontaneous and largely unconscious. This one will have to be a fully conscious operation, guided by the best foresight that science can provide—foresight pushed to its limit. If we actually do it, the undertaking will be absolutely unique in humanity's stay on the earth.

The shape of this undertaking cannot be clearly seen from where we now stand. The conventional image is that of a crossroads: a forced choice of one direction or another that determines the future for some appreciable period. But this does not at all capture the complexity of the current situation. A more appropriate image would be that of a canoeist shooting the rapids: survival depends on continually responding to information by correct steering. In this case the information is supplied by science and economic events; the steering is the work of policy, both governmental and private.

Taking control of the future therefore means tightening the connection between science and policy. We need to understand where the rocks are in time to steer around them. Yet we will not devote the appropriate level of resources to science or accept the policies mandated by science unless we do something else. We have to understand that we are all in the same canoe and that steering toward sustainability is necessary.

Sustainability was the original economy of our species. Preindustrial peoples lived sustainably because they had to; if they did not, if they expanded their populations beyond the available resource base, then sooner or later they starved or had to migrate. The sustainability of their way of life was maintained by a particular consciousness regarding nature: the people were spiritually connected to the animals and plants on which they subsisted; they were part of the landscape, or of nature, not set apart as masters.

The era of this "original sustainability" eventually came to an end. The development of cities and the maintenance of urban populations called for intensive agriculture yielding a surplus. As a population grows, it requires an expansion of production, either by conquest or colonization or improved technique. A different consciousness, also embodied in a structure of myth, sustains this mode of life. The earth and its creatures are considered the property of humankind, a gift from the supernatural. Man stands outside of nature, which is a passive playing field that he dominates, controls and manipulates. Eventually, with industrialization, even the past is colonized: the forests of the Carboniferous are mined to support ever-expanding populations. Advanced technology gives impetus to the basic assumption that there is essentially no limit to humanity's power over nature.

This consciousness, this condition of "transitional unsustainability," is dominant today. It has two forms. In the underdeveloped, industrializing world, it is represented by the drive to develop at any environmental cost. It includes the wholesale destruction of forests, the replacement of sustainable agriculture by cash crops, the attendant exploitation of vulnerable lands by people such cash cropping forces off good land and the creation of industrial centers that are also centers of environmental pollution.

In the industrialized world, unsustainable development has generated wealth and relative comfort for about one fifth of humankind, and among the populations of the industrialized nations the consciousness supporting the unsustainable economy is nearly universal. With a few important exceptions, the environmental-protection movement in those nations, despite its major achievements in passing legislation and mandating pollution-control measures, has not had a substantial effect on the lives of most people. Environmentalism has been ameliorative and corrective—not a restructuring force. It is encompassed within the consciousness of unsustainability.

Although we cannot return to the sustainable economy of our distant ancestors, in principle there is no reason why we cannot create a sustainability consciousness suitable to the modern era. Such a consciousness would include the following beliefs:

1. *The human species is part of nature. Its existence depends on its ability to draw sustenance from a finite natural world; its continuance depends on its ability to abstain from destroying the natural systems that regenerate this world.* This seems to be the major lesson of the current environmental situation as well as being a direct corollary of the second law of thermodynamics.
2. *Economic activity must account for the environmental costs of production.* Environmental regulation has made a start here, albeit a small one. The market has not even begun to be mobilized to preserve the environment; as a consequence an increasing amount of the "wealth" we create is in a sense stolen from our descendants.
3. *The maintenance of a livable global environment depends on the sustainable development of the entire human family.* If 80 percent of the members of our species are poor, we can not hope to live in a world at peace; if the poor nations attempt to improve their lot by the methods we rich have pioneered, the result will eventually be world ecological damage.

This consciousness will not be attained simply because the arguments for change are good or because the alternatives are unpleasant. Nor will exhortation suffice. The central lesson of realistic policy-making is that most individuals and organizations change when it is in their interest to change, either because they derive some benefit from changing or because they incur sanctions when they do not—and the shorter the time between change (or failure to change) and benefit (or sanction), the better. This is not mere cynicism. Although people will struggle and suffer for long periods to achieve a goal, it is not reasonable to expect people or organizations to work against their immediate interests for very long—particularly in a democratic system, where what they perceive to be their interests are so important in guiding the government.

To change interests, three things are required. First, a clear set of values consistent with the consciousness of sustainability must be articulated by leaders in both the public and the private sector. Next, motivations need to be established that will support the values. Finally, institutions must be developed that will effectively apply the motivations. The first is relatively easy, the second much harder and the third perhaps hardest of all.

Values similar to those I described above have indeed been articulated by political leaders throughout the world. In the past year the president and the secretary of state of the U.S., the leader of the Soviet Union, the prime minister of Great Britain and the presidents of France and Brazil have all made major environmental statements. In July the leaders of the Group of Seven major industrialized nations called for "the early adoption, worldwide, of policies based on sustainable development" Most industrialized nations have a structure of national environmental law that to at least some extent reflects such values, and there is even a small set of international conventions that begin to do the same thing.

Mere acceptance of a changed value structure, although it is a prerequisite, does not generate the required change in consciousness, nor does it change the environment. Although diplomats and lawyers may argue passionately over the form of words, talk is not action. In the U.S., which has a set of environmental statutes second to none in their stringency, and where for the past 15 years poll after poll has recorded the American people's desire for increased environmental protection, the majority of the population participates in the industrialized world's most wasteful and most polluting style of life. The values are there; the appropriate motivations and institutions are patently inadequate or nonexistent.

The difficulties of moving from stated values to actual motivations and institutions stem from basic characteristics of the major industrialized nations—the nations that must, because of their economic strength, preeminence as polluters and dominant share of the world's resources, take the lead in any changing of the present order. These nations are market-system democracies. The difficulties, ironically, are inherent in the free-market economic system on the one hand and in democracy on the other.

The economic problem is the familiar one of externalities: the environmental cost of producing a good or service is not accounted for in the price paid for it. As the economist Kenneth E. Boulding has put it: "All of nature's systems are closed loops, while economic activities are linear and assume inexhaustible resources and 'sinks' in which to throw away our refuse." In willful ignorance, and in violation of the core principle of capitalism, we often refuse to treat environmental resources as capital. We spend them as income and are as befuddled as any profligate heir when our checks start to bounce.

Such "commons" as the atmosphere, the seas, fisheries and goods in public ownership are particularly vulnerable to being overspent in this way, treated as either inexhaustible resources or bottomless sinks. The reason is that the incremental benefit to each user accrues exclusively to that user, and in the short term it is a gain. The environmental degradation is spread out among all users and is apparent only in the long term, when the resource shows signs of severe stress or collapse. Some years ago the biologist Garrett Hardin called this the tragedy of the commons.

The way to avoid the tragedy of the commons—to make people pay the full cost of a resource use—is to close the loops in economic systems. The general failure to do this in the industrialized world is related to the second problem, the problem of action in a democracy. Modifying the market to reflect environmental costs is necessarily a function of government. Those adversely affected by such modifications, although they may be a tiny minority of the population, often have disproportionate influence on public policy. In general, the much injured minority proves to be a more formidable lobbyist than the slightly benefited majority.

The Clean Air Act of 1970 in the U.S., arguably the most expensive and far-reaching environmental legislation in the world, is a case in point. Parts of the act were designed not so much to cleanse the air as to protect the jobs of coal miners in high-sulfur coal regions. Utilities and other high-volume consumers were not allowed to substitute low-sulfur coal to meet regulatory requirements but instead had to install scrubbing devices.

Although the act expired seven years ago, Congress found it extraordinarily difficult to develop a revision, largely because of another set of contrary interests involving acid rain. The generalized national interest in reducing the environmental damage attributable to this long-range pollution had to overcome the resistance of both high-sulfur-coal mining interests and the Midwestern utilities that would incur major expenses if they were forced to control sulfur emissions. The problem of conflicting interests is exacerbated by the distance between major sources of acid rain and the regions that suffer the most damage. It is accentuated when the pollution crosses state and national boundaries: elected representatives are less likely to countenance short-term adverse effects on their constituents when the immediate beneficiaries are nonconstituents.

The question, then, is whether the industrial democracies will be able to overcome political constraints on bending the market system toward long-term sustainability. History provides some cause for optimism: a number of contingencies have led nations to accept short-term burdens in order to meet a long-term goal.

War is the obvious example. Things considered politically or economically impossible can be accomplished in a remarkably short time, given the belief that national survival is at stake. World War II mobilized the U.S. population, changed work patterns, manipulated and controlled the price and supply of goods and reorganized the nation's industrial plant.

Another example is the Marshall Plan for reconstructing Europe after World War II. In 1947 the U.S. spent nearly 3 percent of its gross domestic product on this huge set of projects. Although the impetus for the plan came from fear that Soviet influence would expand into Western Europe, the plan did establish a precedent for massive investment in increasing the prosperity of foreign nations.

There are other examples. Feudalism was abandoned in Japan, as was slavery in the U.S., in the 19th century; this century has seen the retreat of imperialism and the creation of the European Economic Community. In each case important interests gave way to new national goals.

If it is possible to change, how do we begin to motivate change? Clearly, government policy must lead the way, since market prices of commodities typically do not reflect the environmental costs of extracting and replacing them, nor do the prices of energy from fossil fuels reflect the risks of climatic change. Pricing policy is the most direct means of ensuring that the full environmental cost of goods and services is accounted for. When government owns a resource, or supplies it directly, the price charged can be made to reflect the true cost of the product. The market will adjust to this as it does to true scarcity: by product substitution and conservation.

Environmental regulation should be refocused to mobilize rather than suppress the ingenuity and creativity of industry. For example, additional gains in pollution control should be sought not simply by increasing the stringency or technical specificity of command-and-control regulation but also by implementing incentive-based systems. Such systems magnify public-sector decisions by tens of thousands of individual and corporate decisions. To be sure, incentive systems are not a panacea. For some environmental problems, such as the use of unacceptably dangerous chemicals, definitive regulatory measures will always be required. Effective policies will include a mixture of incentive-based and regulatory approaches.

Yet market-based approaches will be a necessary part of any attempt to reduce the greenhouse effect. Here the most attractive options involve the encouragement of energy efficiency. Improving efficiency meets the double-benefit standard of insurance: it is good in itself, and it combats global warming by reducing carbon dioxide emissions. If the world were to improve energy efficiency by 2 percent a year, the global average temperature could be kept within one degree Celsius of present levels. Many industrialized nations have maintained a rate of improvement close to that over the past 15 years.

Promoting energy efficiency is also relatively painless. The U.S. reduced the energy intensity of its domestic product by 23 percent between 1973 and 1985 without much notice. Substantial improvement in efficiency is available even with existing technology. Something as simple as bringing all U.S. buildings up to the best world standards could save enormous amounts of energy. Right now more energy passes through the windows of buildings in the U.S. than flows through the Alaska pipeline.

Efficiency gains may nevertheless have to be promoted by special market incentives, because energy prices tend to lag behind increases in income. A "climate protection" tax of $1 per million Btu's on coal and 60 cents per million Btu's on oil is an example of such an incentive. It would raise gasoline prices by 11 cents a gallon and the cost of electricity an average of 10 percent, and it would yield $53 billion annually.

Direct regulation by the setting of standards is cumbersome, but it may be necessary when implicit market signals are not effective. Examples are the mileage standards set in the U.S. for automobiles and the efficiency standards for appliances that were adopted in 1986. The appliance standards will save $28 billion in energy costs by the year 2000 and keep 342 million tons of carbon out of the atmosphere.

Over the long term it is likely that some form of emissions-trading program will be necessary—and on a much larger scale than has been the case heretofore. (Indeed, the President's new Clean Air Act proposal includes a strengthened system of tradeable permits.) In such a program all major emitters of pollutants would be issued permits specifying an allowable emission level. Firms that decide to reduce emissions below the specified level—for example, by investing in efficiency—could sell their excess "pollution rights" to other firms. Those that find it prohibitively costly to retrofit old plants or build new ones could buy such rights or could close down their least efficient plants and sell the unneeded rights.

Another kind of emissions trading might reduce the impact of carbon dioxide emissions. Companies responsible for new greenhouse-gas emissions could be required to offset them by improving overall efficiency or closing down plants, or by planting or preserving forests that would help absorb the emissions. Once the system is established, progress toward further reduction of emissions would be achieved by progressively cranking down the total allowable levels of various pollutants, on both a national and a permit-by-permit basis.

The kinds of programs I have just described will need to be supported by research providing a scientific basis for new environmental-protection strategies. Research into safe, nonpolluting energy sources and more energy-efficient technologies would seem to be particularly good bets. An example: in the mid-1970's the U.S. Department of Energy developed a number of improved-efficiency technologies at a cost of $16 million; among them were a design for compact fluorescent lamps that could replace incandescent bulbs, and window coatings that save energy during both heating and cooling seasons. At current rates of implementation, the new technologies should generate $63 billion in energy savings by the year 2010.

The motivation of change toward sustainability will have to go far beyond the reduction of pollution and waste in the developed countries, and it cannot be left entirely to the environmental agencies in those countries. The agencies whose goals are economic development, exploitation of resources and international trade—and indeed foreign policy in general—must also adopt sustainable development as a central goal. This is a formidable challenge, for it touches the heart of numerous special interests. Considerable political skill will be required to achieve for environmental protection the policy preeminence that only economic issues and national security (in the military sense) have commanded.

But it is in relations with the developing world that the industrialized nations will face their greatest challenges. Aid is both an answer and a perpetual problem. Total official development assistance from the developed to the developing world stands at around $35 billion a year. This is not much money. The annual foreign-aid expenditure of the U.S. alone would be $127 billion if it spent the same proportion of its gross national product on foreign aid as it did during the peak years of the Marshall Plan.

There is no point, of course, in even thinking about the adequacy of aid to the undeveloped nations until the debt issue is resolved. The World Bank has reported that in 1988 the 17 most indebted countries paid the industrialized nations and multilateral agencies $31.1 billion more than they received in aid. This obviously cannot go on. Debt-for-nature swapping has taken place between such major lenders as Citicorp and a number of countries in South America: the bank forgives loans in exchange for the placing of land in conservation areas or parks. This is admirable, but it will not in itself solve the problem. Basic international trading relations will have to be redesigned in order to eliminate, among other things, the ill effects on the undeveloped world of agricultural subsidies and tariff barriers in the industrialized world.

A prosperous rural society based on sustainable agriculture must be the prelude to future development in much of the developing world, and governments there will have to focus on what motivates people to live in an environmentally responsible manner. Farmers will not grow crops when governments subsidize urban populations by keeping prices to farmers low. People will not stop having too many children if the labor of children is the only economic asset they have. Farmers will not improve the land if they do not own it; it is clear that land-tenure reform will have to be instituted.

Negative sanctions against abusing the environment are also missing throughout much of the undeveloped world; to help remedy this situation, substantial amounts of foreign aid could be focused directly on improving the status of the environmental ministries in developing nations. These ministries are typically impoverished and ineffective, particularly in comparison with their countries' economic-development and military ministries. To cite one small example: the game wardens of Tanzania receive an annual salary equivalent to the price paid to poachers for two elephant tusks—one reason the nation has lost two thirds of its elephant population to the ivory trade in the past decade.

To articulate the values and devise the motivations favoring a sustainable world economy, existing institutions will need to change and new ones will have to be established. These will be difficult tasks, because institutions are powerful to the extent that they support powerful interests—which usually implies support of the status quo.

The important international institutions in today's world are those concerned with money, with trade and with national defense. Those who despair of environmental concerns ever reaching a comparable level of importance should remember that current institutions (for example, NATO, the World Bank, multinational corporations) have fairly short histories. They were formed out of pressing concerns about acquiring and expanding wealth and maintaining national sovereignty. If concern for the environment becomes comparably pressing, comparable institutions will be developed.

To further this goal, three things are wanted. The first is money. The annual budget of the United Nations Environment Program (UNEP) is $30 million, a derisory amount considering its responsibilities. If nations are serious about sustainability, they will provide this central environmental organization with serious money, preferably money derived from an independent source in order to reduce its political vulnerability. A tax on certain uses of common world resources has been suggested as a means to this end.

The second thing wanted is information. We require strong international institutions to collect, analyze and report on environmental trends and risks. The Earthwatch program run by the UNEP is a beginning, but there is need for an authoritative source of scientific information and advice that is independent of national governments. There are many nongovernmental or quasi-governmental organizations capable of filling this role; they need to be pulled together into a cooperative network. We need a global institution capable of answering questions of global importance.

The third thing wanted is integration of effort. The world cannot afford a multiplication of conflicting efforts to solve common problems. On the aid front in particular, this can be tragically absurd: Africa alone is currently served by 82 international donors and more than 1,700 private organizations. In 1980, in the tiny African nation Burkina Faso (population about eight million) 340 independent aid projects were under way. We need to form and strengthen coordinating institutions that combine the separate strengths of nongovernmental organizations, international bodies and industrial groups and to focus their efforts on specific problems.

Finally, in creating the consciousness of advanced sustainability, we shall have to redefine our concepts of political and economic feasibility. These concepts are, after all, simply human constructs; they were different in the past, and they will surely change in the future. But the earth is real, and we are obliged by the fact of our utter dependence on it to listen more closely than we have to its messages.

Further Reading

The Global Possible: Resources, Development, and the New Century. Edited by Robert Repetto. Yale University Press, 1985.

Are Today's Institutional Tools up to the Task? Michael Gruber in *EPA Journal,* Vol. 14, No. 7, pages 2–6; November/December, 1988.

State of the World 1989. Lester R. Brown et al. W. W. Norton & Company, February, 1989.

Critical Thinking

1. Which is the biggest obstacle to achieving sustainability, the uncertainty of the science or the potential pain that might be generated by the actions that will be necessary?
2. Explain the statement: "All of nature's systems are closed loops, while economic activities are linear and assume inexhaustible resources and 'sinks' in which to throw away our refuse."
3. It has been nearly 25 years since this article was published. Is there evidence that industrial democracies have been "able to overcome political constraints on bending the market system toward long-term sustainability?"

Abolishing GDP[1]

JEROEN C. J. M. VAN DEN BERGH

1. Introduction

It is not original to criticize GDP (or GNP[2]) as an indicator of welfare or progress. But it turns out to be necessary to repeat the critique, as well as update it to reflect the most recent theoretical and empirical insights. For the critique is only fully accepted within a small circle of academics, while it insufficiently seeps through to economists working in businesses and government, to economic teachers at various levels of education, to policy makers, politicians and journalists. As a result, obvious conclusions and policy implications are not being picked up.

Gross domestic product (GDP) is the monetary, market value of all final goods and services produced in a country over a period of a year. The real GDP per capita (corrected for inflation) is generally used as the core indicator in judging the position of the economy of a country over time or relative to that of other countries. The GDP is thus identified, or considered even synonymous, with social welfare—witness the substituting phrase 'standard of living'. This approach does not follow from a thorough theory about GDP as a welfare measure, but has grown to become like this in the course of time. What is perhaps most striking is that many journalists and politicians, regardless of their political preferences, express critiqueless statements about GDP. Not surprisingly, then, one can observe a strong urge for GDP growth worldwide. This is being reinforced by international organizations like the IMF and the OESO, where (macro)economists play first fiddle. But these same economists should know better than anyone that GDP (per capita) is not an adequate beacon for steering the economy, at least when the ultimate goal is to serve social welfare.

It so happens that there is a quite extensive theoretical and empirical literature in which the use of GDP per head as a measure of welfare and progress is being criticized. Closely related is a growing literature that proposes corrections and alternative indicators. In spite of this, the influence of GDP information on the economy—through the decisions of firms, financial institutions, consumers and governments—has by no means declined. On the contrary, with the formation of the EU GDP growth has become an even more explicit and important goal, witness the unconditional 3 percent growth objective of the Lissabon strategy.

It is pertinent that economists express themselves clearly about the shortcomings of the GDP indicator and the implications of these. For the longstanding critique on GDP as a welfare indicator is either correct, in which case the inevitable conclusions is: we have to get rid of GDP as a welfare measure, because policy and economic decisions guided by it lead to lower than feasible social welfare. Or the critique is incorrect, in which case the counter arguments need to be made explicit and clear. So far, the latter has not occurred. Indeed, the economic literature does not offer any serious efforts to refute the critiques of GDP per capita as a welfare indicator. With this article I intend to invite my fellow economists to arrive at a clear position on what to do with the GDP indicator.

The organization of the remainder of the text is as follows. Section 2 discusses the arguments of the various critiques, which are divided into eight categories. Section 3 illustrates that the influence of GDP information on the economy is easily underestimated. Section 4 analyses the customary arguments in favour of GDP. Section 5 critically reviews the main alternative social welfare or progress indicators that have been proposed. Section 6 derives policy implications of abolishing GDP as a macroeconomic indicator. Section 7 concludes.

2. Shortcomings of the GDP Indicator

Since the 1960s, the implicit and explicit interpretation of GDP (per capita) as a proxy of social welfare has received much criticism. Moreover, criticism has come from some of the most respected economists of the 20th century, including various Nobel laureates. Among the most well-known critics are Kuznets (1941), Galbraith (1958), Samuelson (1961), Mishan (1967), Nordhaus and Tobin (1972), Hueting (1974), Hirsch (1976), Sen (1976), Scitovsky (1976), Daly (1977), Hartwick (1990), Tinbergen and Hueting (1992), Arrow et al. (1995), Vellinga and Withagen (1996), Weitzman and Löfgren (1997), Dasgupta and Mäler (2000), and Dasgupta (2001). The many arguments of the critique are organized into the following eight categories.[3]

2.1 Principles of Proper Accounting

The use and calculation of the GDP indicator is inconsistent with three principles of good bookkeeping: (i) divide clearly between costs and benefits; (ii) correct for changes in stocks and supplies; and (iii) use accurate measures for all social costs

(= private + external costs). If a commercial company were to employ the method that is the basis for calculating GDP, its accounts would not be legally approved. The fact that the GDP calculation method continues to coexist with institutionalised, legal rules for financial accounting of firms is somewhat of a mystery.

Firms employ separate accounts for benefits (revenues) and costs (outlays). The GDP, however, adds benefits and costs together. A company that would function as such, would quickly go broke (countries, however, face another type of competitive environment than firms). According to Stiglitz (2005) "No one would look at just a firm's revenues to assess how well it was doing. Far more relevant is the balance sheet, which shows assets and liabilities. That is also true for a country." In addition, a decline in stocks that represent value or welfare is not taken into account (e.g. natural gas in the earth). An additional shortcoming is that GDP covers the costs of the provision of certain public goods, such as national defence, even though it is evident that the costs of public goods cannot serve as an adequate measure of the benefits associated with these goods. Finally, many private goods show diverging private and social costs because of all kinds of market failure, including imperfect competition, price agreements and technical-physical externalities.

Mishan (1967) and Daly (1977) conclude that GDP must be considered as an estimate of the total cost of all market-related economic activities in a country. Their actual benefits or real welfare effects are unobserved, that is, not measured by means of GDP.[4] As an implication, GDP growth should not be considered as an indicator of progress, but as a reflection of increasing costs of economic change (whether progress or decline). This explains why GDP and welfare growth do not necessarily coincide. At a certain moment, GDP growth creates more costs than benefits, so that an optimal scale of economic activity will be surpassed (Daly, 1992). Economists are happy to argue in favour of cost-benefit analysis as a general method for policy evaluation and support. When it comes to the direction of the economy as a whole, many of them suddenly are satisfied with only information about costs, that is, GDP information.

Finally, the correction of GDP for inflation is required to make estimates comparable over time. This leads to the particular problem that the correction is based on an average consumption basket, which is regarded as representative for the entire population. However, the more skewed is the income distribution or the more heterogeneous in terms of consumption (purchase) behaviour is the population, the more inaccurate and thus less representative this procedure will be.

2.2 Intertemporal Considerations

Macroeconomics, and within it especially economic growth theory, is concerned with the dynamic aspects of the economy as a whole. Macroeconomics does not offer any support for the idea that GDP can serve as a measure of social welfare. Quite the contrary, optimal (normative) growth theory proposes models that explicitly use social welfare as an objective function (based on continuous or overlapping generations), and certainly not a GDP type of criterion.[5]

Apart from this, it should be realized that, although an intertemporal welfare function is usually posed as a truth in theoretical economic growth analyses, it lacks any basis in empirical studies (Section 2.4). One can, of course, claim that people take expected future own-welfare effects of their actions into account. For example, they may respond to uncertainty about the future by creating sufficient financial reserves (wealth). But this is not the same as saying that individuals maximize some intertemporal welfare function, usually utilitarian welfare defined as the aggregation (sum or integral) of a discounted future stream of instantaneous or momentary utilities.

Finally, the fact or belief that GDP growth in certain periods or regions has correlated positively with progress (however measured) should not be confused with the idea that GDP (growth) is a good measure of social welfare (progress) in general. In other words, the correlation may be low or even negative in certain periods and regions. If, by way of thought experiment, one extrapolates a constant tempo of real GDP growth towards the distant future, one will end up with an incredibly high GDP. But it is very unrealistic to suppose that social welfare will reach a comparably high level. Somewhere in time the two need to be de-linked (assuming that they are closely connected during an initial period). To illustrate this, using 2 percent as a conservative estimate of the average yearly GDP growth rate over the past decades, extrapolation of this rate 1000 years into the future gives a GDP that is $(1.02)^{1000} \approx 400$ million times as high as the current GDP. Surely, no one can believe–if only on the basis of introspection—that individual and social welfare can increase to such an extent. This shows that, in the long run, GDP can not serve as a good indicator or even rough approximation of social welfare. Definitely, at some point a de-linking of GDP and welfare must occur. In fact, it is very well possible that such de-linking has already occurred for many rich countries in the world. On the basis of pure theoretical reasoning one can not decisively conclude on this issue. Empirical analysis is required (see Section 2.4).

2.3 Lexicographic Preferences

People have various basic needs, such as air, water, food, sex, shelter, company, respect and freedom. These cannot be traded off against luxury services and material goods—in fact, the latter often serve as a sublimation of the basic needs themselves (e.g. a fancy car to gain respect from peers). In other words, substitution in consumption is very limited. This is the core of the notion of lexicographic preferences, which is closely connected with the Maslow pyramid. Lexicographic preferences can be defined as having two characteristics: (i) individuals have limited needs in certain goods or services, as feelings of satisfaction occur after consumption reaches a certain level; (ii) 'lower' needs (e.g. the removal of thirst and hunger) need to be fulfilled before 'higher' needs (e.g. recreation) can appear. Within this framework, income growth and the associated growth of material consumption, notably in urban and polluted environments, is an imperfect compensation for a lack of satisfaction of basic needs, such as relaxation, space, serenity, clean air and water, and direct access to nature. As a result, one cannot exclude that, despite GDP and individual income growth, (individual and social) welfare remains constant or even declines.

The previous point does not mean to suggest that GDP growth always implies more material consumption. It is quite possible that it comprises an increase in services. What this in turn yields in terms of welfare is difficult to say in general. Sen's (1999) concept of individual 'capabilities' may be useful here. This tries to bring goods and services into a single denominator by emphasizing freedom and opportunities to choose, as well as context-dependent functionality of goods and services, based on taking into account the peculiarities and environments of individuals. Examples of the latter are being disabled versus being perfectly healthy, and living in a dense, busy city versus living in the countryside: different goods and services may be needed to realize the same level of welfare in these alternative circumstances. Income indicators do not correlate well with capabilities and opportunities in these various welfare-relevant dimensions.

2.4 Empirical Analysis of Individual Happiness and Social Welfare

A growing field of subjective well-being analysis on the basis of empirical data has produced many insights about the determinants of welfare and happiness.[6] Studies of this type are being undertaken by economists, psychologists and sociologists. This research has first of all delivered the insight that, somewhere in-between 1950 and 1970, the increase in welfare stagnated or even reversed into a negative trend in most western (OECD) countries, despite a steady pace of GDP growth. Blanchflower and Oswald (2004) offer such an analysis for the UK and the USA.[7] This insight is supported by the 'Eurobarometer surveys', the half-yearly opinion polls of the inhabitants of the EU Member States, as well as by corrections of GDP that seem to point more in the direction of social welfare (e.g. the ISEW indicator of Daly and Cobb, 1989; see Section 5 on this). The income level at which de-linking occurs between GDP and (subjective) social welfare has been estimated to approximate $15,000 (Helliwell, 2003). This has been referred to as a 'threshold hypothesis', reflecting that the costs of growth exceed the benefits (Max-Neef, 1995: p.117): ". . . for every society there seems to be a period in which economic growth (as conventionally measured) brings about an improvement in the quality of life, but only up to a point—the threshold point—beyond which, if there is more economic growth, quality of life may begin to deteriorate." It seems wise to consider any estimates of the threshold point as a rough indication, which may not hold generally for all countries and cultures. Nevertheless, the various empirical findings provide evidence for a stabilization of social welfare in spite of continued GDP growth.

Subjective well-being studies also show that, at the individual level, income does not perfectly correlate with welfare–indeed much less than is often taken for granted–so that individual income is not a good proxy of individual welfare (Easterlin, 2001; Frey and Stutzer, 2002; van Praag and Ferrer-i-Carbonell, 2004; Ferrer-i-Carbonell, 2005). Relative income turns out to be critical (see Section 2.5). In addition, other—income—independent—factors influence individual welfare or happiness. Important ones are: being employed, having a stable family (and having a partner), being healthy, personal freedom (political system), having friends, and belonging to a tight social community. This type of research further shows the relevance of unobservable or not easily observable factors, notably a pessimistic or optimistic attitude towards life in general. A recent empirical study by Ferrer-i-Carbonell and Frijters (2004) concludes, on the basis of an analysis of the effect of this attitude, that the belief that "being rich makes people happy" can better be replaced by "happy people are more likely to be rich". For it appears that optimistic individuals are on average relatively happy and successful in life (ceteris paribus), and on the basis of the latter enjoy a relatively high average income. In addition, this type of research indicates that the well-being of men on average responds differently to income changes than that of women. Responses also differ among income brackets. Now if income does not render a reliable and robust measure of happiness at the micro-level, then it is extremely unlikely that the aggregation of individual incomes in a GDP provides a good indicator of social welfare at the national level.

Another important insight of this literature is that individuals adapt or get used to changed circumstances, such that their subjectively felt well-being does not increase (Frederick and Lowenstein, 1999). This relates to the fact that our senses can only handle a limited amount of stimuli, so that beyond a certain threshold a feeling of satisfaction or boredom arises. A change in circumstances can of course create a one-off or ephemeral welfare effect that quickly fades away. Since people do not realize the phenomenon of adaptation they keep striving for 'more'. Terms like 'addiction', 'hedonic adaptation' (Helson 1964), 'hedonic treadmill' (Brickman and Campbell, 1971) and 'preference drift' (van Praag, 1971) are used in this respect.

Utilizing subjective well-being indicators for a large number of countries—on the basis of World Values Surveys data www.worldvaluessurvey.org—Layard (2005) concludes that, whereas countries with high incomes show little variation in average reported happiness, this is quite different for countries with low incomes. The first group is dominated by countries of Protestant origin, which may point to a religious factor at stake. But, on closer inspection of both groups, it appears that one cannot exclude a serious influence of climate conditions and political systems (communism versus enlightened capitalism) on happiness (see also Boersema, 2004). This illustrates that there is no simple relationship between GDP and happiness or welfare. In addition, the country comparison clarifies that happiness is characterized by diminishing returns of increases in GDP per capita.

Happiness evidently depends on leisure. But leisure is not captured by GDP. Quite the contrary, it has an opportunity cost of not being productive in terms of contributing to GDP. A recent study by the OECD (2006) makes adjustments of GDP by valuing leisure at GDP per hour worked (somewhat debatable), and finds that the result (in per capita terms) leads to a different ranking than according to GDP per capita. In this ranking, The Netherlands scores best of all OECD countries, for two reasons: the inactive part of the working force is relatively large, and part-time working is very common (cf. de Groot et al., 2004).[8]

2.5 Income Distribution, Relative Welfare and Rivalry

Sen (1976, 1979) considers the implicit treatment of income distribution as the main objection against GDP as a measure of welfare. The GDP per capita indicator emphasizes average income. An unequal distribution implies unequal opportunities for personal development and well-being.[9] Furthermore, individuals or families with low incomes benefit relatively much from an income rise, because of the diminishing marginal utility of income. GDP per capita does not, however, distinguish between the expenditures of the poor on basic goods and of the rich on luxury (and often status) goods. In fact, given the higher prices of the latter these implicitly receive a relatively high weight. Of course, GDP growth can occur with a decrease in income inequality, but this is not a general fact. The Kuznets (inverted U) curve is often considered as indicative of the temporal relationship between income level and income inequality as countries undergo economic development (Kuznets, 1934). Nevertheless, it is not so relevant for describing countries beyond a certain income level. Then, complex interactions between economic and political cycles will affect the income distribution. For example, a higher GDP or national income can offer more financial room for public expenditures that redistribute income, such as social security, which in turn contributes to a higher social welfare (ceteris paribus).

A related but more subtle aspect of distribution is that individual welfare cannot be separated from the welfare of other individuals in the relevant social environment, also known as the 'peer group'. Therefore, the term relative welfare or context dependent preferences is used (Tversky and Simonson, 2000). Such preferences are characterized by an urge to compare oneself with others and rivalry ("keeping up with the Jones's") or "reference drift" (Kapteyn et al., 1978). This is (also) a finding of empirical studies on the basis of subjective well-being. The relevant social context of individual welfare does not need to be fixed, but can change over time as a result of information and the media. Globalization means that the media transfer consumption images across the planet, with possible consequences for peer group size and relative welfare of people. Consistent with the notion of relative welfare is the idea that poverty has a relative dimension (here Sen's notion of 'capabilities' is relevant too). Subjective well-being research has shown that poverty often means that individuals are unhappy because they can consume much less than the majority of individuals in their social environment. Consumption surely is not only driven by (basic) needs but also by imitation and search for status.

The striving towards conspicuous consumption (Veblen, 1899), "positional goods" (Hirsch, 1976) and "status goods" (Howarth and Brekke, 2003) are at the core of rivalry in consumption. On the basis of experiments and surveys, Alpizar et al. (2005) find that relative consumption not only plays a role in the case of goods like houses but also holidays and even insurance.[10] Earlier, Solnick and Hemenway (1998) assessed that a majority of respondents would rather opt for being poor in absolute terms and rich in a relative sense than vice versa.[11] Ever since Darwin, biologists have known that the function of conspicuous and extravagant features of animals is to attract sexual partners and repel competitors.[12] Humans are no exception—we are animals after all. Moreover, it is confirmed by studies across time and cultures (e.g. Buss, 1989). The fact that individuals who already have a partner and offspring still keep seeking for status through consumption is the mere result of the automatic nature of this type of behaviour, which became fixed in our genes through repeated sexual selection within our species and its predecessors. Striving towards individual income growth is thus completely understandable but will not necessarily lead to an increase in happiness when others aim at the same goal. In fact, more inequality will tend to lead to a happy few relatively rich people and a large majority of less happy relatively poor ones, suggesting a negative effect on social welfare. The GDP completely omits the relative income aspect of welfare.

Relative welfare is closely related to changes in preferences. Consumer preferences are to a large extent formed by the media that in turn is steered by commercial (business) interests. Here ample use is made of individuals' feelings of rivalry. In other words, advertising (mis)uses our sensitivity to status and imitation or fashion. Children show the utmost sensitivity to advertising aimed at fostering rivalry, but adults do not behave fundamentally differently. The rivalry in the striving towards individual growth of income and consumptive outlays is referred to as the "rat race" and the "affluenza virus" (e.g. Layard, 2005). Income growth almost always goes together with new products and related changes in preferences, but no one guarantees that creating new preferences contributes to people's happiness. Reference drift can then ultimately result from a combination of advertising and comparing with, as well as imitating, others.

The phenomenon of relative welfare does not just explain why humans strive for income rises. It also clarifies that an increase in relative income can improve the welfare of the respective individual, whereas social welfare is not being served by it. The reason is simple: rises in relative income and welfare are a zero-sum game: one individual loses what another gains (Layard, 2005).[13] In other words, you cannot make everyone increase in relative welfare. The relatively rich are generally happier than the relatively poor; this has always been so, and GDP growth will not change it. The rise of the relative income of an individual can be regarded as a negative external effect (external cost) on the welfare of the one whose relative income drops as a result of it. As is well known, externalities are harmful to social welfare, and need to be corrected.[14] The important conclusion of the foregoing for our purpose here is that GDP entirely leaves out considerations of relative welfare and rivalry in consumption.

2.6 Formal Versus Informal Economy

In general, GDP just covers activities and transactions that have a market price and thus completely neglects informal transactions between people that occur outside formal markets. The formal market dimension of human activities can comprise a large or small part of total human activity, depending on whether one observes OECD countries (a large part), economies in transition (medium) or less developed countries (small). The fact that the informal economy is left out of consideration explains why GDP per capita for many countries in the latter group can be so extremely low. At the same time, it can easily give a wrong picture of how (un)happy people really are.[15] This is amplified by the problem that the size of the informal economy relative to that of the formal economy may change considerably over time, both in developing and developed countries. For the Netherlands, for instance, the variation in the estimates of unpaid household services as a percentage of GNP has changed between 1975 and 1990, i.e. over a period of only fifteen years, from [67–108 percent] to [51–91 percent] (Bos, 2006).

Actual GDP growth often comes down to a transfer of existing informal activities (unpaid labour) to the formal market. This means that the benefits were already enjoyed but the market costs were not yet part of GDP. This clearly illustrates the earlier point by Mishan and Daly (Section 2.1) that GDP reflects the costs of reaching a certain, unknown welfare level and not that welfare itself (i.e. the benefits). With transfer of existing activities from the informal to the formal circuit, economic growth means that the costs increase more rapidly than the benefits, and in the worst case only the costs rise. This holds, for instance, when informal activities like subsistence agriculture in developing countries, voluntary work, household work, and child care disappear. Such activities originally took place within the informal family circle and the local community. Transfer from the informal sphere to the formal market also occurs when people are born, die or are nursed in a hospital instead of at home.

The GDP therefore does not recognize the value of all kinds of informal activities and services. As a result, public policy is often aimed at cutting back and discouraging informal activities. This can be interpreted as a strongly normative goal that is not entirely without risks for social welfare. The transition from an informal to a formal economy in itself offers no guarantee for a rise in happiness or welfare. For example, local social contacts—that form the basis for stable and happy lives—are much stronger and occur more frequently within informal than formal economies. In other words, 'society' in the strict meaning of the term has a value that is not captured by GDP.

Obviously, it is not my intention to defend the extreme position that a transition from an informal to a formal economy automatically works out badly for social welfare. It is evident that labour division and specialization can be carried through more extensively in a formal than in an informal economy. It is possible, though not certain, that as a result productivity increases, labour conditions improve and the choice spectrum (diversity of products) for consumers is widened.[16] These advantages do not neutralize the earlier mentioned negative welfare consequences of a transition to a formal market economy. Indeed, many other negative aspects should be taken into account. For example, if the labour market grows in scale, it stimulates commuting (distances), as well as people changing their house for a job. This erodes local community structures, with negative effects on individual happiness. In developing countries, the trajectory towards the formal economy often goes together with a large-scale migration of 'subsistence' farmers with big families to the slums of large cities.

The main message here is that GDP cannot serve as a measure to judge the welfare impact of fundamental changes that involve transitions of the informal to the formal economy. The expansion of markets to include informal activities is not always good for social welfare, even if GDP is raised. One might suspect that a certain combination of informal and formal (market) relationships between people would render the best of both worlds. With the GDP indicator, however, one cannot judge this in any way, since GDP omits the informal dimension of the economy. In the light of the continuous, idealistic public debate on the (un)desirability of expansion of the market domain it is therefore of utmost importance to not rely on GDP but to use adequate, real welfare indicators.

2.7 Environmental Externalities and Depletion of Natural Resources

The previous point is a specific case—through conceptually a very important one—of the more general criticism that GDP omits the value of non-market goods and services. Another example of unpriced effects relates to the natural environment and resources. This involves negative external effects as well as goods and services delivered by nature. The presence of externalities means that the current set of market prices insufficiently reflects the total (private + external) costs, which makes these prices unreliable signals in whatever calculation aimed at producing a social welfare indicator. Moreover, if air, water, or a natural area are being polluted any damage does not enter GDP, but when pollution is being cleaned this increases GDP. In addition, the (capital) depreciation associated with environmental changes (fish stocks, forests, biodiversity) and depletion of resource supplies (fossil energy, metal ores) is missing from the GDP calculation.[17] As a result, we are considering ourselves 'richer' than we really are (Atkinson et al., 1997). It makes sense to define real, meaningful income as sustainable income: namely, as the maximum amount of income (or consumption) in one period without depleting capital or without harming the capacity to generate the same or higher level of income in future periods (Hicks, 1948). Maintaining this capacity requires sustainable capital utilization, regardless of the type of capital: human capital, machinery, and natural capital (natural resources and ecosystem services). Here we focus on sustaining the latter.

A fundamental consequence of neglecting sustainability of natural capital has already been mentioned: namely, that the use of GDP as an indicator of welfare and progress means regarding substitution of basic conditions—like space, serenity, and direct access to nature and water—by market goods–like large houses, roads, cars, sewage systems and water purification, and expensive holidays in exotic locations—as progress. This in turn will unnecessarily stimulate the replacement of 'nature' by the 'market economy'. A correct economic welfare approach would only characterize changes as real progress (welfare improvement) if they are accompanied by a sustainable use of environment and nature. Hueting (1974) already recognized this early on, and his elaboration of a measure of a green or sustainable income is based exactly on this insight (Gerlagh et al., 2002).[18]

2.8 Aggregation of Information

A general shortcoming of GDP as an indicator of social welfare has to do with its aggregated character. Aggregation of information always leads to information loss. The advantage could, of course, be improved overview. Searching for an unambiguous indicator will, however, always be problematic because weights for different components are not always evident. Moreover, it is very ambitious to require any resulting indicator to be usable in all time periods, countries and development phases. Given the discussions in Sections 2.4 and 2.5, a basis for it would have to be looked for in the information and insights generated by studies on subjective well-being, since individuals are to be regarded as the best judges of their own happiness. This would also do justice to the statistical variation in relationships between individual income and happiness. Nevertheless, even though one can believe in empirical (subjective) measures of ordinal or even cardinal well-being, the aggregation of these over individuals to arrive at social welfare seems overly ambitious.

2.9 *Synthesis*

The above list of arguments is surely non-exhaustive.[19,20] Nevertheless, it makes flagrantly clear that defending GDP as a social welfare indicator is futile.[21] An effort to summarize all the arguments provides a quite general picture of GDP per capita growth: namely, as consisting of five main elements. (1) real individual and social welfare growth when unsatisfied basic needs are being fulfilled; (2) a transfer of activities from the informal to the formal (market) circuit, with an often neutral or possibly negative effect on social welfare; (3) adaptation to a higher income; (4) a change in the income distribution, implying a rise in relative welfare for some and a fall for others, and rivalry with regard to individual income growth and consumer expenditures, with largely negative consequences for social welfare; and (5) damage done to the environment and nature, with negative impacts on social welfare. Adaptation and rivalry may together mean that a major part of increases in income (70–80 percent) is not translated into improvements in happiness or well-being (van Praag and Ferrer-i-Carbonell, 2004).

A difficult but important question is: To what extent do elements 2–5 cancel out welfare gains arising from element 1. Corrections of GDP try to answer this question in a sort of empirical manner. A general theoretical outcome is unlikely here. Nevertheless, limits to welfare gains in terms of element 1, largely neutral effects in terms of elements 2–3, neutral to negative effects in terms of element 4, and negative effects in terms of element 5, all together suggests the existence of a maximum level to social welfare. That is, beyond a certain per capita income level, persistent GDP growth will be completely disconnected from trends in welfare. In other words, social welfare will stabilize beyond a certain income threshold.

3. How Serious is the Influence of GDP Information on the Economy?

3.1 *Mechanisms of Influence*

Despite the fundamental and many critiques, GDP is still being considered as an important source of information to measure economic progress. A not unusual response to the critiques is that one should not worry too much about the shortcomings of GDP as it does not actually have so much influence on the real economy. Many signs, however, point to the opposite. In the first place, one can wonder why then do governments invest structurally in calculating and predicting GDP.

Furthermore, one can identify concrete influences of GDP information on economically-relevant decisions. Banks and financial markets have made the prediction of GDP a core indicator of their dealings. Central banks adjust their interest policy when expectations about growth are beyond or below certain thresholds or do not become true. Private companies regard GDP growth as an important element of the general investment climate. And the trust of consumers, which determines their purchasing behaviour, is easily influenced by expectations of GDP growth.

The conclusion is clearly that the influence of GDP information should not be underestimated. It runs through multiple channels—government, politics, private businesses, financial markets and consumers. This influence is reinforced as all kinds of public and private research institutes and advisory councils give much attention to GDP information. The consequence is a large effect of GDP information on consumption, savings and investment decisions, with evident repercussions for economic structure and the social environment.

3.2 *A Self-Fulfilling Prophecy*

GDP growth and economic stability are characterized by a 'self-fulfilling prophecy' mechanism. If everyone believes that GDP has a large influence on economic reality and this belief induces pessimistic and optimistic responses by individuals, firms, and governments to low and high GDP growth (predictions), respectively, then the belief is translated into reality. GDP information in this way creates a pro-cyclic effect. National governments, advisory boards, central banks, and international organizations such as the IMF and the OECD reinforce this phenomenon. The newspaper reader is indoctrinated by the media–and the economics student also by his education–with the idea that GDP growth is relevant. GDP can thus be seen as an abstraction invented by humans without direct physical meaning. The GDP concept is active in the domain of perceptions, theories and beliefs. Only in this way does it influence the real economy.

The 'self-fulfilling prophecy' character of GDP growth resembles the way in which behaviour in financial markets is steered by perceptions. The majority of small investors are responsive mainly to general information about the market and the economy (including GDP) that is publicly available, rather than to private information about specific investment opportunities obtained through their own in-depth research. This herd behaviour causes expectations to become true. Likewise, general information about GDP can lead to large market responses. This is in effect because individuals imitate one another and act on the basis of the same, and unfortunately misleading, information that GDP represents.

Politicians are worried about low GDP growth because they fear negative voter responses.[22] To some extent this is motivated by the belief that insufficient growth will lead to a recession (instability) with much unemployment (see the discussion on 'self-fulfilling prophecy' below). In addition, GDP growth allows for rising tax revenue–as a result of which public expenditures can increase–a nice prospect for politicians in power. But 'no growth' should not be confused with 'no GDP'. There is no reason to fear that without GDP the economy would end up in a recession; indeed, no single study supports this worry. In fact, the chance of recessions is very likely much smaller because the 'self-fulfilling prophecy' of negative GDP growth expectations disappears once GDP is abolished.

4. Does GDP Convey any Useful Information?

Are there advantages of using the GDP indicator that compensate for its discussed shortcomings? Here I review a number of common defences. A possible advantage might be that GDP growth creates trust and economic stability. But the previous discussion has shown that the reverse side of economic stability on the basis of GDP growth expectations is that instability results from negative expectations. Moreover, expectations become reality through the perceptions and information about GDP. Without them, expectations would not be influenced this way so that the specific cause of instability would no longer exist (this does, of course, not remove all causes of economic instability).

That GDP only has an effect on the real economy through the domain of perceptions is consistent with the fact that it is an aggregated, macro-level indicator, while the real economy is the outcome of micro-level processes scaled up. This does not, of course, mean that one cannot design abstract macro-level models in which the GDP indicator occupies a central role, through mechanisms that generate aggregate consumption, savings, investments, trade volumes, and tax revenues. The current critique is not inconsistent with letting GDP play this role of model variable (even though it is contrary to the micro-foundations project). Nevertheless, it is an entirely different issue to assign to GDP the role of a central macro-indicator with an implicit or explicit welfare interpretation.

A regular defence in oral discussions with colleagues is that GDP per capita conveys information about productivity. But this is not correct. It is often thought that average labour productivity of a country is identical to GDP per capita. But a correct productivity measurement needs to be related to the number of hours worked, which varies between countries, as well as over time. GDP per hour is a more useful indicator of productivity than GDP per capita. It might be interpreted as an indication of freedom, power, or even potential welfare (somewhat comparable with Sen's capabilities). Nevertheless, increase of labour productivity cannot be an ultimate goal that supersedes all other goals. Another problem with national productivity indicators is that they hide the diversity of productivity levels among sectors. It is much more useful to know which sectors perform relatively well and poorly, both in a national and international context. Aggregating across very different sectors does not serve any (welfare) purpose.

An often-mentioned advantage of GDP is that it can serve as a basis to roughly estimate tax revenues. This can in turn be useful to predict taxes in the future, to evaluate creditworthiness in the case of providing loans to countries (as done by the IMF and the World Bank), or to determine fair financial contributions of member states to a federation of states (e.g. USA, EU). In this case, GDP functions not as a performance indicator but more modestly as a model variable.[23] Of course, estimates based on disaggregated information (e.g. value added per sector) will be required in order to arrive at a sufficiently accurate estimate of tax revenues. In other words, for such tax calculation purposes, there is no need to aggregate national accounts into a GDP.

In addition, one can point to the importance of economic growth for developing countries. Indeed, one would expect welfare growth here to show a higher correlation with GDP growth than in rich countries (especially because of the arguments in Section 2.3).[24] In fact, however, this correlation turns out to be unstable, meaning that the advantages of growth are not automatically and consistently realized (possibly explained by the arguments presented in Section 2.6).[25] The ultimate goal is, of course, welfare improvement through economic development, not GDP growth itself. There is therefore a need for concrete welfare indicators, notably for poor countries where development aimed at welfare growth is a complex issue.

Finally, one can regard the international standard for national accounts and GDP as a guarantee for uniformity of data on GDP. This contributes to a clear economic comparison of countries. This is, however, a necessary but insufficient condition for useful international comparisons. Indeed, one can search for a lost key in the light of a lantern (= guided by GDP), but when the key lies elsewhere a more effective strategy is to grope in the dark (not guided by GDP) (Meadows et al., 1981). A disadvantage of the international GDP standard is, moreover, that it will not be easy to implement improvements in the GDP calculation method to neutralize the critiques documented here (assuming that such improvements are in principle feasible). Many proposed improvements have met a lot of resistance from various organisations and countries, often for strategic reasons (presently in the EU because all kinds of redistribution decisions are linked to GDP).

The foregoing suggests a number of non-welfare information features of GDP. Of course, GDP can reflect certain aspects of reality in which one may be interested. Examples are the size of the formal versus the informal economy, prognoses of tax revenues, and increases in productivity. None of these, however, even comes close to completely capturing social welfare. Moreover, in many cases better indicators or more disaggregate income type of indicators are available at sector levels. It is thus important to realize that GDP can still serve a useful role in providing information to construct non-welfarist economic indicators, but that for this purpose it does not need to play any central or public role.

5. Alternatives to GDP

It is wise to remove an indicator that is seriously misleading, irrespective of whether an acceptable alternative is available. Hence, the removal of GDP information would be an enormous improvement because a structural information failure is being eliminated that is without precedent. In other words, abolishing GDP should be unconditional. Moreover, one of the conclusions of Section 2.9: namely, that social welfare will stabilize beyond a certain income threshold, suggests that beyond this threshold there is no need to measure social welfare, as the latter will have reached its maximum value. But it is true that until this threshold is reached, welfare improvement in principle is still feasible, so that one may wish to measure it. For this purpose, the main alternatives for GDP as an indicator of social welfare are briefly reviewed in this section.

There are three types of alternative indicators available now. A first type is based on rather pragmatic, accounting adjustments to GDP, such as the Index of Sustainable Economic Welfare (ISEW: Daly and Cobb, 1989), derived indicators like the Genuine Progress Indicator (GPI)[26], and the Sustainable Net Benefit Index (SNBI) (Lawn and Sanders, 1999). These indicators represent a correction of the regular GDP by repairing important deficiencies through adding or subtracting certain partially-calculated money amounts to/from GDP.[27] The ISEW is aimed at measuring the (consumption related) services that directly influence human welfare. This is accomplished by adding to GDP services that it omits, while deleting GDP categories that do not directly render services to consumers. The ISEW can thus be considered as a measure of the benefits of economic activity (see Section 2.1). In addition, the ISEW includes corrections to neutralize income inequality and the unsustainability of production and consumption. In particular, the ISEW approach adapts GDP for non-market goods and services (housework), defensive costs of social and environmental protection and repair (health expenditure, costs of road accidents, costs of urbanization), reduction of future welfare caused by present production and consumption (loss of natural areas, loss of soil, depletion of non-renewable resources, air and water pollution, greenhouse effect), the costs of efforts to obtain the present welfare level (commuting, advertising, duration and intensity of work), and the distribution of income and labour (inequality among workers, between employed and unemployed, between males and females). The GPI deviates slightly from the ISEW in terms of the specific categories of corrections included. Important additional categories that the GPI corrects for are voluntary work, criminality, divorce, (loss of) leisure time, unemployment and damage to the ozone layer.

The ISEW has been calculated—using slightly distinct methods–for a range of regions and countries, including Australia, Austria, Chile, Denmark, Germany, Italy, the Netherlands, Scotland, Sweden, and the UK (an overview of studies is given in Lawn, 2003). The various applications show that, whereas GDP follows a rising trend, the ISEW shows a constant or even decreasing pattern after a certain time. The temporal breakpoint varies with the country, but lies somewhere in-between the late 1960s and the 1980s. Important reasons for this de-linking of GDP and ISEW have been a substitution of informal household production by services provided by the market (e.g. child care), increased inequity, natural resource depletion, and the emergence of global environmental problems (global warming, acid rain, biodiversity loss). For example, the GPI for the USA increased during the 1950s and 1960s, but has declined by about 45 percent since 1970. Moreover, the rate of decline in per capita GPI has increased from an average of 1 percent in the 1970s to 6 percent in the 1990s. Both ISEW and the GPI suggest that the costs of economic growth now outweigh the benefits, leading to "growth that is uneconomic" (paraphrasing Herman Daly).

Neumayer (2000) questions the general findings of the ISEW and GPI studies. Using sensitivity analysis he suggests that the widening gap between ISEW (GPI) and GDP—supporting the 'threshold hypothesis' (Section 2.9)–might be an artefact of debatable methodological assumptions with regard to the valuation of non-renewable resource depletion (resource rent or replacement cost) and cumulative long-term environmental damage. In addition, Neumayer notes that the way inequality (changes) is addressed is ad hoc and should be replaced by making a preference for income equality—or aversion to inequality–explicit. For example, Jackson et al. (1997) use an Atkinson index (Atkinson, 1970). Lawn (2003) emphasizes that ISEW and GPI require more robust monetary valuation in order to arrive at acceptable indicators of social welfare.[28]

A second type of indicator also starts from GDP but focuses entirely on environmental externalities and natural resource depletion. Corrections here give rise to 'sustainable' or 'green(ed)' GDP type of indicators. 'Sustainable income' denotes a level of income that can be sustained, i.e. that is based on a reproducible economic and environmental base. The concepts or indicators of green and sustainable GDP are rooted in welfare economics. Important externalities are noise, air and water pollution, soil erosion, resource exhaustion, desiccation, fragmentation, biodiversity loss, radioactivity, and various health-affecting toxins. Recalculation of a GDP with externalities 'internalized' is not a simple matter, as it implies a completely different set of prices in the economy. It is not surprising, then, that there have been few empirical exercises aimed at calculating a green or sustainable income.

The best known of these is Hueting's Sustainable National Income (SNI), which has been developed for the Netherlands (Gerlagh et al., 2002). It is based on the conceptual work by Hueting (1974) and can be seen to reflect the basic notion of 'sustainable income', as expressed by Hicks (1948). The SNI approach uses a general equilibrium model that calculates the impact on national income of imposing sustainability constraints for the nine most important environmental themes (for the Netherlands): climate change; depletion of the ozone layer; acidification; eutrophication; fine air-borne particles (PM10); volatile organic compounds; dispersion of heavy metals and PAKs/PCBs to water bodies; dessication; and soil contamination.[29] In particular, data on abatement costs associated with these environmental themes are integrated within an existing and somewhat adapted general equilibrium model. This approach not only implies a strong sustainability framework (preservation of all types of natural capital), as it allows for neither trade-offs between environmental themes nor substitution of natural by economic capital. But also the approach comes down to regarding the value of environmental degradation as being equal to the conservation costs. El Serafy (2001) has criticized this, arguing instead in favour of a 'user cost method', which would lead to a higher sustainable income value, where the difference would depend on the speed of natural resource depletion. The (static) general equilibrium approach is required as some of the sustainability constraints on the nine environmental themes are so tight that technical measures alone cannot realize them, so that economic restructuring is inevitable. The policy interpretation of this SNI approach is that an economy is subjected to a strong sustainability policy with a tremendous impact on national income: the calculations for the Netherlands show that the SNI is roughly half the size of GDP (Gerlagh et al., 2002). In itself, this is not very informative. Differential time patterns for SNI and GDP would be of more interest. Hofkes et al. (2004) have therefore analysed the development of SNI for the Netherlands over the period 1990–2000, for 1990–1995 and 1995–2000. They find that not only did SNI increase substantially in this period, but moreover SNI growth rates exceed GDP growth rates for both sub-periods. Over the whole period 1990–2000, the enhanced greenhouse effect appears to be the binding environmental constraint that determined most of the developments for the SNI. Nevertheless, the gap between NNI and SNI remains considerable.

Comparing SNI with ISEW (and GPI), it becomes clear that the first has the advantage of taking into account general equilibrium effects of corrections, but the disadvantage of restricting itself to environmental and natural resource issues. ISEW and GPI correct for a much wider array of GDP imperfections, even though in a partial manner that is likely to involve mutually inconsistent corrections. Furthermore, the SNI results are sensitive to the exact specification of the sustainability condition for each environmental theme, since the marginal abatement costs are sharply rising for low values of pollution or resource use. It is, however, fair to say that the arbitrariness of sustainability conditions also affects the ISEW value, but in a less extreme way.

A third type of indicator relates to distinguishing between measures of current well-being and measures of well-being over time. The latter, however, turn out to be largely theoretical in nature (see also Section 2.2). Dominant approaches here are net present value type or discounted utilitarian intertemporal or multi-generational welfare functions (e.g. Weitzman, 1976), and Rawlsian or fairness-biased maxmin functions (Rawls, 1972; Arrow, 1973; Solow, 1974). A pragmatic indicator that focuses on intertemporal issues is genuine savings (or genuine investment). It means maintaining or increasing wealth, opulence or total capital—the sum of economic, human and natural capital—by sufficiently saving in a broad sense (Hamilton and Clemens, 1999; Dasgupta and Mäler, 2000). Recently, genuine savings (GS) has been adopted as a central indicator by the World Bank, under the name of 'adjusted net savings'. GS can be defined as traditional net savings subject to a number of corrections (Bolt et al., 2002): (i) the value of depletion of natural resources is deducted; (ii) the costs associated with pollution damage, including economic and health effects, are deducted; (iii) expenditures on education are treated not as consumption but as savings/investments in human capital and thus added; (iv) net foreign borrowing is deducted, while net official transfers are added; (v) capital depreciation (capital consumption) is deducted. The result represents a weak sustainability indicator, in that it allows for substitution of nature and natural resources by produced and human capital (Hartwick, 1977). Categories (i) and (ii) are the most difficult to estimate. Nevertheless, the World Bank has produced estimates for most countries in the world. The outcome is that, as a general rule, GS are less than half the gross savings. Moreover, genuine savings are negative for the Middle East and North Africa, and Sub-Saharan Africa regions, positive for OECD countries, and the highest for the East Asia/Pacific region (World Bank, 2006).

A main disadvantage of the GS indicator is that losses of natural capital are not regarded as worrisome as long as they are compensated by economic and human capital (weak sustainability). However, a positive value of GS does not always imply environmental sustainability. The advantage of the GS approach is that it evaluates rapid growth that goes hand-in-hand with consuming, rather than with investing the revenues of unsustainable resource exploitation as negative (i.e. a negative value of GS). But a disadvantage of the approach is that it adopts a partial perspective with respect to time, as it neglects historical contexts. For instance, a country that has depleted all its natural resources can hardly score negative on genuine savings afterwards. At a more fundamental level, one can criticize the approach for assuming that changes in wealth or investment are a good proxy for changes in well-being and social welfare. However, there is no high and stable correlation between wealth and well-being or happiness, apart from the relative income effect discussed earlier in Section 2.5.[30] Against the background of the various criticisms of GDP in Section 2, however, the main shortcoming of the GS approach is that it mainly addresses the problem of capital depreciation (Section 2.7), and may partially cover valuation of informal activities (Section 2.6). Pillarisetti (2005) illustrates that GS is both conceptually and empirically an imperfect indicator for policy, regardless of whether it focuses on environmental sustainability or human well-being. Dasgupta (2001: Section 9.4) shows that neither can net national product (NNP) as the sum of consumption and GS serve as an indicator of welfare.

A fourth and final type of indicator of social welfare is a composite index that combines indicators that are considered to capture relevant aspects of human well-being. Unlike the previous types of indicators, this does not generate a monetary value. The best-known example of this type is the Human Development index (HDI) of the United Nations, which aggregates a number of indicators: GDP per capita (in PPP), life expectancy at birth, adult literacy rate, and combined primary, secondary, and tertiary gross enrolment ratios. The incorporation of GDP reflects, through a log-transformation and a maximum income limit, a decreasing marginal utility of income. This already means an improvement over GDP. Nevertheless, the HDI approach carries an element of arbitrariness, in the sense of selecting arbitrary components, as well as an arbitrary aggregation procedure. The latter generates normalized values for each component based on defined upper and lower bounds, and then calculates an arithmetic mean; this results in an index with a value between 0 and 1. Publications on the HDI argue that potential extensions of HDI with additional components are hampered by measurability problems. But income inequality is in any case measurable and clearly an important criterion for evaluating the position of, and changes in, developing countries. Moreover, it would in principle be feasible to develop quite objective indexes of political freedom[31], time use (work, leisure, commuting), and available public health services. Not surprisingly then, various proposals have been done to extend or adjust the HDI, so as to address some of the omissions (e.g. Hicks, 1997; Noorbakhsh, 1998). In addition, other approaches to aggregate the components of the HDI are available, such as the Human Poverty Index (similar components as the HDI but differently weighted) and the Borda ranking. Dasgupta (2001: chapter 5) uses the latter procedure to extend, for illustrative purposes, the HDI with per capita private consumption and indexes of political and civil rights. In spite of its aforementioned deficiencies, the HDI is considered to be an improvement over GDP, especially for evaluating changes in developing countries. A main disadvantage of the HDI in comparison with the other indicators is a complete neglect of (environmental) sustainability. Dasgupta (2001, Section 5.8) notes that the HDI can be seen as "one-third intertemporal" because of the inclusion of adult literacy; but he adds the shortcoming that, although this reflects a capital asset, the HDI does not cover all relevant types of capital and is therefore inadequate to provide information useful for addressing intertemporal concerns. Neumayer (2001) proposes to combine the HDI and GS indicators to arrive at a more complete picture of sustainable development, notably of poor countries. He does, however, not arrive at a really integrated (composite) indicator.[32]

Comparing the aforementioned alternative indicators of social welfare in light of the main points of criticism of GDP as noted in Section 2, it turns out that, at present, there is no perfect alternative available.[33] All available approaches are far from perfect and do not succeed in systematically repairing the list of shortcomings of GDP as a social welfare indicator noted in Section 2. In particular, the dynamic aspects, lexicographic preferences (basic needs), subjective well-being basis, and relative welfare and rivalry are neglected. Nevertheless, one can expect all of these alternatives to serve as a much better approximation of social welfare than GDP. ISEW (and GPI) are perhaps the most complete in that they try to repair multiple shortcomings as opposed to SNI and GS. A disadvantage of ISEW, however, is that is based on partial corrections. Finally, all alternatives except HDI address environmental (capital) sustainability in one way or another, while ISEW and SNI adopt a strong and GS a weak sustainability perpective. HDI is the least attractive from a methodological viewpoint, and certainly unsuitable to subtly evaluate richer countries.

In conclusion, an ideal indicator of social welfare is not available. This would require an approach that takes its starting point in the findings of research on happiness and subjective well-being. ISEW can be regarded as the most balanced alternative available right now, which is a clear improvement over GDP. Still, if GDP is replaced by ISEW or another measure then there is a risk that growth fetishism—i.e. striving for growth under all circumstances—will be directed at this alternative. Evidently, this would be undesirable if such a measure would still be far from perfect.

6. Policy Implications

6.1 *Removing an Information Failure*

We have seen that GDP not only provides misleading information about social welfare but also exerts a large influence on economic reality, and therefore on the daily life and well-being of all of us. One can frame this phenomenon as a serious form of information failure, which is an instance of the general case of market failures, or given the fact that the government generates GDP information, as an instance of government failure. GDP information influences all agents in the economy: consumers, savers, investors, banks, stock and option markets, private companies, the government, central banks and international organizations.[34] Because of the misleading nature of GDP information economic agents take wrong decisions from the perspective of social welfare. Given the many shortcomings of GDP as a measure of social welfare and the economy-wide effects, one has to expect a large loss of social welfare, certainly in the long run—when repeatedly and cumulatively false information steers economic decisions. Currently, economists are insufficiently aware of this potentially huge cumulative negative impact of GDP over time. In fact, I am inclined to think that there is no larger information failure in the world than that caused by the GDP indicator.

Economists and their schooling play a central role in maintaining the idea that GDP information matters, in a positive sense. Economic studies should pay due attention to the shortcomings of GDP information, and the entire curriculum (notably macroeconomics) should be screened for the use of GDP information. This will inevitably imply a thorough revision of many textbooks.[35] Economists should also inform and convince politicians, banks and financial markets that they should no longer let their decisions be influenced by GDP information. Research and publications by economists should be filtered for use of GDP information. By including GDP indicators in economic studies the myth that GDP matters is kept alive.

The 'self-fulfilling prophecy' character of the influence of GDP information on the economy means that this phenomenon can, in principle, be avoided. Hence, the government can consciously choose to no longer supply aggregate GDP information, without threatening concrete economic mechanisms. Of course, this view does not mean the refutation of useful economic growth, that is, welfare growth, quite the opposite. But GDP growth not offer any guarantees for this. Moreover, unlimited welfare growth is very unlikely, if only because of the phenomenon of relative welfare and rivalry in income and consumption.

6.2 *A Better World without GDP*

Without the availability of a GDP indicator decisions will be more aimed at welfare improvement, since the systematic error resulting from economic behavioural responses to misleading GDP information will be gone. Such a systematic error means that the economy follows a trend away from a situation or path that is desirable from a social welfare perspective. The removal of GDP information will more likely lead to white noise type of errors (random drift), rather than systematic, cumulative and trend-like errors. Panic responses and recessions due to the threat of stagnating GDP growth are no longer possible. In addition, certain aspects of public policy will have to be adapted. Monetary policy, for example, can focus on more useful things than GDP growth. One will less dogmatically deal with stimulating developing countries to enter a transition to a formal economy (for that matter, the World Bank and UNDP has already moved a long way in this direction but do not seem to dare taking the step to discard GDP entirely). There will also be less resistance against policies—notably environmental policies–which improve welfare (partly of future generations) at the cost of GDP growth. This is especially relevant to the case of climate policy. All current economic studies focus entirely on the development of GDP under alternative climate (policy) scenarios, and therefore on the trade-off between GDP growth and climate change related risks (Kelly and Kolstad, 1999). But this trade-off is misplaced as GDP is not a good welfare measure. Moreover, this trade-off concerns a period of 50 to 100 years in the future, during which GDP of the rich countries will certainly have grown far beyond any welfare-maximizing level.[36] See further Frank (1985), Ireland (2001) and Layard (2005) for a number of examples of alterations in economic policy that are in line with replacing the GDP indicator by information obtained from the subjective-empirical welfare literature: e.g. extra taxation of working overtime, extra taxes on status goods, limiting commercial advertising; and restricting flexible labour contracts. Although from an economic growth perspective these look like bad measures, they will be more positively evaluated from a real happiness perspective.

6.3 Abolishing GDP Should Not Be Confused with 'Anti-Growth', 'Anti-Innovation' or 'Anti-Accounting'

Many people responding to an early draft of this paper have concluded that I must be against GDP growth. But they seem to confuse 'no GDP indicator' with 'no GDP growth'. It is important to realize that, without a GDP indicator, GDP growth cannot be measured at all. This merely reflects the irrelevance of GDP growth rather than anti-growth. Given that GDP is not a good welfare indicator, one should not be in favour of 'always GDP growth' (which is not the same as being against growth). To strive under all circumstances for GDP growth ('growth fetishism') puts an unnecessary constraint on the space within which we search for welfare growth. In fact, it means that GDP growth cannot be traded-off against something else (better). History, moreover, shows that, with a GDP indicator, structurally striving for (GDP) growth is not just a risk but inevitable. Evidently, the temptation is too large.[37]

My plea to abolish GDP as a macro-indicator should not be confused with a plea to restrict individual income growth. I am very well aware of the fact that income growth can mean satisfaction of more (basic) needs or more happiness due to an improved relative income position (Sections 2.3 and 2.5). From an individual perspective it can therefore make sense to strive towards income growth. Moreover, even if individuals do not get happier from individual income growth (e.g. through adaptation, or a constant or decreasing relative income position), then still status-seeking, imitation and habits make them strive for more. However, as argued in Section 2, from a social perspective continued income growth ultimately results in what is at best a zero sum game and—due to social and environmental problems—at worst a negative sum game. For this reason, amongst other things, a society and its government should not uncritically and unconditionally foster economic growth.

In addition, it is good to realize that abolishing GDP does not imply any position against innovation. It is true, however, that it will result in attaching more value to innovations that as a net effect improve or at least do not harm social welfare than to innovations that promote or are strongly correlated with GDP growth pur sang. Innovations with a postitive welfare effect and negative growth effect are perhaps the most interesting to consider. Possibly, examples can be found in the areas of renewable energy, energy-efficient houses, car-sharing systems, and energy-efficient technologies. It is of course difficult *ex ante* to provide evidence for the positive net welfare effect of certain innovations. What is important is that this does not need to be planned or regulated. Instead, removal of GDP growth incentives is likely to shift net innovations in the direction of welfare improvement. Moreover, less 'growth-fetishism' should go together with increasing attention and incentives for welfare-enhancing innovations.

Finally, in response to my critique on GDP as a welfare indicator, various commentators have noted that the system of national accounting serves a useful role. It should therefore be stressed that my critique of GDP—as a welfare indicator and how it is used in public debates and policy preparation—should not be misinterpreted as a critique of the system of national accounts. Perhaps the most important distinction that can clarify this is the level of aggregation. Whereas GDP is the most aggregated description of the economy, the national accounting system provides a detailed, disaggregated picture of the flows of goods and services along with complementary monetary flows. As a result, the national accounts can support economic modelling, analysing productivity growth of sectors, and financial planning by the government. Interestingly, accounts are being transformed and extended to eliminate various shortcomings related to informal markets, natural resources and environmental damage. On the other hand, the method of GDP calculation has remained largely the same.

6.4 Different Implications For Developing, Middle Income, and Rich Countries

Does GDP per capita serve a more useful function in evaluating processes and policies in developing countries–defined by the World Bank as having an average GDP per capita of less than US$6000—than in developed countries? It is often believed that it is especially poor countries which require GDP growth in order to improve the well-being of their citizens. In Section 2.6 it was suggested, however, that in many cases income growth here just represents a shifting of activities from the informal to the formal sector. Particularly in very poor countries this may be accompanied by a loss of local community and subsistence agriculture, as well as migration to urban slums, with predictable negative consequences for food availability, health and quality of life. Moreover, income inequality may increase during the process (the initial part of the famous Kuznets curve). Such very poor countries dominate among those that the World Bank classifies as 'low income countries' and the UN as 'low human development countries' (according to the HDI). Once countries have moved onto a track leading to a formal market economy, there does not seem to be a way back. Middle income countries may then often see a positive correlation between income growth and welfare growth. However, negative impacts on welfare and health may result there from severe environmental pollution and resource degradation. This is illustrated by the current development of China. Finally, moving on to the high income countries, the results reported in Section 2.4 support the idea that GDP growth there does not contribute much to welfare growth (the 'threshold hypothesis': Section 2.9). Indeed, welfare seems to have stagnated for the richest countries in the world, despite continuing GDP per capita growth. This is not surprising given that in these countries all basic needs are more than satisfied, so that the consumers in these countries are mainly involved in a zero-sum rivalry game of income and status consumption (Section 2.5). Combining all development stages, a non-monotonic relationship appears between well-being and GDP per capita, where for low incomes GDP growth is accompanied by decreasing welfare, for middle incomes with increasing welfare, and for high incomes with—at best—stabilized welfare.

7. Conclusions

Economists are in strong support of the idea that public policy should be based on rational arguments. They should therefore no longer delude themselves and others by allowing simplistic indicators like GDP to exert any serious influence on economic reality, and public policy in particular. In this article I have tried to make clear that the elimination of GDP from the set of macroeconomic indicators is rational. Economists can improve the world by pleading in favour of this, and thus clear the way for economic policies aimed at improvements in human happiness instead of assuming that GDP growth is necessary and sufficient for this.

Economists who support the GDP indicator or do not want to abolish it need to realize that they in fact make the assumption that there exists a structurally positive (and high) correlation between GDP and social welfare. This is an (unrealistic) assumption because there is no study which presents convincing evidence of such robust correlation. In fact, on the basis of theoretical and empirical arguments I have shown that GDP growth is compatible with decreasing or constant social welfare, that is, a negative or zero correlation. Moreover, given the satisfaction of (basic) needs, adaptation (preference drift) and a zero-sum game type of rivalry in consumption beyond a certain income (reference drift), one can only hope for a correlation that is on average, if positive, very low. All things considered, a rigid GDP growth objective will often act as a constraint on realizing social welfare growth.

The most common, almost instinctive response of macroeconomists to advice to remove GDP from the set of macroeconomic indicators has been to point to the uniform, consistent character of the GDP calculation worldwide, which allows the 'economic performance' of countries to be compared (of course, all measurements can then easily be consistently wrong). This pragmatic view raises the important question: How many substantial points of criticism are required to counter balance a pragmatic 'consistency of measurement' argument? Suppose that GDP was not measured at all but that the welfare-economic criticism on a hypothetical GDP indicator was known (published). Who would then be willing to support an initiative to implement such an indicator, knowing that it was going to play the role of a central welfare or progress indicator? A supporter would have to be heroic, as he would very likely be accused of exemplifying bad economics.

Another not uncommon response is that happiness and well-being are beyond economics. This opinion is often motivated with the suggestion that aiming at maximum happiness through public policy is overly ambitious. At the same time, economists do not worry about giving a central place to utility and welfare (maximization) in positive and normative theories of economic agents. However, utility and welfare are nothing more than the economist's jargon for happiness.

It is a well-known and often repeated fact that the GDP indicator was never developed for the purpose of welfare measurement.[38] In the absence of a better indicator, it has taken up this role. Every well-trained economist should be a warm supporter of removing market failures, in this case misleading information, from the public sphere. This holds especially when such information is generated on a structural basis and has a large influence on the real economy. From a microeconomic welfarist perspective, GDP information represents a situation of persistent market failure, namely on the grounds of imperfect information. In addition, a 'self-fulfilling prophecy' holds, that is, GDP information influences the economy only because people believe it matters. No (other) micro-level physical cause-effect mechanisms play a role. This means that the role of GDP information in the modern economy should not be considered as inevitable (a law of nature). The positive policy lesson is that there are no fundamental barriers against undoing the current type of market failure.

Evidently, it is easy to underestimate the resistance against such a scenario. Despite the fact that many respected economists have expressed or support the severe criticism of GDP as a welfare indicator, the majority of economists, journalists, investors, civil servants and politicians are not concerned at all about the imperfections of GDP information. Moreover, the usual arguments in defence of the GDP indicator consistently come down to the unproven belief that GDP structurally correlates strongly and positively with social welfare. The support for the GDP indicator thus turns out to be rather dogmatic instead of well thought over and reasoned. This is the result of subtle indoctrination by endless repetition in (economic) education and the mass media. In addition, an immense intellectual effort over time has gone into perfecting a standardized GDP calculation suitable for all the countries in the world. This has created large vested interests to continue calculation of GDP by (national) statistical offices and related governmental activities, even though many national income accountants will accept the shortcomings and misuse of GDP as a welfare indicator.

Because of these realities, we are in fact facing a situation known as 'lock-in' of a non-optimal configuration, in this case of the erroneous idea that GDP growth means progress. By definition, it is extremely difficult to escape from a lock-in situation. At least a large shock is needed. Economists could cause such a shock, by pleading together for the removal of the GDP indicator from the public sphere. Such a strategy is evidently not a plea against welfare growth, quite the contrary. Neither should it be confused with being against economic growth under all circumstances or against innovation. Indeed, abolishing GDP would imply being disinterested in whether GDP grows or not.[39] For innovation it would mean that innovations which increase GDP but not welfare will receive less support, while innovations which increase welfare (regardless of their effect on GDP) would receive more support.

With regard to alternatives for GDP, it is good to repeat that adding (welfare) indicators, such as in the case of the HDI, does not solve the problem of misleading information that GDP represents. The replacing of GDP by a corrected GDP or another (either or not monetized) aggregate welfare indicator means effectively the elimination of GDP as such. We should not, however, wait until a perfect alternative welfare indicator is available. It is unlikely that a single indicator can be constructed to undo the long list of objections against GDP. It is, however, true that each well-thought alternative will represent a better approximation of social welfare than GDP. It would therefore be a good strategy to first strive towards less misleading information and subsequently magnify the amount of correct and useful information.

My proposal here was to discard GDP regardless of a good alternative being available. Nevertheless, if we choose to use a single index of aggregate human well-being–which is not necessarily required—then we should try to come up with much better alternatives than have been proposed thus far, and better sooner than later. For the moment, the ISEW and derived indicators seem to offer the best starting point in terms of the coverage of items that need correction. Nevertheless, their calculation methods should be much improved, notably to undo the partiality and inconsistency of corrections. In order to neutralize the critiques relating to lexicographical preferences, relative welfare, status and rivalry it seems inevitable that a basis is sought in the literature on subjective well-being and happiness (see Sections 2.4 and 2.5). In particular, subjective indicators obtained from (international) studies and comparisons of happiness via surveys can form the basis for social welfare indicators. This would suggest a role for (economic) psychologists in macroeconomic policy preparation and advice. Given the findings of the happiness literature, one should also be prepared to accept that welfare can reach a maximum, so that welfare growth has its limits. Beyond any point of stable welfare, (costly) measurement of welfare aimed at assessing or discovering welfare improvements will then, of course, become quite useless.

It is easy to discard my plea as unorthodox. However, given the consistency of my position with mainstream microeconomics and welfare theory, and in view of the long list of illustrious economists that have criticized the GDP indicator (see especially the first paragraph of Section 2), my position should really be regarded as entirely orthodox. The problem is that the majority of economists have up till now been too silent, pragmatic or defensive on this issue, and therefore unwilling to draw the evident conclusion: it is perfectly rational to abolish the GDP indicator from the public sphere.

References

Alpizar, F., F. Carlsson and O. Johansson-Stenman (2005). How much do we care about absolute versus relative income and consumption. *Journal of Economic Behaviour and Organization* 56: 405–421.

Aronsson, T., P-O. Johansson and K-G. Löfgren (1997), *Welfare Measurement, Sustainability and "Green" Accounting: A Growth Theoretical Approach.* Edward Elgar, Cheltenham.

Arrow, K.J. (1973). Some ordinalist-utilitarian notes on Rawl's Theory of Justice. *Journal of Philosophy* 70: 245–263.

Arrow, K.J., B. Bolin, R. Costanza, P. Dasgupta, C. Folke, C. S. Holling, B.-O. Jansson, S. Levin, K.-G. Mäler, C. Perrings and D. Pimentel (1995). Economic growth, carrying capacity, and the environment. *Science* 268(April 28): 520–21.

Arrow, K., P. Dasgupta, L. Goulder, G. Daly, P. Ehrlich, G. Heal, S. Levin, K.-G. Mäler, S. Schneider, D. Starrett and B. Walker (2004). Are we consuming too much? *Journal of Economic Perspectives* 18: 147–172.

Asheim, G. (1994). Net national product as an indicator of sustainability. *Scandinavian Journal of Economics* 96: 257–265.

Asheim, G. (2000). Green national accounting: why and how? *Environment and Development Economics* 5: 25–48.

Atkinson, A.B. (1970). On the measurement of inequality. *Journal of Economic Theory* 2: 244–263.

Atkinson, G. (1995). Measuring sustainable economic welfare: A critique of the UK ISEW. Working Paper GEC 95-08. Centre for Social and Economic Research on the Global Environment, Norwich and London.

Atkinson, G., R. Dubourg, K. Hamilton, M. Munasinghe, D. Pearce and C. Young (1997). *Measuring Sustainable Development: Macroeconomics and the Environment.* Edward Elgar, Cheltenham.

Azar, C., and S.H. Schneider (2002). Are the economic costs of stabilising the atmosphere prohibitive? *Ecological Economics* 42: 73–80.

Blanchflower, D.G., and A.J. Oswald (2004). Well-being over time in Britain and the USA. *Journal of Public Economics* 88(7–8): 1359–86.

Bleys, B. (2006). The index of sustainable economic welfare: case study for Belgium–first attempt and preliminary results. March 2006, Vrije Universiteit Brussel, Belgium.

Boersema, J. (2004). Het goede leven is te weinig groen. Mimeo, Vrije Universiteit.

Bolt, K., M. Matete and M. Clemens (2002). Manual for calculating adjusted net savings. Environment Department, World Bank, September 2002.

Bos, F. (2006). The development of the Dutch national accounts as a tool for analysis and policy. *Statistica Neerlandica* 60(2): 225–258.

Brickman, P., and D.T. Campbell (1971). Hedonic relativism and planning the good society. In: M.H. Apley (ed.), *Adaptation-Level Theory: A Symposium.* Academic Press, New York, pp. 287–302.

Buss, D.M. (1989). Sex differences in human mate preferences: evolutionary hypotheses tested in 37 cultures. *Behavioral and Brain Sciences* 12: 1–49.

Bruni, L., and P.L. Porta (eds) (2005). *Economics and Happiness: Framing the Analysis.* Oxford University Press, Oxford.

Charmes, J. (2000). The contribution of the informal sector to GDP in developing countries: assessment, estimates, methods, orientations for the Future. 4th Meeting of the Delhi Group on Informal sector Statistics, Geneva 28–30 August 2000. C3ED, University of Versailles, Saint Quentin and Yvelines.

Clark, D., J.R. Kahn and H. Ofek (1988). City size, quality of life, and the urbanization deflator of the GNP: 1910–1984. *Southern Economic Journal* 54(3): 701–714.

de Groot, H.L.F., R. Nahuis and P.J.G. Tang (2004). Is the American model miss World? Choosing between the Anglo-Saxon model and a European-style alternative. CPB Discussion Paper 40, Centraal Planbureau, Den Haag.

Daly, H.E. (1977). *Steady-State Economics.* W.H. Freeman, San Francisco.

Daly, H.E. (1992). Allocation, distribution, and scale: towards an economics that is efficient, just and sustainable. *Ecological Economics* 6: 185–193.

Daly, H.E., and J. Cobb (1989). *For the Common Good: Redirecting the Economy Toward Community, the Environment, and a Sustainable Future.* Beacon Press, Boston, MA.

Dasgupta, P. (2001). *Human Well-Being and the Natural Environment.* Oxford University Press, Oxford.

Dasgupta, P., and K.-G. Mäler (2000). Net national product, wealth, and social well-being. *Environment and Development Economics* 5(1–2): 69–93.

Dougan, W.R. (1991). The cost of rent seeking: Is GNP negative? *Journal of Political Economy* 99(3): 660–664

Duesenberry, J.S. (1949). *Income, Saving and the Theory of Consumer Behavior.* Harvard University Press, Cambridge, MA.

Easterlin, R.A. (1974). Does economic growth improve the human lot? Some empirical evidence. In: P.A. David and M.W. Reder (eds), *Nations and Households in Economic Growth: Essays in Honour of Moses Abramowitz,* Academic Press, New York.

Easterlin, R.A. (2001). Income and happiness: Towards a unified theory. *The Economic Journal* 111: 465–484.

El Serafy, S.E. (1989). The proper calculation of income from depletable natural resources. In: Ahmad, Y.J., S.E. Serafy and E. Lutz (eds). *Environmental Accounting for Sustainable Development: A UNDP-World Bank Symposium.* The World Bank, Washington D.C., pp. 10–18.

El Serafy, S. (2001). Steering the right compass: the quest for a better assessment of the national product. In: E.C. van Ierland, J. van der Straaten and H.R.J. Vollebergh (eds). *Economic Growth and Valuation of the Environment.* Edward Elgar, Cheltenham, UK, pp. 189–210.

Ferrer-i-Carbonell, A. (2005). Income and well-being: an empirical analysis of the comparison income effect. *Journal of Public Economics* 89(5–6): 997–1019.

Ferrer-i-Carbonell, A., and P. Frijters (2004). How important is methodology for the estimates of the determinants of happiness? *The Economic Journal* 114(July): 641–659.

Frank, R.H. (1985). *Choosing the Right Pond: Human Behavior and the Quest for Status.* Oxford University Press, New York.

Frederick, S., and G. Lowenstein (1999). Hedonic adaptation. In: D. Kahneman, E. Diener and N. Schwartz (eds), *Well-Being: The Foundations of Hedonic Psychology.* Russell Sage Foundation, New York, pp. 302–329.

Frey, B.S, and A. Stutzer (2002). What can economists learn from happiness research? *Journal of Economic Literature* 40: 402–435.

Galbraith, J.K. (1958). *The Affluent Society.* Houghton Mifflin Company, Boston.

Gerlagh, R., R.B. Dellink, M.W. Hofkes and H. Verbruggen (2002). A measure of sustainable national income for the Netherlands. *Ecological Economics* 41: 157–174.

Hamilton, K., and M.A. Clemens (1999). Genuine savings rates in developing countries. *World Bank Economic Review* 13(2): 333–356.

Hartwick, J. (1977). Intergenerational equity and the investing of rents from exhaustible resources. *American Economic Review* 67: 972–974.

Hartwick, J. M. (1990). Natural resources, national accounting and economic depreciation. *Journal of Public Economics* 43: 291–304.

Helliwell, J. (2003). How's life? Combining individual and national variations to explain subjective well-being. *Economic Modelling* 20: 331–360.

Helson, H. (1964). *Adaptation-level Theory.* Harper and Row, New York.

Hicks, D.A. (1997). The inequality-adjusted human development index: a constructive proposal. *World Development* 25: 1283–1298.

Hicks, J.R. (1948). *Value and Capital* (2nd ed.), Clarendon Press, Oxford.

Hirsch, F. (1976). *Social Limits to Growth.* Harvard University Press, Cambridge, MA.

Hofkes, M., R. Gerlagh and V. Linderhof (2004). Sustainable National Income: A Trend Analysis for the Netherlands for 1990–2000. Report R-04/02, Institute for Environmental Studies, Free University, Amsterdam.

Hopkins, E., and T. Kornienko (2004). Running to keep in the same place: consumer choice as a game of status. *American Economic Review* 94: 1085–1107.

Howarth, R.B., and K.A. Brekke (2003). *Status, Growth and the Environment: Goods As Symbols in Applied Welfare Economics* Edward Elgar Publishing, Cheltenham.

Hueting, R. (1974). *Nieuwe Schaarste and Economische Groei.* Elsevier, Amsterdam. (English edition 1980, *New Scarcity and Economic Growth,* North-Holland, Amsterdam).

Hueting, R. (1996). Three persistent myths in the environmental debate. *Ecological Economics* 18(2): 81–88.

Ireland, N.J. (2001). Optimal income tax in the presence of status effects. *Journal of Public Economics* 81: 193–212.

Iyengar, S.S., and M. Lepper (2000). When choice is demotivating: Can one desire too much of a good thing? *Journal of Personality and Social Psychology* 76: 995–1006.

Jackson, T., F. Laing, A. MacGillivray, N. Marks, J. Ralls and S. Stymne (1997). An index of sustainable economic welfare for the UK 1950–1996. Centre for Environmental Strategy, University of Surrey, Guildford.

Johansson, C. (2004). The human development indices. Human Development Report Office, United Nations Development Programme, Presentation, Oxford, September 14 2004.

Kapteyn, A., B.M.S. van Praag, and F.G. van Herwaarden (1978). Individual welfare functions and social preference spaces. *Economics Letters* 1: 173–177.

Kelly, D.L. and C.D. Kolstad (1999). Integrated assessment models for climate change control. In: H. Folmer and T. Tietenberg (eds), *The International Yearbook of Environmental and Resource Economics 1999/2000.* Edward Elgar, Cheltenham.

Kenny, C. (2005). Why are we worried about income? Nearly everything that matters is converging. *World Development* 33(1): 1–19.

Kuznets, S. (1934). *National Income, 1929–32,* 1934. National Bureau of Economic Research, New York.

Kuznets, S. (1941). *National income and its composition 1919–1938.* National Bureau of Economic Research, New York.

Laband, D.N., and J.P. Sophocleus (1988). The social cost of rent-seeking: first estimates. *Public Choice* 58: 269–275.

Lawn, P., and R. Sanders (1999). Has Australia surpassed its optimal macroeconomic scale: finding out with the aid of 'benefit' and 'cost' accounts and a sustainable net benefit index. *Ecological Economics* 28: 213–229.

Lawn, P.A. (2003). A theoretical foundation to support the Index of Sustainable Economic Welfare (ISEW), Genuine Progress Indicator (GPI), and other related indexes. *Ecological Economics* 44(1): 105–118.

Layard, R. (2005). *Happiness: Lessons from A New Science.* Penguin, London.

Max-Neef, M. (1995). Economic growth and quality of life: A threshold hypothesis. *Ecological Economics* 15: 115–118.

Meadows, D.H., D.L. Meadows, J. Randers, and W.W. Behrens, III (1981). *Groping in the Dark: The First Decade of Global Modeling.* Wiley, New York

Mishan, E.J. (1967). *The Cost of Economic Growth.* Staples Press, London.

Neumayer, E. (2000). On the methodology of ISEW, GPI and related measures: Some constructive Comments and some doubt on the threshold hypothesis. *Ecological Economics* 34(3): 347–361.

Neumayer, E. (2001). The Human Development Index and sustainability: A constructive proposal. *Ecological Economics* 39(1): 101–114.

Noorbakhsh, F. (1998). A modified human development index. *World Development* 26: 517–528.

Nordhaus, W.D. and J. Tobin (1972). Is growth obsolete? Economic Growth. 50th anniversary colloquium V. Columbia University Press for the National Bureau of Economic Research. New York. Reprinted in: Milton Moss (ed.), *The Measurement of Economic and Social Performance,* Studies in Income and Wealth, Vol. 38, National Bureau of Economic Research, 1973.

OECD (2006). *Going for Growth.* OECD, Paris.

Pillarisetti, J.R. (2005). The World Bank's 'genuine savings' measure and sustainability. *Ecological Economics* 55: 599–609.

Rawls, J. (1972). *A Theory of Justice.* Clarendon Press, Oxford.

Scitovsky, T. (1976). *The Joyless Economy.* Oxford University Press, New York.

Samuelson, P.A. (1961). The evaluation of social income: capital formation and wealth. In: F. Lutz and D. Hague (eds). *The Theory of Capital.* St. Martin's Press, New York.

Schorr, J. (1998). *The Overspent American.* Basic Books, New York.

Sen, A. (1976). Real national income. *Review of Economic Studies* 43(1): 19–39.

Sen, A. (1979). The welfare basis of real income comparisons. *Journal of Economic Literature* 17(1): 1–45.

Sen, A. (1999). *Commodities and Capabilities.* Oxford University Press, Oxford.

Sen, A. (2000). *Development as Freedom.* Oxford University Press, Oxford.

Solnick, S., and D. Hemenway (1998). Is more always better? A survey on positional concerns. *Journal of Economic Behaviour and Organization* 37: 373–383.

Solow, R.M. (1974). Intergenerational equity and exhaustible resources. *Review of Economic Studies* (symposium) 41: 29–45.

Stiglitz, J.E. (2005). The Ethical Economist–A review of "The Moral Consequences of Economic Growth". By B.M. Friedman (Knopf, 2005). *Foreign Affairs,* November/December 2005.

Tinbergen, J., and R. Hueting (1992). GNP and market prices: wrong signals for sustainable economic success that mask environmental destruction. In: R. Goodland, H. Daly and S. El Serafy (eds). *Population, Technology and Lifestyle: The Transition to Sustainability.* Island Press, Washington D.C.

Tversky, A., and I. Simonson (2000). Context-dependent preferences. In: D. Kahneman and A. Tversky (eds), *Choices, Values and Frames.* Cambridge University Press, Cambridge, pp. 518–527.

van Praag, B.M.S. (1971). The welfare function of income in Belgium: an empirical investigation. *European Economic Review* 2: 337–369.

van Praag, B., and A. Ferrer-i-Carbonell (2004). *Happiness Quantified: A Satisfaction Calculus Approach.* Oxford University Press, Oxford.

Veblen, T. (1899). *The Theory of the Leisure Class: An Economic Study in the Evolution of Institutions.* Macmillan, New York.

Vellinga, N., and C. Withagen (1996). On the concept of green national income. *Oxford Economic Papers* 49(4): 499–514.

Weil, D.N. (2005). *Economic Growth.* Pearson Education, Addison-Wesley, Boston.

Weitzman, M.L. (1976). On the welfare significance of national product in a dynamic economy. *Quarterly Journal of Economics* 90: 156–162.

Weitzman, M.L,. and K.-G. Löfgren (1997). On the welfare significance of green accounting as taught by parable. *Journal of Environmental Economics and Management* 32: 139–153.

World Bank (2006). *Where is the Wealth of Nations.* The World Bank, Washington D.C.

Endnotes

1. I have benefited from discussions with and comments by E.J. Bartelsman, J. Boersema, F. Bos, F.A.G. den Butter, H.E. Daly, A. Ferrer-i-Carbonell, R. Gerlagh, H.L.F. de Groot, R. Hueting, J.W. Gunning, F. van der Lecq, G. Lettinga, J. Pen, P. Rietveld, A.B.T.M. van Schaik, E. Tellegen, H. van Tuinen, W.J. van den Berg, H. Verbruggen, P. Victor and C.A.A.M. Withagen. The usual disclaimer applies.
2. Gross domestic product (GDP) is the market output generated within a country's boundaries by both its citizens and foreigners. Gross national product (GNP) is the output that is generated by the citizens of a country, irrespective of where production takes place. For most countries the difference between GDP and GNP is not very large (a notable exception is Ireland). Moreover, the conceptual difference does not matter for our discussion here: all the weaknesses of GDP as a welfare indicator discussed hereafter apply equally to GNP.

3. Some of the criticism of the GDP indicator coincides with criticism of economic growth, which is not surprising as this growth is traditionally equated with an increase of GDP over time. Nevertheless, for the sake of clarity, it should be noted that the rejection of the GDP indicator does not imply a rejection of GDP growth in general. This article can be interpreted as drawing attention to the problem that the correlation between GDP growth and welfare growth is not generally positive and high.
4. Daly (1977) has proposed the notions of "ultimate means" and "ultimate ends". He considers economic activity as an intermediate end (or an intermediate means), so that it is best regarded as a cost factor from the perspective of ultimate ends.
5. Weitzman (1976) has shown that, under certain restrictions, the national product can serve as a proxy (stationary equivalent) of a utilitarian intertemporal welfare function formulated as a net present value of future consumption flows. Vellinga and Withagen (1996) have generalized this result. Three objections can be raised against the approaches adopted by these authors. In the first place, a specific intertemporal function is posed, without any empirical support, as a suitable representation of social welfare (an egalitarian or Rawlsian welfare function, for example, would render an entirely different outcome). Furthermore, the instantaneous utility function is approximated by consumption (Weitzman) or by a Taylor series in the arguments of utility–consumption and capital (Vellinga and Withagen). Last but not least, it needs to be assumed that there is no pure time dependence in the form of, for instance, exogenous technical change or exogenously changing world prices; time can only enter in the form of a time preference rate. All these choices imply a serious deviation from actual social welfare.
6. No sharp distinction is made here between notions like utility, welfare, well-being and happiness. As a rule, they are used to denote roughly the same thing. Disciplines such as economics, sociology, and psychology seem to have developed their own jargon in this respect. More importantly, empirical research does not (and cannot) make a sharp distinction between these various notions, and, not surprisingly, one finds conflicting interpretations of them. As is clarified later, this is not meant to deny the existence of different interpretations of happiness.
7. The original influential study is Easterlin (1974).
8. For a discussion of different interpretations of happiness (short-term feelings of euphoria or long-term feelings of satisfaction), see, for example, the introductory chapter in Bruni and Porta (2005). They note a fundamental distinction between subjective hedonism ("seeking pleasure and avoiding pain") and objective eudaimonia ("striving for perfection that represents the realization of one's true potential").
9. Stiglitz (2005) notes that median rather than average GDP serves a better job in capturing inequality. He emphasizes that average GDP per capita in the US has been steadily rising whereas median (household) income has been falling over the last decades.
10. The study also assesses an absolute welfare effect. A derived question, the answer to which is not easy, can be formulated as: To what extent does technological change, in particular product innovation, contribute to happiness through an increase of absolute (as opposed to relative) welfare of consumers? The relationship between GDP growth, technical change and welfare or happiness deserves critical examination. But even if technological change turned out to always be welfare enhancing (which has not been proven), one should not *ex ante* assume that the best technological and therefore welfare-enhancing strategy would be 'GDP growth no matter what'.
11. The analytical-theoretical work starts with Duesenberry (1949). Recent contributions are Arrow et al. (2004) and Hopkins and Kornienko (2004).
12. In biology the fact that only individuals with a high physical or mental quality can carry the costs of extravagance is denoted by the term "handicap principle". That one can waste means to seemingly useless–but certainly not 'fitnessless' or 'function-less'–consumption provides a signal of superiority and quality, which increases the social status, as a result of which the probability of finding a suitable sexual partner and therefore fitness increase.
13. Empirical studies even suggest asymmetry in the sense that the 'poor' lose more happiness than the 'rich' gain (Ferrer-i-Carbonell, 2005). This would imply a *negative*-sum game. This asymmetry was already suggested a long time ago (Duesenberry, 1949).
14. In addition, the attention for 'market consumption', including search, gathering information about (new) products, spending much time in shops and malls, and buying things that are hardly ever used, can divert adults as well as children from activities that contribute much more to happiness, such as more leisure (less working, less spending), playing with your kids, taking time for the extended family (grandparents), friends and neighbours, etc. (Schorr, 1998).
15. For a large number of African and a small number of Asian countries, GDP has been regularly subject to corrections and adaptations based on estimates of the value added of the informal agricultural sector. Such corrections are less common for Latin-American countries (Charmes, 2000).
16. According to recent field and experimental studies by psychologists a larger choice set may have a direct negative effect on individual motivation and welfare because of "choice overload", caused by incomplete information, search costs, and too much dependence on information offered by experts (Iyengar and Lepper, 2000). Herbert Simon's notion of bounded rationality, i.e. limited brain capacity to process information and 'satisficing' behaviour , offers an additional, complementary explanation of this phenomenon (Simon's famous statement ". . . a wealth of information creates a poverty of attention . . ." is also relevant here).
17. Non-renewable resource depletion presents a special problem. It is widely agreed upon that the method proposed by El Serafy (1989) should be followed here. This transforms the finite income stream from a non-renewable resource stock into a (lower) infinite stream of income from other types of capital (manufactured and human capital). The method computes the difference between these two income streams and deducts the resulting cost. The latter is referred to as the 'user cost' of non-renewable resource extraction. This approach thus assumes that (and works only if) one can sustain an infinite stream of income by substituting the non-renewable resource by other types of capital. Hartwick (1977) provides the theoretical basis for this line of thought.
18. For that matter, the growth of GDP is chiefly generated by the ± 30 percent economic activities that cause the major part of total environmental pressure (Hueting, 1996).

19. For example, Dasgupta (2001) suggests that expenditures on defence, a significant part of GDP, in many countries contributes to GDP overestimating social welfare. He notes: "But in poor countries the machinery for warfare is all too frequently used by governments against their own people . . . This means we can ignore defence expenditure . . ." (p.53). Laband and Sophocleus (1988) and Dougan (1991) further draw attention to the enormous rent-seeking costs of modern economies which boost GDP (e.g. through the activities of lawyers). Clark et al. (1988) use hedonic price techniques (based on income compensation) to assess the impact of urbanization on social welfare. Their findings suggest a GNP-as-welfare deflator in the order of 6 to 7 percent, which steadily increases at a rate of half a percent per decade.
20. Neither have I tried to be complete in terms of studies into the correction of GDP. It is unlikely that improvements which would neutralize the above critique will come available as part of the international standard of GDP construction at short notice.
21. Even *The Economist* (11 February 2006, p.70) characterises GDP as "badly flawed as a guide to a nation's economic well-being".
22. A common defence against the critique of GDP is that it is just one of various goals of macroeconomic policy (besides stable prices, low unemployment, an acceptable income distribution, etc.). However, adding indicators cannot undo or compensate the imperfect nature of the GDP indicator. The same objection holds, for example, for the Human Development Index (HDI), which incorporates GDP as a component, as well as life expectation and education (Dasgupta, 2001). See further Section 5 below on this aspect.
23. A corrected GDP—to accommodate the critiques documented here–would lack the pure financial (cash flow) interpretation ('taxable income'), which would be disadvantageous from the perspective of this tax revenue prediction goal (unless taxable income were redefined in line with the corrections undertaken).
24. Even so, I have been unable to find a thorough study that supports this doubtlessly widely held belief.
25. Kenny (2005: p.10) concludes on the basis of empirical data: "There has been convergence across a wide range of indicators of the quality of life. Given that there has not been convergence in the standard income indicator, this may suggest that income is only one among a number of factors in determining quality of life outcomes. In turn, this suggests some hope that improvements can be sustained even in the absence of sustained income growth."
26. See www.rprogress.org.
27. A predecessor of this approach is Nordhaus and Tobin (1972). A theoretical basis was created by Hartwick (1977, 1990) and Asheim (1994). For an overview, see Aronsson et al. (1997) and Asheim (2000). The theoretical literature shows that there are many fundamental problems with calculating a welfare measure based on GDP. Not only does it require valuation of non-market goods and services, but also it runs into fundamental problems regarding technological progress and changing prices due to open economies (international trade and relocation of activities). Another fundamental issue, neglected in most of this literature, is that status goods and rivalry in consumption cannot be addressed. This in itself would suggest a welfare-maximizing level of income rather than unlimited welfare growth. The calculation of income-based social welfare indicators is further hampered by the difficulty of translating certain basic needs into individual willingness-to-pay or accounting prices.
28. Atkinson (1995) and Neumayer (2000) also express criticism.
29. Gerlagh et al. (2002) note that the number of environmental themes might be extended. For the Netherlands, land use and waste disposal seems of high relevance as well. They further observe that the list of chosen environmental themes is biased to 'sink' (rather than 'source') functions of the environment. If the calculations were repeated for other countries, one might want to consider including source functions related to forests, mineral deposits, topsoil, fish stocks and water resources.
30. The World Bank has recently published *Where is the Wealth of Nations* (2006), which suggests that it regards wealth as important, but mainly as a basis for future welfare. In other words, wealth then serves as an indicator of potential future welfare.
31. In fact, the UN published a Human Freedom Index (HFI) in 1991 and a Political Freedom Index (PFI) in 1992 (Johansson, 2004).
32. Sen (2000: p.318, note 41) states: "Indeed, getting public attention has clearly been a part of UNDP's objective, particularly in its attempt to combat the overconcentration on the simple measure of GNP per head, which often serves as the only indicator of which the public takes any notice. To compete with the GNP, there is a need for another–broader–measure with the same level of crudeness as the GNP. This need is partly met by the use of the HDI . . .". He adds that the HDI has attracted much more attention than often more informative, less aggregated information on diversity at the micro level.
33. Other alternatives are mentioned in the literature, but these have not proceeded beyond the stage of conceptualization. Bleys (2006) offers an overview.
34. What I sketch here is not a 'straw man', as one critical reader suggested to me. And if one is unwilling to accept that GDP has so much influence on our economic reality, in spite of the arguments offered in Section 3, then one can hardly object to the removal of the GDP indicator from the public domain.
35. To take just one example: the textbook by Weil (2005), which is entirely devoted to the theme of economic (GDP) growth, does not contain a single reference to the many criticiques of GDP as an indicator of (welfare) progress. Only on the very last pages (pp.508-510) a box is presented in which the question is addressed: "Will growth make us happy". The answer given is twofold: "Income is not the only determinant of happiness, but clearly happiness rises with income . . ." and "Thus, although growth will not make us as happy as we expect it to, it will still make us happier than we would be if there were no growth". Unfortunately, neither statement is convincingly supported with data or arguments.
36. Azar and Schneider (2002) show that the costs of reaching what the IPCC considers "safe" concentrations of CO_2 in the atmosphere, for the world as a whole, fall in the range US$1 tot US$20 trillion. Although these are impressive figures, they imply less than 3 years delay of reaching a certain income level 100 years from now (given 2 percent GDP growth). Interpreted this way, the costs of a stringent climate policy are marginal in economic terms in the long run.

37. But neither is there a good reason to be generally against GDP growth. Growth surely does correlate positively with welfare growth in many situations. In fact, growth critics should seriously consider shifting to oppose GDP as an indicator of progress instead of to oppose GDP growth in general (i.e. under all circumstances, in all countries, at all times).
38. In 1665 Sir William Petty produced the first estimate of a national income: namely, for England. His work aimed to determine which outlays on warfare could be supported by means of tax revenues. Work by Nobel Laureates Simon Kuznets (for the USA) and Richard Stone and James Meade (for the UK) in the early twentieth century allowed for the rapid diffusion of the GDP indicator in economic research and politics. Again, war acted a stimulating factor, as with so many innovations. For there was a need to determine the production capacity of the Allied Forces just before and during World War II.
39. Neither can we expect GDP to converge to a close to perfect welfare indicator through a series of marginal improvements. One reason is that in deciding about which changes to incorporate in the GDP calculation procedure, countries may act strategically. For example, currently, EU members may be motivated to show such strategic behaviour, given that their contributions to the EU budget are to a large extent determined by the size of their economy, measured in terms of GDP. Countries that would see their GDP go up (relative to other countries) as a result of an altered GDP calculation method might therefore vote against a proposal to implement such an alteration.

Critical Thinking

1. Will it be possible to achieve sustainability if GDP continues to be the primary mechanism by which to judge the "position of the economy of a county over time relative to that of other countries?"
2. Which of the alternatives to the GDP identified in this article do you believe is the most promising?
3. What can the average citizen do regarding an issue of this magnitude?

The Efficiency Dilemma

If our machines use less energy, will we just use them more?

DAVID OWEN

In April, the federal government adopted standards for automobiles requiring manufacturers to improve the average fuel economy of their new-car fleets thirty percent by 2016. The *Times,* in an editorial titled "Everybody Wins," said the change would produce "a trifecta of benefits." Those benefits were enumerated last year by Steven Chu, the Secretary of Energy: a reduction in total oil consumption of 1.8 billion barrels; the elimination of nine hundred and fifty million metric tons of greenhouse-gas emissions; and savings, for the average American driver, of three thousand dollars.

Chu, who shared the Nobel Prize in Physics in 1997, has been an evangelist for energy efficiency, and not just for vehicles. I spoke with him in July, shortly after he had conducted an international conference called the Clean Energy Ministerial, at which efficiency was among the main topics. "I feel very passionate about this," he told me. "We in the Department of Energy are trying to get the information out that efficiency really does save money and doesn't necessarily mean that you're going to have to make deep sacrifices."

Energy efficiency has been called "the fifth fuel" (after coal, petroleum, nuclear power, and renewables); it is seen as a cost-free tool for accelerating the transition to a green-energy economy. In 2007, the United Nations Foundation said that efficiency improvements constituted "the largest, the most evenly geographically distributed, and least expensive energy resource." Last year, the management-consulting firm McKinsey & Company concluded that a national efficiency program could eliminate "up to 1.1 gigatons of greenhouse gases annually." The environmentalist Amory Lovins, whose thinking has influenced Chu's, has referred to the replacement of incandescent light bulbs with compact fluorescents as "not a free lunch, but a lunch you're paid to eat," since a fluorescent bulb will usually save enough electricity to more than offset its higher purchase price. Tantalizingly, much of the technology required to increase efficiency is well understood. The World Economic Forum, in a report called "Towards a More Energy Efficient World," observed that "the average refrigerator sold in the United States today uses three-quarters less energy than the 1975 average, even though it is 20 percent larger and costs 60 percent less"—an improvement that Chu cited in his conversation with me.

But the issue may be less straightforward than it seems. The thirty-five-year period during which new refrigerators have plunged in electricity use is also a period during which the global market for refrigeration has burgeoned and the world's total energy consumption and carbon output, including the parts directly attributable to keeping things cold, have climbed. Similarly, the first fuel-economy regulations for U.S. cars—which were enacted in 1975, in response to the Arab oil embargo—were followed not by a steady decline in total U.S. motor-fuel consumption but by a long-term rise, as well as by increases in horsepower, curb weight, vehicle miles travelled (up a hundred percent since 1980), and car ownership (America has about fifty million more registered vehicles than licensed drivers). A growing group of economists and others have argued that such correlations aren't coincidental. Instead, they have said, efforts to improve energy efficiency can more than negate any environmental gains—an idea that was first proposed a hundred and fifty years ago, and which came to be known as the Jevons paradox.

Great Britain in the middle of the nineteenth century was the world's leading military, industrial, and mercantile power. In 1865, a twenty-nine-year-old Englishman named William Stanley Jevons published a book, "The Coal Question," in which he argued that the bonanza couldn't last. Britain's affluence, he wrote, depended on its endowment of coal, which the country was rapidly depleting. He added that such an outcome could not be delayed through increased "economy" in the use of coal—what we refer to today as energy efficiency. He concluded, in italics, *"It is wholly a confusion of ideas to suppose that the economical use of fuel is equivalent to a diminished consumption. The very contrary is the truth."*

He offered the example of the British iron industry. If some technological advance made it possible for a blast furnace to produce iron with less coal, he wrote, then profits would rise, new investment in iron production would be attracted, and the price of iron would fall, thereby stimulating additional demand. Eventually, he concluded, "the greater number of furnaces will more than make up for the diminished consumption of each." Other examples of this effect abound. In a paper published in 1998, the Yale economist William D. Nordhaus estimated the cost of lighting throughout human history. An ancient Babylonian, he calculated, needed to work more than forty-one hours to acquire enough lamp oil to provide a thousand lumen-hours of light—the equivalent of a seventy-five-watt incandescent bulb burning for about an hour. Thirty-five hundred years later, a contemporary of Thomas Jefferson's could buy the same amount of illumination, in the form of tallow candles, by working for about five hours and twenty minutes. By 1992, an average American, with access to compact fluorescents, could do the same in less than half a second. Increasing the energy efficiency of illumination is nothing new; improved lighting has been "a lunch you're paid to eat" ever since humans upgraded from cave fires (fifty-eight hours of labor for our early Stone Age ancestors). Yet our efficiency gains haven't reduced the energy we expend on illumination or shrunk our energy consumption over all. On the contrary, we now generate light so extravagantly that darkness itself is spoken of as an endangered natural resource.

Jevons was born in Liverpool in 1835. He spent two years at University College, in London, then went to Australia, where he had been offered a job as an assayer at a new mint, in Sydney. He left after five years, completed his education in England, became a part-time college instructor, and published a well-received book on gold markets. "The Coal Question" made him a minor celebrity; it was admired by John Stuart Mill and William Gladstone, and it inspired the government to investigate his findings. In 1871, he published "The Theory of Political Economy," a book that's still considered one of the founding texts of mathematical economics. He drowned a decade later, at the age of forty-six, while swimming in the English Channel. In 1905, John Maynard Keynes, who was then twenty-two and a graduate student at Cambridge University, wrote to Lytton Strachey that he had discovered a "thrilling" book: Jevons's "Investigations in Currency and Finance." Keynes wrote of Jevons, "I am convinced that he was one of *the* minds of the century."

Jevons might be little discussed today, except by historians of economics, if it weren't for the scholarship of another English economist, Len Brookes. During the nineteen-seventies oil crisis, Brookes argued that devising ways to produce goods with less oil—an obvious response to higher prices—would merely accommodate the new prices, causing energy consumption to be higher than it would have been if no effort to increase efficiency had been made; only later did he discover that Jevons had anticipated him by more than a century. I spoke with Brookes recently. He told me, "Jevons is very simple. When we talk about increasing energy efficiency, what we're really talking about is increasing the productivity of energy. And, if you increase the productivity of anything, you have the effect of reducing its implicit price, because you get more return for the same money—which means the demand goes up."

Nowadays, this effect is usually referred to as "rebound"—or, in cases where increased consumption more than cancels out any energy savings, as "backfire." In a 1992 paper, Harry D. Saunders, an American researcher, provided a concise statement of the basic idea: "With fixed real energy price, energy efficiency gains will increase energy consumption above where it would be without these gains."

In 2000, the journal *Energy Policy* devoted an entire issue to rebound. It was edited by Lee Schipper, who is now a senior research engineer at Stanford University's Precourt Energy Efficiency Center. In an editorial, Schipper wrote that the question was not whether rebound exists but, rather, "how much the effect appears, how rapidly, in which sectors, and in what manifestations." The majority of the *Energy Policy* contributors concluded that there wasn't a lot to worry about. Schipper, in his editorial, wrote that the articles, taken together, suggested that "rebounds are significant but do not threaten to rob society of most of the benefits of energy efficiency improvements."

I spoke with Schipper recently, and he told me that the Jevons paradox has limited applicability today. "The key to understanding Jevons," he said, "is that processes, products, and activities where energy is a very high part of the cost—in this country, a few metals, a few chemicals, air travel—are the only ones whose variable cost is very sensitive to energy. That's it." Jevons wasn't wrong about nineteenth-century British iron smelting, he said; but the young and rapidly growing industrial world that Jevons lived in no longer exists.

Most economists and efficiency experts have come to similar conclusions. For example, some of them say that when you increase the fuel efficiency of cars you lose no more than about ten percent of the fuel savings to increased use. And if you look at the whole economy, Schipper said, rebound effects are comparably trivial. "People like Brookes would say—they don't quite know how to say it, but they seem to want to say the extra growth is more than the saved energy, so it's like a backfire. The problem is, that's never been observed on a national level."

But troublesome questions have lingered, and the existence of large-scale rebound effects is not so easy to dismiss. In 2004, a committee of the House of Lords invited a number of experts to help it grapple with a conundrum: the United Kingdom, like a number of other countries, had spent heavily to increase energy efficiency in an attempt to reduce its greenhouse emissions. Yet energy consumption and carbon output in Britain—as in the rest of the world—had continued to rise. Why?

Most economic analyses of rebound focus narrowly on particular uses or categories of uses: if people buy a more efficient clothes dryer, say, what will happen to the energy they use as they dry clothes? (At least one such study has concluded that, for appliances in general, rebound is nonexistent.) Brookes dismisses such "bottom-up" studies, because they ignore or understate the real consumption effects, in economies as a whole.

A good way to see this is to think about refrigerators, the very appliances that the World Economic Forum and Steven Chu cited as efficiency role models for reductions in energy use. The first refrigerator I remember is the one my parents owned when I was little. They acquired it when they bought their first house, in 1954, a year before I was born. It had a tiny, uninsulated freezer compartment, which seldom contained much more than a few aluminum ice trays and a burrow-like mantle of frost. (Frost-free freezers stay frost-free by periodically heating their cooling elements—a trick that wasn't widely in use yet.) In the sixties, my parents bought a much improved model—which presumably was more efficient, since the door closed tight, by means of a rubberized magnetic seal rather than a mechanical latch. But our power consumption didn't fall, because the old refrigerator didn't go out of service; it moved into our basement, where it remained plugged in for a further twenty-five years—mostly as a warehouse for beverages and leftovers—and where it was soon joined by a stand-alone freezer. Also, in the eighties, my father added an icemaker to his bar, to supplement the one in the kitchen fridge.

This escalation of cooling capacity has occurred all over suburban America. The recently remodelled kitchen of a friend of mine contains an enormous side-by-side refrigerator, an enormous side-by-side freezer, and a drawer-like under-counter mini-fridge for beverages. And the trend has not been confined to households. As the ability to efficiently and inexpensively chill things has grown, so have opportunities to buy chilled things—a potent positive-feedback loop. Gas stations now often have almost as much refrigerated shelf space as the grocery stores of my early childhood; even mediocre hotel rooms usually come with their own small fridge (which, typically, either is empty or—if it's a minibar—contains mainly things that don't need to be kept cold), in addition to an icemaker and a refrigerated vending machine down the hall.

The steadily declining cost of refrigeration has made eating much more interesting. It has also made almost all elements of food production more cost-effective and energy-efficient: milk lasts longer if you don't have to keep it in a pail in your well. But there are environmental downsides, beyond the obvious one that most of the electricity that powers the world's refrigerators is generated by burning fossil fuels. James McWilliams, who is the author of the recent book "Just Food," told me, "Refrigeration and packaging convey to the consumer a sense that what we buy will last longer than it does. Thus, we buy enough stuff to fill our capacious Sub-Zeros and, before we know it, a third of it is past its due date and we toss it." (The item that New Yorkers most often throw away unused, according to the anthropologist-in-residence at the city's Department of Sanitation, is vegetables.) Jonathan Bloom, who runs the Website wastedfood.com and is the author of the new book "American Wasteland," told me that, since the mid-nineteen-seventies, per-capita food waste in the United States has increased by half, so that we now throw away forty percent of all the edible food we produce. And when we throw away food we don't just throw away nutrients; we also throw away the energy we used in keeping it cold as we lost interest in it, as well as the energy that went into growing, harvesting, processing, and transporting it, along with its proportional share of our staggering national consumption of fertilizer, pesticides, irrigation water, packaging, and landfill capacity. According to a 2009 study, more than a quarter of U.S. freshwater use goes into producing food that is later discarded.

Efficiency improvements push down costs at every level—from the mining of raw materials to the fabrication and transportation of finished goods to the frequency and intensity of actual use—and reduced costs stimulate increased consumption. (Coincidentally or not, the growth of American refrigerator volume has been roughly paralleled by the growth of American body-mass index.) Efficiency-related increases in one category, furthermore, spill into others. Refrigerators are the fraternal twins of air-conditioners, which use the same energy-hungry compressor technology to force heat to do something that nature doesn't want it to. When I was a child, cold air was a far greater luxury than cold groceries. My parents' first house—like eighty-eight percent of all American homes in 1960—didn't have air-conditioning when they bought it, although they broke down and got a window unit during a heat wave, when my mom was pregnant with me. Their second house had central air-conditioning, but running it seemed so expensive to my father that, for years, he could seldom be persuaded to turn it on, even at the height of a Kansas City summer, when the air was so humid that it felt like a swimmable liquid. Then he replaced our ancient Carrier unit with a modern one, which consumed less electricity, and our house, like most American houses, evolved rapidly from being essentially un-air-conditioned to being air-conditioned all summer long.

Modern air-conditioners, like modern refrigerators, are vastly more energy efficient than their mid-twentieth-century predecessors—in both cases, partly because of tighter standards established by the Department of Energy. But that efficiency has driven down their cost of operation, and manufacturing efficiencies and market growth have driven down the cost of production, to such an extent that the ownership percentage of 1960 has now flipped: by 2005, according to the Energy Information Administration, eighty-four percent of all U.S. homes had air-conditioning, and most of it was central. Stan Cox, who is the author of the recent book "Losing Our Cool," told me that, between 1993 and 2005, "the energy efficiency of residential air-conditioning equipment improved twenty-eight percent, but energy consumption for A.C. by the average air-conditioned household rose thirty-seven percent." One consequence, Cox observes, is that, in the United States, we now use roughly as much electricity to cool buildings as we did for all purposes in 1955.

As "Losing Our Cool" clearly shows, similar rebound effects permeate the economy. The same technological gains that have propelled the growth of U.S. residential and commercial cooling have helped turn automobile air-conditioners, which barely existed in the nineteen-fifties, into standard equipment on even the least luxurious vehicles. (According to the National Renewable Energy Laboratory, running a mid-sized car's air-conditioning increases fuel consumption by more than twenty percent.) And access to cooled air is self-reinforcing: to someone who works in an air-conditioned office, an un-air-conditioned house quickly becomes intolerable, and vice versa. A resident of Las Vegas once described cars to me as "devices for transporting air-conditioning between buildings."

In less than half a century, increased efficiency and declining prices have helped to push access to air-conditioning almost all the way to the bottom of the U.S. income scale—and now those same forces are accelerating its spread all over the world. According to Cox, between 1997 and 2007 the use of air-conditioners tripled in China (where a third of the world's units are now manufactured, and where many air-conditioner purchases have been subsidized by the government). In India, air-conditioning is projected to increase almost tenfold between 2005 and 2020; according to a 2009 study, it accounted for forty percent of the electricity consumed in metropolitan Mumbai.

All such increases in energy-consuming activity can be considered manifestations of the Jevons paradox. Teasing out the precise contribution of a particular efficiency improvement isn't just difficult, however; it may be impossible, because the endlessly ramifying network of interconnections is too complex to yield readily to empirical, mathematics-based analysis. Most modern studies of energy rebound are "bottom-up" by necessity: it's only at the micro end of the economics spectrum that the number of mathematical variables can be kept manageable. But looking for rebound only in individual consumer goods, or in closely cropped economic snapshots, is as futile and misleading as trying to analyze the global climate with a single thermometer.

Schipper told me, "In the end, the impact of rebound is small, in my view, for one very key reason: energy is a small share of the economy. If sixty percent of our economy were paying for energy, then anything that moved it down by ten percent would liberate a huge amount of resources. Instead, it's between six and eight percent for primary energy, depending on exactly what country you're in." ("Primary energy" is the energy in oil, coal, wind, and other natural resources before it's been converted into electricity or into refined or synthetic fuels.) Schipper believes that cheap energy is an environmental problem, but he also believes that, because we can extract vastly more economic benefit from a ton of coal than nineteenth-century Britons did, efficiency gains now have much less power to stimulate consumption. This concept is closely related to one called "decoupling," which suggests that the growing efficiency of machines has weakened the link between energy use and economic activity, and also to the idea of "decarbonization," which holds that, for similar reasons, every dollar we spend represents a shrinking quantity of greenhouse gas.

These sound like environmentally valuable trends—yet they seem to imply that the world's energy and carbon challenges are gradually solving themselves, since decoupling and decarbonization, like increases in efficiency, are nothing new. One problem with decoupling, as the concept is often applied, is that it doesn't account for energy use and carbon emissions that have not been eliminated but merely exported out of the region under study (say, from California to a factory in China). And there's a more fundamental problem, described by the Danish researcher Jørgen S. Nørgård, who has called energy decoupling "largely a statistical delusion." To say that energy's economic role is shrinking is a little like saying, "I have sixteen great-great-grandparents, eight great-grandparents, four grandparents, and two parents—the world's population must be imploding." Energy production may account for only a small percentage of our economy, but its falling share of G.D.P. has made it more important, not less, since every kilowatt we generate supports an ever larger proportion of our well-being. The logic misstep is apparent if you imagine eliminating primary energy from the world. If you do that, you don't end up losing "between six and eight percent" of current economic activity, as Schipper's formulation might suggest; you lose almost everything we think of as modern life.

Blake Alcott, an ecological economist, has made a similar case in support of the existence of large-scale Jevons effects. Recently, he told me, "If it is true that greater efficiency in using a resource means less consumption of it—as efficiency environmentalists say—then less efficiency would logically mean more consumption. But this yields a reductio ad absurdum: engines and smelters in James Watt's time, around 1800, were far less efficient than today's, but is it really imaginable that, had technology been frozen at that efficiency level, a greater population would now be using vastly more fossil fuel than we in fact do?" Contrary to the argument made by "decouplers," we aren't gradually reducing our dependence on energy; rather, we are finding ever more ingenious ways to leverage B.T.U.s. Between 1984 and 2005, American electricity production grew by about sixty-six percent—and it did so despite steady, economy-wide gains in energy efficiency. The increase was partly the result of population growth; but per-capita energy consumption rose, too, and it did so even though energy use per dollar of G.D.P. fell by roughly half. Besides, population growth itself can be a Jevons effect: the more efficient we become, the more people we can sustain; the more people we sustain, the more energy we consume.

The Model T was manufactured between 1908 and 1927. According to the Ford Motor Company, its fuel economy ranged between thirteen and twenty-one miles per gallon. There are vehicles on the road today that do worse than that; have we really made so little progress in more than a hundred years? But focussing on miles per gallon is the wrong way to assess the environmental impact of cars. Far more revealing is to consider the productivity of driving. Today, in contrast to the early nineteen-hundreds, any American with a license can cheaply travel almost anywhere, in almost any weather, in extraordinary comfort; can drive for thousands of miles with no maintenance other than refuelling; can easily find gas, food, lodging, and just about anything else within a short distance of almost any road; and can order and eat meals without undoing a seat belt or turning off the ceiling-mounted DVD player.

A modern driver, in other words, gets vastly more benefit from a gallon of gasoline—makes far more economical use of fuel—than any Model T owner ever did. Yet motorists' energy consumption has grown by mind-boggling amounts, and, as the productivity of driving has increased and the cost of getting around has fallen, the global market for cars has surged. (Two of the biggest road-building efforts in the history of the world are currently under way in India and China.) And developing small, inexpensive vehicles that get a hundred miles to the gallon would only exacerbate that trend. The problem with efficiency gains is that we inevitably reinvest them in additional consumption. Paving roads reduces rolling friction, thereby boosting miles per gallon, but it also makes distant destinations seem closer, thereby enabling people to live in sprawling, energy-gobbling subdivisions far from where they work and shop.

Chu has said that drivers who buy more efficient cars can expect to save thousands of dollars in fuel costs; but, unless those drivers shred the money and add it to a compost heap, the environment is unlikely to come out ahead, as those dollars will inevitably be spent on goods or activities that involve fuel consumption—say, on increased access to the Internet, which is one of the fastest-growing energy drains in the world. (Cox writes that, by 2014, the U.S. computer network alone will each year require an amount of energy equivalent to the total electricity consumption of Australia.) The problem is exactly what Jevons said it was: the economical use of fuel is not equivalent to a diminished consumption. Schipper told me that economy-wide Jevons effects have "never been observed," but you can find them almost anywhere you look: they are the history of civilization.

Jevons died too soon to see the modern uses of oil and natural gas, and he obviously knew nothing of nuclear power. But he did explain why "alternative" energy sources, such as wind, hydropower, and biofuels (in his day, mainly firewood and whale oil), could not compete with coal: coal had replaced them, on account of its vastly greater portability, utility, and productivity. Early British steam engines were sometimes used to pump water to turn water wheels; we do the equivalent when we burn coal to make our toothbrushes move back and forth.

Decreasing reliance on fossil fuels is a pressing global need. The question is whether improving efficiency, rather than reducing total consumption, can possibly bring about the desired result. Steven Chu told me that one of the appealing features of the efficiency discussions at the Clean Energy Ministerial was that they were never contentious. "It was the opposite," he said. "No one was debating about who's responsible, and there was no finger-pointing or trying to lay blame." This seems encouraging in one way but dismaying in another. Given the known level of global disagreement about energy and climate matters, shouldn't there have been *some* angry table-banging? Advocating efficiency involves virtually no political risk—unlike measures that do call for sacrifice, such as capping emissions or putting a price on carbon or increasing energy taxes or investing heavily in utility-scale renewable-energy facilities or confronting the deeply divisive issue of global energy equity. Improving efficiency is easy to endorse: we've been doing it, globally, for centuries. It's how we created the problems we're now trying to solve.

Efficiency proponents often express incredulity at the idea that squeezing more consumption from less fuel could somehow carry an environmental cost. Amory Lovins once wrote that, if Jevons's argument is correct, "we should mandate inefficient equipment to save energy." As Lovins intended, this seems laughably illogical—but is it? If the only motor vehicle available today were a 1920 Model T, how many miles do you think you'd drive each year, and how far do you think you'd live from where you work? No one's going to "mandate inefficient equipment," but, unless we're willing to do the equivalent—say, by mandating costlier energy—increased efficiency, as Jevons predicted, can only make our predicament worse.

At the end of "The Coal Question," Jevons concluded that Britain faced a choice between "brief greatness and longer continued mediocrity." His preference was for mediocrity, by which he meant something like "sustainability." Our world is different from his, but most of the central arguments of his book still apply. Steve Sorrell, who is a senior fellow at Sussex University and a co-editor of a recent comprehensive book on rebound, called "Energy Efficiency and Sustainable Consumption," told me, "I think the point may be that Jevons has yet to be disproved. It is rather hard to demonstrate the validity of his proposition, but certainly the historical evidence to date is wholly consistent with what he was arguing." That might be something to think about as we climb into our plug-in hybrids and continue our journey, with ever-increasing efficiency, down the road paved with good intentions.

Critical Thinking

1. If the Jevons paradox is real, what will be required to achieve sustainability?
2. Explain the statement: "Energy production may account for only a small percentage of our economy, but its falling share of G.D.P. has made it more important, not less, since every kilowatt we generate supports an ever larger population of our well-being."
3. It is suggested in this article that "continued mediocrity" is synonymous with "sustainability." Do you agree?

Consumption, Not CO_2 Emissions: Reframing Perspectives on Climate Change and Sustainability

ROBERT HARRISS AND BIN SHUI

A stunning documentary film titled *Mardi Gras: Made in China* provides an insightful and engaging perspective on the globalization of desire for material consumption. Tracing the life cycle of Mardi Gras beads from a small factory in Fuzhou, China, to the streets of the Mardi Gras celebration in New Orleans, the viewer grasps the near-universal human tendency to strive for an affluent lifestyle. David Redmon, an independent filmmaker, follows the beads' genealogy back to the industrial town of Fuzhou, and to the factory that is the world's largest producer of Mardi Gras beads and related party trinkets. He explores how these frivolous and toxic products affect the people who make them and those who consume them. Redmon captures the harsh daily reality of working in this Chinese facility. Members of its workforce—approximately 500 young female workers and a handful of young male workers—live like prisoners in a fenced-in compound. These young people, often working 16-hour days, are constantly exposed to styrene, a chemical known to cause cancer—all for about 10 cents an hour. In addition to the indoor pollution, the decrepit coal-fired factory is also symbolic of China's fast rise to the world's top producer of carbon dioxide (CO_2) emissions.[1] The process of industrialization and modernization in China is happening at an unprecedented rate and scale.

The filming of Mardi Gras celebrations in New Orleans provides a startling contrast to the Fuzhou factory, showing indulgent, affluent Americans engaging in obnoxious exhibitionism. When questioned by the filmmaker, the partygoers are unaware of the origins of the Mardi Gras beads. In an early morning scene, after a night of Mardi Gras celebrations, the party crowd has disappeared and sanitation vehicles are seen sweeping up mounds of discarded beads for disposal in a New Orleans landfill. The party beads will ultimately decay, producing CO_2 and other pollutants—an invisible and unintended consequence from the perspective of both the Chinese factory workers and the American party crowd. The transformation of the Mardi Gras beads from objects of desire to trash in a matter of hours illustrates the complexities associated with human perceptions of sufficiency, conspicuous consumption, and present and future well-being.[2]

A Matter of Scale and Focus

For many people the transition from the human scale, as depicted in *Mardi Gras: Made in China,* to issues of global consumption and climate change poses a daunting cognitive challenge. Making the connection between the ongoing growth of a global consumer culture and global climate change has proven to be both an intellectual and institutional quagmire. The devil is in transitioning from knowing the local to imagining the global.

Reducing global emissions of CO_2 and other factors that contribute to climate change has been at the center of highly politicized and publicized global climate policy negotiations for almost two decades. Ongoing international negotiations under the auspices of the UN Framework Convention on Climate Change (UNFCCC) and the Kyoto Protocol have largely failed, especially in the area given greatest attention: mitigation of industrial CO_2 emissions.[3] The recent Copenhagen Accord was only able to get agreement on acknowledging the scientific view that the increase in global temperature should be kept below 2°C, but without adequate commitments by nations to mitigate CO_2 emissions. While some experts remain cautiously positive about the UNFCCC process,[4] others think it is time to move beyond the intense focus on climate change as a physical threat.[5] A compelling case has been made for why it might be more productive to address the sociocultural dimensions that contribute to why we disagree on climate change.[6]

This article reviews evidence for a growing influence of international trade on global CO_2 emissions. We conclude that economic globalization as currently practiced will undermine future progress toward achieving the goals of the UNFCCC and post-Kyoto negotiations on reducing the growth of global CO_2 emissions and potential impacts of climate change. Our analysis adds to the growing evidence that a reframing of the climate change policy debate is urgently needed. We recommend a broader dialogue on strategies for a societal transition to long-term sustainability, recognizing that global warming is not the primary concern of many nations.

Economic Globalization and CO_2 Emissions

For the past several decades, growth in international trade has outpaced the growth of global gross domestic product (GDP), energy consumption, and world population. This surge of economic globalization has resulted in a dynamic shifting in the geographic patterns of production and consumption of consumer goods, and consequently the fossil fuels and CO2 emissions needed to make them.

Economic globalization reflects the logic of increasing the production of consumer goods at the lowest possible costs while maintaining the qualities and quantities that buyers demand. Estimating the net benefits and costs of economic globalization is a contentious and widely debated topic. An unintended consequence of economic globalization has been a shifting of the burden of additional CO_2 emissions and other environmental pollutants from developed consumer to developing producer countries. This process is also known as "offshoring" the emissions of CO_2 and other pollutants by wealthy countries. This large-scale geographical separation of material production and consumption has raised fundamental policy questions concerning responsibility for CO_2 emissions.

Scientific comparisons of production-based versus consumption-based national CO_2 emission inventories have illustrated that economic globalization is undermining the validity of using the national emissions inventory methodologies as the sole basis allocating responsibility for CO_2 emissions.[7] The UNFCCC/Kyoto Protocol process has based negotiations on CO_2 emissions that originate within national boundaries (i.e., national emissions inventories). Developing countries that are large CO_2 emitters, like China, have recently argued that national emission inventories do not represent a true measure of a country's consumption, the fundamental culprit driving global climate change. They are recommending the use of consumption-based measures of CO_2 in the cases where emissions were generated during the manufacturing of a commodity in a developing country and the commodity was subsequently exported for use or consumption in a developed country.

Chinese officials have noted the "common but differentiated responsibility" criteria declared in Article 3 of the UNFCCC as a basis for their concerns about the allocation of responsibility for CO_2 emissions. China's President Hu Jintao, who spoke at the G-8 meeting held in summer 2008 in Japan, stated that "as a result of changes in international division of labor and manufacturing location, China faces mounting pressure of international transferred emissions."[8] At a meeting in Washington, D.C., in March 2010, Dr. Gao Li, who heads the climate change department of the Chinese National Development and Reform Commission, met with top U.S. policymakers and their counterparts from the European Union (EU), Japan, and Mexico. In his message to the gathering, he said that his country was "at the low end of the production line for the global economy . . . We produce products and these products are consumed by other countries, especially the developed countries." Li estimated that the CO_2 emitted in China during the manufacture of exports to the United States and other countries accounted for some 15 to 25 percent of his country's total emissions. He submitted that "[T]his share of emissions should be taken by the consumers, not the producers." He then predicted that this would be a "very important item" in reaching a fair post-Kyoto global agreement on greenhouse gas reductions.[9]

Measuring Embodied Carbon in International Trade

Recent advances in consumption-based accounting provide an opportunity to quantitatively determine the importance of international trade as a factor in shifting the burden of CO_2 emissions from developed to developing nations. For example, if a computer manufactured in China resulted in one ton of CO_2 emissions and the computer is exported and sold in the United States, which country should be responsible for this ton of CO_2? In a consumption-based accounting methodology, the one ton of CO_2 emissions associated with manufacturing the computer, and the emissions produced by the international transport of the computer from China to a U.S. port of entry, would be the American buyer's responsibility. Consumption-based accounting is focused on the consumer as the driver of emissions. The widely employed IPCC national inventory methodologies focus on emissions generated within countries' territorial boundaries. In the extreme, a wealthy country with an economy based on financial and similar relatively nonpolluting services could purchase all of its manufactured goods from a developing country at low cost, thereby avoiding the industrial pollution.

CO_2 emission associated with the production and export of goods in international trade is most commonly characterized as "carbon embodied in trade." The terms "embedded carbon in trade" and "virtual carbon in trade" have also been used in the same context.

Consumption-based accounting methodologies capable of estimating CO_2 emissions associated with manufacturing a product were initially of particular interest to scientists and engineers advancing the science of life-cycle analysis and "green" design and manufacturing. A pioneering example of life-cycle analysis software is the Economic Input–Output Life Cycle Assessment (EIO-LCA) methodology for assessing the environmental impacts of products developed at Carnegie Mellon University in the 1990s by researchers at the Green Design Institute. A public Website provides a comprehensive overview of the EIO-LCA methods and access to online tools and guidance.[10]

Briefly, conventional economic input–output (EIO) tables map the monetary values of basic materials or goods traded between countries. The life-cycle assessment (LCA) of a product estimates emissions associated with the entire cycle of going from raw materials to a finished product. In a comprehensive assessment, the LCA also includes environmental impacts from product uses and any human or environmental health factors associated with a product from its production origins to final disposal.

Currently, there is no agreed-upon standard methodology for estimating embodied carbon in internationally traded goods. Input–output approaches and emerging multi-region input–output (MRIO) models are research tools that provide a state-of-the-art methodological framework for estimating embodied carbon in trade at national and supranational scales.[11] Further improvements are needed in data availability and quality and in assessing the precision and accuracy of MRIO modeling.[12] There is little doubt that if consumption-based accounting attains official status as a methodology for the estimation of embodied carbon in international trade, the MRIO models will become an important methodology. However, the data requirements and complexity of training a wide range of international users in both the private and public sectors will be a challenge.

How Important Is Embodied Carbon in International Trade?

A state-of-the-art analysis of embodied carbon in international trade published in the *Proceedings of the National Academy of Sciences* (PNAS) reports that the embodied carbon in goods and services imported for consumption in the United States was equivalent to transferring about 11 percent of U.S. national CO_2 inventory emissions to the exporting countries, which is approximately 2.4 tons of CO_2 per American citizen.[13] In other words, the American gets the benefit of the purchased goods while the exporting country gets credited with the CO_2 emissions produced during manufacturing. This transaction has currently little economic significance because the United States and its major trading partners are not fully engaged in an international agreement that places a price on carbon emissions, but it illustrates that the magnitude of embodied carbon in international trade is certainly not trivial. Japan's imported goods were equivalent to nearly 18 percent of domestic emissions, and European nations reduced their national CO_2 emissions 20 to 50 percent as a result of importing goods rather than manufacturing the goods within their national territories.

The PNAS study used published international trade data to create a global model of the flow of products, and estimated embodied CO_2 emissions across 57 industry sectors and 113 countries or regions. Most of the imports to wealthy countries were produced in developing countries. Small wealthy nations, such as Switzerland, avoided the largest quantities of CO_2 emissions by importing most of their manufactured goods. On the flip side, nearly 25 percent of China's CO_2 emissions, for example, were dedicated to making goods for export and consumption in other countries.

"We produce products and these products are consumed by other countries, especially the developed countries."

An estimated 23 percent of total global CO_2 emissions—or 6.8 billion tons (6.2 billion metric tons) of CO_2—was associated with international trade in 2004, with most of the exported goods originating from low-income countries and being consumed in middle- and high-income countries. China was identified as the largest exporter of embodied carbon in exported goods, followed by Russia, the Middle East, South Africa, Ukraine, and India. The largest trade flows of embodied carbon were from China to the United States, Europe, and Japan. The embodied carbon flows from Russia to Europe and from countries in the Middle East to the United States and the European Union were also significant, as was the trade between the United States and the European Union.

Imports from Russia, China, and India were significantly higher in CO_2 per U.S. dollar spent than imports from European countries. The reasons for these differences can arise from a combination of the larger fraction of coal in the producer country energy mix, the lower energy efficiency of manufacturing, and the market valuation of the products being exported.

Emerging Policy Perspectives on Embodied Carbon

The issue of embodied carbon in goods traded internationally was first raised well over a decade ago. A study of the carbon embodied in the manufactured goods imported by the six largest Organization for Economic Cooperation and Development (OECD) countries between 1984 and 1986 warned as early as 1994 that policies predicated on the reduction of greenhouse gas emissions at home might not be effective if imports were contributing significantly to domestic consumption.[14]

This brief review of recent advances in methodologies and published case studies indicates that consumption-based emission inventories can provide reliable scientific assessments of embodied carbon in international trade. The policy relevance of embodied carbon in trade is a more contentious issue. There are valid arguments that policy applications of consumption-based emissions would reduce concerns about carbon leakage, provide a quantitative basis for reducing emission responsibilities for some developing countries, increase options for mitigation, support the design of financial penalties based on environmental externalities, and encourage the international diffusion of low-carbon technologies.[15] On the other hand, national emission inventories based on the IPCC methodologies are relatively well accepted, especially in the case of CO_2 emissions. There would undoubtedly be resistance in the UNFCCC to changing a fundamental technical procedure, given the tenuous nature of the ongoing COP (Conference of the Parties) negotiations.

As would be expected, the early proponents for using embodied carbon in trade and consumption-based emission inventories as important metrics in the ongoing climate change negotiations are China and other major exporting countries in the developing world. The opponents of a consumption-based emissions approach, which include the EU's chief climate negotiator, Artur Runge-Metzger, doubt that asking importers to accept responsibility for embodied carbon in purchased goods would work. In addition to the logistical difficulties involved in regulating embodied CO_2 emissions in the country of destination, Runge-Metzger has noted, importing countries would then "like to have jurisdiction and legislative powers in order to control and limit emissions in the exporting country and I'm not sure whether my Chinese colleagues would agree on that particular point."[9]

In what appears to be a defensive move, some importing countries are discussing a border tax on the carbon content of imported goods from China and other exporting countries that are major emitters of CO_2. This notion assumes that manufacturing goods with high-carbon fuels like soft coal offers an economic competitive advantage. A policy research working paper issued from the World Bank reinforced this idea, stating that "a border tax adjustment based on carbon content in domestic production, especially if it applies to both imports and exports, would broadly address the competitiveness concerns of producers in high income countries and less seriously damage developing country trade."[16] A working paper recently published by the Stockholm Environment Institute contradicts the World Bank results.[17] This paper concludes that "China's success in trade is based on low labor costs, not on embodied carbon emissions; there is literally no correlation between the amount of CO_2 emissions emitted per unit of product and revealed comparative advantage within the Chinese economy today." It is also likely that the use of border taxes to penalize developing countries with high CO_2 emissions could escalate to the point of undermining the effectiveness of the World Trade Organization and the importance of trade to the reduction of global poverty.

Economic globalization also has important and well–documented environmental consequences on air quality, water quality, and land use at local and regional environmental scales in developing countries.[18] The United States and other developed countries should view China and other developing countries as important markets and research opportunities for advancing environmental technologies. Actions to improve air quality often have the co-benefit of reducing CO_2 emissions as a result of fuel switching from coal to natural gas, nuclear power, or renewable energy sources. This "stealth" technology-sharing approach to reducing developing country embodied carbon is likely to gain far more political traction in many developing countries than a penalty approach (e.g., border tax). Given current knowledge, it is now certain that the fastest and most effective path to concurrently reducing most environmental pollutants is to accelerate the transition to clean energy technologies. However, the transitions of major technologies have historically taken 50 to 100 years. The transition from fossil fuels to clean energy technologies will require a wide range of new infrastructures, regulatory frameworks, and enormous financial investments. The challenge of achieving the focus and scale necessary to avert serious consequences of global warming will remain daunting.

What's Next?

The UNFCC COP-16 will convene in Mexico in December 2010. The modest gains achieved in the Copenhagen Accord do not bode well for the future of UNFCCC-COP negotiations on climate change. Each disappointing COP meeting has enhanced the broader perception of the process being in a state of "slow-motion failure," and heading for an eventual "multilateral zombie" outcome or "death by climatocracy."[19] The zombie scenario would have the process stagger along piteously, never making much progress, while never quite dying either. The more likely scenario of death by climatocracy is potentially more dangerous in imagining success on reaching an agreement that subsequently fails due to inadequate attention paid to institutions necessary for effective implementation.

The recent failure of the U.S. Congress to take action on climate and energy legislation and China's lack of interest in discussing binding commitments are clear signals that further negotiations on the mitigation of climate change will be wasted time. A focus on a selected group of issues that concern both developed and developing countries is urgently needed to break the current gridlock. Some progress may be possible on issues related to rebuilding trust in the IPCC science process, the reducing of emissions from deforestation and forest degradation (REDD), increasing research on the deployment of low carbon technologies, and international support for climate adaptation actions in developing countries.

The increasingly relevant question is how long the UNFCCC-COP process can survive without substantial progress on a realistic agenda for international accountability at a scale appropriate to the global climate change problem. The COP process has become locked into a classic free-rider problem where each country wants everyone else to do the "right thing" while that country benefits from being the exception. Unfortunately, the history of these negotiations gives the appearance that there are few disincentives for failure. The important question is what comes next if the UNFCCC process fails.

Prosperity Without Conspicuous Consumption

The prosperity of the United States has benefited from more than a century of its status as a world leader in the manufacturing and production of goods based, in part, on access to cheap fossil fuels and ignorance of the climate consequences of emitting CO_2. The consumption of goods and services now accounts for more than two-thirds of United States economic activity. The Chinese and American economies together accounted for a third of global economic output and two-fifths of worldwide economic growth from 1998 to 2007. As a result of the "Chimerican" symbiosis, China quadrupled its gross domestic product from 2000 to 2008, increased exports by a factor of five, imported Western technology, and created tens of millions of manufacturing jobs for the rural poor.[20] American overconsumption meant that from 2000 to 2008, the United States consistently outspent its national income, leading to an unsustainable increase in the national debt. Goods imported from China accounted for about a third of that overconsumption.

Given the magnitude and trajectory of China's likely continued economic expansion, we are experiencing only the initial phase of this nation's potential impact on the global environment, geopolitics, and society at large. India and other developing nations are not far behind China with similar aspirations to improve the well-being of their millions of impoverished people. Indeed, it seems likely that we are in the midst of an acceleration of globalization and consumption of considerable historic importance.

The pursuit of a consumption-based approach to measuring CO_2 emissions reveals questions central to climate change and sustainable development. For example, what are the geopolitical implications of China and other emerging economies becoming increasingly formidable competitors in world markets and in the competition for energy and other strategic resources? And, perhaps most importantly, what will the emergence of China, India, and other developing countries as the world's largest consumer economies mean for an already fragile global environment?

We see the UNFCCC and IPCC as being too narrowly focused on climate change. A comprehensive and integrated climate and sustainability strategy that acknowledges the need to account for both the impacts of conspicuous consumption in wealthy countries and unmet basic needs in developing countries is urgently needed. Various publications have appeared in recent months that offer innovative ideas for reframing the global change narrative.[21]

One certainty is that consumption-based accounting of carbon, nitrogen, water, and other environment factors associated with international trade will be important to addressing both climate change and sustainable development challenges that lie ahead. The information derived from consumption-based accounting, together with attention to physical and cultural needs, will provide a framework for dialogues on sufficiency versus conspicuous consumption. This approach would also better integrate climate change into the larger suite of issues associated with sustainable development.

Sustainable development, as reflected in the United Nations Millennium Development Goals, proposes a broad set of policy challenges for stabilizing the world's population growth, narrowing the well-being gaps between the rich and poor, and protecting the environment.[22] While it is obvious that climate change and sustainable development are intimately intertwined, a sustainability strategy will be more likely to gain wide acceptance among all nations by focusing initially on the moral basis and practical pathways to a future world based on principles of sufficiency in meeting basic material needs, non-violence, and global common goods.

Notes

1. D. Guan, K. Hubacek, C. L. Weber, G. P. Peters, and D. M. Reiner, "The Drivers of Chinese CO_2 Emissions from 1980 to 2030," *Global Environmental Change* 18 no. 4 (2008): 626–634; N. Zeng, Y. Ding, J. Pan, H. Wang, and J. Gregg, "Climate Change—The Chinese Challenge," *Science* 319: no. 5864 (8 February 2008): 730–731; G. P. Peters, C. L. Weber, D. Guan, and K. Hubacek, "China's Growing CO_2 Emissions—A Race Between Increasing Consumption and Efficiency Gains," *Environmental Science & Technology* 41, no. 17 (2007): 5939–5944.
2. L. Cohen, *A Consumers' Republic,* (New York: Vintage Books, 2003); E. R. Shell, *Cheap: The High cost of Discount Culture,* (New York: Penguin, 2009).
3. D. J. Hoffmann, J. H. Butler, and P. P. Tans, "A New Look at Atmospheric Carbon Dioxide," *Atmospheric Environment* 43, no. 12 (2009): 2084–2086.
4. R. N. Stavins, and R. C. Stowe, "What Hath Copenhagen Wrought? A Preliminary Assessment," *Environment* 52 no. 3 (2010): 8–14.
5. M. Hulme, "Moving Beyond Climate Change," *Environment* 52, no. 3 (2010): 15–19.
6. M. Hulme, *Why We Disagree About Climate Change,* (Cambridge, UK: Cambridge University Press, 2009).
7. G. P. Peters, "From Production-Based to Consumption-Based National Inventories," *Ecological Economics* 65, no. 1 (2008): 13–23.
8. Xinhua News Agency, "President Hu Elaborates on China's Stance on Climate Change," *Peoples Daily Online,* http:// english.peopledaily.com.cn (accessed 7 January 2009)
9. Anonymous, "Who Should Pay for Embedded Carbon," *International Centre for Trade and Sustainable Development News and Analysis,* 13, no. 1 (2009).
10. Economic Input–Output Life Cycle Assessment (EIO-LCA), www.eiolca.net (accessed 18 July 2010).
11. T. Wiedmann, "A Review of Recent Multi-Region Input-Output Models Used for Consumption-Based Emission and Resource Accounting," *Ecological Economics* 69 no. 2 (2009): 211–222; T. Wiedmann, M. Lenzen, K. Turner, and J. Barrett, "Examining the Global Environmental Impact of Regional Consumption Activities—Part 2: Review of Input-Output Models for the Assessment of Environmental Impacts Embodied in Trade," *Ecological Economics* 61 no. 1 (2007): 15–26.
12. A. Druckman, P. Bradley, E. Papathanasopoulou, T. Jackson, "Measuring Progress Towards Carbon Reduction in the UK," *Ecological Economics* 66 no. 4 (2008): 594–604.
13. S. Davis and K. Caldeira, "Consumption-Based Accounting of CO_2 Emissions," *Proceedings of the National Academy of Sciences USA* 107 no. 12 (2010): 5687–5692.
14. A. Wyckoff and J. Roop, "The Embodiment of Carbon in Imports of Manufactured Products. Implications for International Agreements on Greenhouse Gas Emissions," *Energy Policy* 22 no. 3 (1994): 187–194.
15. G. P. Peters and E. Hertwick, "Post-Kyoto Greenhouse Gas Inventories: Production versus Consumption," *Climatic Change* 86 no. 1 (2008): 51–66.
16. A. Mattoo, A. Subramanian, D. van der Mensbrugghe, and J. He, "Reconciling Climate Change and Trade Policy," Policy Research Working Paper WPS5123, The World Bank, Washington, D.C, November 2009.
17. F. Ackerman, "Carbon Embedded in China's Trade," Working Paper WP-US-0906, Stockholm Environment Institute, Stockholm, Sweden, 16 June 2009.
18. K. Gallagher, "Economic Globalization and the Environment," *Annual Review of Environment Resources* 34 (2009): 279–304.
19. A. Evans and D. Steven, "Hitting Reboot, Where Next for Climate After Copenhagen?," Brookings Institution, 21 December 2009, www.brookings.edu/papers/2009/1221_climate_evans_steven.aspx (accessed 4 August 2010).
20. N. Ferguson, "The Great Wallop," *Finance & Economics* November 2009, www.niallferguson.com/site/FERG/Templates/ArticleItem.aspx?pageid=221 (accessed 18 July 2010).
21. M. Hulme, *Why We Disagree About Climate Change* (Cambridge: Cambridge University Press, 2009); N. Birdsall, A. Subramanian, D. Hammer, and K. Ummel, "Energy Needs and Efficiency, Not Emissions: Re-Framing the Climate

Change Narrative," Working Paper 187, Center for Global Development, World Bank, Washington, D.C., November, 2009; The Hartwell Paper: A New Direction for Climate Policy After the Crash of 2009, http://eprints.lse.ac.uk/27939 (accessed 18 July 2010); R. Pielke, Jr., *The Climate Fix: What Scientists and Politicians Won't Tell You About Global Warming,* (New York: Basic Books, 2010).

22. United Nations Millennium Development Goals, www.un.org/millenniumgoals, (accessed 31 July 2010).

Critical Thinking

1. Should CO_2 emissions be the responsibility of the producer or consumer of a product that causes them?
2. How would alternatives to the GDP account for CO_2 emissions?
3. Could the arguments provided in this article be applied to other emissions? If so, which ones?

Robert Harriss is president of the Houston Advanced Research Center, a nonprofit research and education organization dedicated to sustainability science, engineering, and policy. He holds adjunct professorships at Texas A&M University—Galveston and the University of Houston. He has published extensively on the role of human actions in driving regional and global environmental change. **Bin Shui** is a scientist at the Joint Global Change Research Institute located in College Park, Maryland. Dr. Shui's research is focused on human dimensions of global change, including environmental impacts of household consumption, embodied carbon in trade, and urban transportation.

The contents of this article reflect the views of the authors and do not necessarily reflect the official views or policies of their respective institutions.

The Rise of Vertical Farms

Growing crops in city skyscrapers would use less water and fossil fuel than outdoor farming, eliminate agricultural runoff and provide fresh food

DICKSON DESPOMMIER

Together the world's 6.8 billion people use land equal in size to South America to grow food and raise livestock—an astounding agricultural footprint. And demographers predict the planet will host 9.5 billion people by 2050. Because each of us requires a minimum of 1,500 calories a day, civilization will have to cultivate another Brazil's worth of land—2.1 billion acres—if farming continues to be practiced as it is today. That much new, arable earth simply does not exist. To quote the great American humorist Mark Twain: "Buy land. They're not making it any more."

Agriculture also uses 70 percent of the world's available freshwater for irrigation, rendering it unusable for drinking as a result of contamination with fertilizers, pesticides, herbicides and silt. If current trends continue, safe drinking water will be impossible to come by in certain densely populated regions. Farming involves huge quantities of fossil fuels, too—20 percent of all the gasoline and diesel fuel consumed in the U.S. The resulting greenhouse gas emissions are of course a major concern, but so is the price of food as it becomes linked to the price of fuel, a mechanism that roughly doubled the cost of eating in most places worldwide between 2005 and 2008.

Some agronomists believe that the solution lies in even more intensive industrial farming, carried out by an ever decreasing number of highly mechanized farming consortia that grow crops having higher yields—a result of genetic modification and more powerful agrochemicals. Even if this solution were to be implemented, it is a short-term remedy at best, because the rapid shift in climate continues to rearrange the agricultural landscape, foiling even the most sophisticated strategies. Shortly after the Obama administration took office, Secretary of Energy Steven Chu warned the public that climate change could wipe out farming in California by the end of the century.

What is more, if we continue wholesale deforestation just to generate new farmland, global warming will accelerate at an even more catastrophic rate. And far greater volumes of agricultural runoff could well create enough aquatic "dead zones" to turn most estuaries and even parts of the oceans into barren wastelands.

As if all that were not enough to worry about, foodborne illnesses account for a significant number of deaths worldwide—salmonella, cholera, *Escherichia coli* and shigella, to name just a few. Even more of a problem are life-threatening parasitic infections, such as malaria and schistosomiasis. Furthermore, the common practice of using human feces as a fertilizer in most of Southeast Asia, many parts of Africa, and Central and South America (commercial fertilizers are too expensive) facilitates the spread of parasitic worm infections that afflict 2.5 billion people.

Clearly, radical change is needed. One strategic shift would do away with almost every ill just noted: grow crops indoors, under rigorously controlled conditions, in vertical farms. Plants grown in high-rise buildings erected on now vacant city lots and in large, multistory rooftop greenhouses could produce food year-round using significantly less water, producing little waste, with less risk of infectious diseases, and no need for fossil-fueled machinery or transport from distant rural farms. Vertical farming could revolutionize how we feed ourselves and the rising population to come. Our meals would taste better, too; "locally grown" would become the norm.

The working description I am about to explain might sound outrageous at first. But engineers, urban planners and agronomists

Problem

Feeding the Future: Not Enough Land

Growing food and raising livestock for 6.8 billion people require land equal in size to South America. By 2050 another Brazil's worth of area will be needed, using traditional farming; that much arable land does not exist.

- Present (6.8 billion people) = Uses cropland the size of South America
- 2050 (9.5 billion people) = Size of South America + Would require added cropland the size of Brazil

who have scrutinized the necessary technologies are convinced that vertical farming is not only feasible but should be tried.

Do No Harm

Growing our food on land that used to be intact forests and prairies is killing the planet, setting up the processes of our own extinction. The minimum requirement should be a variation of the physician's credo: "Do no harm." In this case, do no further harm to the earth. Humans have risen to conquer impossible odds before. From Charles Darwin's time in the mid-1800s and forward, with each Malthusian prediction of the end of the world because of a growing population came a series of technological breakthroughs that bailed us out. Farming machines of all kinds, improved fertilizers and pesticides, plants artificially bred for greater productivity and disease resistance, plus vaccines and drugs for common animal diseases all resulted in more food than the rising population needed to stay alive.

That is until the 1980s, when it became obvious that in many places farming was stressing the land well beyond its capacity to support viable crops. Agrochemicals had destroyed the natural cycles of nutrient renewal that intact ecosystems use to maintain themselves. We must switch to agricultural technologies that are more ecologically sustainable.

As the noted ecologist Howard Odum reportedly observed: "Nature has all the answers, so what is your question?" Mine is: How can we all live well and at the same time allow for ecological repair of the world's ecosystems? Many climate experts—from officials at the United Nations Food and Agriculture Organization to sustainable environmentalist and 2004 Nobel Peace Prize winner Wangari Maathai—agree that allowing farmland to revert to its natural grassy or wooded states is the easiest and most direct way to slow climate change. These landscapes naturally absorb carbon dioxide, the most abundant greenhouse gas, from the ambient air. Leave the land alone and allow it to heal our planet.

Examples abound. The demilitarized zone between South and North Korea, created in 1953 after the Korean War, began as a 2.5-mile-wide strip of severely scarred land but today is lush and vibrant, fully recovered. The once bare corridor separating former East and West Germany is now verdant. The American dust bowl of the 1930s, left barren by overfarming and drought, is once again a highly productive part of the nation's breadbasket. And all of New England, which was clear-cut at least three times since the 1700s, is home to large tracts of healthy hardwood and boreal forests.

The Vision

For many reasons, then, an increasingly crowded civilization needs an alternative farming method. But are enclosed city skyscrapers a practical option?

Yes, in part because growing food indoors is already becoming commonplace. Three techniques—drip irrigation, aeroponics and hydroponics—have been used successfully around the world. In drip irrigation, plants root in troughs of lightweight, inert material, such as vermiculite, that can be used for years, and small tubes running from plant to plant drip nutrient-laden water precisely at each stem's base, eliminating the vast amount of water wasted in traditional irrigation. In aeroponics, developed in 1982 by K. T. Hubick, then later improved by NASA scientists, plants dangle in air that is infused with water vapor and nutrients, eliminating the need for soil, too.

Agronomist William F. Gericke is credited with developing modern hydroponics in 1929. Plants are held in place so their roots lie in soilless troughs, and water with dissolved nutrients is circulated over them. During World War II, more than eight million pounds of vegetables were produced hydroponically on South Pacific islands for Allied forces there. Today hydroponic greenhouses provide proof of principles for indoor farming: crops can be produced year-round, droughts and floods that often ruin entire harvests are avoided, yields are maximized because of ideal growing and ripening conditions, and human pathogens are minimized.

Most important, hydroponics allows the grower to select where to locate the business, without concern for outdoor environmental conditions such as soil, precipitation or temperature profiles. Indoor farming can take place anywhere that adequate water and energy can be supplied. Sizable hydroponic facilities can be found in the U.K., the Netherlands, Denmark, Germany, New Zealand and other countries. One leading example is the 318-acre Eurofresh Farms in the Arizona desert, which produces large quantities of high-quality tomatoes, cucumbers and peppers 12 months a year.

Most of these operations sit in semirural areas, however, where reasonably priced land can be found. Transporting the food for many miles adds cost, consumes fossil fuels, emits carbon dioxide and causes significant spoilage. Moving greenhouse farming into taller structures within city limits can solve these remaining problems. I envision buildings perhaps 30 stories high covering an entire city block. At this scale, vertical farms offer the promise of a truly sustainable urban life: municipal wastewater would be recycled to provide irrigation water, and the remaining solid waste, along with inedible plant matter, would be incinerated to create steam that turns turbines that generate electricity for the farm. With current technology, a wide variety of edible plants can be grown indoors. An adjacent aquaculture center could also raise fish, shrimp and mollusks.

Start-up grants and government-sponsored research centers would be one way to jump-start vertical farming. University partnerships with companies such as Cargill, Monsanto, Archer Daniels Midland and IBM could also fill the bill. Either approach would exploit the enormous talent pool within many agriculture, engineering and architecture schools and lead to prototype farms perhaps five stories tall and one acre in footprint. These facilities could be the "playground" for graduate students, research scientists and engineers to carry out the necessary trial-and-error tests before a fully functional farm emerged. More modest, rooftop operations on apartment complexes, hospitals and schools could be test beds, too. Research installations already exist at many schools, including the University of California, Davis, Pennsylvania State University, Rutgers University, Michigan State University, and schools in Europe and Asia. One

Growing Techniques

Three technologies would be exploited in vertical farms.

Aeroponics

Plants are held in place so their roots dangle in air that is infused with water vapor and nutrients. Good for root crops (potatoes, carrots).

Hydroponics

Plants are held in place so their roots lie in open troughs; water with dissolved nutrients is continually circulated over them. Good for many vegetables (tomatoes, spinach) and berries.

Drip Irrigation

Plants grow in troughs of lightweight, inert material, such as vermiculite, reused for years. Small tubing on the surface drips nutrient-laden water precisely at each stem's base. Good for grains (wheat, corn).

of the best known is the University of Arizona's Controlled Environment Agriculture Center, run by Gene Giacomelli.

Integrating food production into city living is a giant step toward making urban life sustainable. New industries will grow, as will urban jobs never before imagined—nursery attendants, growers and harvesters. And nature will be able to rebound from our insults; traditional farmers would be encouraged to grow grasses and trees, getting paid to sequester carbon. Eventually selective logging would be the norm for an enormous lumber industry, at least throughout the eastern half of the U.S.

Practical Concerns

In recent years I have been speaking regularly about vertical farms, and in most cases, people raise two main practical questions. First, skeptics wonder how the concept can be economically viable, given the often inflated value of properties in cities such as Chicago, London and Paris. Downtown commercial zones might not be affordable, yet every large city has plenty of less desirable sites that often go begging for projects that would bring in much needed revenue.

Maximum Yield

On most floors of a vertical farm, an automated conveyor would move seedlings from one end to the other, so that the plants would mature along the way and be at the height of producing grain or vegetables when they reached a harvester. Water and lighting would be tailored to optimize growth at each stage. Inedible plant material would drop down a chute to electricity-generating incinerators in the basement.

High-Rise Crops

A 30-story vertical farm would exploit different growing techniques on various floors. Solar cells and incineration of plant waste dropped from each floor would create power. Cleansed city wastewater would irrigate plants intead of being dumped into the environment. The sun and artificial illumination would provide light. Incoming seeds would be tested in a lab and germinate in a nursery. And a grocery and restaurant would sell fresh food directly to the public.

In New York City, for example, the former Floyd Bennett Field naval base lies fallow. Abandoned in 1972, the 2.1 square miles scream out for use. Another large tract is Governors Island, a 172-acre parcel in New York Harbor that the U.S. government recently returned to the city. An underutilized location smack in the heart of Manhattan is the 33rd Street rail yard. In addition, there are the usual empty lots and condemned buildings scattered throughout the city. Several years ago my graduate students surveyed New York City's five boroughs; they found no fewer than 120 abandoned sites waiting for change, and many would bring a vertical farm to the people who need it most, namely, the underserved inhabitants of the inner city. Countless similar sites exist in cities around the world. And again, rooftops are everywhere.

Simple math sometimes used against the vertical farm concept actually helps to prove its viability. A typical Manhattan block covers about five acres. Critics say a 30-story building would therefore provide only 150 acres, not much compared with large outdoor farms. Yet growing occurs year-round. Lettuce, for example, can be harvested every six weeks, and even a crop as slow to grow as corn or wheat (three to four months from planting to picking) could be harvested three to four times annually. In addition, dwarf corn plants, developed for NASA, take up far less room than ordinary corn and grow to a height of just two or three feet. Dwarf wheat is also small in stature but high in nutritional value. So plants could be packed tighter, doubling yield per acre, and multiple layers of dwarf crops could be grown per floor. "Stacker" plant holders are already used for certain hydroponic crops.

Combining these factors in a rough calculation, let us say that each floor of a vertical farm offers four growing seasons, double the plant density, and two layers per floor—a multiplying factor of 16 ($4 \times 2 \times 2$). A 30-story building covering one city block could therefore produce 2,400 acres of food (30 stories $\times$ 5 acres $\times$ 16) a year. Similarly, a one-acre roof atop a hospital or school, planted at only one story, could yield 16 acres of victuals for the commissary inside. Of course, growing could be further accelerated with 24-hour lighting, but do not count on that for now.

Other factors amplify this number. Every year droughts and floods ruin entire counties of crops, particularly in the American Midwest. Furthermore, studies show that 30 percent of what is harvested is lost to spoilage and infestation during storage and transport, most of which would be eliminated in city farms because food would be sold virtually in real time and on location as a consequence of plentiful demand. And do not forget that we will have largely eliminated the mega insults of outdoor farming: fertilizer runoff, fossil-fuel emissions, and loss of trees and grasslands.

The second question I often receive involves the economics of supplying energy and water to a large vertical farm. In this regard, location is everything (surprise, surprise). Vertical farms in Iceland, Italy, New Zealand, southern California and some parts of East Africa would take advantage of abundant geothermal energy. Sun-filled desert environments (the American Southwest, the Middle East, many parts of Central Asia) would actually use two- or three-story structures perhaps 50 to 100 yards wide but miles long, to maximize natural sunlight for growing and photovoltaics for power. Regions gifted with steady winds (most coastal zones, the Midwest) would capture that energy. In all places, the plant waste from harvested crops would be incinerated to create electricity or be converted to biofuel.

One resource that routinely gets overlooked is very valuable as well; in fact, communities spend enormous amounts of energy and money just trying to get rid of it safely. I am referring to liquid municipal waste, commonly known as blackwater. New York City occupants produce one billion gallons of wastewater every day. The city spends enormous sums to cleanse it and then dumps the resulting "gray water" into the Hudson River. Instead that water could irrigate vertical farms. Meanwhile the solid by-products, rich in energy, could be incinerated as well. One typical half-pound bowel movement contains 300 kilocalories of energy when incinerated in a bomb calorimeter. Extrapolating to New York's eight million people, it is theoretically possible to derive as much as 100 million kilowatt-hours of electricity a year from bodily wastes alone, enough to run four, 30-story farms. If this material can be converted into useful water and energy, city living can become much more efficient.

Upfront investment costs will be high, as experimenters learn how to best integrate the various systems needed. That expense is why smaller prototypes must be built first, as they are for any new application of technologies. Onsite renewable energy production should not prove more costly than the use of expensive fossil fuel for big rigs that plow, plant and harvest crops (and emit volumes of pollutants and greenhouse gases). Until we gain operational experience, it will be difficult to predict how profitable a vertical farm could be. The other goal, of couse, is for the produce to be less expensive than current supermarket prices, which should be attainable largely because locally grown food does not need to be shipped very far.

Desire

It has been five years since I first posted some rough thoughts and sketches about vertical farms on a Website I cobbled together (www.verticalfarm.com). Since then, architects, engineers, designers and mainstream organizations have increasingly taken note. Today many developers, investors, mayors and city planners have become advocates and have indicated a strong desire to actually build a prototype high-rise farm. I have been approached by planners in New York City, Portland, Ore., Los Angeles, Las Vegas, Seattle, Surrey, B.C., Toronto, Paris, Bangalore, Dubai, Abu Dhabi, Incheon, Shanghai and Beijing. The Illinois Institute of Technology is now crafting a detailed plan for Chicago.

All these people realize that something must be done soon if we are to establish a reliable food supply for the next generation. They ask tough questions regarding cost, return on investment, energy and water use, and potential crop yields. They worry about structural girders corroding over time from humidity, power to pump water and air everywhere, and economies of scale. Detailed answers will require a huge input from engineers, architects, indoor agronomists and businesspeople. Perhaps budding engineers and economists would like to get these estimations started.

Because of the Website, the vertical farm initiative is now in the hands of the public. Its success or failure is a function only of those who build the prototype farms and how much time and

Hurdles

Several roadblocks could stifle the spread of urban farms, but all can be solved.

- Reclaim enough abandoned city lots and open rooftops as sites for indoor agriculture.
- Convert municipal wastewater into usable irrigation water.
- Supply inexpensive energy to circulate water and air.
- Convince city planners, investors, developers, scientists and engineers to build prototype farms where practical issues could be resolved.

More to Explore

- **Our Ecological Footprint: Reducing Human Impact on the Earth.** Mathis Wackernagel and William Rees. New Society Publishers, 1996.
- **Cradle to Cradle: Remaking the Way We Make Things.** William McDonough and Michael Braungart. North Point Press, 2002.
- **Growing Vertical.** Mark Fischetti in *Scientific American Earth 3.0,* Vol. 18, No. 4, pages 74–79; 2008.
- University of Arizona Controlled Environment Agricultural Center: http://ag.arizona.edu/ceac
- **Vertical Farm: The Big Idea That Could Solve the World's Food, Water and Energy Crises.** Dickson Despommier. Thomas Dunne Books/ St. Martin's Press (in press).

effort they apply. The infamous Biosphere 2 closed-ecosystem project outside Tucson, Ariz., first inhabited by eight people in 1991, is the best example of an approach not to take. It was too large of a building, with no validated pilot projects and a total unawareness about how much oxygen the curing cement of the massive foundation would absorb. (The University of Arizona now has the rights to reexamine the structure's potential.)

If vertical farming is to succeed, planners must avoid the mistakes of this and other non-scientific misadventures. The news is promising. According to leading experts in ecoengineering such as Peter Head, who is director of global planning at Arup, an international design and engineering firm based in London, no new technologies are needed to build a large, efficient urban vertical farm. Many enthusiasts have asked: "What are we waiting for?" I have no good answer for them.

Critical Thinking

1. Do you agree with the assertion "that much new, arable Earth simpley does not exist"? Why or why not?
2. Are there alternatives to "wholesale deforestation" to "generate new farmland"? If so, what are they?
3. What do you think are the arguments against vertical farming?

Climate 2030

A National Blueprint for a Clean Energy Economy

UNION OF CONCERNED SCIENTISTS

Reducing oil dependence. Strengthening energy security. Creating jobs. Tackling global warming. Addressing air pollution. Improving our health. The United States has many reasons to make the transition to a clean energy economy. What we need is a comprehensive set of smart policies to jump-start this transition without delay and maximize the benefits to our environment and economy. *Climate 2030: A National Blueprint for a Clean Energy Economy* ("the Blueprint") answers that need.

Recent rapid growth of the wind industry (developers have installed more wind power in the United States in the last two years than in the previous 20) and strong sales growth of hybrid vehicles show that the U.S. transformation to a clean energy economy is already under way. However, these changes are still too gradual to address our urgent need to reduce heat-trapping emissions to levels that are necessary to protect the well-being of our citizens and the health of our environment.

Global warming stems from the release of carbon dioxide and other heat-trapping gases into the atmosphere, primarily when we burn fossil fuels and clear forests. The problems resulting from the ensuing carbon overload range from extreme heat, droughts, and storms to acidifying oceans and rising sea levels. To help avoid the worst of these effects, the United States must play a lead role and begin to cut its heat-trapping emissions today—and aim for at least an 80 percent drop from 2005 levels by 2050.

The Blueprint Cuts Carbon Emissions and Saves Money

Blueprint policies lower U.S. heat-trapping emissions to meet a cap set at 26 percent below 2005 levels in 2020, and 56 percent below 2005 levels in 2030. The actual year-by-year emissions reductions differ from the levels set in the cap because firms have the flexibility to over-comply with the cap in early years, bank allowances, and then use them to meet the cap requirements in later years.

To meet the cap, the cumulative *actual* emissions must equal the cumulative tons of emissions set by the cap. In 2030, we achieve this goal.

The nation achieves these deep cuts in carbon emissions while saving consumers and businesses $465 billion annually by 2030. The Blueprint also builds $1.7 trillion in net cumulative savings between 2010 and 2030.[1]

Blueprint policies stimulate significant consumer, business, and government investment in new technologies and measures by 2030. The resulting savings on energy bills from reductions in electricity and fuel use more than offset the costs of these additional investments. The result is net annual savings for households, vehicle owners, businesses, and industries of $255 billion by 2030.[2]

Climate 2030 Blueprint shows that deep emissions cuts can be achieved while saving U.S. consumers and businesses $465 billion in 2030.

We included an additional $8 billion in government-related costs to administer and implement the policies. However, auctioning carbon allowances will generate $219 billion in revenues that is invested back into the economy.[3] This brings annual Blueprint savings up to $465 billion by 2030.[4]

Under the Blueprint, every region of the country stands to save billions. Households and businesses—even in coal-dependent regions—will share in these savings.

The Blueprint keeps carbon prices low. Under the Blueprint, the price of carbon allowances starts at about $18 per ton of CO_2 in 2011, and then rises to $34 in 2020, and to $70 in 2030 (all in 2006 dollars). Those prices are well within the range that other analyses find, despite our stricter cap on economywide emissions.

The Climate 2030 Approach

This report analyzes the economic and technological feasibility of meeting stringent targets for reducing global warming emissions, with a cap set at 26 percent below 2005 levels by 2020, and 56 percent below 2005 levels by 2030. Meeting this cap means the United States would limit total emissions—the crucial measure for the climate—to 180,000 million metric tons carbon dioxide equivalent ($MMTCO_2eq$) from 2000 to 2030.[6]

The nation's long-term carbon budget for 2000 to 2050—as defined in a previous UCS analysis (Luers et al. 2007)—is 160,000 to 265,000 $MMTCO_2eq$. The 2000–2030 carbon budget in our analysis would put us on track to reach the mid-range of that long-term budget by 2050, if the nation continues to cut emissions steeply.

To reach the 2020 and 2030 cap and carbon budget targets, the Blueprint proposes a comprehensive policy approach (the "Blueprint policies") that combines an economywide cap-and-trade program with complementary policies. This approach finds cost-effective ways to reduce fossil fuel emissions throughout our economy—including in industry, buildings, electricity, and transportation—and to store carbon through agricultural activities and forestry.

Our analysis relies primarily on a modified version of the U.S. Department of Energy's National Energy Modeling System (referred to as UCS-NEMS). We supplemented that model with an analysis of the impact of greater energy efficiency in industry and buildings by the American Council for an Energy Efficient Economy. We also worked with researchers at the University of Tennessee to analyze the potential for crops and residues to provide biomass energy. We then combined our model with those studies to capture the dynamic interplay between energy use, energy prices, energy investments, and the economy while also considering competition for limited resources and land.

Our analysis explores two main scenarios. The first—which we call the Reference case—assumes no new climate, energy, or transportation policies beyond those already in place as of October 2008.[7] The second—the Blueprint case—examines an economywide cap-and-trade program, plus a suite of complementary policies to boost energy efficiency and the use of renewable energy in key economic sectors: industry, buildings, electricity, and transportation. Our analysis also includes a third "sensitivity" scenario that strips out the policies targeted at those sectors, which we refer to as the No Complementary Policies case.

Our analysis shows that the technologies and policies pursued under the Blueprint produce dramatic changes in energy use and cuts in carbon emissions. The analysis also shows that consumers and businesses reap significant net savings under the comprehensive Blueprint approach, while the nation sees strong economic growth.

In addition, the Blueprint achieves much larger cuts in carbon emissions *within the capped sectors* because of the tighter limits that we set on "offsets"[5] and because of our more realistic assumptions about the cost-effectiveness of investments in energy efficiency and renewable energy technologies.

The economy grows by at least 81 percent by 2030 under the Blueprint. U.S. gross domestic product (GDP) expands by 81 percent between 2005 and 2030 under our approach—virtually the same as in the Reference case, which shows the U.S. economy growing by 84 percent. In fact, our model predicts that the Blueprint will slow economic growth by less than 1.5 percent in 2030—equivalent to only 10 months of economic growth over the 25-year period.[8]

The Blueprint also shows practically the same employment trends as the Reference case. In fact, nonfarm employment is slightly higher under the Blueprint than in the Reference case (170 million jobs versus 169.4 million in 2030).

We should note that there are significant limitations in the way NEMS accounts for the GDP and employment effects of the Blueprint policies. NEMS does not fully consider the economic growth that would arise from investments in clean technology, or from the spending of the money consumers and businesses saved on energy due to these investments. And the Reference case does not include the costs of global warming itself.

The Blueprint cuts the annual household cost of energy and transportation by $900 in 2030. The average U.S. household would see net savings on electricity, natural gas, and oil of $320 per year compared with the Reference case, after paying for investments in new energy efficiency and low-carbon technologies.

Transportation expenses for the average household would fall by about $580 per year in 2030. Those savings take into account the higher costs of cleaner cars and trucks, new fees used to fund more public transit, and declining use of gasoline.

Businesses save nearly $130 billion in energy-related expenses annually by 2030 under the Blueprint. Neither the energy nor the transportation savings account for the revenue from auctioning carbon allowances that will be invested back into the economy, lowering consumer and business costs (or increasing consumer and business savings) even further.

The Blueprint Changes the Energy We Use

Blueprint policies reduce projected U.S. energy use by one-third by 2030. Significant increases in energy efficiency across the economy and reductions in car and truck travel drive down energy demand and carbon emissions.

Climate 2030 Blueprint Policies

Climate Policies

- Economywide cap-and-trade program with:
    - Auctioning of all carbon allowances
    - Recycling of auction revenues to consumers and businesses[9]
    - Limits on carbon "offsets" to encourage "decarbonization" of the capped sectors
    - Flexibility for capped businesses to over-comply with the cap and bank excess carbon allowances for future use

Industry and Buildings Policies

- An energy efficiency resource standard requiring retail electricity and natural gas providers to meet efficiency targets
- Minimum federal energy efficiency standards for specific appliances and equipment
- Advanced energy codes and technologies for buildings
- Programs that encourage more efficient industrial processes
- Wider reliance on efficient systems that provide both heat and power
- R&D on energy efficiency

Electricity Policies

- A renewable electricity standard for retail electricity providers
- R&D on renewable energy
- Use of advanced coal technology, with a carbon-capture-and-storage demonstration program

Transportation Policies

- Standards that limit carbon emissions from vehicles
- Standards that require the use of low-carbon fuels
- Requirements for deployment of advanced vehicle technology
- Smart-growth policies that encourage mixed-use development, with more public transit
- Smart-growth policies that tie federal highway funding to more efficient transportation systems
- Pay-as-you-drive insurance and other per-mile user fees

Carbon-free electricity and low-carbon fuels together make up more than one-third of the remaining U.S. energy use by 2030. A significant portion of U.S. reductions in carbon emissions in 2030 comes from a 25 percent increase in the use of renewable energy from wind, solar, geothermal, and bioenergy under the Blueprint. Carbon emissions are also kept low because the use of nuclear energy and hydropower—which do not directly produce carbon emissions—remain nearly the same as in the Reference case.

In 2030, the Blueprint cuts the use of petroleum products by six million barrels a day—as much as we are currently importing from OPEC countries.

The Blueprint reduces U.S. dependence on oil and oil imports. By 2030, the Blueprint cuts the use of oil and other petroleum products by 6 million barrels per day, compared with 2005. That is as much oil as the nation now imports from the 12 members of OPEC (the Organization of Petroleum Exporting Countries). Those reductions will help drop imports to less than 45 percent of the nation's oil needs, and cut projected expenditures on those imports by more than $85 billion in 2030, or more than $160,000 per minute.

Smart Energy and Transportation Policies Are Essential for the Greatest Savings

Many of the Blueprint's complementary policies have a proven track record at state and federal levels. These policies include emission standards for vehicles and fuels, energy efficiency standards for appliances, buildings, and industry, and renewable energy standards for electricity. The Blueprint also relies on innovative policies to reduce the number of miles people travel in their cars and trucks.

These policies are essential to delivering significant consumer and business savings under the Blueprint. Our No Complementary Policies case shows that if we remove these policies from the Blueprint, consumers and businesses will save much less money.[10] Excluding the complementary policies we recommend for the energy and transportation sectors would reduce net cumulative consumer and business savings through 2030 from a total of $1.7 trillion to $0.6 trillion.

Our No Complementary Policies case also shows that excluding the policies we recommend for the energy and transportation sectors will double the price of carbon allowances.

Where the Blueprint Cuts Emissions and Saves Money

Five sectors of the U.S. economy account for the majority of the nation's heat-trapping emissions: electricity, transportation, buildings (commercial and residential), industry, and land use. Blueprint policies ensure that each of these sectors contributes to the drop in the nation's net carbon emissions.

Blueprint policies reduce projected U.S. energy use by one-third by 2030.

The electricity sector—with help from efficiency improvements in industry and buildings—leads the way by providing more than half (57 percent) of the needed cuts in heat-trapping emissions by 2030. Transportation delivers the next-largest cut (16 percent). Carbon offsets provide 11 percent of the overall cuts in carbon emissions by 2030. Reduced emissions of heat-trapping gases other than carbon dioxide (non-CO_2 emissions) deliver another 7 percent of the cuts. Savings in direct fuel use in the residential, commercial, and industrial sectors are the final pieces, contributing 3 percent, 2 percent, and 4 percent, respectively, of the reductions in emissions.

National savings on annual energy bills (the money consumers save on their monthly electricity bills or gasoline costs, for example) total $414 billion in 2030. As noted, these savings more than cover the costs of carbon allowances that utilities and fuel providers pass through to households and businesses in higher energy prices. The incremental costs of energy investments (expenditures on energy-consuming products such as homes, appliances, and vehicles) reach $160 billion. The result is net annual savings of $255 billion for households and businesses in 2030.

Households and businesses that rely on the transportation sector see nearly half of the net annual savings ($119 billion) in 2030. However, Blueprint policies ensure that consumers and businesses throughout the economy save money on energy expenses. Lower electricity costs for industrial, commercial, and residential customers are responsible for $118 billion in net annual savings.

The Blueprint Cuts Emissions in Each Sector

Blueprint policies dramatically reduce carbon emissions from power plants. Under the Blueprint, carbon emissions from power plants are 84 percent below 2005 levels by 2030. Sulfur dioxide (SO_2), nitrogen oxides (NO_X), and mercury pollution from power plants are also significantly lower, improving air and water quality and providing important public health benefits.

The Blueprint cuts carbon emissions from power plants by 84 percent below 2005 levels by 2030.

Most of these cuts in emissions come from reducing the use of coal to produce electricity through greater use of energy efficiency and renewable energy technologies. For example, energy efficiency measures—such as advanced buildings and industrial processes—and high-efficiency appliances, lighting, and motors reduce demand for electricity by 35 percent below the Reference case by 2030. The use of efficient combined-heat-and-power systems that rely on natural gas in the commercial and industrial sectors more than triples over current levels, providing 16 percent of U.S. electricity by 2030. And largely because of a national renewable electricity standard, wind, solar, geothermal, and bioenergy provide 40 percent of the remaining electricity.

Hydropower and nuclear power continue to play important roles, generating slightly more carbon-free electricity in 2030 than they do today. Efforts to capture and store carbon from advanced coal plants, and new advanced nuclear plants, play a minor role, as our analysis shows they will not be economically competitive with investments in energy efficiency and many renewable technologies. However, carbon capture and storage and advanced nuclear power could play a more significant role both before and after 2030 if their costs decline faster than expected, or if the nation does not pursue the vigorous energy efficiency and renewable energy policies and investments we recommend.

Industry and buildings cut fuel use through greater energy efficiency. By 2030, a drop in direct fuel used in industry and buildings accounts for 9 percent of the cuts in carbon emissions from non-electricity sources under the Blueprint.

Transportation gets cleaner, smarter, and more efficient. Under the Blueprint, carbon emissions from cars and light trucks are 40 percent below 2005 levels by 2030. Global warming emissions from freight trucks hold steady despite a more than 80 percent growth in the nation's economy. However, carbon emissions from airplanes continue to grow nearly unchecked, pointing to the need for specific policies targeting that sector. Overall, carbon emissions from the transportation sector fall 19 percent below 2005

Beyond the Climate 2030 Blueprint—Technologies for Our Future

Our analysis did not include several renewable energy and transportation sector technologies that are at an early stage of development, but offer promise. These include:

- Thin film solar
- Biopower with carbon capture and storage
- Advanced geothermal energy
- Wave and tidal power
- Renewable energy heating and cooling
- Advanced storage and smart grid technologies
- Dramatic expansion of all-electric cars and trucks
- High-speed electric rail
- Expanded public transit-oriented development
- Breakthroughs in third-generation biofuels

levels by 2030—and more than 30 percent below the Reference case.

Much of the improvement in this sector comes from greater vehicle efficiency and the use of the lowest-carbon fuels, such as ethanol made from plant cellulose. Measures to encourage more efficient travel options—such as per-mile insurance and congestion fees, and more emphasis on compact development linked to transit—also provide significant reductions. Renewable electricity use in advanced vehicles such as plug-in hybrids begins to grow significantly by 2030.

These advances represent the second half of an investment in a cleaner transportation system that began with the 2007 Energy Independence and Security Act.[11] These investments provide immediate benefits and will be essential to dramatically cutting carbon emissions from the transportation sector by 2050.

Blueprint Cuts Are Conservative and Practical

The Blueprint includes only technologies that are commercially available today, or that will very likely be available within the next two decades. our analysis excludes many promising technologies, or assumes they will play only a modest role by 2030. We also did not analyze the full potential for storing more carbon in U.S. agricultural soils and forests, although studies show that such storage could be significant.

Our estimates of cuts in carbon emissions are therefore conservative. More aggressive policies and larger investments in clean technologies could produce even deeper U.S. reductions.

Recommendations: Building Blocks for a Clean Energy Future

Given the significant savings under the Blueprint, building a clean energy economy not only makes sense for our health and well-being and the future of our planet, but is clearly also good for our economy. However, the nation will only realize the benefits of the Climate 2030 Blueprint if we quickly put the critical policies in place—some as soon as 2010. All these policies are achievable, but near-term action is essential.

An important first step is science-based legislation that would enable the nation to cut heat-trapping emissions by at least 35 percent below 2005 levels by 2020,[12] and at least 80 percent by 2050. Such legislation would include a well-designed cap-and-trade program that guarantees the needed emission cuts and does not include loopholes, such as "safety valves" that prevent the free functioning of the carbon market.

Impact of the Blueprint Policies in 2020

A central insight from the Blueprint analysis is that the nation has many opportunities for making cost-effective cuts in carbon emissions in the next 10 years (through 2020). Our analysis shows that firms subject to the cap on emissions find it cost-effective to cut emissions more than required—and to bank carbon allowances for future years. Energy efficiency, renewable energy, reduced vehicle travel, and carbon offsets all contribute to these significant near-term reductions.

By 2020, we find that the United States can:

- Achieve, and go beyond, the cap requirement of a 26 percent reduction in emissions below 2005 levels, at a net annual savings of $243 billion to consumers and businesses. The reductions in excess of the cap are banked by firms for their use in later years to comply with the cap and lower costs.
- Reduce annual energy use by 17 percent compared with the Reference case levels.
- Cut the use of oil and other petroleum products by 3.4 million barrels per day compared with 2005, reducing imports to 50 percent of our needs.
- Reduce annual electricity generation by almost 20 percent compared with the Reference case while producing 10 percent of the remaining electricity with combined heat and power and 20 percent with renewable energy sources, such as wind, solar, geothermal, and bioenergy.
- Rely on complementary policies to deliver cost effective energy efficiency, conservation, and renewable energy solutions. Excluding those energy and transportation sector policies from the Blueprint would reduce net cumulative consumer savings through 2020 from $795 billion to $602 billion.

Equally important, policy makers should require greater energy efficiency and the use of renewable energy in industry, buildings, and electricity. Policy makers should also require and provide incentives for cleaner cars, trucks, and fuels and better alternatives to car and truck travel.

U.S. climate policy must also have an international dimension. That dimension should include funding the preservation of tropical forests, sharing energy efficiency and renewable energy technologies with developing nations, and helping those nations adapt to the unavoidable effects of climate change.

Cleaner vehicles, better transportation choices, and low-carbon fuels cut transportation emissions by 30 percent.

Conclusion

We are at a crossroads. The Reference case shows that we are on a path of rising energy use and heat-trapping emissions. We are already seeing significant impacts from this carbon overload, such as rising temperatures and sea levels and extreme weather events. If such emissions continue to climb at their current rate, we could reach climate "tipping points" and face irreversible changes to our planet.

In 2007 the Intergovernmental Panel on climate change (IPCC) found it "unequivocal" that the Earth's climate is warming, and that human activities are the primary cause (IPCC 2007). The IPCC report concludes that unchecked global warming will only create more adverse impacts on food production, public health, and species survival.

The climate will not wait for us. More recent studies have shown that the measured impacts—such as rising sea levels and shrinking summer sea ice in the Arctic—are occurring more quickly, and often more intensely, than IPCC projections (Rosenzweig et al. 2008; Rahmstorf et al. 2007; Stroeve et al. 2007).

The most expensive thing we can do is nothing. One study also estimates that if climate trends continue, the total cost of global warming in the United States could be as high as 3.6 percent of GDP by 2100 (Ackerman and Stanton 2008).

The most expensive thing we can do is nothing.

The climate 2030 Blueprint demonstrates that we can choose to cut our carbon emissions while maintaining robust economic growth and achieving significant energy-related savings. While the Blueprint policies are not the only path forward, a near-term comprehensive suite of climate, energy, and transportation policies is essential if we are to curb global warming in an economically sound fashion. These near-term policies are also only the beginning of the journey toward achieving a clean energy economy. The nation can and must expand these and other policies beyond 2030 to ensure that we meet the mid-century reductions in emissions that scientists deem necessary to avoid the worst consequences of global warming.

References

Ackerman, F., and E.A. Stanton. 2008. *The cost of climate change: What we'll pay if global warming continues unchecked.* New York, NY: Natural Resources Defense Council. Online at www.nrdc.org/globalWarming/cost/cost.pdf.

Intergovernmental Panel on Climate Change (IPCC). 2007. *Climate change 2007: The physical science basis.* Contribution of Working Group I to the Fourth Assessment Report of the Intergovernmental Panel on Climate Change, edited by S. Solomon, D. Qin, M. Manning, Z. Chen, M. Marquis, K.B. Averyt, M. Tignor, and H.L. Miller. Cambridge, UK: Cambridge University Press.

Luers, A.L., M.D. Mastrandrea, K. Hayhoe, and P.C. Frumhoff. 2007. *How to avoid dangerous climate change: A target for U.S. emissions reductions.* Cambridge, MA: Union of Concerned Scientists.

Rahmstorf, S., A. Cazenave, J.A. Church, J.E. Hansen, R.F. Keeling, D.E. Parker, and R.C.J. Somerville. 2007. Recent climate observations compared to projections. *Science* 316:709.

Rosenzweig, C., D. Karoly, M. Vicarelli, P. Neofotis, Q. Wu, G. Casassa, A. Menzel, T.L. Root, N. Estrella, B. Seguin, P. Tryjanowski, C. Liu, S. Rawlins, and A. Imeson. 2008. Attributing physical and biological impacts to anthropogenic climate change. *Nature* 453:353—357.

Sperling, D., and D. Gordon. 2009. *Two billion cars: Driving toward sustainability.* New York: Oxford University Press.

Stroeve, J., M. Serreze, S. Drobot, S. Gearheard, M. Holland, J. Maslanik, W. Meier, and T. Scambos. 2008. Arctic sea ice extent plummets in 2007. *Eos, Transactions, American Geophysical Union* 89(2):13—20.

Notes

1. Unless otherwise noted, all amounts are in 2006 dollars, and cumulative figures are discounted using a 7 percent real discount rate.
2. Net savings include both energy bills (the direct cost of energy such as diesel, electricity, gasoline, and natural gas) and the cost of purchasing more efficient energy-consuming products such as appliances and vehicles. The cost of carbon allowances passed through to consumers and businesses is also included in their energy bills.
3. We could not model a targeted way of recycling these revenues. The preferred approach would be to target revenues from auctions of carbon allowances toward investments in energy efficiency, renewable energy, and protection for tropical forests, as well as transition assistance to consumers, workers, and businesses in moving to a clean energy economy. However, limitations in the NEMS model prevented us from directing auction revenues to specific uses. Instead, we could only recycle revenues in a general way to consumers and businesses.
4. Values may not sum properly due to rounding.
5. In a cap-and-trade system, rather than cutting their emissions directly, capped companies can "offset" them by paying uncapped third parties to reduce their emissions instead. The cap-and-trade program we modeled includes offsets from storing carbon in domestic soils and vegetation—set at a maximum of 10 percent of the emissions cap, to encourage "decarbonization" of the capped sectors—and from investing in reductions in other countries, mainly from preserving tropical forests, set at a maximum of 5 percent of the emissions cap.
6. This amount is equivalent to the emissions from nearly 1 billion of today's U.S. cars and trucks over the same 30-year period. The nation now has some 230 million cars and trucks, and more than 1 billion vehicles are on the road worldwide. Given today's trends, we can expect at least 2 billion vehicles by 2030 (Sperling and Gordon 2009).

7. Our analysis includes the tax credits and incentives for energy technologies included in the October 2008 Economic Stimulus Package (H.R. 6049), as well as the transportation and energy policies in the 2007 Energy Independence and Security Act. However, the timing of the February 2009 American Recovery and Reinvestment Act did not allow us to incorporate its significant additional incentives.
8. This means that under the Blueprint the economy reaches the same level of economic growth in October 2030 as the Reference case reaches in January 2030.
9. See endnote 3.
10. Some or all of the economic benefits of the complementary policies could also occur if policy makers effectively use the revenues from auctioning carbon allowances to fund the technologies and measures included in these policies. Our study did not address that approach.
11. Because our Reference case includes the policies in the 2007 legislation, the Blueprint's 30 percent reduction from that case in 2030 represents benefits beyond those delivered from the fuel economy standards and renewable fuel standard in the act. If our Reference case did not include the provisions in the act, Blueprint transportation policies would deliver nearly a 40 percent reduction compared with the Reference case.
12. Note that this recommendation encompasses more possibilities for reducing emissions than we were able to model in UCS-NEMS. For example, investments in reducing emissions from tropical deforestation could help meet this 2020 target. The Blueprint reductions can and should be supplemented by these and other sources of emissions reductions.

Critical Thinking

1. Does the U.S. need to "transition to a clean energy economy" or an economy that simply requires less energy?
2. What do think will be the impact on the Blueprint if auctioning carbon allowances does not "generate $219 billion in revenues?"
3. What is your reaction to the Union of Concerned Scientists using GDP in its analysis?

The Power of Green

What does America need to regain its global stature? Environmental leadership.

THOMAS L. FRIEDMAN

I.

One day Iraq, our post-9/11 trauma and the divisiveness of the Bush years will all be behind us—and America will need, and want, to get its groove back. We will need to find a way to reknit America at home, reconnect America abroad and restore America to its natural place in the global order—as the beacon of progress, hope and inspiration. I have an idea how. It's called "green."

In the world of ideas, to name something is to own it. If you can name an issue, you can own the issue. One thing that always struck me about the term "green" was the degree to which, for so many years, it was defined by its opponents—by the people who wanted to disparage it. And they defined it as "liberal," "tree-hugging," "sissy," "girlie-man," "unpatriotic," "vaguely French."

Well, I want to rename "green." I want to rename it geostrategic, geoeconomic, capitalistic and patriotic. I want to do that because I think that living, working, designing, manufacturing and projecting America in a green way can be the basis of a new unifying political movement for the 21st century. A redefined, broader and more muscular green ideology is not meant to trump the traditional Republican and Democratic agendas but rather to bridge them when it comes to addressing the three major issues facing every American today: jobs, temperature and terrorism.

How do our kids compete in a flatter world? How do they thrive in a warmer world? How do they survive in a more dangerous world? Those are, in a nutshell, the big questions facing America at the dawn of the 21st century. But these problems are so large in scale that they can only be effectively addressed by an America with 50 green states—not an America divided between red and blue states.

Because a new green ideology, properly defined, has the power to mobilize liberals and conservatives, evangelicals and atheists, big business and environmentalists around an agenda that can both pull us together and propel us forward. That's why I say: We don't just need the first black president. We need the first green president. We don't just need the first woman president. We need the first environmental president. We don't just need a president who has been toughened by years as a prisoner of war but a president who is tough enough to level with the American people about the profound economic, geopolitical and climate threats posed by our addiction to oil—and to offer a real plan to reduce our dependence on fossil fuels.

After World War II, President Eisenhower responded to the threat of Communism and the "red menace" with massive spending on an interstate highway system to tie America together, in large part so that we could better move weapons in the event of a war with the Soviets. That highway system, though, helped to enshrine America's car culture (atrophying our railroads) and to lock in suburban sprawl and low-density housing, which all combined to get America addicted to cheap fossil fuels, particularly oil. Many in the world followed our model.

Today, we are paying the accumulated economic, geopolitical and climate prices for that kind of America. I am not proposing that we radically alter our lifestyles. We are who we are—including a car culture. But if we want to continue to be who we are, enjoy the benefits and be able to pass them on to our children, we do need to fuel our future in a cleaner, greener way. Eisenhower rallied us with the red menace. The next president will have to rally us with a green patriotism. Hence my motto: "Green is the new red, white and blue."

The good news is that after traveling around America this past year, looking at how we use energy and the emerging alternatives, I can report that green really has gone Main Street—thanks to the perfect storm created by 9/11, Hurricane Katrina and the Internet revolution. The first flattened the twin towers, the second flattened New Orleans and the third flattened the global economic playing field. The convergence of all three has turned many of our previous assumptions about "green" upside down in a very short period of time, making it much more compelling to many more Americans.

But here's the bad news: While green has hit Main Street—more Americans than ever now identify themselves as greens, or what I call "Geo-Greens" to differentiate their more muscular and strategic green ideology—green has not gone very far down Main Street. It certainly has not gone anywhere near the distance required to preserve our lifestyle. The dirty little secret is that we're fooling ourselves. We in America talk like we're already "the greenest generation," as the business writer Dan Pink once called it. But here's the really inconvenient truth: We have not even begun to be serious about the costs, the effort and the scale of change that will be required to shift our country, and eventually the world, to a largely emissions-free energy infrastructure over the next 50 years.

II.

A few weeks after American forces invaded Afghanistan, I visited the Pakistani frontier town of Peshawar, a hotbed of Islamic radicalism. On the way, I stopped at the famous Darul Uloom Haqqania, the biggest madrasa, or Islamic school, in Pakistan, with 2,800 live-in students. The Taliban leader Mullah Muhammad Omar attended this madrasa as a younger man. My Pakistani friend and I were allowed to observe a class of young boys who sat on the floor, practicing their rote learning of the Koran from texts perched on wooden holders. The air in the Koran class was so thick and stale it felt as if you could have cut it into blocks. The teacher asked an 8-year-old boy to chant a Koranic verse for us, which he did with the elegance of an experienced muezzin. I asked another student, an Afghan refugee, Rahim Kunduz, age 12, what his reaction was to the Sept. 11 attacks, and he said: "Most likely the attack came from Americans inside America. I am pleased that America has had to face pain, because the rest of the world has tasted its pain." A framed sign on the wall said this room was "A gift of the Kingdom of Saudi Arabia."

Sometime after 9/11—an unprovoked mass murder perpetrated by 19 men, 15 of whom were Saudis—green went geostrategic, as Americans started to realize we were financing both sides in the war on terrorism. We were financing the U.S. military with our tax dollars; and we were financing a transformation of Islam, in favor of its most intolerant strand, with our gasoline purchases. How stupid is that?

Islam has always been practiced in different forms. Some are more embracing of modernity, reinterpretation of the Koran and tolerance of other faiths, like Sufi Islam or the populist Islam of Egypt, Ottoman Turkey and Indonesia. Some strands, like Salafi Islam—followed by the Wahhabis of Saudi Arabia and by Al Qaeda—believe Islam should be returned to an austere form practiced in the time of the Prophet Muhammad, a form hostile to modernity, science, "infidels" and women's rights. By enriching the Saudi and Iranian treasuries via our gasoline purchases, we are financing the export of the Saudi puritanical brand of Sunni Islam and the Iranian fundamentalist brand of Shiite Islam, tilting the Muslim world in a more intolerant direction. At the Muslim fringe, this creates more recruits for the Taliban, Al Qaeda, Hamas, Hezbollah and the Sunni suicide bomb squads of Iraq; at the Muslim center, it creates a much bigger constituency of people who applaud suicide bombers as martyrs.

The Saudi Islamic export drive first went into high gear after extreme fundamentalists challenged the Muslim credentials of the Saudi ruling family by taking over the Grand Mosque of Mecca in 1979—a year that coincided with the Iranian revolution and a huge rise in oil prices. The attack on the Grand Mosque by these Koran-and-rifle-wielding Islamic militants shook the Saudi ruling family to its core. The al-Sauds responded to this challenge to their religious bona fides by becoming outwardly more religious. They gave their official Wahhabi religious establishment even more power to impose Islam on public life. Awash in cash thanks to the spike in oil prices, the Saudi government and charities also spent hundreds of millions of dollars endowing mosques, youth clubs and Muslim schools all over the world, ensuring that Wahhabi imams, teachers and textbooks would preach Saudi-style Islam. Eventually, notes Lawrence Wright in "The Looming Tower," his history of Al Qaeda, "Saudi Arabia, which constitutes only 1 percent of the world Muslim population, would support 90 percent of the expenses of the entire faith, overriding other traditions of Islam."

Saudi mosques and wealthy donors have also funneled cash to the Sunni insurgents in Iraq. The Associated Press reported from Cairo in December: "Several drivers interviewed by the A.P. in Middle East capitals said Saudis have been using religious events, like the hajj pilgrimage to Mecca and a smaller pilgrimage, as cover for illicit money transfers. Some money, they said, is carried into Iraq on buses with returning pilgrims. 'They sent boxes full of dollars and asked me to deliver them to certain addresses in Iraq,' said one driver. . . . 'I know it is being sent to the resistance, and if I don't take it with me, they will kill me.' "

No wonder more Americans have concluded that conserving oil to put less money in the hands of hostile forces is now a geostrategic imperative. President Bush's refusal to do anything meaningful after 9/11 to reduce our gasoline usage really amounts to a policy of "No Mullah Left Behind." James Woolsey, the former C.I.A. director, minces no words: "We are funding the rope for the hanging of ourselves."

No, I don't want to bankrupt Saudi Arabia or trigger an Islamist revolt there. Its leadership is more moderate and pro-Western than its people. But the way the Saudi ruling family has bought off its religious establishment, in order to stay in power, is not healthy. Cutting the price of oil in half would help change that. In the 1990s, dwindling oil income sparked a Saudi debate about less Koran and more science in Saudi schools, even experimentation with local elections. But the recent oil windfall has stilled all talk of reform.

That is because of what I call the First Law of Petropolitics: The price of oil and the pace of freedom always move in opposite directions in states that are highly dependent on oil exports for their income and have weak institutions or outright authoritarian governments. And this is another reason that green has become geostrategic. Soaring oil prices are poisoning the international system by strengthening antidemocratic regimes around the globe.

Look what's happened: We thought the fall of the Berlin Wall was going to unleash an unstoppable tide of free markets and free people, and for about a decade it did just that. But those years coincided with oil in the $10-to-$30-a-barrel range. As the price of oil surged into the $30-to-$70 range in the early 2000s, it triggered a countertide—a tide of petroauthoritarianism—manifested in Russia, Iran, Nigeria, Venezuela, Saudi Arabia, Syria, Sudan, Egypt, Chad, Angola, Azerbaijan and Turkmenistan. The elected or self-appointed elites running these states have used their oil windfalls to ensconce themselves in power, buy off opponents and counter the fall-of-the-Berlin-Wall tide. If we continue to finance them with our oil purchases, they will reshape the world in their image, around Putin-like values.

You can illustrate the First Law of Petropolitics with a simple graph. On one line chart the price of oil from 1979 to the present; on another line chart the Freedom House or Fraser Institute freedom indexes for Russia, Nigeria, Iran and Venezuela for the same years. When you put these two lines on the same graph you see something striking: the price of oil and the pace of freedom are inversely correlated. As oil prices went down in the early 1990s, competition, transparency, political participation and accountability of those in office all tended to go up in these countries—as measured by free elections held, newspapers opened, reformers elected, economic reform projects started and companies privatized. That's because their petroauthoritarian regimes had to open themselves to foreign investment and educate and empower their people more in order to earn income. But as oil prices went up around 2000, free speech, free press, fair elections and freedom to form political parties and NGOs all eroded in these countries.

The motto of the American Revolution was "no taxation without representation." The motto of the petroauthoritarians is "no representation without taxation": If I don't have to tax you, because I can get all the money I need from oil wells, I don't have to listen to you.

It is no accident that when oil prices were low in the 1990s, Iran elected a reformist Parliament and a president who called for a "dialogue of civilizations." And when oil prices soared to $70 a barrel, Iran's conservatives pushed out the reformers and ensconced a president who says the Holocaust is a myth. (I promise you, if oil prices drop to $25 a barrel, the Holocaust won't be a myth anymore.) And it is no accident that the first Arab Gulf state to start running out of oil, Bahrain, is also the first Arab Gulf state to have held a free and fair election in which women could run and vote, the first Arab Gulf state to overhaul its labor laws to make more of its own people employable and the first Arab Gulf state to sign a free-trade agreement with America.

People change when they have to—not when we tell them to—and falling oil prices make them have to. That is why if we are looking for a Plan B for Iraq—a way of pressing for political reform in the Middle East without going to war again—there is no better tool than bringing down the price of oil. When it comes to fostering democracy among petroauthoritarians, it doesn't matter whether you're a neocon or a radical lib. If you're not also a Geo-Green, you won't succeed.

The notion that conserving energy is a geostrategic imperative has also moved into the Pentagon, for slightly different reasons. Generals are realizing that the more energy they save in the heat of battle, the more power they can project. The Pentagon has been looking to improve its energy efficiency for several years now to save money. But the Iraq war has given birth to a new movement in the U.S. military: the "Green Hawks."

As Amory Lovins of the Rocky Mountain Institute, who has been working with the Pentagon, put it to me: The Iraq war forced the U.S. military to think much more seriously about how to "eat its tail"—to shorten its energy supply lines by becoming more energy efficient. According to Dan Nolan, who oversees energy projects for the U.S. Army's Rapid Equipping Force, it started last year when a Marine major general in Anbar Province told the Pentagon he wanted alternative energy sources that would reduce fuel consumption in the Iraqi desert. Why? His air-conditioners were being run off mobile generators, and the generators ran on diesel, and the diesel had to be trucked in, and the insurgents were blowing up the trucks.

"When we began the analysis of his request, it was really about the fact that his soldiers were being attacked on the roads bringing fuel and water," Nolan said. So eating their tail meant "taking those things that are brought into the unit and trying to generate them on-site." To that end Nolan's team is now experimenting with everything from new kinds of tents that need 40 percent less air-conditioning to new kinds of fuel cells that produce water as a byproduct.

Pay attention: When the U.S. Army desegregated, the country really desegregated; when the Army goes green, the country could really go green.

"Energy independence is a national security issue," Nolan said. "It's the right business for us to be in. . . . We are not trying to change the whole Army. Our job is to focus on that battalion out there and give those commanders the technological innovations they need to deal with today's mission. But when they start coming home, they are going to bring those things with them."

III.

The second big reason green has gone Main Street is because global warming has. A decade ago, it was mostly experts who worried that climate change was real, largely brought about by humans and likely to lead to species loss and environmental crises. Now Main Street is starting to worry because people are seeing things they've never seen before in their own front yards and reading things they've never read before in their papers—like the recent draft report by the United Nations's 2,000-expert Intergovernmental Panel on Climate Change, which concluded that "changes in climate are now affecting physical and biological systems on every continent."

I went to Montana in January and Gov. Brian Schweitzer told me: "We don't get as much snow in the high country as we used to, and the runoff starts sooner in the spring. The river I've been fishing over the last 50 years is now warmer in July by five degrees than 50 years ago, and it is hard on our trout population." I went to Moscow in February, and my friends told me they just celebrated the first Moscow Christmas in their memory with no snow. I stopped in London on the way home, and I didn't need an overcoat. In 2006, the average temperature in central England was the highest ever recorded since the Central England Temperature (C.E.T.) series began in 1659.

Yes, no one knows exactly what will happen. But ever fewer people want to do nothing. Gov. Arnold Schwarzenegger of California summed up the new climate around climate when he said to me recently: "If 98 doctors say my son is ill and needs medication and two say 'No, he doesn't, he is fine,' I will go with the 98. It's common sense—the same with global warming. We go with the majority, the large majority. . . . The key thing now is that since we know this industrial age has created it, let's get our act together and do everything we can to roll it back."

But how? Now we arrive at the first big roadblock to green going down Main Street. Most people have no clue—no clue—how huge an industrial project is required to blunt climate change. Here are two people who do: Robert Socolow, an engineering professor, and Stephen Pacala, an ecology professor, who together lead the Carbon Mitigation Initiative at Princeton, a consortium designing scalable solutions for the climate issue.

People change when they have to, and falling oil prices make them have to. That is why if we are looking for a Plan B for Iraq there is no better tool than briging down the price of oil.

They first argued in a paper published by the journal *Science* in August 2004 that human beings can emit only so much carbon into the atmosphere before the buildup of carbon dioxide (CO_2) reaches a level unknown in recent geologic history and the earth's climate system starts to go "haywire." The scientific consensus, they note, is that the risk of things going haywire—weather patterns getting violently unstable, glaciers melting, prolonged droughts—grows rapidly as CO_2 levels "approach a doubling" of the concentration of CO_2 that was in the atmosphere before the Industrial Revolution.

"Think of the climate change issue as a closet, and behind the door are lurking all kinds of monsters—and there's a long list of them," Pacala said. "All of our scientific work says the most damaging monsters start to come out from behind that door when you hit the doubling of CO_2 levels." As Bill Collins, who led the development of a model used worldwide for simulating climate change, put it to me: "We're running an uncontrolled experiment on the only home we have."

So here is our challenge, according to Pacala: If we basically do nothing, and global CO_2 emissions continue to grow at the pace of the last 30 years for the next 50 years, we will pass the doubling level—an atmospheric concentration of carbon dioxide of 560 parts per million—around midcentury. To avoid that—and still leave room for developed countries to grow, using less carbon, and for countries like India and China to grow, emitting double or triple their current carbon levels, until they climb out of poverty and are able to become more energy efficient—will require a huge global industrial energy project.

To convey the scale involved, Socolow and Pacala have created a pie chart with 15 different wedges. Some wedges represent carbon-free or carbon-diminishing power-generating technologies; other wedges represent efficiency programs that could conserve large amounts of energy and prevent CO_2 emissions. They argue that the world needs to deploy any 7 of these 15 wedges, or sufficient amounts of all 15, to have enough conservation, and enough carbon-free energy, to increase the world economy and still avoid the doubling of CO_2 in the atmosphere. Each wedge, when phased in over 50 years, would avoid the release of 25 billion tons of carbon, for a total of 175 billion tons of carbon avoided between now and 2056.

Here are seven wedges we could chose from: "Replace 1,400 large coal-fired plants with gas-fired plants; increase the fuel economy of two billion cars from 30 to 60 miles per gallon; add twice today's nuclear output to displace coal; drive two billion cars on ethanol, using one-sixth of the world's cropland; increase solar power 700-fold to displace coal; cut electricity use in homes, offices and stores by 25 percent; install carbon capture and sequestration capacity at 800 large coal-fired plants." And the other eight aren't any easier. They include halting all cutting and burning of forests, since deforestation causes about 20 percent of the world's annual CO_2 emissions.

"There has never been a deliberate industrial project in history as big as this," Pacala said. Through a combination of clean power technology and conservation, "we have to get rid of 175 billion tons of carbon over the next 50 years—and still keep growing. It is possible to accomplish this if we start today. But every year that we delay, the job becomes more difficult—and if we delay a decade or two, avoiding the doubling or more may well become impossible."

IV.

In November, I flew from Shanghai to Beijing on Air China. As we landed in Beijing and taxied to the terminal, the Chinese air hostess came on the P.A. and said: "We've just landed in Beijing. The temperature is 8 degrees Celsius, 46 degrees Fahrenheit and the sky is clear."

I almost burst out laughing. Outside my window the smog was so thick you could not see the end of the terminal building. When I got into Beijing, though, friends told me the air was better than usual. Why? China had been host of a summit meeting of 48 African leaders. *Time* magazine reported that Beijing officials had "ordered half a million official cars off the roads and said another 400,000 drivers had 'volunteered' to refrain from using their vehicles" in order to clean up the air for their African guests. As soon as they left, the cars returned, and Beijing's air went back to "unhealthy."

Green has also gone Main Street because the end of Communism, the rise of the personal computer and the diffusion of the Internet have opened the global economic playing field to so many more people, all coming with their own versions of the American dream—a house, a car, a toaster, a microwave and a refrigerator. It is a blessing to see so many people growing out of poverty. But when three billion people move from "low-impact" to "high-impact" lifestyles, Jared Diamond wrote in "Collapse," it makes it urgent that we find cleaner ways to fuel their dreams. According to Lester Brown, the founder of the Earth Policy Institute, if China keeps growing at 8 percent a year, by 2031 the per-capita income of 1.45 billion Chinese will be the same as America's in 2004. China currently has only one car for every 100 people, but Brown projects that as it reaches American income levels, if it copies American consumption, it will have three cars for every four people, or 1.1 billion vehicles. The total world fleet today is 800 million vehicles!

That's why McKinsey Global Institute forecasts that developing countries will generate nearly 80 percent of the growth in world energy demand between now and 2020, with China representing 32 percent and the Middle East 10 percent. So if Red China doesn't become Green China there is no chance we will keep the climate monsters behind the door. On some days, says the U.S. Environmental Protection Agency, almost 25 percent of the polluting matter in the air above Los Angeles comes from China's coal-fired power plants and factories, as well as fumes from China's cars and dust kicked up by droughts and deforestation around Asia.

The good news is that China knows it has to grow green—or it won't grow at all. On Sept. 8, 2006, a Chinese newspaper reported that China's E.P.A. and its National Bureau of Statistics had re-examined China's 2004 G.D.P. number. They concluded that the health problems, environmental degradation and lost workdays from pollution had actually cost China $64 billion, or 3.05 percent of its total economic output for 2004. Some experts believe the real number is closer to 10 percent.

Thus China has a strong motivation to clean up the worst pollutants in its air. Those are the nitrogen oxides, sulfur oxides and mercury that produce acid rain, smog and haze—much of which come from burning coal. But cleaning up is easier said than done. The Communist Party's legitimacy and the stability of the whole country depend heavily on Beijing's ability to provide rising living standards for more and more Chinese.

So, if you're a Chinese mayor and have to choose between growing jobs and cutting pollution, you will invariably choose jobs: coughing workers are much less politically dangerous than unemployed workers. That's a key reason why China's 10th five-year plan, which began in 2000, called for a 10 percent reduction in sulfur dioxide in China's air—and when that plan concluded in 2005, sulfur dioxide pollution in China had increased by 27 percent.

But if China is having a hard time cleaning up its nitrogen and sulfur oxides—which can be done relatively cheaply by adding scrubbers to the smokestacks of coal-fired power plants—imagine what will happen when it comes to asking China to curb its CO_2, of which China is now the world's second-largest emitter, after America. To build a coal-fired power plant that captures, separates and safely sequesters the CO_2 into the ground before it goes up the smokestack requires either an expensive retrofit or a whole new system. That new system would cost about 40 percent more to build and operate—and would produce 20 percent less electricity, according to a recent M.I.T. study, "The Future of Coal."

China—which is constructing the equivalent of two 500-megawatt coal-fired power plants every week—is not going to pay that now. Remember: CO_2 is an invisible, odorless, tasteless gas. Yes, it causes global warming—but it doesn't hurt anyone in China today, and getting rid of it is costly and has no economic payoff. China's strategy right now is to say that CO_2 is the West's problem. "It must be pointed out that climate change has been caused by the long-term historic emissions of developed countries and their high per-capita emissions," Jiang Yu, a spokeswoman for China's Foreign Ministry, declared in February. "Developed countries bear an unshirkable responsibility."

So now we come to the nub of the issue: Green will not go down Main Street America unless it also goes down Main Street China, India and Brazil. And for green to go Main Street in these big developing countries, the prices of clean power alternatives—wind, biofuels, nuclear, solar or coal sequestration—have to fall to the "China price." The China price is basically the price China pays for coal-fired electricity today because China is not prepared to pay a premium now, and sacrifice growth and stability, just to get rid of the CO_2 that comes from burning coal.

"The 'China price' is the fundamental benchmark that everyone is looking to satisfy," said Curtis Carlson, C.E.O. of SRI International, which is developing alternative energy technologies. "Because if the Chinese have to pay 10 percent more for energy, when they have tens of millions of people living under $1,000 a year, it is not going to happen." Carlson went on to say: "We have an enormous amount of new innovation we must put in place before we can get to a price that China and India will be able to pay. But this is also an opportunity."

V.

The only way we are going to get innovations that drive energy costs down to the China price—innovations in energy-saving appliances, lights and building materials and in non-CO_2-emitting power plants and fuels—is by mobilizing free-market capitalism. The only thing as powerful as Mother Nature is Father Greed. To a degree, the market is already at work on this project—because some venture capitalists and companies understand that clean-tech is going to be the next great global industry. Take Wal-Mart. The world's biggest retailer woke up several years ago, its C.E.O. Lee Scott told me, and realized that with regard to the environment its customers "had higher expectations for us than we had for ourselves." So Scott hired a sustainability expert, Jib Ellison, to tutor the company. The first lesson Ellison preached was that going green was a whole new way for Wal-Mart to cut costs and drive its profits. As Scott recalled it, Ellison said to him, "Lee, the thing you have to think of is all this stuff that people don't want you to put into the environment is waste—and you're paying for it!"

So Scott initiated a program to work with Wal-Mart's suppliers to reduce the sizes and materials used for all its packaging by five percent by 2013. The reductions they have made are already paying off in savings to the company. "We created teams to work across the organization," Scott said. "It was voluntary—then you had the first person who eliminated some packaging, and someone else started showing how we could recycle more plastic, and all of a sudden it's $1 million a quarter." Wal-Mart operates 7,000 huge Class 8 trucks that get about 6 miles per gallon. It has told its truck makers that by 2015, it wants to double the efficiency of the fleet. Wal-Mart is the China of companies, so, explained Scott, "if we place one order we can create a market" for energy innovation.

For instance, Wal-Mart has used its shelves to create a huge, low-cost market for compact fluorescent bulbs, which use about a quarter of the energy of incandescent bulbs to produce the same light and last 10 times as long. "Just by doing what it does best—saving customers money and cutting costs," said Glenn Prickett of Conservation International, a Wal-Mart adviser, "Wal-Mart can have a revolutionary impact on the market for green technologies. If every one of their 100 million customers in the U.S. bought just one energy-saving compact fluorescent lamp, instead of a traditional incandescent bulb, they could cut CO_2 emissions by 45 billion pounds and save more than $3 billion."

Those savings highlight something that often gets lost: The quickest way to get to the China price for clean power is by becoming more energy efficient. The cheapest, cleanest, non-emitting power plant in the world is the one you don't build. Helping China adopt some of the breakthrough efficiency programs that California has adopted, for instance—like rewarding electrical utilities for how much energy they get their customers to save rather than to use—could have a huge impact. Some experts estimate that China could cut its need for new power plants in half with aggressive investments in efficiency.

Yet another force driving us to the China price is Chinese entrepreneurs, who understand that while Beijing may not be ready to impose CO_2 restraints, developed countries are, so this is going to be a global business—and they want a slice. Let me introduce the man identified last year by Forbes Magazine as the seventh-richest man in China, with a fortune now estimated at $2.2 billion. His name is Shi Zhengrong and he is China's leading manufacturer of silicon solar panels, which convert sunlight into electricity.

Clean-tech plays to America's strength, because making things like locomotives lighter and smarter takes a lot of knowledge—not cheap labor. Embedding clean-tech into everything we design can revive America as a manufacturing power.

"People at all levels in China have become more aware of this environment issue and alternative energy," said Shi, whose company, Suntech Power Holdings, is listed on the New York Stock Exchange. "Five years ago, when I started the company, people said: 'Why do we need solar? We have a surplus of coal-powered electricity.' Now it is different; now people realize that solar has a bright future. But it is still too expensive. . . . We have to reduce the cost as quickly as possible—our real competitors are coal and nuclear power."

Shi does most of his manufacturing in China, but sells roughly 90 percent of his products outside China, because today they are too expensive for his domestic market. But the more he can get the price down, and start to grow his business inside China, the more he can use that to become a dominant global player. Thanks to Suntech's success, in China "there is a rush of business people entering this sector, even though we still don't have a market here," Shi added. "Many government people now say, 'This is an industry!' " And if it takes off, China could do for solar panels what it did for tennis shoes—bring the price down so far that everyone can afford a pair.

VI.

All that sounds great—but remember those seven wedges? To reach the necessary scale of emissions-free energy will require big clean coal or nuclear power stations, wind farms and solar farms, all connected to a national transmission grid, not to mention clean fuels for our cars and trucks. And the market alone, as presently constructed in the U.S., will not get us those alternatives at the scale we need—at the China price—fast enough.

Prof. Nate Lewis, Caltech's noted chemist and energy expert, explained why with an analogy. "Let's say you invented the first cellphone," he said. "You could charge people $1,000 for each one because lots of people would be ready to pay lots of money to have a phone they could carry in their pocket." With those profits, you, the inventor, could pay back your shareholders and plow more into research, so you keep selling better and cheaper cellphones.

But energy is different, Lewis explained: "If I come to you and say, 'Today your house lights are being powered by dirty coal, but tomorrow, if you pay me $100 more a month, I will power your house lights with solar,' you are most likely to say: 'Sorry, Nate, but I don't really care how my lights go on, I just care that they go on. I won't pay an extra $100 a month for sun power. A new cellphone improves my life. A different way to power my lights does nothing.'

"So building an emissions-free energy infrastructure is not like sending a man to the moon," Lewis went on. "With the moon shot, money was no object—and all we had to do was get there. But today, we already have cheap energy from coal, gas and oil. So getting people to pay more to shift to clean fuels is like trying to get funding for NASA to build a spaceship to the moon—when Southwest Airlines already flies there and gives away free peanuts! I already have a cheap ride to the moon, and a ride is a ride. For most people, electricity is electricity, no matter how it is generated."

If we were running out of coal or oil, the market would steadily push the prices up, which would stimulate innovation in alternatives. Eventually there would be a crossover, and the alternatives would kick in, start to scale and come down in price. But what has happened in energy over the last 35 years is that the oil price goes up, stimulating government subsidies and some investments in alternatives, and then the price goes down, the government loses interest, the subsidies expire and the investors in alternatives get wiped out.

The only way to stimulate the scale of sustained investment in research and development of non-CO_2 emitting power at the China price is if the developed countries, who can afford to do so, force their people to pay the full climate, economic and geopolitical costs of using gasoline and dirty coal. Those countries that have signed the Kyoto Protocol are starting to do that. But America is not.

Up to now, said Lester Brown, president of the Earth Policy Institute, we as a society "have been behaving just like Enron the company at the height of its folly." We rack up stunning profits and G.D.P. numbers every year, and they look great on paper "because we've been hiding some of the costs off the books." If we don't put a price on the CO_2 we're building up or on our addiction to oil, we'll never nurture the innovation we need.

Jeffrey Immelt, the chairman of General Electric, has worked for G.E. for 25 years. In that time, he told me, he has seen seven generations of innovation in G.E.'s medical equipment business—in devices like M.R.I.s or CT scans—because health care market incentives drove the innovation. In power, it's just the opposite. "Today, on the power side," he said, "we're still selling the same basic coal-fired power plants we had when I arrived. They're a little cleaner and more efficient now, but basically the same."

The one clean power area where G.E. is now into a third generation is wind turbines, "thanks to the European Union," Immelt said. Countries like Denmark, Spain and Germany imposed standards for wind power on their utilities and offered sustained subsidies, creating a big market for wind-turbine manufacturers in Europe in the 1980s, when America abandoned wind because the price of oil fell. "We grew our wind business in Europe," Immelt said.

As things stand now in America, Immelt said, "the market does not work in energy." The multibillion-dollar scale of investment that a company like G.E. is being asked to make in order to develop new clean-power technologies or that a utility is being asked to make in order to build coal sequestration facilities or nuclear plants is not going to happen at scale—unless they know that coal and oil are going to be priced high enough for long enough that new investments will not be undercut in a few years by falling fossil fuel prices. "Carbon has to have a value," Immelt emphasized. "Today in the U.S. and China it has no value."

I recently visited the infamous Three Mile Island nuclear plant with Christopher Crane, president of Exelon Nuclear, which owns the facility. He said that if Exelon wanted to start a nuclear plant today, the licensing, design, planning and building requirements are so extensive it would not open until 2015 at the earliest. But even if Exelon got all the approvals, it could not start building "because the cost of capital for a nuclear plant today is prohibitive."

That's because the interest rate that any commercial bank would charge on a loan for a nuclear facility would be so high—because of all the risks of lawsuits or cost overruns—that it would be impossible for Exelon to proceed. A standard nuclear plant today costs about $3 billion per unit. The only way to stimulate more nuclear power innovation, Crane said, would be federal loan guarantees that would lower the cost of capital for anyone willing to build a new nuclear plant.

The 2005 energy bill created such loan guarantees, but the details still have not been worked out. "We would need a robust loan guarantee program to jump-start the nuclear industry," Crane said—an industry that has basically been frozen since the 1979 Three Mile Island accident. With cheaper money, added Crane, CO_2-free nuclear power could be "very competitive" with CO_2-emitting pulverized coal.

Think about the implications. Three Mile Island had two reactors, TMI-2, which shut down because of the 1979 accident, and TMI-1, which is still operating today, providing clean electricity with virtually no CO_2 emissions for 800,000 homes. Had the TMI-2 accident not happened, it too would have been providing clean electricity for 800,000 homes for the last 28 years. Instead, that energy came from CO_2-emitting coal, which, by the way, still generates 50 percent of America's electricity.

Similar calculations apply to ethanol production. "We have about 100 scientists working on cellulosic ethanol," Chad Holliday, the C.E.O. of DuPont, told me. "My guess is that we could double the number and add another 50 to start working on how to commercialize it. It would probably cost us less than $100 million to scale up. But I am not ready to do that. I can guess what it will cost me to make it and what the price will be, but is the market going to be there? What are the regulations going to be? Is the ethanol subsidy going to be reduced? Will we put a tax on oil to keep ethanol competitive? If I know that, it gives me a price target to go after. Without that, I don't know what the market is and my shareholders don't know how to value what I am doing. . . . You need some certainty on the incentives side and on the market side, because we are talking about multiyear investments, billions of dollars, that will take a long time to take off, and we won't hit on everything."

Summing up the problem, Immelt of G.E. said the big energy players are being asked "to take a 15-minute market signal and make a 40-year decision and that just doesn't work. . . . The U.S. government should decide: What do we want to have happen? How much clean coal, how much nuclear and what is the most efficient way to incentivize people to get there?"

He's dead right. The market alone won't work. Government's job is to set high standards, let the market reach them and then raise the standards more. That's how you get scale innovation at the China price. Government can do this by imposing steadily rising efficiency standards for buildings and appliances and by stipulating that utilities generate a certain amount of electricity from renewables—like wind or solar. Or it can impose steadily rising mileage standards for cars or a steadily tightening cap-and-trade system for the amount of CO_2 any factory or power plant can emit. Or it can offer loan guarantees and fast-track licensing for anyone who wants to build a nuclear plant. Or—my preference and the simplest option—it can impose a carbon tax that will stimulate the market to move away from fuels that emit high levels of CO_2 and invest in those that don't. Ideally, it will do all of these things. But whichever options we choose, they will only work if they are transparent, simple and long-term—with zero fudging allowed and with regulatory oversight and stiff financial penalties for violators.

The politician who actually proved just how effective this can be was a guy named George W. Bush, when he was governor of Texas. He pushed for and signed a renewable energy portfolio mandate in 1999. The mandate stipulated that Texas power companies had to produce 2,000 new megawatts of electricity from renewables, mostly wind, by 2009. What happened? A dozen new companies jumped into the Texas market and built wind turbines to meet the mandate, so many that the 2,000-megawatt goal was reached in 2005. So the Texas Legislature has upped the mandate to 5,000 megawatts by 2015, and everyone knows they will beat that too because of how quickly wind in Texas is becoming competitive with coal. Today, thanks to Governor Bush's market intervention, Texas is the biggest wind state in America.

President Bush, though, is no Governor Bush. (The Dick Cheney effect?) President Bush claims he's protecting

American companies by not imposing tough mileage, conservation or clean power standards, but he's actually helping them lose the race for the next great global industry. Japan has some of the world's highest gasoline taxes and stringent energy efficiency standards for vehicles—and it has the world's most profitable and innovative car company, Toyota. That's no accident.

The politicians who best understand this are America's governors, some of whom have started to just ignore Washington, set their own energy standards and reap the benefits for their states. As Schwarzenegger told me, "We have seen in California so many companies that have been created that work just on things that have do with clean environment." California's state-imposed efficiency standards have resulted in per-capita energy consumption in California remaining almost flat for the last 30 years, while in the rest of the country it has gone up 50 percent. "There are a lot of industries that are exploding right now because of setting these new standards," he said.

VII.

John Dineen runs G.E. Transportation, which makes locomotives. His factory is in Erie, Pa., and employs 4,500 people. When it comes to the challenges from cheap labor markets, Dineen likes to say, "Our little town has trade surpluses with China and Mexico."

Now how could that be? China makes locomotives that are 30 percent cheaper than G.E.'s, but it turns out that G.E.'s are the most energy efficient in the world, with the lowest emissions and best mileage per ton pulled—"and they don't stop on the tracks," Dineen added. So China is also buying from Erie—and so are Brazil, Mexico and Kazakhstan. What's the secret? The China price.

"We made it very easy for them," said Dineen. "By producing engines with lower emissions in the classic sense (NOx [nitrogen oxides]) and lower emissions in the future sense (CO_2) and then coupling it with better fuel efficiency and reliability, we lowered the total life-cycle cost."

The West can't impose its climate or pollution standards on China, Dineen explained, but when a company like G.E. makes an engine that gets great mileage, cuts pollution and, by the way, emits less CO_2, China will be a buyer. "If we were just trying to export lower-emission units, and they did not have the fuel benefits, we would lose," Dineen said. "But when green is made green—improved fuel economies coupled with emissions reductions—we see very quick adoption rates."

One reason G.E. Transportation got so efficient was the old U.S. standard it had to meet on NOx pollution, Dineen said. It did that through technological innovation. And as oil prices went up, it leveraged more technology to get better mileage. The result was a cleaner, more efficient, more exportable locomotive. Dineen describes his factory as a "technology campus" because, he explains, "it looks like a 100-year-old industrial site, but inside those 100-year-old buildings are world-class engineers working on the next generation's technologies." He also notes that workers in his factory make nearly twice the average in Erie—by selling to China!

The bottom line is this: Clean-tech plays to America's strength because making things like locomotives lighter and smarter takes a lot of knowledge—not cheap labor. That's why embedding clean-tech into everything we design and manufacture is a way to revive America as a manufacturing power.

"Whatever you are making, if you can add a green dimension to it—making it more efficient, healthier and more sustainable for future generations—you have a product that can't just be made cheaper in India or China," said Andrew Shapiro, founder of GreenOrder, an environmental business-strategy group. "If you just create a green ghetto in your company, you miss it. You have to figure out how to integrate green into the DNA of your whole business."

Ditto for our country, which is why we need a Green New Deal—one in which government's role is not funding projects, as in the original New Deal, but seeding basic research, providing loan guarantees where needed and setting standards, taxes and incentives that will spawn 1,000 G.E. Transportations for all kinds of clean power.

Bush won't lead a Green New Deal, but his successor must if America is going to maintain its leadership and living standard. Unfortunately, today's presidential hopefuls are largely full of hot air on the climate-energy issue. Not one of them is proposing anything hard, like a carbon or gasoline tax, and if you think we can deal with these huge problems without asking the American people to do anything hard, you're a fool or a fraud.

Being serious starts with reframing the whole issue—helping Americans understand, as the Carnegie Fellow David Rothkopf puts it, "that we're not 'post-Cold War' anymore—we're pre-something totally new." I'd say we're in the "pre-climate war era." Unless we create a more carbon-free world, we will not preserve the free world. Intensifying climate change, energy wars and petroauthoritarianism will curtail our life choices and our children's opportunities every bit as much as Communism once did for half the planet.

Equally important, presidential candidates need to help Americans understand that green is not about cutting back. It's about creating a new cornucopia of abundance for the next generation by inventing a whole new industry. It's about getting our best brains out of hedge funds and into innovations that will not only give us the clean-power industrial assets to preserve our American dream but also give us the technologies that billions of others need to realize their own dreams without destroying the planet. It's about making America safer by breaking our addiction to a fuel that is powering regimes deeply hostile to our values. And, finally, it's about making America the global environmental leader, instead of laggard, which as Schwarzenegger argues would "create a very powerful side product." Those who dislike America because of Iraq, he explained, would at least be able to say, "Well, I don't like them for the war, but I do like them because they show such unbelievable leadership—not just with their blue jeans and hamburgers but with the environment. People will love us for that. That's not existing right now."

In sum, as John Hennessy, the president of Stanford, taught me: Confronting this climate-energy issue is the epitome of what John Gardner, the founder of Common Cause, once described as "a series of great opportunities disguised as insoluble problems."

Am I optimistic? I want to be. But I am also old-fashioned. I don't believe the world will effectively address the climate-energy challenge without America, its president, its government, its industry, its markets and its people all leading the parade. Green has to become part of America's DNA. We're getting there. Green has hit Main Street—it's now more than a hobby—but it's still less than a new way of life.

Why? Because big transformations—women's suffrage, for instance—usually happen when a lot of aggrieved people take to the streets, the politicians react and laws get changed. But the climate-energy debate is more muted and slow-moving. Why? Because the people who will be most harmed by the climate-energy crisis haven't been born yet.

"This issue doesn't pit haves versus have-nots," notes the Johns Hopkins foreign policy expert Michael Mandelbaum, "but the present versus the future—today's generation versus its kids and unborn grandchildren." Once the Geo-Green interest group comes of age, especially if it is after another 9/11 or Katrina, Mandelbaum said, "it will be the biggest interest group in history—but by then it could be too late."

An unusual situation like this calls for the ethic of stewardship. Stewardship is what parents do for their kids: think about the long term, so they can have a better future. It is much easier to get families to do that than whole societies, but that is our challenge. In many ways, our parents rose to such a challenge in World War II—when an entire generation mobilized to preserve our way of life. That is why they were called the Greatest Generation. Our kids will only call us the Greatest Generation if we rise to our challenge and become the Greenest Generation.

Critical Thinking

1. Do you agree "living, working, designing, manufacturing and projecting America in a green way can be the basis for a new unifying political movement for the 21st century?" Why or why not?
2. If people "change when they have to" not when they are told to, what will it take to make them change in the direction of sustainability?
3. Explain the statement "Green will not go down Main Street America unless it also goes down Main Street China, India and Brazil."

Thomas L. Friedman is a columnist for *The New York Times* specializing in foreign affairs.

Test-Your-Knowledge Form

We encourage you to photocopy and use this page as a tool to assess how the articles in *Annual Editions* expand on the information in your textbook. By reflecting on the articles you will gain enhanced text information. You can also access this useful form on a product's book support website at www.mhhe.com/cls.

NAME: DATE:

TITLE AND NUMBER OF ARTICLE:

BRIEFLY STATE THE MAIN IDEA OF THIS ARTICLE:

LIST THREE IMPORTANT FACTS THAT THE AUTHOR USES TO SUPPORT THE MAIN IDEA:

WHAT INFORMATION OR IDEAS DISCUSSED IN THIS ARTICLE ARE ALSO DISCUSSED IN YOUR TEXTBOOK OR OTHER READINGS THAT YOU HAVE DONE? LIST THE TEXTBOOK CHAPTERS AND PAGE NUMBERS:

LIST ANY EXAMPLES OF BIAS OR FAULTY REASONING THAT YOU FOUND IN THE ARTICLE:

LIST ANY NEW TERMS/CONCEPTS THAT WERE DISCUSSED IN THE ARTICLE, AND WRITE A SHORT DEFINITION:

We Want Your Advice

ANNUAL EDITIONS revisions depend on two major opinion sources: one is our Advisory Board, listed in the front of this volume, which works with us in scanning the thousands of articles published in the public press each year; the other is you—the person actually using the book. Please help us and the users of the next edition by completing the prepaid article rating form on this page and returning it to us. Thank you for your help!

ANNUAL EDITIONS: Sustainability 12/13

ARTICLE RATING FORM

Here is an opportunity for you to have direct input into the next revision of this volume.
We would like you to rate each of the articles listed below, using the following scale:

1. **Excellent: should definitely be retained**
2. **Above average: should probably be retained**
3. **Below average: should probably be deleted**
4. **Poor: should definitely be deleted**

Your ratings will play a vital part in the next revision.
Please mail this prepaid form to us as soon as possible.
Thanks for your help!

RATING	ARTICLE
________	1. World Scientists' Warning to Humanity
________	2. Population and the Environment: The Global Challenge
________	3. Ecosystems and Human Well-Being: Summary for Decision-Makers
________	4. The Anthropocene: Are Humans Now Overwhelming the Great Forces of Nature?
________	5. The State of the Nation's Ecosystems 2008: What the Indicators Tell Us
________	6. Global Biodiversity Outook 3: Executive Summary
________	7. Top 10 *Myths* About Sustainability
________	8. The Century Ahead: Searching for Sustainability
________	9. The Invention of Sustainability
________	10. The Future of Sustainability: Re-thinking Environment and Development in the Twenty-First Century
________	11. Sustainable Co-evolution
________	12. Framing Sustainability
________	13. Synthesis
________	14. Ecosystem Services: Benefits Supplied to Human Societies by Natural Ecosystems
________	15. How Have Ecosystems Changed?
________	16. How Have Ecosystems Services and Their Uses Changed?
________	17. The Competitive Exclusion Principle
________	18. The Historical Roots of Our Ecological Crisis
________	19. The Cultural Basis for Our Environmental Crisis
________	20. The Tragedy of the Commons
________	21. The Narcotizing Dysfunction
________	22. Mind the Gap: Why Do People Act Environmentally and What are the Barriers to Pro-environmental Behavior?
________	23. Do Global Attitudes and Behaviors Support Sustainable Development?
________	24. The Latest On Trends In: Nature-Based Outdoor Recreation
________	25. New Consumers: The Influence of Affluence on the Environment
________	26. Human Domination of Earth's Ecosystems
________	27. Human Alteration of the Global Nitrogen Cycle: Causes and Consequences
________	28. The Story of Phosphorus: Global Food Security and Food for Thought
________	29. Biodiversity Loss Threatens Human Well-Being
________	30. Soil Diversity and Land Use in the United States
________	31. Can Selfishness Save the Environment?
________	32. Toward A Sustainable World
________	33. Abolishing GDP
________	34. The Efficiency Dilemma
________	35. Consumption, Not CO_2 Emissions: Reframing Perspectives on Climate Change and Sustainability
________	36. The Rise of Vertical Farms
________	37. Climate 2030: A National Blueprint for a Clean Energy Economy
________	38. The Power of Green

NO POSTAGE
NECESSARY
IF MAILED
IN THE
UNITED STATES

BUSINESS REPLY MAIL
FIRST CLASS MAIL PERMIT NO. 551 DUBUQUE IA

POSTAGE WILL BE PAID BY ADDRESSEE

McGraw-Hill Contemporary Learning Series
501 BELL STREET
DUBUQUE, IA 52001

ABOUT YOU

Name Date

Are you a teacher? ❒ A student? ❒
Your school's name

Department

Address City State Zip

School telephone #

YOUR COMMENTS ARE IMPORTANT TO US!

Please fill in the following information:
For which course did you use this book?

Did you use a text with this ANNUAL EDITION? ❒ yes ❒ no
What was the title of the text?

What are your general reactions to the Annual Editions concept?

Have you read any pertinent articles recently that you think should be included in the next edition? Explain.

Are there any articles that you feel should be replaced in the next edition? Why?

Are there any World Wide Websites that you feel should be included in the next edition? Please annotate.

May we contact you for editorial input? ❒ yes ❒ no
May we quote your comments? ❒ yes ❒ no

NOTES

NOTES

NOTES

NOTES

NOTES

NOTES

NOTES

NOTES